Soaps and Detergents

A Theoretical and Practical Review

Soaps and Detergents
A Theoretical and Practical Review

Editor

Luis Spitz
L. Spitz, Inc.
Skokie, Illinois

AOCS
PRESS

Champaign, Illinois

Library of Congress Cataloging-in-Publication Data

Soaps and detergents: a theoretical and practical review/editor,
  Luis Spitz.
      p. cm.
    Includes bibliographical references and index.
    ISBN 0-935315-72-1
    1. Soap. 2. Detergents. I. Spitz, Luis.
  TP991.S688 1996
  668'.12—dc20   96-17566
                                                        CIP

Printed in the United States of America with vegetable oil-based inks.
00 99 98 97 96 95 5 4 3 2 1

Preface

The first seminar dedicated entirely to soaps was held in May 1989 in Cincinnati, Ohio, and was attended by a record number of 375 participants from around the world. The book based on that seminar, *Soap Technology for the 1990's*, published in 1990, has become the largest-selling AOCS Press technical book. Comments and requests received in the following years showed a genuine interest for a new, expanded, combination soaps and detergents seminar. The 1994 Conference drew a record number of 730 attendees from 49 countries, with one-third of the registrants coming from Central and South America. The two goals of the three-day Conference were to present an updated overview of Soap and Detergent subjects of key importance, and to follow up with the publication of a comprehensive reference book.

Soaps and Detergents: A Theoretical and Practical Review contains practically all the subjects presented at the Conference with additional material added for publication. As the book's subtitle indicates, theoretical coverage of the various subjects is combined with many practical, easy-to-use isometric drawings, flow sheets, sample calculations, definitions, and reference tables.

The book also includes important material presented for the very first time. The subjects range from raw materials pretreatment, classification, selection, synthesis, and uses to product formulations, processing systems for the production of soaps, glycerine and powered detergents, quality control, analysis, performance evaluation, packaging, process control, computerization, fragrance technology, and market trends.

I extend my deepest appreciation to the entire AOCS organization for their accepting, supporting, promoting, and organizing this first-of-a-kind project, which became the largest and most successful technical conference in AOCS history. During the past three years, I have had the privilege and honor to work very closely with a number of most competent and friendly AOCS staff members.

To all the contributors of this book, I send my heartfelt thanks for their professionalism, expertise, and for sharing their knowledge to enhance our own.

I hope that this publication will be a valuable reference book for a wide range of professionals, from newcomers to those already practicing in the soap and detergent industry, who will also find new information in areas other than their own specialty. I feel fortunate to work in an industry that preserves and enhances people's health, well-being, and beauty.

May the soap and detergent industry "bubble" and keep "bubbling" with activities and opportunities for new and exciting products as we move towards the technologically challenging 21st Century.

Luis Spitz

Soaps and Detergents

A Love Story

SUNLIGHT had just come up on the horizon. **ALL** was quiet.
She took an early morning walk along the **COAST**, her **SPIRIT** high.
With outstretched arms, as if she was about to **CARESS** the beauty
    about her, she looked around.
The **TIDE** was low, a **SURF** board and a **LIFEBUOY** lay abandoned.
The **TONE** of a **DOVE** sounded like the **WHISK** of a whip.
As she waited for him to arrive she needed to **GAIN** confidence
    to talk about her **FAB**ulous wedding plans and a **LUX**urious
    honeymoon in Ireland during the fresh **IRISH SPRING**
With her bare toes she outlined a **DIAL** on the pure clean **IVORY** colored sand.
Feeling the sand's warmth she realized that it is time to **SHIELD** herself
    from the bright sun.
JERGENS would arrive soon and she knew that he **DUZ** not approve of her being
    alone on the deserted beach.
The **BASIS** of his **ZEST** and his **BOLD** nature were derived from his
    OLD DUTCH ancestry.
To him it was **ESSENTIAL** to remember his heritage.
He became **CHEER**ful seeing his beloved **SWEETHEART**
    waving in the distance.

Written by Ms. Rachael Smoler, a dear family friend of many years and presented at the
opening remarks of the Conference on October 12, 1994.

Contents

Chapter 1

Raw Materials and Their Pretreatment for Soap Production

Rafael A. Corredoira[a] and Antonio R. Pandolfi[b]

[a]Compañía Bāo S.A., Montevideo, Uruguay

[b]Consultant, Montevideo, Uruguay

Introduction

This chapter examines fats and oils as the basic raw materials used in soap manufacturing, their selection according to formulation needs, and how they can be upgraded by pretreatment to achieve the desired soap standards of quality.

A classification of raw materials will be presented, stating which ones are used in diverse regions and how they must be treated to be adapted for use in soap manufacturing. The effect of the quality of raw materials on the quality of the finished product will also be analyzed. The reasons behind the need for pretreatment will be stated; different approaches will be presented for the different methods used to produce the base soap.

Fats and Oils Composition and Soap Performance

Natural triglycerides are generally formed with mixes of fatty acids, and their structure need not concern soap manufacturers. What is of interest is which fatty acids can be found in different raw materials, and in what proportion, because soaps made from different fatty acids have different properties.

Soaps composed of fatty acids with chains from 12 to 18 carbons are those of interest in soap production. Those with shorter chains do not have the soapy effect and are skin irritants. Chains longer than 20 carbons have very little solubility and are used only in mixes, with great care taken in their formulation.

For example, sodium laureate provides quick and profuse foam but inferior detergency. Sodium palmitate and sodium stearate are very good detergents at high temperatures, whereas sodium oleate provides good foam, good detergency, and solubility but is poor in firmness. Sodium myristate yields soap with properties nearing the ideal.

Triglycerides can be saponified to obtain soaps and glycerol, hydrolyzed to obtain a mix of fatty acids and glycerol, or esterified with methyl alcohol to produce a mix of methyl esters and glycerol. (See Fig. 1.1.)

The fatty acids are neutralized with soda or sodium carbonate to obtain soap. The methyl esters are saponified to make soap (Fig. 1.2). No matter which method

1

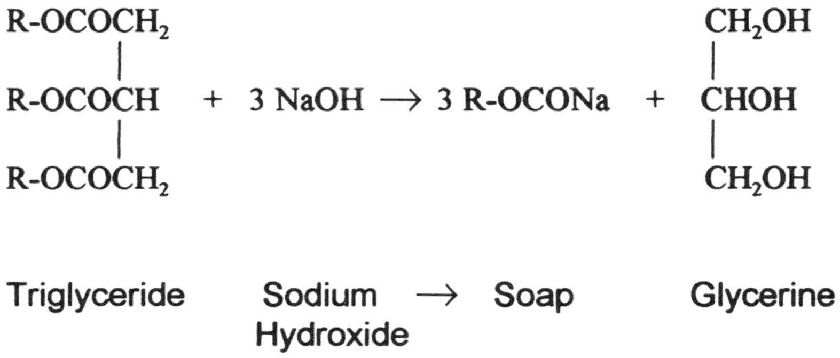

$$\begin{array}{ccccc}
\text{R-OCOCH}_2 & & & & \text{CH}_2\text{OH} \\
| & & & & | \\
\text{R-OCOCH} & + \ 3 \ \text{NaOH} \rightarrow 3 \ \text{R-OCONa} & + & \text{CHOH} \\
| & & & & | \\
\text{R-OCOCH}_2 & & & & \text{CH}_2\text{OH}
\end{array}$$

Triglyceride   Sodium $\rightarrow$ Soap   Glycerine
Hydroxide

Fig. 1.1. Saponification of triglycerides.

is employed, it must always be remembered that the raw materials for soap are the fatty acids that make up the triglyceride, and quality of the former depends on the latter.

Raw Materials

Classification

Triglycerides, the main raw materials in soap production, can be divided into four groups: animal fats (tallow and greases), lauric oils (coconut oil, babassu oil, and palm kernel oil), nonlauric oils (palm oil and cottonseed oil), and marine oils (anchovy and sardine).

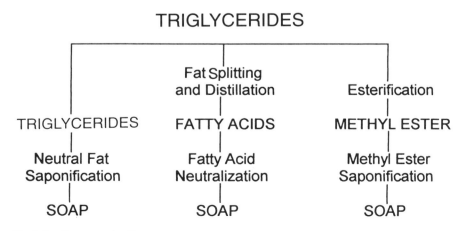

Fig. 1.2. Soap production routes.

Animal Fats

Tallow is the most widely used animal fat in soap manufacturing. Its quality depends above all on the kind of treatment it may have received. During the natural decomposition process that begins with slaughter, both the color and content of fatty acids increase, to the detriment of the soap quality. For this reason, all the care taken during rendering, handling, and storage of the fat is of vital importance because it will prevent degradation before soap production.

Tallow is obtained by putting animal tissue containing fat through a process called rendering (1). This process consists of the separation of tissue from water using heat and pressure. There are two rendering methods: the wet method, preferred for edible-quality fats, and the dry method, used when the fat is of nonedible quality and the taste and odor of the product are of secondary importance.

In the United States there is a commercial classification that defines types of fats (2). This classification is of special interest because of the degree of influence the United States has in fat trade. The tallows of commercial quality of most importance to the soap industry are edible grade (ET), top white (TWT), fancy bleachable (BFT), and choice white grease (CWG). The American Fats and Oils Association (AFOA) specifications for these qualities can be found in Table 1.1.

Edible grade includes lard and tallow. Edible qualities have very low color and free fatty acid levels. These grades have been used to produce food products for human consumption. Due to a change in consumer preference, the consumption of animal fats has diminished. The premium that is paid for these grades over the price of the soap grades has gone down. The quality obtained from soaps made from these tallows is the best.

Top white and bleachable fancy are, to the soap manufacturer, of quality similar to that of edible grade. Their acidity and color levels allow the production of very-high-quality soap, and their price is slightly lower than that of edible grade.

TABLE 1.1

American Fats and Oils Association Specifications for Different Grades of Animal Fat (1)

Grade	Titer, min C	FFA, max	FAC, max	R&B, max	MIU
Edible tallow	41.0	0.5	3	none	a
Lard (edible)	38.0	0.5	b	none	a
Top white tallow	41.0	2	5	0.5	1
All beef packer tallow	42.0	2	none	0.5	1
Extra-fancy tallow	41.0	3	5	none	1
Fancy tallow	40.5	4	7	none	1
Bleachable fancy tallow	40.5	6	none	1.5	1
Prime tallow	40.5	6	13–11B	none	1.5
Choice White Grease	36.0	4	13–11B	none	1
Yellow Grease	c	15	39	none	2

[a]Moisture maximum 0.20%; insoluble impurities maximum 0.05%.
[b]Lovibond color 5 1/4-inch cell—maximum 1.5 red. Lard peroxide value maximum 4.0 ME/k.
[c]Titer minimum, when required, to be negotiated between buyer and seller on a contract-by-contract basis.

Choice white grease differs from BFT in the lowest titer (36°C min.). It can be obtained from hogs or fowl. It is more unstable because of its highly unsaturated fatty acid content.

In toilet soap manufacturing, the product must be as white as possible. The better the tallow, the brighter the color of the soap and the lesser the quantity of white colorant (such as titanium dioxide) and fluorescent whitening agent needed, along with lesser perfume interference. Greases are rarely used for toilet soaps because of their instability and odor. Their main use is for powder and industrial laundry soaps.

Lauric Oils

Coconut oil has the distinction of being the most important vegetable oil in soap manufacturing. It is obtained from the copra (coconut pulp) and produced mainly in the Philippines. Coconut oil is a white fat that melts at around 25°C.

For the purpose of soap manufacturing, an oil with an iodine index from 9 to 13 and an acidity index between 2% and 6% are used, usually blended with tallow, fractions of palm oil, or hydrogenated oils.

Babassu oil is of Brazilian origin and can be used, along with palm kernel oil, as a substitute for coconut oil. These oils have very similar lauric oil contents. One basic difference lies in the lower levels of oleic acid found in coconut oil. Their main contribution to the formulas is lauric acid, which noticeably improves the solubility and foam properties of soap.

Nonlauric Oils

This category includes those vegetable oils in which lauric acid is not the main component. Among the nonlauric oils used in soap manufacturing, palm oil and some of its fractions are undergoing important growth. Thus stearin, subproduct of the fractionation of palm oil usually used in edible oil, is becoming an important substitute for tallow in soap manufacturing.

The use of edible oils in soap manufacturing is not very widespread. However, the use of soapstock obtained from the refining of crude oils is quite common. Soapstock contains soaps derived from alkali refining, neutral oil, organic substances that are not fatty acids, and water. Compared to neutral oil, its price is lower based on its fatty acid content. The proportion of the different fatty acids contained in soapstock is the same as in crude oil.

Soap manufacturing based on soapstock of edible oil processing is of great importance. This is the way in which many plants get rid of refining by-products. Soapstock is normally acidulated with sulfuric acid to change soaps into free fatty acids, or dried to diminish its volume and transportation costs.

Soapstock can be saponified with soda by a boiling or semiboiling process and washed with brine to obtain industrial and laundry soaps. Fatty acids obtained by acidifying soapstock can be distilled to produce either higher-quality soaps or, through hydrogenation, stearic acid.

Marine Oils

Oils obtained from different fish and marine mammals are sometimes used in soap manufacturing after being hydrogenated. The compositions of the marine oils vary according to the species from which they have been obtained. In general, they contain important quantities of polyunsaturated long-chain fatty acids (C20, C22, and so on). Hydrogenating to iodine values in the range of 50 to 60 achieves stability, preventing the reappearance of odor during storage.

The long-chain fatty acids saturated during marine oil hydrogenation are not adequate for soap manufacturing by themselves, but they can be used with shorter-chain fatty acids such as those found in lauric oils.

Raw Materials Used in Different Regions

Tallow is used as the basic fat raw material in modern soap formulas all over the world. It was available in those regions where industrial production of soap was developed. To improve the quality of soap, coconut oil was added, obtaining a more agreeable result with respect to foam and performance during use.

In those regions where tallow is not available, fatty substances of local provenance have been used as substitutes. Thus palm oil and the palm stearin replace tallow in tropical regions, whereas in others it is replaced by vegetable oils from local species that are partially or totally hydrogenated and, in some cases, split and distilled.

India is of special interest because it has done significant research in the production, purification, and use of different local varieties of oils.

Due to the shortage of edible oils, the Indian government has promoted the use of nonedible oils for soap and paint production as a means of diminishing imports. Boosted by government incentives, manufacturers have developed technology for the improvement of different local varieties of oils.

Rice bran is among those oils used for soap manufacturing in India. An important portion of the nonedible oil is destined for soap production.

Another oil whose main use is in soap production is castor oil. This oil has 85% ricinoleic acid. It is hydrogenated and dehydrated for its use in soap manufacturing.

Other oils of lesser volume in India are karanja oil and neem oil. The former has a very peculiar odor and nonglyceride that provokes a change of color in soap made from this oil. To prevent this inconvenience, two pretreatment technologies have been developed: (1) alcoholic alkali extraction, deodorization, alkali refining, and bleaching; and (2) physical refining to remove odor, splitting, and steam distillation of the fatty acids to improve the color.

The only profitable process is physical refining.

In the United States the fats used for quality soap production, depending on their availability, are as follows:

- Tallow and coconut oil
- Marine oils and their hydrogenated fractions, palm oils and their fractions, babassu oil, and palm kernel oil.

Modification of Raw Materials

Sometimes the raw materials available for soap manufacturing have either a fatty acid composition that does not allow mixes to reach adequate proportions, or too many polyunsaturated fatty acids that must be transformed into more stable fatty acids. Hydrogenation is the process that makes these transformations possible.

Hydrogenation

Fat and oil hydrogenation consists of adding hydrogen to the double bonds of the fatty acids in the triglyceride. An improvement in the resistance of fats and oils to oxidation and odor reversion is achieved by this process.

For the reaction to take place, an intimate contact between the fat or oil, hydrogen, and catalyst must be maintained, at an adequate temperature. To this end, reactors keep a suspension of catalyst in the oil under a hydrogen atmosphere (3). The variables that are handled to obtain different results are hydrogen pressure, reaction temperature, agitation, quantity of catalyst, and the quantity of hydrogen added to the fat or oil.

Hydrogenation can be carried out selectively. Conditions must be controlled so that the hydrogen reacts first with the double bonds of the most unsaturated fatty acids (4). The reaction speeds increase with the degree of unsaturation.

Effect of the Process Conditions

Temperature
Hydrogenation speeds up by an increase in temperature, but this effect is barely noticeable with low agitations. The rise in temperature also increases selectivity.

Pressure
A pressure increase speeds up the reaction but diminishes selectivity. Operating pressures vary between 0, 2, and 3 bar in the case of neutral fats and can reach 20 bar in the case of fatty acids.

Agitation
An increase in the agitation provokes a noticeable increase in the reaction speed with a great reduction of selectivity.

Catalyst
An increase in catalyst concentration provokes a notable increase in reaction speed and diminishes selectivity very little. At present, catalysts are commercialized with different selectivity levels. Selectivity is controlled by means of porosity and poisons. Selective catalysts are catalysts poisoned mainly with sulfur. Poisoned catalysts require the use of larger doses due to their decreased activity .

Hydrogenation Applied to Soap Manufacturing

Hydrogenation applied to soap manufacturing improves the firmness and stability properties of soap.

Before hydrogenation, it is economical to treat the triglycerides to eliminate the possibility of catalyst poisoning. The costs of removing impurities through washing, neutralizations, and bleaching are lower than the costly catalysts used in hydrogenation.

The selection of the type of hydrogenation method used depends on the raw materials available. When dealing with highly saturated fats, selective hydrogenation of an oil to obtain an important input of monosaturated fatty acids might be convenient. However, if highly unsaturated fats are used, they must be hydrogenated until saturated fatty acids are obtained. This will ensure good stability and allow for the production of soaps with adequate in-use properties.

Paths in Soap Production

There are two important paths in soap production: saponification of neutral fats and neutralization of fatty acids.

Saponification of Neutral Fats

Saponification of neutral fats saponifies the triglyceride directly by boiling, by semiboiling, or by continuous processes. The chosen pretreatment processes are carried out before saponification based on the type of soap to be produced and the quality of the raw materials used.

Neutralization of Fatty Acids

Fatty acids obtained from natural triglycerides by hydrolysis and distillation are neutralized to make soap.

A third path, which has not had commercial importance, is fatty methyl ester saponification.

Pretreatment of Raw Materials

Pretreatment is necessary because raw materials undergo a natural degradation process. This deterioration is what forces pretreatment to improve the overall quality of the raw materials used for soap making.

Raw material production is sometimes seasonal, making storage during shortage periods necessary. If the raw materials are not treated and stored with the proper care, they deteriorate very quickly.

There are raw materials of poor quality whose prices justify the costs of the pretreatment. In other cases, when very good color or stability levels are expected in the soap bases, the fats must be pretreated to reach these levels.

Pretreatment Methods and Selection

Washing, alkali refining, bleaching, and deodorization are the various pretreatment methods that may be used when saponifying neutral fats and oils. Washing, splitting, and distilling are the pretreatment options when neutralizing fatty acids.

Special care has to be taken when processing palm and palm kernel fats and fatty acids due to high color problems.

When oils with a high polyunsaturated fatty acid content such as linseed or fish are available, they must undergo hydrogenation to produce a base soap with stable color and odor.

Raw materials that are very poor in color, with odoriferous substances, or unbleachable by simple bleaching, demand fatty acid splitting and distillation processes.

Distilled fatty acids in those cases result in the most elaborated but costly pretreatment for bad-quality raw materials.

Pretreatment for Neutral Fats Saponification

Storage

Pretreatment can be avoided or minimized at least to ensure that the effects will last as long as possible.

The most economical way to have good-quality fat is to avoid deterioration from bad handling. The way to achieve this is through proper storage. Remember the following basic points:

- Deterioration is caused by several chemical reactions—some of them autocatalytic—and their speed has a exponential dependence on temperature. It can be estimated that a 10°C rise in temperature will double the reaction speed.
- There are catalysts, metals, water, enzymes, and others that may speed up some of these reactions.
- Deterioration is produced in part by the reaction between oxygen in the air and fat.

Bearing all of this in mind, the following is recommended:

- Avoid overheating the materials and exposing them, while hot, to the oxygen in the air.
- Avoid bringing the materials in contact with copper or iron parts, especially if they are hot.
- Avoid storing humid materials or those that have impurities.
- Keep the fats at the lowest possible temperature for the longest possible time.
- Keep materials under an inert atmosphere (for example, nitrogen).

Hot water is preferable for heating. Even though it requires a larger heat exchange surface, it ensures that all the surfaces are in contact with the materials at temperatures below 100°C. In heating with steam, the use of pressures below 3 bar is advised, as long as the fats are not above 80°C. When melting the fat is necessary, do not raise the temperature more than 10°C above its melting point. Use stainless steel heating surfaces.

When raw materials must be stored for prolonged periods of time, the air must previously be displaced from the tank in which the materials will be placed while an inert atmosphere is maintained. Hot fats and those that are under vacuum must be discharged in nitrogen atmospheres, thus keeping the oxygen in the air from coming into contact with the raw materials, eliminating a deteriorating agent.

All these safeguards become more and more necessary as fats with higher unsaturated fatty acid contents are used. In any case, it is more economical to prevent the deterioration of the fats than to have to treat them to improve their quality. When storing tallow, it is important to eliminate those substances that contribute to deterioration. Among these are enzymes, proteins, and water.

One way to eliminate them is by denaturalizing proteins with heat. This can be done during bleaching, separating them from the tallow during filtering, or during cleaning by centrifuging using steam instead of water to coagulate proteins, before entering the centrifuge.

Washing

Washing with hot water is a very effective cleaning treatment for different fats. A soft surfactant may be added, separating the fat from water and impurities by means of batch settling or centrifuging. With good-quality tallow, those improvements made in crude color and impurity content may be enough.

In this treatment, the hot water hydrates some of the impurities, such as proteins, which pass into the water phase, absorbing bodies that give the tallow color. The solids are then separated by settling in a batch process or continuously using a centrifuge. Working with water close to its boiling point improves washing because it coagulates part of the proteins, making their hydration easier.

The impurity hydration process is also enhanced by trisodium phosphate when added to the water, making a dilute 10% solution.

For batch washing, water or brine is added to enhance impurity settling. After the tank has been allowed to decant for several hours it is purged, sending the accumulated solids to the sewers. The water is then treated and the tallow is rendered as a finished product. The advantages of this method are as follows:

- Zero investment in equipment (all that is needed are settling tanks with a conical bottom, a water spray system, and good heating)
- Simplicity in the cleaning operation
- Good quality level for most tallow in regard to moisture and impurities

The following are among the inconveniences this method presents:

- Risk of reducing tallow quality because of the prolonged period of time it must be kept hot in contact with water and impurities
- Loss of tallow by occlusion in the water along with the impurities
- High tank and shutdown time requirements for large hourly productions
- Significant wastewater contamination due to settling tank cleansing and occlusions in the water

In addition, the impurities present in fats may form very stable emulsions that are practically impossible to decant. In these cases, the fats must be subjected to higher temperatures for longer contact time during washing.

In continuous washing, the fat is mixed with water that is very close to its boiling point (it may be a weak surfactant solution) and then conduced through a system of centrifuges that can be a decanter and self-cleaning type. The former is used to separate solids, whereas the latter permits drying and reduces the level of impurities to a minimum.

The advantages of using a centrifuge are:

- It allows continuous processing.
- It diminishes losses by making continuous control of the operation easier.
- It reduces labor.
- It eliminates washing of settling tanks, thus annulling a source of drainage contamination.

The disadvantages are that it requires a larger investment in equipment and a large maintenance cost.

For example, tallow with acidity under 1%, moisture under 0.5%, and impurities under 0.1%, stored in clean tanks at temperatures below 24°C for periods of 6 months, did not noticeably increase its color nor its acidity.

Alkali Refining

Alkali refining is one of the ways to improve quality. This process increases product losses, raises costs, and often provides only limited color improvement benefits.

Alkali refining consists of the neutralization of a fat through treatment with an approximately 13% concentration caustic soda solution. The technique used is the same as in the case of edible oils.

Among the disadvantages of using alkali refining are the important increase of losses. Those caused by this process are estimated at three times the fatty acid content when it is less than 3% before refining and more if is higher.

The soapstock obtained is a raw material of poor quality for the soap maker, with high recovery costs. Soapstock is composed of soap formed by free fatty acids and caustic soda, neutral fat matter, glycerine in the form of mono and diglycerides, and other impurities such as proteins and color-generating bodies. In order to use the fatty acids and the neutral fat matter left in the soapstock, they can be acidified. This increases the recovery costs for a soap raw material of poor quality.

Another option is to saponify the soapstock as is, by a kettle process, for the production of lower-grade laundry-type soap. In case there is no lower-grade laundry-type soap required, alkali refining for a soap producer should be avoided.

Due to the increase in processing costs and losses that alkali refining implies as a step in quality improvement, the soapmaker must carefully analyze the fat to be purchased. Sometimes the savings obtained in the sales price are lost several times in the treatments to which the fat must be subjected to reach the desired quality.

Improvements in color, for example, in tallow of 10 red can reach values of 1 or 2 red after washing and 0.5 or 1 after refining. Such small color differences do not justify alkali refining, but overall product quality should be kept in mind.

Bleaching

The most important method of improving quality is bleaching. After washing, it becomes a very attractive option because it permits considerable improvements with low costs.

Bleaching consists of treating the fat with acid-activated earth at high temperatures and then filtering it through conventional methods. Acid activated earths retain the pigments present in many fats. The presence of water in quantities of about 1/2 to 1% along with the clay improves its pigment retention.

In the soap industry, bleaching is usually performed on batch equipment with agitator-equipped tanks and heating systems with a small vacuum. There is also continuous equipment for bleaching, but it is used mostly by refineries.

A decrease in the odor of the fat or oil may be achieved by bleaching, but this does not imply that the benefits obtained through deodorization will be realized, because they include the reduction of prooxidant substances that are catalysts of the deterioration of the fat or oil.

Use of Silica

Silica appeared in the market near 10 years ago, and its use was suggested to complement the clarifying effects of activated bleaching earths (ABEs). Actually, the field of application has broadened to a complete change in fats and oils refining operation, with or without the use of activated bleaching earths.

In the case of pretreatment for tallow, Grace Davison (Grace Argentina S.A), expresses the following (5):

A very good quality soap can be made with any kind of tallow available in the market. Low quality tallow can be upgraded through adequate processing. The total cost to achieve a good quality toilet soap is a combination of the raw material cost plus the cost of upgrading process. In our territory (Argentina) most customers or prospects use medium to low quality tallow, so bleaching is a key procedure. In this general concept Trisyl has proved as a valuable tool in tallow bleached soapstock for soap bar production, [e]specially when low quality tallow[s] are used.

Those tallows have not only high red color (over 8), but also high blue color (over 2). Color level is referred to sap color measured with Lovibond 5 1/4 (sap color = Background color + pigment color + alkali-condensable color).

Experience demonstrates that sequential use of Trysil and activated bleaching earths helps to remove carotene, chlorophyll, and other derivatives from hemoglobin and bilis (see Table 1.2).

We have industrial experience in pretreatment for tallow based on Trisyl use. Initially a phosphoric acid treatment is made (adding 1 L/ton at 80°C over 10 min), and then enough soda caustic solution is added to neutralize the acid, with an excess to produce 600 to 1,000 PPM soap into the product. This soap occlude impurities.

When the moisture content must be reduced to 0.1%, the tallow should not be heated above 80°C to avoid fixing the color bodies. If very light colors are required, add 0.1/0.2% silica, mix over 10 minutes and bleach under normal conditions.

This procedure produces excellent results for low-quality tallow, with very low losses processed both in batches and continuously.

Theory on Bleaching

Bleaching is the adsorption of color-generating pigments in the adsorbent, diminishing the color of the fat. According to this theory, the Langmuir isotherm may be applied to relate the amounts of adsorbent, adsorbate, and unabsorbed substance. The Langmuir isotherm links the extent of a gas adsorption with the pressure at a

TABLE 1.2
Tallow Pretreatment

Tallow	Control	Trisyl	Trisyl 300
Batch volume (tons)	16.7	16.8	16.8
Temperature (°C)	105	104	103
Moisture (%)	0.1	0.1	0.1
SAP color (5¼")	6 R / 2 Bl	8 R / 2 Bl	8 R / 2.7 Bl
Raw color bleached at lab (2.52% clay)	4.5 R	6.0 R	4.0 R
Adsorbents			
Clay (kg)	420	210	245
Trisyl (kg)	—	40	—
Trisyl 300 (kg)	—	—	40
Clay (%)	2.5	1.25	1.46
Trisyl (%)	—	0.24	—
Trisyl 300 (%)	—	—	0.24
Final color (5¼")	5.1 R	5.0 R	4.0 R
Filtration (min)	227	100	125
Throughput (tons/hr)	4.4	10.1	6.1

Courtesy of Grace S.A.Argentina.

constant temperature. This equation was used by Freundlich to explain the results he encountered studying the solute adsorption from the solution. For this reason the following equation is known as the Freundlich isotherm, adapted for practical use to determine the degree of adsorption of a pigment by an adsorbent:

$$\frac{x}{m} = kc^n$$

where

x = amount of substance adsorbed

m = amount of adsorbent

c = concentration of substance remaining unabsorbed

k = constant linked with the general capacity of that adsorbent for that adsorbate

n = constant relating the manner in which the efficiency of adsorption changes as it progresses from higher to lower color

It follows that

$$\log \frac{x}{m} = \log k + n \log c$$

The following dependencies can be deduced from the equation:

- The higher k, the lower the amount of adsorbent is needed to reach a certain color value.
- The value of n shows a variation in the efficiency of the adsorbent, depending on whether it is used in solutions with high or low concentrations of adsorbate.

In Fig. 1.3 the action of three adsorbents used in oil bleaching can be seen. The slope of the line is n and the intercept is k, plotted on a log-log scale. From this figure it may be concluded that activated carbon is very efficient when adsorbing pigments to reach high values of residue color, but this efficiency descends abruptly when the residue color is low. That Fuller's earth has a lower n can also be observed, indicating that its efficiency is not as affected by the final residue color and also that it is more efficient in obtaining low-residue colors than activated carbon.

Bleaching Plants and Equipment

The types of bleachers used are batch, semicontinuous, and continuous.

Batch Bleachers

Batch equipment, the most commonly used equipment in soap manufacturing, has not changed significantly in the past 30 years. It consists of cylindrical tanks with capacities ranging from 5 to 60 tons, agitation systems, and coil heating and cool-

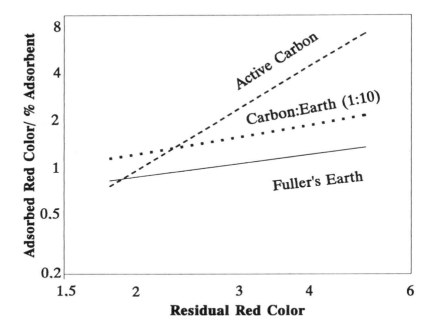

Fig. 1.3. Bleaching of refined vegetable oil (25Y-5.5R).

ing. In vacuum bleaching equipment, the tank and vacuum system are designed to maintain 50 torr of absolute pressure.

Semicontinuous Bleachers
Semicontinuous equipment operates on an automated system, where the passage from one stage to another is performed by a variable timing system and the oil, along with the earth, is pumped continuously from a final buffer section, obtaining in this way continuous service (6). One of the most popular semicontinuous systems is the Pellerin-Zenith. In the case of continuous equipment, the De Smet is of note. It doses the bleaching earth prior to entering the first stage. The slurry enters a vacuum area where it is heated with a coil and agitated with open steam. It overflows from that first compartment into a second, which gives it the retention time necessary for the operation. In the final stage, a pump sends the slurry to a filter bank that permits continuous operation. Although the Pellerin-Zenith and De Smet systems have been designed for edible fat and oil bleaching, both are useful for soap manufacturing, where excellent results are obtained.

Continuous Bleachers
Semicontinuous and continuous equipment have the advantage inherent to all the continuous processes in the industry. Among the continuous bleaching plants, the

Mazzoni SGA system has been specifically designed for use in soap production. The SGA plant (Fig. 1.4) permits the selection of the proper bleaching temperature and residence time in the bleacher, according to the characteristics of processed fatty matter. The operation is continuous and automatic. The heating is done with external preheaters. The fat and earth are deaerated and dehydrated before the earth and part of the fat are fed to the mixer tank. Then the rest of the fat and the fat and earth mix are fed into the continuous multistage bleacher. From the bleacher the slurry is sent through a cooler to a leaf filter bank that permits continuous operation.

Filters

A filter system is used to separate the clays used in bleaching from the fat or oil. For many years plate-and-frame, and recessed plate press filters have been used. They use cloths of different sizes as filtering material. Their operation is heavy and dirty when they are made of iron or steel. They are lighter when made of wood, aluminum, or certain nonmetallic materials. Cleaning has been made easier through automatization, and the plates and frames can be moved mechanically in all directions. This equipment is cheap, robust, and widely applicable, and it does not take up much space. It requires a great deal of labor and has prolonged shutdown times. Filter automatization cuts down on labor, shutdown, times and consumption of filter cloths.

There are closed self-cleaning filters with horizontal and vertical leaves. Both types have important advantages. They are versatile, use less labor, and prevent

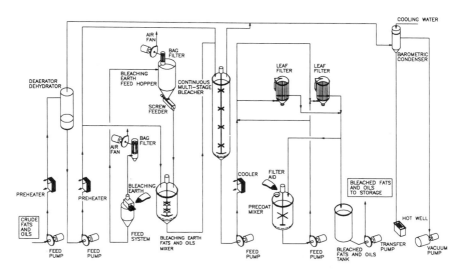

Fig. 1.4. Mazzoni SGA continuous fat and oil bleaching plant. (Courtesy of Mazzoni LB, S.p.A.)

contact between the oil and the air. They have larger flow rates per square meter than the press types, perform automatic cleaning in less time, and have gauzes that last longer than cloths. They are more expensive than the press-type filters per square meter, require high-quality maintenance, and have several moving parts, and the gauzes are more expensive than cloths.

When choosing a filter, the following points must be analyzed:

- Space available for the filter system
- Frequency of product changes and volumes in each run
- Labor cost
- Filtration rate (kg/m^2/h), including cleaning and discharge times
- Total investment including building, filter, filtering materials used in its operation, and the like

Conditions Affecting Operation

There are several conditions that affect the bleaching results. As indicated, it is an adsorption process that depends on the characteristics of the adsorbent and the adsorbate. Under some conditions the interaction is enhanced, whereas in others it is prevented. Temperature, agitation, clay doses, degree of activation, contact time, and the presence of moisture and oxygen are the main factors to be taken into account.

There must be enough agitation to ensure good contact between the adsorbent and adsorbate, but not so much as to incorporate air. The contact time must generally be under 30 minutes. The major part of the adsorption takes place in the first 15 minutes, whereas times longer than 30 minutes favor color reversion.

There are several important things to bear in mind about dosage. The efficiency of a bleaching earth depends on the requirements to which it is subjected. Adsorbents are selective of certain impurities; in this way silica, which is very effective in adsorbing gums and soaps, is useless for removing chlorophyll. For these reasons, the correct selection of an adsorbent is of vital importance in order to obtain the best results with a minimum doses.

The doses depend as much on the adsorbent as on the quality of the oil to be bleached; it is a variable that needs permanent adjustment in practice. In a year´s production, important wastes (of clay, oil in the clay, and labor) may be caused by inadequate doses and faulty clay adaption to the quality of the raw materials available.

Another important factor in bleaching results is the temperature at which it is performed (see Figs. 1.5 and 1.6). Several effects that take place during bleaching are dependent on it. The adsorption of impurities is favored by temperatures below 100°C; chemical pigment destruction at temperatures above 120°C is catalyzed by the activated earths. If the temperature exceeds 160°C, structural modifications of the unsaturated fatty acids are found. Superior temperatures cause thermic decomposition of carotenoid pigments. High temperatures in the presence of activated

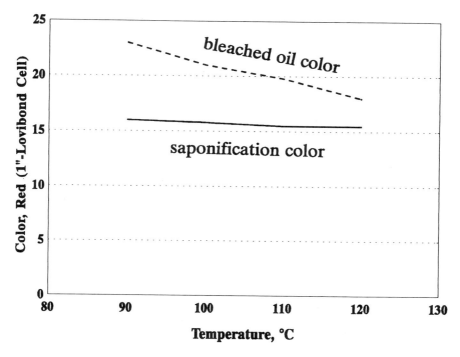

Fig. 1.5. Bleaching of palm oil with 6.1% FFA. Operating conditions: 2.0% bleaching earth, 30 min atmospheric. (Courtesy Química Sumex S.A. de C.V.)

earths provoke the formation and fixation of new pigments due to their catalytic oxidant characteristics.

The comparison of bleaching under vacuum or atmospheric bleaching is controversial.

The results obtained by one method or the other depend on the quality of the feed material. In the case of low-quality tallow, the pigment-generation effect, caused by the oxidation and the action of the adsorbent, results in a better final color than in the case of bleaching obtained by the action of the adsorbent in vacuum. On the other hand, the saponification color stability of those soaps made from fats bleached in vacuum is better than those that have undergone atmospheric bleaching, as can be seen from Table 1.3.

In the case of good tallow (see Fig. 1.7), because there are no pigments to be oxidized to improve color, the final color obtained with atmospheric bleaching (transforming colorless substances into colored substances by means of oxidation) would be worse than that obtained in vacuum bleaching.

Atmospheric bleaching also has a more harmful effect on the anisidine value than vacuum bleaching, which makes it rise only slightly (see Fig. 1.8).

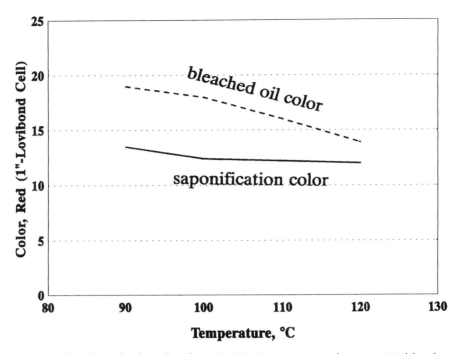

Fig. 1.6.   Bleaching of palm oil with 6.1% FFA. Operating conditions: 2.0% bleaching earth, 30 min, 50 torr vacuum. (Courtesy Química Sumex S.A. de C.V.)

Due to the effect of the quality of raw materials used, the results obtained with lower-quality palm oil (greater acidity) are poorer under the same temperature, pressure, and contact time conditions, even though it has similar initial color (see Figs. 1.9 and 1.10).

Bleaching Recommendations

The practices recommended for bleaching are the same as those recommended for storage: avoid oxidation opportunities and operate at the lowest possible temperature. When bleaching any fat and oil, it is advisable to do the following:

TABLE 1.3
Saponification Color Stability, Vacuum *vs.* Atmospheric Bleaching

	Peroxide Value	Saponification Color		
		1 Hour	1 Week	3 Months
Bleached tallow (vacuum)	1	2.2	2.4	3.5
Bleached tallow (atmospheric)	13	2.7	3.2	5.5

Original peroxide value 14 ME/k.

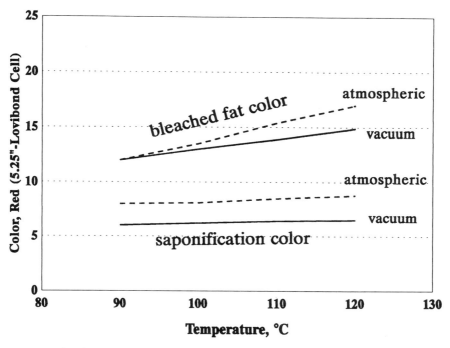

Fig. 1.7. Bleaching of tallow with 3.6% FFA. Operating conditions: 2.5% bleaching earth, 30 min. (Courtesy Química Sumex S.A. de C.V.)

- Bleach in a vacuum or inert atmosphere.
- Keep the bleaching time to 30 minutes or less if possible.
- Feed fats with a soap content under 50 ppm to the bleaching process. Soap competes with other impurities for places in the adsorbents.
- Use smaller doses of highly activated earths, diminishing losses by retention in the cake.
- Bleach separately those raw materials that require different treatment.

Tallow

Highly activated earths in quantities of 0.3% or mildly activated earths of 1% are used for high-quality tallow. The recommended contact time is 20 minutes at a temperature of 95° to 100°C.

Coconut Oil

Lightly activated earths, in doses of 0.2 to 0.4%, are used to bleach industrial-grade coconut oil. They are sufficient to achieve adequate bleaching for soap

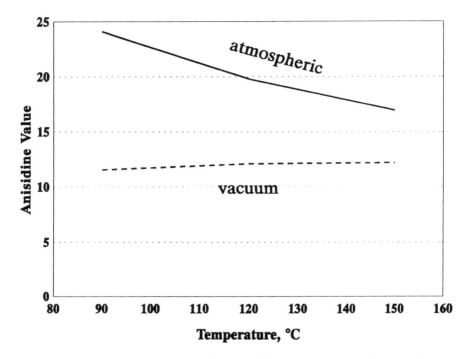

Fig. 1.8. Bleaching of palm stearin with 3.6% FFA. Operating conditions: 0.4% bleaching earth, 30 min. (Courtesy Química Sumex S.A. de C.V.)

manufacturing. Better results are obtained using 0.4% activated carbon, added 10 minutes after the earth, keeping up agitation for 45 minutes more at a temperature of 90° to 95°C.

Palm Oil

Satisfactory bleaches may be performed with the most highly activated clays with doses of 1 to 1.5% and, in some cases, 2%. The temperature may rise to 140°C, with a contact time of 20 to 30 minutes.

Another bleaching method used for technical purposes is a high-temperature method. Highly activated earths are used with doses of 0.2 to 1.0%. The oil and earth are heated under vacuum of 20 mm Hg absolute minimum to 180°C and held at this temperature for 120 minutes (7).

Palm oil olein is more easily bleachable, requiring moderately activated earths in a 1.5% doses for 30 minutes at 140°C. Palm oil stearin is the most difficult fraction to treat. It demands highly activated earths in doses of 2% with contact times of 30 minutes followed by thermic treatment at 260° to 270°C.

Table 1.4 is a comparison between bleaching conditions for several fats and oils. For more information about bleaching refer to (8).

These three analyses—crude color, refined bleached color, and free fatty acids—measure three extremely important parameters that determine the final quality of the soap produced from a fat or oil.

MIU (AOCS Methods Ca 2a-45, Ca 3a-46, Ca 6a-40)

This analysis indicates the sum of the moisture, unsaponifiables, and insoluble matter, generally traces of proteins not extracted in the rendering process. The MIU measures the amount of fat or oil that cannot be turned into soap. The presence of water provokes hydrolysis of the fat with the ensuing increase of free fatty acid content and loss of glycerine yield. A high MIU value implies that the percentage of impurities will not allow the transformation into soap; and also implies an important increase in the pretreatment costs of the fat in order to obtain a high-quality soap.

Other important raw material analyses for soap formulation are the iodine value (AOCS Method Cd 1-25) and the titer (AOCS Method Cc 12-59). The iodine value measures the degree of unsaturation of the fat, oil, or fatty acid, whereas the titer measures the solidification point of fats and fatty acids. High titer values imply hard fats that produce hard soaps, which are more stable. High iodine values imply a high unsaturated fatty acid content, which produces soft soaps but with higher instability due to the reaction of the fatty acids with the oxygen in the air, causing rancidity and generating undesirable color and odor.

Peroxide value (AOCS Method Cd 8-53) and anisidine value (AOCS Method Cd 19-90) show the degree of deterioration and ill treatment a fat may have suffered.

The combination of these two indexes, the total oxidation (TOTOX), measures the presence of components of primary and secondary decomposition of the fat. It allows us to infer the fat's previous history. This is of great importance in evaluating the fat's future behavior.

Saponification color (French Norm NF T 60-216) measures the color of the fatty acids after saponification and can be correlated with the soap color.

For more details on the listed and other AOCS methods, please refer to Chapter 3 in this book.

Effect on the Base Soap

No matter what method is chosen to make the base soap, its quality and stability depend on the quality of the raw materials that have been used.

The effect of the quality of the tallow on the color of the base soap is shown in Table 1.5. The Hunter method used in this table for color valuation measures the reflection of standard light on solid base soap. The Hunter Whiteness Index correlates with the color perceived by the human eye.

As can be seen in Table 1.5, the best base soap color is reached when edible tallow is used. When distilled palm oil stearin fatty acids are used, good quality is achieved. As the quality of the tallow goes down (TWT and BFT) the corresponding products also decrease in quality. Through refining, bleaching, and deodorization we can obtain from raw materials of inferior quality soaps as good as those obtained from the best tallow.

TABLE 1.5
Effect of Raw Material Quality on the Base Soap

Raw Material Type	Raw Material Color (Lovibond Red)	Base Soap Color (Hunter Whiteness)
Edible tallow	3	90[a]
Top white (TWT)	7–8	85[a]
Bleachable fancy (BFT)	50–100	72[a]
Bleached TWT	2	89[a]
Bleached BFT	10–20	80[a]
Packer beef—refined and bleached	<1	93[a]
TWT—hydrogenated	2–3	90[a]
Distilled palm oil stearin fatty acids	<3	93.2[b]

[a]Data based on Appleby, D.B., and Halloran, K.A. (1990) in *Soap Technology for the 1990's,* Edited by Spitz, L., p. 103, AOCS, Champaign, Illinois.
[b]Measured on 23% water soap. Data based on Ainie, K., and Hamirin, K. (1994) in *Selected Readings on Palm Oil and Its Uses,* edited by Abdullah, A., et al., p. 189, Palm Oil Research Institute of Malaysia, Kuala Lumpur.

Thus it may be concluded that the selection of a raw material for soap production must be subjected to a double analysis: its price, and costs that must be incurred in order to reach the established quality goals.

Conclusions

The selection of raw materials and pretreatment consists of a search for the most economical way to obtain a product of the required quality. To this end, both the costs of the raw materials and their necessary pretreatment must be considered to reach the quality objectives.

The raw material quality improvement methods discussed in this chapter are those used today all over the world. The objective is the correct selection of the necessary and sufficient pretreatment based on the raw materials available. The adequacy of this choice will reflect on the base soap with regards to quality, at the lowest cost and with the greatest benefit for the manufacturer.

References

1. Krishnamurthy, R.G., in *Bailey's Industrial Oil and Fat Products,* 4th edn., edited by D. Swern, Wiley, New York, 1982, vol. 2, pp. 188–194.
2. National Renderers Association, *Pocket Information Manual,* National Renderers Association, Alexandria, Virginia, 1993, pp. 11–17.
3. Allen, R.R., in *Bailey's Industrial Oil and Fat Products,* 4th edn., edited by D. Swern, Wiley, New York, 1982, vol. 2, pp. 27–34.
4. Albright, L.F., in *Hydrogenation: Proceedings of an AOCS Colloquium,* edited by R. Hastert, American Oil Chemists' Society, Champaign, Illinois, 1987, p. 11–26.
5. Grace Davison, personal communication, 1995.

6. Patterson, H.B.W., *Bleaching and Purifying Fats and Oils, Theory and Practice,* American Oil Chemists' Society, Champaign, Illinois, 1993, p. 165.
7. PORIM Techno. Palm Oil Res. Inst. Malaysia No. 1, 1984, p. 14.
8. Patterson, H.B.W., *Bleaching and Purifying Fats and Oils, Theory and Practice,* American Oil Chemists' Society, Champaign, Illinois, 1993.
9. Sonntag, N.O.V., in *Bailey's Industrial Oil and Fat Products,* 4th edn., edited by D. Swern, Wiley, New York, 1982, vol. 2, p. 99.
10. Stage, H., *J. Amer. Oil Chem. Soc. 61:* 204–214 (1984).

Chapter 2

The Formulation of Bar Soaps

Edmund D. George[a] **and Joseph A. Serdakowski**[b]

[a]Original Bradford Soap Works, Inc., West Warwick, Rhode Island, U.S.A.

[b]Chemical Engineering Consultants, East Greenwich, Rhode Island, U.S.A.

Introduction

This chapter discusses the inclusion of functional additives in various traditional and synthetic soap bases. By *traditional* we mean the sodium or potassium salts of triglycerides and fatty acids, notably from beef tallow and coconut oil. These include opaque, translucent, transparent, and specialty bases, such as shaving and nonpowdering opaque soaps. The making of these soap bases has been well documented (1–15) and will not be discussed further. We consider syndet soap bases (16–24) to consist of synthetic detergents plasticized with other ingredients to yield a solid composition that can be formed into a bar. Syndet bases can be combined in various ratios with traditional soaps to form combination, or combo, soaps that have properties of both bases. Typically two methods are used for base production. The cold method involves the addition of ingredients into a mixer or blender with no heat; meltable compounds are melted separately and added to the mixer with other ingredients. After proper mixing, the material is released to the production line. The hot, or molten, method involves blending all ingredients into a melt tank that is maintained at elevated temperature and mixing until uniform. The batch is then flaked on a chill roll, and the resulting flake is then processed as any other soap base. The mode of manufacture is largely dependent on formulation and availability of equipment.

The formulation of soap bars has become more complex over the years due to the need to incorporate more additives to an ever-increasing number of soap bases. This is true because consumers have become accustomed to multifunctional products such as conditioning shampoos; deo-antiperspirants; and sunscreens in lotions, creams, and other cosmetics. Traditionally, soaps were designed for cleaning skin and clothes. But over time soaps began to be used as a delivery system for perfumes and superfatting agents. Today, the cleansing aspect appears to be secondary to the effects that various types and amounts of additives can deliver from the soap system. This apparent contradiction leaves the formulator with the task of being familiar with the myriad of compounds that are available for cosmetic and drug use and deciding how they can be applied to the soap and detergent systems used today. The CTFA International Cosmetic Ingredient Dictionary (25) alone lists more than 5,300 cosmetic ingredients manufactured worldwide.

As with any drug or cosmetic, matrix effects must be considered when developing soap formulas. Among these are the following:

- Base-additive interactions
- pH effects
- Additive-additive interactions
- Fragrance effects
- Processing effects

Any combination of the above may positively or negatively affect the physical and aesthetic characteristics of the final product. Sometimes it is difficult to predict every possible consequence of matrix effects. Only with time and experience can the formulator begin to understand these interactions.

Base-Additive Interactions

It is important to be familiar with the types of additives from information in vendor product data sheets and specifications. The chemical and physical properties must be understood before the additive is considered for use. For instance, additives may interact with soap bases by changing the chemical or physical characteristics of the base. Traditional alkaline soap bases may be influenced by the addition of compounds that are acidic enough to break down the soap into fatty acids, which renders the soap base ineffective. This may not be immediately noticeable because the soap base does not have sufficient water to behave like a solution, but it can occur in time after processing and storage. With syndet bases, the opposite may take place, especially if the base contains fatty acids such as stearic acid, which could be neutralized by alkaline additives. This can seriously affect the processing characteristics of the formula.

pH Effects

Much of the stability problems related to pH seems to lie with traditional soaps; the majority of soap additives can be found in cosmetic and personal care products that have an acid pH. Certain compounds such as some quaternary compounds and some fragrance ingredients are unstable under the pH conditions found in traditional soaps. In addition, some OTC active ingredients such as salicylic acid and benzoyl peroxide have greater stability in syndet and combo systems that have neutral to acid pHs.

Additive-Additive Interactions

Similar interactions between additives must be considered in the same light as base-additive interactions (see above).

Fragrance Effects

Fragrances are developed from many compounds, such as aliphatic and aromatic acids, esters, ketones, and glycols. They can profoundly effect the processing characteristics by increasing the softness and tackiness of the soap or, in the case of translucent and transparent soap, influence the clarity of the system. For instance, fragrance solvents such as some glycols and esters appear to soften and/or cloud translucent soaps, making an already difficult base to run more difficult. Vanillin is known to cause severe browning of soaps due to a chemical reaction in the alkaline pH range. We recommend working closely with fragrance suppliers to optimize these fragrance selections prior to production.

Processing Considerations

As described above, chemical interactions can influence the outcome of the finished product, but processing effects can also play a role in the results. Process parameters such as temperatures, shear, recycle of scrap, viscosity, vacuum, die refrigeration, size and type of equipment, milling *vs.* refining, and plodder speed must be monitored and controlled. Although this subject is out of the scope of this chapter, we believe this to be an important aspect of a successful product.

Properties of Soap Base

We offer a brief review of the properties of soap and detergent systems, which will aid in further discussions on formulations.

Chemical Properties

Table 2.1 describes some typical chemical properties of various soap and syndet systems. As can be seen from the table, there is a wide range of chemical properties, especially in transparent, syndet, and combo systems. This is influenced by the variations in the starting materials in the base formulas. For instance, transparent bases can be made from detergents, and fats and oils using combinations of sodium hydroxide, potassium hydroxide, and alkanolamines such as triethanolamine.

TABLE 2.1
Typical Values of Soap Bases

Base Type	pH	% Free Alkalinity as NaOH	% Free Fatty Acid as Stearic Acid	% Water	% Salt as Sodium Chloride
Tallow/coconut	10.0	0.1	0.5	12.0–14.0	<0.5
Translucent	10.0	0.0	0.5–1.5	12.0–14.0	<0.5
Transparent	8.0–10.0	0.1	1.0	10.0–20.0	0.0
Syndet	5.0–6.0	0.0	0–50.0	5.0–10.0	0–2.0
Combo	7.0	0.0	0–40.0	5.0–10.0	0–2.0

Syndet systems can be plasticized with saturated fatty acids, waxes, fatty alcohols, or combinations of these. The formulator must be familiar with these properties when developing additives packages that are to be stable and functional.

Physical Characteristics

The following are some of the more important physical characteristics that will influence the amount and type of additives that will be incorporated into the final formula. It must be noted that the following characteristics should be established with each formulation to generate a complete product profile.

Wear Rate

Wear rate essentially describes the lasting power of the soap bar under use conditions. It is influenced by the various solubilities of the bases, which are determined by the titre of the fat and oils, the type of alkali used, and the amount of water for traditional soap to the type of plasticizer and solubility of the detergent in syndet systems. For instance, transparent soaps have relatively high wear rates because they are formulated with high solvent contents such as glycols, water, and alcohols to maintain clarity. They also may contain a surfactant system that has high solubility to maintain clarity. This combination tends to let the bar "melt away."

On the other hand, syndet systems need plasticizers and binders to hold the bar together. They are typically waxes, starches, fatty acids, and fatty alcohols that have very limited water solubility, making the system less sensitive to wear rate than other soap types.

Crack Resistance and Sloughing

Crack resistance relates to the tendency of soap bars to crack and/or disintegrate when subjected to repeated wet-dry cycles. This is achieved in the laboratory by submersion of one-half the bar in ambient water from 4 to 24 h, then air drying until completely dry. Cracks will appear if the system is prone to cracking. Also, the amount of sloughing can be determined by a similar method of submersion, calculating the weight loss after drying. Our experience indicates that translucent soaps traditionally have poor crack resistance. One theory suggests that translucent systems lack the "grain" that traditional opaque soaps have. This grain, or crystallinity, allows the soap to hydrate and dehydrate uniformly, whereas translucent soap lacks sufficient structure to stabilize the effect. Additives such as sucrose can promote cracking, as well as certain solvents in fragrances. Syndet systems appear to be less prone to cracking but can produce a high rate of sloughing (19). This is the tendency to form gel-like material (mush) when hydrated, depending on the formulation, which is a major drawback of syndet-combo systems.

Wash Down

The feel of a bar during use can be determined by a wash-down test. This is usually performed at lower temperatures (85–90°F) than normal to determine whether any

grit, drag, or sandiness is present in the bar. Syndet and combo systems traditionally are prone to this problem. Some of the causes include improper processing at the
base and bar-making stages or hard particles in the surfactant system. Formulas
containing sodium cocoyl isethionate appear to be prone to this problem. The
concern arises when the bar is cold, before use; the grit may feel unpleasant to the
user, despite the fact that the grit disappears at normal use temperatures. Sandiness
may occur in traditional opaque soap base due to excessively dry particles formed
during the vacuum drying process. Drying the soap to a higher water range
(12%–14%) will typically reduce this problem.

Foaming

Although lathering does not necessarily equate with detergency, consumers associate quick, copious foam with quality and cleaning. Foaming characteristics can be
influenced by many factors. Among them are fat and oil type and ratio or, in the
case of syndets, the type of surfactants and plasticizers. Many additives that are oily
in nature will tend to act as defoamers if incorporated at high levels, such as in
superfatted bars. Traditional soaps will lather poorly in hard water and seawater,
whereas syndets, if properly formulated, perform well. Standard foam height tests
should be performed as part of the product profile.

Color

Soap bases tend to yellow, so that the color of final bar formulations will also
change. Coupled with additive and fragrance instability, this can produce color
variations in a short period of time. Accelerated stability testing in oven, sunlight,
and florescent light will help predict the color stability of the system. The soaps
may remain in these stations for weeks or months depending on the need. One
problem that this presents, however, is the difficulty in maintaining color standards
from run to run. Color chips, photographs, artist drawings, and the like have been
used to preserve the original color intentions, with limited success. We have been
using a reflectance colorimeter to record the color of a sample in a computer database. This instrument mathematically calculates the color as the human eye sees it,
allowing the computer to recall the color at any time in the future as a reference
standard. We have used this method for all types of soaps, including translucent and
transparent, and the yellowness of soap bases. We have found this to be a reliable
tool for color characterization.

Odor

Olfactory evaluation of soap base and finished product is as important as any other
physical characteristic. Consumers tend to view fragrance perceptions as being as
important as any other characteristic. Therefore it is important that fragrances be
formulated for soap use to ensure as much stability as possible. Evaluation for
chemical stability can be carried out under similar conditions as color stability.
Olfactory evaluation is usually reviewed by trained technicians or panelists.

Colorants

Colorants can essentially be divided into three categories:

- Color additives subject to certification (certified)
- Color additives not subject to certification (noncertifiable)
- Color additives not certified (noncertified)

The use of colorants very much depends on the type of product that is being produced. The distinction usually lies in the application of the Federal Food, Drug, and Cosmetic (FD&C) Act (26) and the Fair Packaging and Labeling (FP&L) Act, where the definition and labeling requirements for drugs and cosmetics are made, and soap is exempt. Drugs must use certified colors, whereas cosmetics can use certified and noncertifiable colors. Any soap that makes drug and/or cosmetic claims must be ingredient labeled accordingly with the appropriate colorants. Soaps, however, can use any combination of colors as long as the product meets the definition of *soap* and is claimed as such. Tables 2.2 and 2.3 list some of the current certified and noncertifiable colorants.

TABLE 2.2
Color Additives Subject to Certification (March 1990)

Color Additives	Use in Cosmetics
FD&C Blue No. 1	
FD&C Green No. 3	Except in eye area
FD&C Red No. 4	Externally except in eye area
FD&C Red No. 40	
FD&C Yellow No. 5	
FD&C Yellow No. 6	Except in eye area
D&C Blue No. 4	Externally except in eye area
D&C Brown No. 1	Externally except in eye area
D&C Green No. 5	
D&C Green No. 6	Externally except in eye area
D&C Green No. 8	Externally except in eye area (0.01% max.)
D&C Orange No. 4	Externally except in eye area
D&C Orange No. 5	Externally except in eye area Lip products (5% max.), Mouthwashes Dentifrices (GMP)
D&C Orange No. 10	Externally except in eye area
D&C Orange No. 11	Externally except in eye area
D&C Red No. 6	Except in eye area
D&C Red No. 7	Except in eye area
D&C Red No. 17	Externally except in eye area
D&C Red No. 21	Except in eye area
D&C Red No. 22	Except in eye area
D&C Red No. 27	Except in eye area
D&C Red No. 28	Except in eye area
D&C Red No. 30	Except in eye area

(continued)

TABLE 2.2 (continued)
Color Additives Subject to Certification (March 1990)

Color Additives	Use in Cosmetics
D&C Red No. 31	Externally except in eye area
D&C Red No. 33	Externally except in eye area
	Lip products (3% max.), Mouthwashes
	Dentifrices (GMP)
D&C Red No. 34	Externally except in eye area
D&C Red No. 36	Externally except in eye area
	Lip products (3% max.)
D&C Violet No. 2	Externally except in eye area
D&C Yellow No. 7	Externally except in eye area
D&C Yellow No. 8	Externally except in eye area
D&C Yellow No. 10	Except in eye area
D&C Yellow No. 11	Externally except in eye area

TABLE 2.3
Color Additives Exempt from Certification (March 1990)

Color Additives	Use in Cosmetics
Aluminum powder	Externally including in eye area
Annetto	No restrictions
Bismuth citrate	Scalp hair dye only
Bismuth oxychloride	No restrictions
Bronze powder	No restrictions
Caramel	No restrictions
Carmine	No restrictions
Carotene, beta	No restrictions
Chromium hydroxide green	Externally including in eye area
Chromium oxide greens	Externally including in eye area
Copper powder	No restrictions
Dihydroxyacetone	Externally except in eye area
Disodium ETDA–copper	Cosmetic shampoo only
Ferric ammonium ferrocyanide	Externally including in eye area
Ferric ferrocyanide	Externally including in eye area
Guanine	No restrictions
Guaiazulene	Externally except in eye area
Henna	Scalp hair dye only
Iron Oxides	No restrictions
Lead acetate	Scalp hair dye only
	(0.6% Pb w/v max.)
Manganese violet	No restrictions
Mica	No restrictions
Potassium sodium copper chlorophyllin	
(chlorophyllin–copper complex)	Cosmetic dentfrices (0.1% max.)
Pyrophyllite	Externally except in eye area
Silver	Nail polish only (1% max.)
Titanium dioxide	No restrictions
Ultramarines (blue, green, pink, red. violet)	Externally except in eye area
Zinc oxide	No restrictions

Certified Colorants

Listed in Table 2.4 are some of the more common colorants found in soap products. The certified colorants may be water soluble, oil soluble, or oil dispersible and may also include the corresponding metal lakes. The solutions or dispersions are typically made at the 1%–2% level, but higher loads (30%–50% pigment) may be obtained from vendors who have specialized equipment for grinding and dispersions. Solubility and dispersion tables should be consulted to determine the optimum concentrations and dispersant. As a general rule, we recommend making concentrations that are of sufficient strength to keep the amount added to soap batches relatively low, so as not to add too much dispersant to the batch.

Certified colorants tend to have the lowest stability of the three categories. Factors include pH, day and florescent light, heat, and additive interaction. For instance, FD&C Green No. 3 renders a green color below pH 7 but royal blue above pH 7. Stability stations should be utilized to determine as overall stability in the surfactant system.

Noncertifiable Colorants

These colorants tend to be very stable compounds and are used extensively in cosmetics as eye shadows, mascaras, and facial makeup. In soap products they can be used as primary or secondary colorants in which they tend to stabilize color drifting due to certified color use. These compounds should be dispersed or wet out before use to maximize their color value. We have found that a 30%–40% potassium cocoate solution gives good dispersability to the colorants. The formulator can develop a color by approximating the amount of dry pigment and then dispersing the powder in the cocoate solution. A "blooming" effect, in which large pigment particles cause a slow migration of color into the soap system, can occur if the colorant is not dispersed properly. Unfortunately, this can occur within hours or days of production causing a blotched color in the finished product. Newer

TABLE 2.4
Common Soap Colorants

Certified	Noncertifiable
FD&C Green No. 3	Caramel
FD&C Red No. 4	Chromium hydroxide green
FD&C Red No. 40	Chromium oxide greens
FD&C Yellow No. 5	Iron oxides
D&C Green No. 5	Mica
D&C Orange No. 4	Titanium dioxide
D&C Red No. 17	Ultramarines (blue, green, pink, red, violet)
D&C Red No. 30	Zinc oxide
D&C Red No. 33	
D&C Violet No. 2	
D&C Yellow No. 10	

materials on the market contain predispersed colorants (Eastman Chemical Corporation) for ease of use. Opacifiers such as titanium dioxide and zinc oxide are used to give uniformity to the color system. Without them, the soap bar may appear to have light- and dark-colored areas caused by various compression areas from the processing and pressing operations. Titanium dioxide is offered in both rutile and anatase forms, with the water-dispersible, anatase USP grade used most often. Rutile types typically would not give the brightness or hiding power that the anatase gives. However, a new rutile version (Kerr-McGee Corporation) appears to be comparable to the anatase type, giving equal opacity and somewhat superior brightness.

Noncertified Colorants

Noncertified pigments are often used in bar soaps as long as they conform to the definition and labeling requirements of soap. These colorants tend to be very stable and less sensitive to pH and provide a wide range of brightness and hues. They are often supplied as dispersions with concentrations of 25% to 50%. Relatively small amounts of these pigments are required to achieve dark colors in most soap bases. Table 2.5 lists some of the common types of pigments.

Fragrances

There are endless possibilities as to the types of fragrances that can be used in soap. Virtually any scent can be created to fit a product profile or marketing concept. All fragrances should be developed for use with a particular base type to ensure proper stability and fragrance quality. Use levels will vary depending on composition, desired impact, price, compatibility with the soap system, and effect on processing, with a low range of 0.25%–0.50% for masking purposes to 3.0%–4.0% for prestige fragrance bars. At high use levels it is important to use vacuum in the extrusion process to minimize surface bubbles, or blistering, caused by the fragrance components. A very small amount of fragrance will be lost through the vacuum system, but the overall effect will be a uniform extrusion.

TABLE 2.5
Common Noncertified Colorants in Soap

Pigment Name	Pigment Type	Color Index Number (CI)
Pthalo Blue RS	Blue 15	74160
Pthalo Blue	Blue 15:1	74250
Pthalo Blue GS	Blue 15:3	74160
Pthalo Green	Green 7	74260
Quinacridone Magenta	Red 122	73915
Napthol Red ITR	Red 5	12490
Carbozole Violet	Violet 23	51319
Hancock Yellow G	Yellow 1	11680

Fragrances will affect processing of soaps to the extent that they will not permit the soap to be extruded or pressed (stamped) within desired process rates. This is seen in syndet and combo systems, but more dramatically in translucent systems where clarity and firmness are critical. We have found that fragrances that contain certain solvents such as dipropylene glycol (DPG) and diethylpthlate (DEP) tend to make translucent systems sticky to the point where the soap does not extrude or press at all. When these solvents were removed from the fragrance (so-called nondiluant formulation), the soap met all process parameters. Certain resonoid compounds used in fragrances may cloud translucent systems because of solubility and particulate concerns. Other components may produce similar effects and should be investigated through experimentation.

In short, every effort should be made to design and verify fragrance compositions that are suitable for the type of soap base being used.

Additives

As mentioned previously, the addition of compounds to achieve a desired functional and/or marketing position has become the prime focus of soap formulators. Below are described several categories that are important in achieving the goals of various product profiles. A more detailed description of functional ingredients can be found in the CTFA Cosmetic Ingredient Handbook (27) and McCutheon's Functional Materials (28). Several important categories are as follows:

- Emollients
- Humectants/moisturizers
- Occlusive agents
- Dermabrasive agents
- Drug components
- Anti-irritants
- Foam boosters
- Miscellaneous compounds

A difficult task for the formulator is the evaluation of additive benefits in soap bases. Most often, expert panels are used to evaluate performance and aesthetic perceptions. A model of a simplified evaluation log that might be used to screen formulations is shown in Fig. 2.1.

More sophisticated and highly technical techniques can be employed, such as transepidermal water loss (TEWL), skin elasticity, tracer studies for residual ingredients after wash-off, spectroscopic and florescence studies, and human skin models, to establish efficacy. Kajs and Garstein (29) further review the use of biophysical instrumentation in evaluating cleansing products.

IN-HOUSE PANEL STUDY

Product Catagory_____   Male   ☐

Bar Shape_____   Female   ☐

Product Code_____

<u>**RATING**</u>

1= Extremely Poor   10= Outstanding

Performance Evaluation:	1	2	3	4	5	6	7	8	9	10
Foaming How well does the product foam?	☐	☐	☐	☐	☐	☐	☐	☐	☐	☐
Foam Stability How stable is the foam during use?	☐	☐	☐	☐	☐	☐	☐	☐	☐	☐

Foam Type
What is the type of foam?

☐ Flat   ☐ Loose   ☐ Creamy   ☐ Big bubbled   ☐ Soap-like

| Detergency
How well does the product clean? | ☐ | ☐ | ☐ | ☐ | ☐ | ☐ | ☐ | ☐ | ☐ | ☐ |
| Wear Rate
How does the product wear down? | ☐ | ☐ | ☐ | ☐ | ☐ | ☐ | ☐ | ☐ | ☐ | ☐ |

Mushing
How does the product appear in the soap dish between use?

☐ Mushy   ☐ Slimy   ☐ Cracked   ☐ Firm   ☐ Soap-like

Aesthetics Evaluation:

Afterfeel
What kind of afterfeel does the product leave you?

☐ Irritating   ☐ Dry   ☐ Residue   ☐ Smooth   ☐ Silky

Usage Feel
How does the product feel during usage?

☐ Grit   ☐ Drag   ☐ Too slippery   ☐ Smooth   ☐ Soap -like

Fig. 2.1. A model of a simplified evaluation log.

It must be remembered that is difficult to incorporate the right type and amount of functional materials in surfactant systems that have a real or perceived benefit. This is somewhat intuitive because the main purpose of cleansers is to remove (clean) material from the skin, not add it.

Emollients

Emollients are compounds that are used to impart and maintain softness and pliability to the skin, and generally to improve the skin's overall appearance. Fatty esters, alkoxylated ethers, and alkoxylated alcohols compose the bulk of the ingredients in this category. Typical use levels are 1%–3%. These compounds are stable under normal bar soap conditions. Table 2.6 describes the general categories and typical emollients. This list is far from complete but serves as an example of common ingredients that are used in cosmetics that may be used in soap products to obtain a desired benefit.

Care should be taken to review the physical properties to determine the best way to incorporate the ingredients into the base, as well as any stability considerations (such as pH and temperature).

Humectants/Moisturizers

Humectants are skin-conditioning agents that increase the moisture to the skin (see Table 2.7). They tend to function as hygroscopic compounds that attract water and thus reduce the amount of water leaving the skin. Their effectiveness is dependent on the humidity and environment in which they operate. Dahlgren et al. (30) reports on instrumental assessments as well as perceived skin benefits of glycerine in various surfactant systems. The results indicate that high levels of glycerine (that is, 10%) provides improved skin feel and softness. Use levels vary widely from 0.10% to 10%.

It should be noted that monosaccharides (simple sugars), such as fructose and glucose, darken under alkaline conditions; this is accelerated by the small amount of free alkali that is present in traditional soap bases. The neutralization of the free alkali by the addition of a fatty acid, such as coconut or stearic acid, will help to delay the darkening effect. The discoloration will vary depending on the type and amount of the sugar. Disaccharides such as sucrose are relatively stable under mild alkaline conditions.

Water droplet formation, or "sweating," may occur with the use of high levels (10% and higher) of glycols and alcohols, such as glycerine and sorbitol, under humid conditions. This is a persistent problem with transparent soaps that employ high levels of multiple humectants that serve also as solvents and solubilizers. This hygroscopic effect appears to be less of a problem with other types of soap systems that employ lower levels (< 5%) of these materials.

Occlusive Agents

Occlusive agents include ingredients designed to prevent moisture evaporation from the skin, thereby helping to maintain soft, smooth skin (see Table 2.8). They typically are lipid in nature and may be blended to achieve a desired effect in so-called dry-skin products. Use levels are in the 1%–10% range. The higher levels may cause extrusion and pressing problems by making the soap excessively soft and sticky, but this may be reduced by incorporating the ingredient(s) in the base-making stage or premilling/refining into the base prior to extruding.

TABLE 2.6
General Emollient Groupings and Common Ingredients

Esters	Ethers	Alcohols	Oils and Oil Type	Glycols	Lanolin Derivatives	Silicone Derivatives
Butyl myristate	Methyl gluceth-10	Cetearyl alcohol	Castor oil	Glycerine	Acetylated lanolin	Dimethicone Copolyol
Cetyl palmitate	Methyl gluceth-20	Cetyl alcohol	Jojoba oil	Propylene glycol	Lanolin	Silicone fluid
Glycerol laurate	PPG-10 methyl glucose ether	Stearyl alcohol	Mineral oil		Lanolin esters	
Glycerol stearate	PPG-20 cetyl ether		Mink oil		Lanolin alcohols	
Isopropyl myristate	PPG-20 lanolin alcohol ether		PEG-glycerides		Lanolin fatty acids	
Isopropyl palmitate	PPG-20 methyl glucose ether		Petrolatum		PEG-lanolin	
Octyl palmitate	PPG-20 oleyl ether		Shea butter			
Tridecyl neopentanoate	PPG-3 myristyl ether		Sweet almond oil, wheat germ oil, olive oil			

TABLE 2.7
Common Humectants

Saccharides	Ethers	Alcohols	Glycols	Miscellaneous Compounds
Fructose	Methyl gluceth-10	Mannitol	Glycerine	Acetamide MEA
Glucose	Methyl gluceth-20	Sorbitol	Propylene glycol	Hydrogenated
Honey	PPG-10 propylene			starch hydrosylate
Lactose	glycol			PCA
Sucrose				Sodium PCA
Xylitol				Urea

Dermabrasive/Exfoliating Agents

Abrasive agents work in conjunction with a cleanser to remove the outermost layer of the stratum cornea by scrubbing, thereby producing a smoother feel to the skin. Loden et al. (31) has described a method to measure the effects of abrasive cleansers on the stratum cornea and verified that there is a perception of smoothness after the use of these agents. Currently, natural components such as bran, loofa, oatmeal, jojoba beads, and seeds are popular because they can be blended to achieve a desired look in the cleanser and a desired feel on the skin in addition to a "natural" claim. Polyethylene beads of various sizes are also used effectively where a natural or visual component is not needed.

These agents often can be added to the amalgamator and processed under normal conditions with some precautions. When refiners are used, the screen sizes should be large enough to allow the scrub agents to pass through; otherwise they will be filtered out. Likewise, when using roll mills, the milling action should not break down the component into fines, which will reduce the scrubbing effect. The

TABLE 2.8
Occlusive Agents

Castor oil
Coconut oil
Dimethicone
Hydrogenated oils
Jojoba oil
Jojoba wax
Mineral oil
Natural and synthetic waxes
Olive oil
Petrolatum
Sesame oil
Shea butter
Sweet almond oil
Tallow
Wheat germ oil

proper gapping of the rolls and the use of low-shear mills will help reduce the problem. If this is not possible, the roll mill is by-passed and the scrub agent is fed into the Duplex Vacuum Plodder. In this case, it is important to use either a very coarse refining screen or no screen at all in the first stage of the Duplex Vacuum Plodder. This problem occurs mostly with the larger-size natural components such as loofa and oatmeal and to a much lesser extent with polyethylene beads.

Drug Components

Soap bar formulations in categories such as antidandruff, astringent, and skin protectorant are not widely known, although antimicrobial and acne soap products have become over-the-counter (OTC) drugs. In the United States, each drug category is governed by monographs published by the federal Food and Drug Administration (FDA). Each monograph, in turn, sets the conditions under which a product that is claimed in a particular category may be sold. The formulator should consult these monographs to obtain specific information of the category prior to development work.

These products need to be controlled according to Good Manufacturing Procedures (GMP) because they are drugs and subject to FDA audits and inspections.

Antimicrobials

Triclosan and Triclocarban are two compounds that are most widely used in antimicrobial bar soaps. Typical use levels are 0.30%–1.0% for Triclosan and 1.0%–1.5% for Triclocarban. They are incorporated at the amalgamator stage and may be dispersed or dissolved in a suitable solvent, such as a perfume, prior to addition. As with all drug products, the soap line should be validated to ensure that the finished product is homogeneous and that the proper level of antimicrobial is in the soap .

Acne

The most common approved acne ingredients used in soap are salicylic acid (0.5%–2%) and sulfur (3%–10%). They have been formulated both in traditional and syndet/combo bases. Sulfur is stable in both alkaline and acid bases. Salicylic acid, however, is stable with acid bases but may destabilize traditional alkaline bases. Therefore a mild alkali such as triethanolamine is sometimes added to stabilize the salicylic acid.

Iron contamination should be avoided because both ingredients will remove iron from equipment into the soap bar, resulting in a "blooming" of the iron particle in the soap. This will cause soap discoloration and may lead to rancidity in traditional soap.

Both ingredients are typically added at the amalgamator stage and processed under normal conditions. As stated above, GMP procedures for this product category should be in force.

Anti-irritants

Several ingredients are on the market that claim to have anti-irritant properties. Among these are sucrose esters (1%–3%), bisabolol (0.1%–0.2%), lactylates (2%–5%), amphoterics (2%–5%), and ethoxylated vegetable oils (2%–5%).

Most of these compounds have been evaluated in shampoo systems in which skin and eye irritation levels are reportedly reduced. The overall mechanism is not fully understood but is believed to involve several factors including binding or complexation of irritants, blocking sites that are prone to irritation, and prophylaxis that covers the skin, reducing or preventing irritant contact.

The soap chamber irritation assay, as described by Frosch and Kligman (32), is a common *in vivo* test method to determine irritation levels of surfactant systems. It involves 5-day exposures of 8% surfactant solutions to the ventral side of the forearm, which is evaluated for scaling, redness, and fissuring. The Zein test, as described by Gotte (33), is an *in vitro* method that measures the ability of surfactants to solubilize the vegetable protein zein. The higher the solubility, the greater are the defatting and irritancy of the surfactant. Newer *in vitro* methods are being developed that utilize human skin factors in a benchtop assay that will be useful to the formulator for screening various additive packages.

Secondary Surfactants

Secondary surfactants are often used to increase the performance of the soap bar resulting in improved skin feel, reduction of irritancy from the primary surfactant, improved solubility, or improved quantity and quality of foam. Typically they are added at relatively low levels (under 5%) as an adjunct to the primary surfactant. Theoretically, most surfactants found in shampoos and liquid soap may be used in bar soap systems as long as they are compatible with the system and are stable at the given pH. However, many of them are pastes or liquids that tend to make soap systems tacky, resulting in increased process problems. This may be due not only to their physical state but also to their tendency to lower the viscosity of soap systems, causing softening of the system. Furthermore, this may be complicated by additives and fragrances, as previously discussed.

However, there are several types of surfactants that are particularly useful in soap systems. Table 2.9 lists some of the more common ingredients. These ingredients may be added to the amalgamator and processed under normal conditions As with all powders, care should be taken in handling dust associated with these surfactants. An alternative method of introduction in traditional bases is to add the material to molten soap in a crutcher and then dry, ensuring maximum uniformity. For syndet and combo bases, these surfactants may be added during the normal compounding process. Rieger (34) has developed an excellent overview of all surfactant types that are useful to the formulator.

TABLE 2.9
Secondary Surfactants

Type	Example	Chemical Type	Form
Acyl isethionates	Sodium cocoyl isethionate	Anionic	Solid
Amphoterics	Disodium cocoamphodiproprionate	Amphoteric	Paste
Sarcosinates	Sodium cocoyl sarcosinate	Anionic	Liquid–solid
Sulfosuccinates	Disodium lauryl sulfosuccinate	Anionic	Liquid–solid
Sulfoacetates	Sodium lauryl sulfoacetate	Anionic	Solid

Miscellaneous

Optical brighteners are sometimes used to shift the appearance of a product from a yellow tone to a bluer tone. Disodium distyrylbiphenyl disulfonate is particularly effective for this purpose at approximately 0.03% in traditional soap base for products labeled as soap. Currently, this compound is viewed as a noncertified colorant in the United States and is thus not acceptable in products labeled as cosmetics. It is approved, however, in the EEC (European Economic Community) for use in cosmetics.

The challenge in using encapsulated products in the soap industry is ensuring the survival of the capsule during the processing while having a small enough particle size to minimize a drag or grit feeling in washing. Recent encapsulation technology (3M Corporation) has allowed the incorporation of colorants, fragrances, additives, and active ingredients in the soap-making process without destroying the capsules. This allows the delivery of process- or formula-sensitive materials during washing for enhanced effectiveness.

In conclusion, the formulator must take a balanced approach in developing surfactant systems. Consideration must be given to base type and function, which will dictate acceptable additives for the product category. In turn, process considerations must be addressed to manufacture a quality product. Surfactant systems can be overwhelmed trying to achieve higher and higher functionality for a world of ever-changing cultures.

References

1. Thomsenn, E.G., and C.R. Kemp, *Modern Soap Making,* MacNair-Dorland, New York, 1937.
2. Davidsohn, J., E.J. Better, and A. Davidsohn, *Soap Manufacture,* vol. 1, Interscience, New York, 1953.
3. Swern, D., ed., *Bailey's Industrial Oil and Fat Products,* vol. 1, 4th edn., J. Wiley and Sons, New York, 1979.
4. Woolatt, E., *The Manufacture of Soaps, Other Detergents and Glycerine,* Halstead Press, New York, 1985.
5. George, E., and J. Serdakowski, Computer Modeling in the Full Boil Soap Making Process, *HAPPI, 24(1):* 34–47 (1987).
6. Chambers, J., T. Instone, and B. Stuart, Eur. Pat. Appl. 385,796 (1990).

7. Chambers, J., and T. Instone, Eur. Pat. Appl. 350,306 (1990).
8. Verite, C., and A. Caudet, Eur. Pat. 336,803 (1981).
9. Wood-Rethwill, J., R. Jaworski, and G. Myers, U.S. Pat. 4,897,063 (1989).
10. Jungerman, E., T. Hassapis, R. Scott, and M. Wortzman, U.S. Pat. 4,758,370 (1988).
11. Dawson, G., and G. Ridley, Eur. Pat. Appl. 311,343 (1989).
12. Joshi, D., U.S. Pat. 4,493,786 (1985).
13. Nagashima, T., Y. Usuba, T. Ogawa, and M. Takehara, U.S. Pat. 4,273,684 (1981).
14. O'Neill, J., J. Komor, T. Babcock, R. Edmundson, and E. Shay, U.S. Pat. 3,793,214 (1974).
15. Deweever, E., and T. Carroll, U.S. Pat. 3,903,008 (1975).
16. Crudden, J.J., U.S. Pat. 5,186,855 (1993).
17. Kutny, Y., F. Osmer, J. Podgorsky, D. Richardson, and K. Rys, U.S. Patent 5,041,233 (1991).
18. Lee, R., T. Instone, and J. Chambers, Eur. Pat. Appl. 434,460 A1 (1989).
19. Hollstein, M., and L. Spitz, Manufacture and Properties of Synthetic Toilet Soaps, *Soap and Cosmetic Specialties, 59(1):* 29–34, 51 (1983).
20. Barker, G., L. Safrin, and M. Barabash, U.S. Pat. 4,335,025 (1981).
21. Kamen, M., and I. Ugelow, U.S. Pat. 3,562,167 (1971).
22. Chemical Services (Proprietary) Ltd., British Patent 1,153,303 (1969).
23. Geitz, R., U.S. Pat. 2,894,912 (1959).
24. Dederen, J.C., Skin Cleansing and Mildness: A Comparison, *Cosmetics and Toiletries Manufacture Worldwide,* 1993, pp. 110–124.
25. Nikitakis, J.M., G.N. McEwen, and J.A. Wenninger, *CTFA International Cosmetic Dictionary,* 4th edn., The Cosmetic, Toiletry, and Fragrance Association, Washington, D.C., 1991.
26. Title 21, U.S. Code of Federal Regulations (21CFR).
27. Wenniger, J.A., and G.N. McEwen, *CTFA Cosmetic Ingredient Handbook,* 2nd edn., The Cosmetic, Toiletry, and Fragrance Association, Washington, D.C., 1992.
28. *McCutheon's Vol. 2: Functional Materials,* North American edn., The Manufacturing Confections Publishing Co., Glen Rock, New Jersey, 1993.
29. Kajs, T.M., and V. Gartsein, Review of the Instrumental Assessment of Skin: Effects of Cleansing Products, *Journal of the Society of Cosmetic Chemists 42(4):* 249–279 (1991).
30. Dahlgren, R.M., M.F. Lukacovic, S.E. Michaels, and M.O. Visscher, "Effects of Bar Soap Constituents on Product Mildness," in *Second World Congress on Detergents,* edited by A.R. Baldwin, Amercian Oil Chemists' Society, Champaign, IL, 1987.
31. Loden, M., and A. Anders, Mechanical Removal of the Superficial Portion of the Stratum Corneum by a Scrub Cream: Methods for the Objective Assessment of the Effects., *J. Soc. Cosmet. Chem. 41:* 111–121 (1990).
32. Frosch, P.J., and A.M. Kligman, The Soap Chamber Test, *J. Am. Acad. Dermatol. 1:* 35–41 (1979).
33. Gotte E., *Asthet Medzin 15:* 313 (1966).
34. Rieger, M.M., Surfactant Encyclopedia, Allured Publishing Co., Wheaton, Illinois, 1993.

Chapter 3

Quality Control and Evaluation of Soap and Related Materials

Thomas E. Wood

Valley Products Company, Memphis, Tennessee, U.S.A.

Introduction

The first part of this chapter provides a review of the various chemical, physical, and instrumental test methods that are commonly used in evaluating fats, oils, and fatty acids for soap making. The relationships among the key attributes are discussed along with their relevance to the quality and physical properties of both the raw materials and the soap products. Also included is a review of the typical test methods used for quantifying minor residual components and certain functional ingredients found in soap raw materials and soap products. The second part of this chapter includes a review of current chromatographic methods that are applicable to fats, oils, fatty acids, and key ingredients used in soap products. The third part of this chapter presents a review of the test methods prescribed for use in evaluating the physical attributes and performance characteristics of finished soap bars.

Analytical Methods for Soap and Related Materials

The analytical methods available to the soap chemist can be thought of as consisting of two types, including the classical physical and chemical analytical procedures that are found in standard reference manuals of analytical methods and the modern instrumental methods including gas chromatography and high-performance liquid chromatography. The two most important reference sources for analytical methods for soap and soap raw materials include the *Official Methods and Recommended Practices of the American Oil Chemists' Society* (1) and the *Annual Book of ASTM Standards* of the American Society for Testing and Materials, Volume 15.04 (2). The instrumental methods are generally found in publications such as the *Journal of the American Oil Chemists' Society*, the *Journal of Chromatography*, the *Journal of Liquid Chromatography*, and other similar publications. (Please refer to Table 3.4 at the end of this section for a concise list of references to selected AOCS official test methods for soap and soap raw materials.)

Chemical and Physical Characteristics of Soap Raw Materials

Among the chemical properties of fats, oils, and fatty acids that are important to the soap chemist, three of the most important are acid value, saponification value, and

iodine value. Additionally, the *trans* isomer content can be an important consideration whenever hydrogenated stocks are involved. Other important characteristics
include the titer, free fatty acid, raw color, bleached color, moisture, insoluble
impurities, and unsaponifiable matter.

Acid Value

Acid value is defined as the number of milligrams of potassium hydroxide required
to neutralize the free acids in 1 g of sample.

For samples of fats and oils, the determination of the acid value involves the
simple titration, in alcohol, of an appropriately sized sample to the phenolphthalein
end point with essentially no sample preparation required. In this case, the acid
value is virtually a measure of the percentage of free fatty acid present and can be
expressed as such by simple mathematical conversion. Typically, in the case of fats
and oils, the level of free fatty acid should be very low, ranging from near zero to
several tenths of a unit for higher-grade material to a few whole units for lower-
grade oils. Please refer to AOCS Official Method Cd 3d-63 for more information
on acid value determination for fats and oils.

For samples of fatty acid stock, the acid value determination is also a direct
titration of the sample in alcohol to the phenolphthalein end point without any
special sample preparation. Because fatty acid stocks by their nature are essentially
all free fatty acid, the acid value will range above 200 units for most of the common
fatty acid blends used in soap making. For fatty acid stocks, the acid value will
approach the saponification value. Please refer to AOCS Official Method Te 1a-64
for the details for acid value determination of fatty acid blends.

For soap products, the acid value is determined on the total fatty acids present
including the free acids, which may range from none to several percent, plus the
major portion that is combined with the cation as soap. Consequently, the acid value
determination for soap requires an initial sample preparation where the combined
fatty acids are liberated by acidulation with an excess of sulfuric acid, recovered,
and then dried. Following this sample preparation, the acid value is determined on
the fatty acids as just described for fatty acid blends. Again, an acid value of more
than 200 units would be expected for the fatty acids associated with most ordinary
soaps. Please see AOCS Official Method Da 14-48 or ASTM Standard Method D
460, Sections 48 and 49, for more details on acid value determination of soap.

The acid value is a key indicator of the fatty acid composition of soap because
it is inversely and linearly related to the average molecular weight and chain length
of the fatty acids in the blend. For example, tallow fatty acid has an approximate
acid value of 204, whereas coconut fatty acid has an approximate acid value of 268.
The usefulness of the acid value determination lies in helping to establish the composition of soap fatty acid blends. In the case of a fatty acid blend that is derived
from an 80:20 tallow:coconut oil blend, for example, the expected acid value should
be about 216 based on the weighted average of the acid values for the blend. If the
ratio were to shift toward higher tallow content and lower coconut oil content, a

lower acid value would result. For example, a 90:10 tallow:coconut oil blend would result in an approximate acid value of 210. Conversely, a lower tallow:coconut oil ratio would result in a higher acid value. A 70:30 tallow:coconut oil blend would have an acid value near 223.

The acid value is frequently included in specifications for soap with only a lower limit indicated. This is done for reasons of both quality and economics because tallow, with the lower acid value, is the lower-cost commodity with the poorer performance characteristics.

Saponification Value

Saponification value is defined as the number of milligrams of potassium hydroxide required to saponify 1 g of sample.

The procedure is carried out directly on a sample of the stock for both commercial whole oils and fatty acid blends. In the case of soap samples, the fatty acids must first be prepared by acidulating a sample of the soap, recovering the liberated fatty acids, and drying the recovered fatty acids. The saponification value determination is carried out by refluxing the sample of oil or fatty acid with an excess of potassium hydroxide in alcoholic solution for 30 to 60 min. Along with each sample, or set of samples, a blank is also run. After the reaction is completed and cooled, the excess potassium hydroxide is titrated with standardized 0.5N hydrochloric acid. With appropriate calculation, the difference between the titrations of the blank and the sample is then reported as the saponification value. For more information, see AOCS Official Methods Cd 3-25 for fats and oils, Tl 1a-64 for fatty acids, and Da 16-48 for soap and soap products.

Note that any difference between results of the saponification value and the acid value on a given sample is called the *ester value*. It is not unusual for the saponification value to be one or two units higher than the acid value on a given sample of soap fatty acid. This results from the reaction of trace amounts of naturally occurring ester-like compounds that are reactive under the conditions of the saponification value determination but not under the milder conditions of the acid value determination.

Iodine Value

Iodine value is defined as the number of centigrams of iodine absorbed by 1 g of sample.

For samples of fats, oils, and fatty acids, the iodine value determination is performed directly on an appropriately sized sample of the material that has been liquefied by melting and filtered to remove any trace impurities, including moisture. For samples of soap, the fatty acids must first be liberated with sulfuric acid, recovered, and then dried. The iodine value is then determined on the prepared fatty acids.

The procedure is carried out by reacting the sample of oil or fatty acid with an excess of iodine monochloride solution (Wij's solution) and titrating the excess iodine with standardized sodium thiosulfate solution using a starch indicator for the end-point determination. Each sample, or set of samples, is done with a blank to determine the quantity of iodine consumed by the sample. The difference between the blank and the sample is attributable to the iodine absorption by the sample. With appropriate calculation, the difference is reported as the iodine value for the sample. See AOCS Official Methods Cd 1-25 for fats and oils, Tg 1a-64 for fatty acids, and Da 15-48 for soap and soap products for detailed information on the procedures. Also, for soap and soap products, refer to ASTM Standard Method D 460, Sections 50 to 52, for detailed procedures.

The iodine value is directly related to the degree of unsaturation present in the fat, oil, or fatty acid. The iodine value also serves as an indicator of the relative hardness of fats and of the derived fatty acids and soap product for otherwise comparable fat stocks. In general, higher levels of unsaturation, as indicated by higher iodine values, indicate a tendency toward softer fat stocks and softer soap product. This relationship is clearly evident when comparing natural animal fats such as tallow and grease that have roughly comparable average chain length distributions and average molecular weights, as indicated by their saponification values, but that have markedly different amounts of unsaturated fatty acids. Tallow, with its lower iodine value indicating less unsaturation, will melt at a higher temperature and will result in firmer soap compared with grease.

The underlying basis for the physical effects of lower melting point and softer material associated with natural animal fats that have a higher degree of unsaturation is found in the predominance of *cis*-unsaturated fatty acids in naturally occurring fats. Due to the geometry around the *cis* double bonds, the molecules with the *cis* configuration have a distinctive bend in the carbon chain at the site of the carbon-to-carbon double bond. This results in looser packing of molecules in the solid, causing reduced intermolecular forces and, consequently, lower melting points. The effects on the properties of fatty acids and soap resulting from varying the *cis* and *trans* isomer content is discussed in a later section.

Titer

Titer is defined as the temperature, expressed in degrees Celsius, at which fatty acids solidify.

Generally, titers of fatty acids will vary inversely with iodine values. Titer is frequently used as a quality control and process control measure in soap making because it is a good indicator of the processing characteristics of the resultant soap at the bar-finishing stage of manufacturing. Soap made from higher-titer fatty acid blends tend to be firmer and vice versa. The effects of *trans* isomer content on this relationship are discussed in a later section.

The typical inverse relationship between titer and iodine value for otherwise comparable fats is illustrated in Table 3.1. Both lard and tallow are similar in average molecular weights, but lard, with the significantly higher iodine value, has a much lower titer and will form a softer soap.

TABLE 3.1
Comparative Data for Lard and Tallow

Property	Lard	Tallow
Saponification value	190–202	190–200
Iodine value	53–77	35–48
Titer (°C)	32–43	40–46

For tallow, it has been reported by Grompone (3) that titer is fundamentally dependent on the stearic/oleic acid ratio. The relationship between titer of tallow and stearic acid content is direct and linear. The relationship between titer and oleic acid concentration is inverse and linear. It appears that titer is not influenced significantly by the various levels of palmitic acid.

The titer test is always performed on the sample in the form of free fatty acids. For fatty acid blends, the procedure requires no special sample preparation. For fats and oils, a sample of the stock to be evaluated must first be saponified, followed by acidulation and then recovery and drying of the fatty acids. For soap samples, the sample needs to be acidulated with subsequent recovery and drying of the fatty acids. Please refer to AOCS Official Methods Cc 12-59 for fats and oils, Tr 1a-64 for fatty acids, and Da 13-48 for soap and soap products. Also, for soap and soap products, see ASTM Standard Method D 460, Sections 46 and 47.

Effects of Trans Isomers on Fatty Acid and Soap Properties
Variation in the *trans* isomer content of fatty acid blends affects the titer, soap characteristics, and the relationship between titer and iodine value for otherwise similar fatty acid stocks. Although the iodine value can serve as a useful indicator of hardness for naturally occurring fats where *cis* unsaturation is prevalent and *trans* unsaturation is low and consistent, this reliability can significantly diminish whenever hydrogenation is involved. If hydrogenation is carried out with temperatures that are higher than normal, hydrogen feed that is lower than normal, or poor catalyst, a significant amount of rearrangement from *cis* to *trans* isomers can occur rather than the reduction of the double bonds. During hydrogenation of fatty acids, the reduction of *cis*-unsaturated fatty acids, with their characteristically nonlinear structure, to saturated fatty acids, which are essentially linear molecules, will predictably result in the formation of harder fatty acid with a lower iodine value. However, the rearrangement of *cis* to *trans* isomers, insofar as it occurs, will not contribute to the lowering of the iodine value but will result in harder fatty acid stock with a higher melting point.

The relative amount of *cis* and *trans* isomers not only has an impact on the melting point and hardness of the fatty acid, but also affects the hardness and plasticity of the resulting soap. The *trans* isomer content will influence titer independently of the degree of saturation. Again, this phenomenon can become important when fatty acids used in soap making are subjected to hydrogenation. The conversion of a significant amount of *cis* to *trans* isomers will result in harder than expected soap for a given iodine value. Under these conditions, titer may be a more useful process control measure for the soap maker than iodine value.

Some interesting relationships between melting points and iodine values can be seen in Table 3.2 (4), in which analogous sets of fatty acids that have various degrees of unsaturation and *cis/trans* ratios are listed. For the four groups of isomers listed, both the iodine values and melting points are given. In the first pair, both isomers have a double bond at carbon atom 6; the first is a *cis* structure and the second is a *trans* structure. Both have an iodine value of 90. Note, however, the difference in melting points; the *trans* isomer has a melting point 21C° higher than the *cis* isomer. The other two pairs of isomers listed here show the same kind of relationship between isomeric structure and melting point. In the last set, each of the three fatty acid isomers contains three double bonds. Note that in the first structure shown, all three double bonds are *cis;* in the second case, one is *cis* and the other two are trans; and in the third case, all three are trans. All three of these isomers have the same iodine value of 274. However, there is a dramatic increase in melting point with increasing *trans* isomer content.

Trans Isomer Measurement
The classical method of analysis for iodine value will establish the relative amounts of saturated and unsaturated fatty acids present but will not differentiate between the *cis* and *trans* isomers. The trans-acid content of fatty acid blends can be quantified by infrared spectroscopy. The method is based on the absorption at 966 cm^{-1} by the *trans* double bonds in the sample. The absorption is directly proportional to the concentration of *trans* double bonds present in the sample. The absorption is thought to be independent of the position and number of *trans* double bonds present. Conjugated *trans* double bonds in excess of 5% may interfere with this method. The presence of such interference is indicated by absorption at 987 cm^{-1}. The details of this method are found in AOCS Tentative Method Cd 14-61.

Color Evaluation of Soap Raw Materials
The following sections will review the various methods for evaluating the color of soap raw materials and of the resulting neat soap product. The color of raw fats and

TABLE 3.2
Comparative Iodine Values and Melting Points

Systematic Name	Formula	IV	MP (°C)
6c-octadecenoic acid	$C_{18}H_{34}O_2$	90	33
6t-octadecenoic acid	$C_{18}H_{34}O_2$	90	54
9c-octadecenoic acid	$C_{18}H_{34}O_2$	90	16.3
9t-octadecenoic acid	$C_{18}H_{34}O_2$	90	45
9c,12c-octadecadienoic acid	$C_{18}H_{32}O_2$	181	−5.0
9t,12t-octadecadienoic acid	$C_{18}H_{32}O_2$	181	28–29
9c,12c,15c-octadecatrienoic acid	$C_{18}H_{30}O_2$	274	−11
9c,11t,13t-octadecatrienoic acid	$C_{18}H_{30}O_2$	274	49
9t,11t,13t-octadecatrienoic acid	$C_{18}H_{30}O_2$	274	71.5

oils stock is an indicator of the degree of abuse and degradation to which the material has been subjected. To that extent, it is a measure of the degree of additional processing that will be required to produce white soap and of the potential for loss of fat in the soap-making process. In the case of fatty acid stocks, the color is also an indicator of the quality of the material and will have a direct bearing on the color of the neat soap made from the fatty acid stock. If the color is high, it may also be an indicator of degradation that will affect the odor of the resulting soap.

Raw Color of Fats and Oils by Colorimeter

Probably the most widely used method of color measurement for fats and oils in the United States is the Wesson method that uses the Lovibond AOCS Tintometer, Model AF710, that is manufactured by The Tintometer, Ltd., Salisbury, UK, and is distributed by HF Scientific, Inc., Fort Myers, Florida.

This method determines the color of fats, oils, and fatty acids by comparison of a column of the oil with standard glass slides under specific controlled conditions. The procedure is also commonly used for commercial fatty acid blends where, in the absence of any turbidity, the material can be evaluated while molten.

In the Lovibond method, a 5¼-inch column of material is placed in a standard color tube that is specially designed for use with the AOCS Tintometer. The tube is inserted into the port of the lighted cabinet where the color of the material in the sample tube can be compared with the standard red and yellow glass slides that have numerical scale values. Usually, both the red and yellow values are specified. Frequently, only the red value is specified and reported, particularly for tallow and coconut oil. The details of this method are found in AOCS Official Method Cc 13b-45.

The accepted international standard, based on the British Standards Institute method BS 684, is the Lovibond method. The Model E AF900 Lovibond Tintometer that is also made by The Tintometer, Ltd., is suitable for this method. Please refer to the AOCS Official Method Cc 13e-92 for details of this method.

Color of Fats, Oils, and Fatty Acids by Photometric Measurement

Another common method of measuring the color of fats and oils utilizes visible spectrophotometry. The photometric measurement of the color of fats and oils measures the absorbance at 460, 550, 620, and 670 nm. The *photometric color* is expressed as follows:

$$\text{Photometric color} = 1.29A_{460} + 69.7A_{550} + 41.2A_{620} - 56.4A_{670}$$

Please refer to AOCS Official Method Cc 13c-50 for the details of this method.

Frequently, the color of fatty acids is also determined by photometric measurement. Either the absorbance or transmission is measured at 440 nm and 550 nm. The results are reported either as *% transmission at 440/550 nm* or in terms of the *photometric index* that is expressed as follows: $100 \times A_{440}$ and $100 \times A_{550}$. Please refer to AOCS Official Method Td 2a-64 for details of this method.

Refined and Bleached Color of Fats and Oils

The refined and bleached color test can be used to predict the color after processing of intermediate and lower grades of tallow that will require pretreatment before use in soap making.

The tallow sample is refined by neutralizing the free fatty acids with caustic soda. The neutral fat is then separated by filtration through cheesecloth or filter paper to remove any soap formed during the refining step. The neutral fat is then treated with AOCS Official Activated Filter Earth that may be purchased from the American Oil Chemists' Society, Champaign, Illinois. The mixture is then filtered, followed by the reading of the Lovibond color of the clear filtered fat. Please refer to AOCS Official Method Cc 8d-55 for the details of this procedure.

Direct Bleach Test for Fats and Oils

The direct bleach test for tallow is useful when working with higher grades of inedible tallow that are low in free fatty acid content. The method uses 300 g of fat that is heated to 110°C, followed by the addition of 15 g of AOCS Official Activated Filter Earth. The mixture is then agitated for 10 min at 80°C or higher, followed by filtration through a heated funnel at 70°C. After the bleaching and filtration, the Lovibond color is determined. All supplies used for this procedure are the same as those specified for the refined and bleach test above. The results of this test indicate the potential for making white soap from fats that will be pretreated by bleaching without refining.

Alcoholic Saponification Color for Fats and Oils

In AOCS Recommended Practice Cc 13g-94, a method is presented for saponification color determination of high-quality tallow and coconut oil intended for use in soap making. In this method, a sample of the untreated oil is reacted with an excess of alcoholic KOH solution. Upon completion of the saponification reaction, a sample of the soap solution is placed in a 5¼-inch tube color evaluation using the Wesson method for Lovibond color measurement found in AOCS Official Method Cc 13b-45.

In AOCS Recommended Practice Cc 13f-94, methods are presented for refined and bleached color and for saponification color of lower-grade tallows and greases intended for use in soap making. First, the sample is subjected to a caustic refining and bleaching operation. Following the refining and bleaching, a portion of the sample is reacted with an excess of alcoholic KOH solution. After saponification is completed, a sample of the soap solution is transferred to a 5¼-inch tube for color reading using the procedure in AOCS Official Method Cc 13b-45.

Saponification Color of Fatty Acids

The color of fatty acids for soap making can be evaluated by measuring the alcoholic saponification color that determines the color of a potassium soap solution. This method is applicable to normal soap fatty acid blends with a saponification value in the range of 214 to 220.

In this procedure, a sample of the fatty acid is dissolved in ethanol, followed by neutralization with aqueous 39% potassium hydroxide solution. The percent transmission at 440 and 550 nm is determined for the potassium soap solution. This method serves as a good indicator for the expected color of finished soap to be made from the fatty acid blend. Due to the relatively short period of time required to run this procedure, it can be useful as a receiving quality control method for bulk fatty acids.

Approximate Conversions for Various Color Scales

Figure 3.1 gives the approximate equivalents of various visual color scales commonly used for fats and oils including Gardner, APHA, FAC, Lovibond, and percent transmission. Additional correlations for several methods of measuring the color of fats and oils are given in AOCS Official Method Cc 13b-45.

Free Fatty Acid Content of Fats and Oils

For determining the free fatty acid content of fats and oils, a weighed sample is dissolved in ethanol, followed by titration with standardized sodium hydroxide solution to the phenolphthalein end point. The result is calculated and conventionally reported as percent oleic acid for most samples. Please refer to AOCS Official Method Ca 5a-40 for details of this method.

The presence of elevated levels of free fatty acid can result from the storage of fats and oils at elevated temperatures in the presence of water, causing hydrolysis of a portion of the triglycerides. Also, prolonged enzymatic activity in the case of animal fats can result in undesirable levels of free fatty acid. The presence of excessive free fatty acid will indicate a decreased level of available glycerin for recovery and an increased potential for color and odor degradation of the stock.

Moisture Content of Fats, Oils, and Fatty Acids

The moisture content of tallow and other nonlauric oils is generally determined by the 130°C air oven method found in AOCS Method Ca 2c-25. The air oven method is not recommended for coconut and palm kernel oils. For the standard Karl Fischer titration method that is recommended for lauric oils and other fat stocks, please see AOCS Method Ca 2e-84. For the Karl Fischer method for fatty acids, see AOCS Method Tb 2-64.

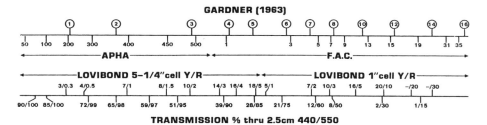

Fig. 3.1. Approximate equivalents of various visual color scales.

Insoluble Impurities in Fats and Oils

This procedure determines dirt, bone meal, and other insoluble impurities present, under the conditions of the test, in all commercial fats and oils.

A properly sized sample of the fat is dried to constant weight in an oven; the residue from the standard air oven moisture determination can be used if available. The dried sample is dissolved in a warm 50-mL portion of kerosene and filtered through a dried, tared Gooch crucible with a glass fiber filter. The filter is washed with five 10-mL portions of hot kerosene, followed by a thorough washing with petroleum ether. The crucible and contents are dried to constant weight in a 101°C air oven, cooled under desiccation, and weighed. The weight of residue is reported as the percent insoluble impurities. See AOCS Official Method Ca 3a-46 for more information.

Unsaponifiable Matter in Fats and Oils

Unsaponifiable matter can be composed of such things as cholesterol, fatty alcohols, and denaturants in fats. These materials do not react with sodium hydroxide to form soap and are extractable with fat solvents. The level of unsaponifiable material in fats and oils can range as high as a few tenths of a percent for better-grade materials. Apart from the obvious economic consideration associated with the purchasing of fats for soap making, the unsaponifiable matter can also have a negative impact on soap performance. It has been reported that one unit of unsaponifiable matter in soap can reduce the detergency action of at least three times its weight in soap (5).

The unsaponifiable matter content of fats is determined by vigorously reacting a sample with potassium hydroxide, followed by petroleum ether extraction of the unreactive organic matter. Several petroleum ether extractions are collected and dried, with the amount of residue determined gravimetrically. Unsaponifiable matter content can be determined by AOCS Method Ca 6a-40 in fats and oils, Tk 1a-64 in fatty acids, and Da 11-42 in soap. Also, for soap and soap products, see ASTM Standard Method D 460, Sections 36 to 38.

MIU Content of Fats and Oils

The sum of the moisture, insoluble matter, and unsaponifiable matter present in commercial fats and oils is typically part of the material specification, the MIU, and may serve as the basis for a price adjustment if the total exceeds specified limits.

Analysis of Soap and Minor Ingredients

The following sections will review the principal test methods used for analyzing soap and soap products. Included here are methods for anhydrous soap, total fatty acid and real soap, moisture, glycerin, chlorides, free fatty acid and free alkalinity, and certain functional ingredients.

Anhydrous Soap Content

In the anhydrous soap determination, a precisely weighed soap sample is treated with mineral acid to liberate the fatty acids. The fatty acids are recovered and reacted with sodium hydroxide solution to form soap. The resulting soap is dried and weighed to establish the anhydrous soap content of the original sample. Please refer to either AOCS Official Method Da 8-48 or ASTM Method D 460, Sections 24 to 25, for details of this procedure.

Total Fatty Acid and Real Soap Content

In the total fatty acid determination (Valley Products Company Analytical Method 401), a precisely weighed soap sample is treated with mineral acid, the liberated fatty acids are recovered by a series of petroleum ether extractions, dried, and weighed. The real soap content can be derived by multiplying the percent TFA by the gravimetric factor for the type of soap.

Moisture Content of Soap

The moisture content of ordinary sodium soap can be reliably determined by the 105°C air oven method as described in AOCS Method Da 2a-48 and by ASTM Method D 460 Section 14. This method is not reliable in the presence of elevated glycerin and free fatty acid levels.

Another useful method that is specific for water is the Karl Fischer titration. This method provides for a rapid determination of water content and is especially useful when water is present in small quantities. It may be used on glycerin, fats and oils, fatty acids, soap, and soap product. Among the automated systems available for Karl Fischer moisture analysis is the Photovolt Aquatest 10, manufactured by Seradyn, Inc., of Indianapolis, Indiana, that generates Karl Fischer reagent on demand based on the sample. Because Karl Fischer reagent is generated electrolytically, no volumetric measurement of reagent is required and no standardization of the solution is needed. Instead, the water content of the sample is determined and computed by the instrument based on the equivalence of 1 coulomb of electricity to 186.53 micrograms of water.

Free Glycerin Content of Soap

This method determines the free glycerin content of soap by way of oxidation of the glycerin with periodic acid. The sample of soap containing glycerin is mixed with chloroform and glacial acetic acid and then quantitatively transferred to a volumetric flask. Distilled water is added and the sample is dissolved with heating, if necessary, to completely dissolve the sample. The flask is filled to volume with distilled water and stoppered, followed by agitation to effect thorough mixing. The water and chloroform layers are then allowed to separate. An aliquot of the aqueous layer containing the glycerin is withdrawn and added to a beaker containing an excess of periodic acid reagent where the oxidation-reduction reaction is allowed to proceed for 30 min. Two blanks are prepared for each batch of samples being analyzed. After the reaction period has passed, potassium iodide is added to each beaker to liberate the excess iodine. Each sample and blank is then diluted with deionized water and titrated with standardized 0.1N sodium thiosulfate solution. The titration difference between the blank and the sample is converted with appropriate calculation and reported as percent free glycerin in the soap sample. See AOCS Official Method Da 23-56 and ASTM Standard Method D 460, Sections 82 to 84.

Another useful procedure for the determination of glycerin content of soap is a modification of the sodium metaperiodate method in AOCS Official Method Ea 6-51. In this procedure (Valley Products Company Analytical Method 408), a 40.0 g ± 0.001 g sample (or other sample size based on the expected results found in the table in AOCS Ea 6-51) is weighed into a 250-mL Erlenmeyer flask. Add 100 mL of deionized water and heat on a steam bath to dissolve. Once dissolved, the solution is acidified to a slight excess with 1:4 sulfuric acid using methyl orange indicator. The flask is covered with a watch glass and heated to clarify the fatty acid layer. The solution is filtered through a qualitative filter paper into a 400-mL beaker, retaining the fatty acids on the filter paper. The fatty acids are washed with several small portions of hot deionized water until the washings are neutral to methyl orange. Cool and neutralize to methyl orange with 50% NaOH very carefully. Adjust the pH of the solution so that it is definitely acidic to methyl orange. Add a few glass beads and boil for 5 min to expel any CO_2. Boil long enough to reduce the volume to 75 mL. Some glycerin may be volatilized if the solution volume goes below 50 mL. Cool to room temperature. Prepare a blank of distilled water and process with the sample solutions in an identical manner. Buffer the pH meter using pH 7.0 and pH 10.0 buffer solutions. Neutralize the sample and blank to a pH of 8.1 ± 0.1 with NaOH. (The strength of NaOH used depends on the acidity of the sample.) Final adjustment should be made with 0.125N or weaker NaOH or with 0.1N or weaker H_2SO_4. The volume of the pH-adjusted solution should not be more than 100 mL. Pipet 50 mL of sodium metaperiodate solution to the pH-adjusted samples. Swirl the beaker to ensure thorough mixing, cover with a watch glass, and immediately place in a dark cupboard at room temperature for 30 min. Remove from the cupboard, add 10 mL of 50% ethylene glycol solution by graduated cylinder, swirl gently, and allow to stand for 20 min. Titrate using 0.125N NaOH to a pH of 8.1 ± 0.1. Dropwise addition of NaOH should be made as the 8.1 pH is approached. The results are calculated as follows:

$$\% \text{ glycerin} = \frac{(A - B)\,(N)\,(9.209)}{\text{Sample weight in grams}}$$

where

A = mL NaOH used for sample titration

B = mL NaOH used for blank titration

Chlorides in Soap.

The soap sample, usually 5 g, is dissolved in about 300 mL of chloride-free deionized water, with boiling as needed. The soap is then reacted with an excess of magnesium nitrate to form insoluble magnesium soaps, filtered, and washed with chloride-free deionized water. The filtrate is then titrated with a standardized 0.1N silver nitrate solution with potassium chromate indicator. The result is calculated and usually reported as percent sodium chloride. See AOCS Official Method Da 9-48 and ASTM Standard Method D 460, Sections 53 to 55.

Free Fatty Acid and Free Alkalinity in Soap

The free fatty acid or free alkalinity in soap is determined by titration, with standard alkali or acid as appropriate, to the phenolphthalein end point. For most soaps, which contain no significant amount of alkaline salts, the procedure involves the titration of the sample, usually 10 or 20 g dissolved in neutralized ethanol, with either standardized sodium hydroxide solution (0.1N or 0.25N) for acidic samples or sulfuric acid (0.1N or 0.25N) for alkaline samples to the phenolphthalein end point. The results for alkaline soaps are usually reported as either percent NaOH or Na_2O for sodium soaps and percent KOH or K_2O for potassium soaps. For acidic soaps, the results are typically reported as percent oleic acid, coconut acid, or lauric acid.

Note that the titration is always performed in neutralized ethanol rather than water due to the hydrolysis of soap in water that would buffer the solution and interfere with the end-point determination. See AOCS Official Method Da 4a-48 and ASTM Standard Method D 460, Section 21, for more detail.

Triclocarban and Triclosan in Soap by Ultraviolet Absorbance

Two important and widely used additives found as the active ingredients in many antibacterial soaps and as the functional ingredients in many deodorant soaps are triclocarban and triclosan. The chemical structures of these two compounds make them amenable to quantitation by ultraviolet (UV) absorbance spectroscopy methods (Valley Products Company Analytical Method #802 for triclocarban and #804 for triclosan). Triclocarban absorbs at 265 nm and triclosan at 282 nm.

Note that interference may be encountered due to absorbance by fragrance components or other soap ingredients at these same wavelengths. This interference can be nullified by preparing a calibration curve for each formulation of soap product. By preparing stock solutions of the product matrix without the active ingredient and a separate stock solution of the active ingredient, various concentrations of the active ingredient over the concentration range of interest can be prepared for UV absorbance measurement.

Listed in Table 3.3 are the UV absorbance values obtained for triclocarban concentrations over a range from 0.0 to 1.0 mg/100 mL in a typical soap formulation. Because the UV absorbance is a direct linear function of the analyte's concentration, the relationship between UV absorbance and concentration can be expressed by an equation in the form of straight line, $A_{CORR} = mC + b$, where A_{CORR} is the UV absorbance and C is the triclocarban (or triclosan) concentration in mg/100 mL. The slope of the line, m, and the y-intercept, b, can be derived by treatment of the concentration and absorbance data using linear least-squares equations (6) as follows:

$$m = \frac{N\Sigma C_i A_i - (\Sigma C_i)\,(\Sigma A_i)}{N\Sigma C_i^2 - (\Sigma C_i)^2} = \frac{(5)\,(3.228) - (2.8)\,(4.27)}{(5)\,(2.16) - (7.84)} = 1.41$$

$$b = \frac{(\Sigma A_i)\,(\Sigma C_i^2) - (\Sigma C_i)\,(\Sigma C_i A_i)}{N\Sigma C_i^2 - (\Sigma C_i)^2} = \frac{(4.27)\,(2.16) - (2.8)\,(3.228)}{(5)\,(2.16) - (7.84)} = 0.0624$$

Thus the linear equation for the data in this example would be

$$A_{CORR} = (1.41)\,(C) + 0.0624$$

Upon rearrangement, the equation becomes

$$C = \frac{A_{CORR} - 0.0624}{1.41}$$

The equation obtained in this manner can be applied at any time in the future during routine analysis of product of this same formulation. By preparing a test solution of the product and determining its UV absorbance at the specified wavelength, the triclocarban (or triclosan) concentration in the test solution can readily be determined. By dividing the triclocarban (or triclosan) concentration obtained from the soap sample by the soap sample concentration in the test solution, the analyte's concentration in the product can then be expressed. The y-intercept value, 0.0624 in this example, represents the correction for interference from other formula components such as fragrance in the product matrix. In Fig. 3.2 the equation that was derived from the data in Table 3.3 is represented graphically.

Chromatographic Methods for Soap Raw Materials, Soap Products, and Minor Ingredients

At this point we will review several gas chromatography and high-performance liquid chromatography methods that are of interest to the soap chemist. Gas chromatography (GC) and high-performance liquid chromatography (HPLC) are two of the fastest-growing areas of analytical chemistry. Their growth can be attributed to the high degree of reproducible qualitative and quantitative data that can be acquired with these two powerful techniques. Their application is constantly

TABLE 3.3
UV Absorbance *vs.* Triclocarban Concentration for a Typical Bar Soap

Concentration (mg/100 mL)	A_{265}	A_{320}	A_{CORR}
0.0	0.17	0.11	0.06
0.4	0.73	0.12	0.61
0.6	1.03	0.10	0.93
0.8	1.33	0.11	1.22
1.0	1.49	0.14	1.45

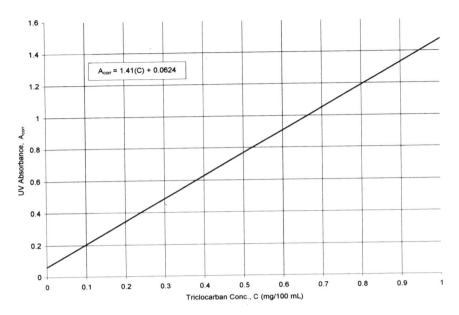

Fig. 3.2. UV absorbance *vs.* triclocarban concentration for a typical antibacterial bar soap.

expanding due to the development of more highly efficient instruments and columns, as well as more application-specific detectors and stationary phases. Many GC and HPLC methods have been developed for analysis of materials used in the soap industry, a few of which will be briefly described here. It can be assumed that both GC and HPLC will continue to play a significant role in the field of analytical chemistry for the soap industry.

Triglyceride Analysis by HPLC
Some work is reported in the literature for the separation of triglycerides using HPLC. Plattner et al. (7) have performed triglyceride separation by chain length and degree of unsaturation using a C_{18} μ-Bondapak column with an acetonitrile-acetone mobile phase and a differential refractometer detector. Waters Associates has a specialty column for triglyceride analysis (Waters, #84346), which is used with a 50:50 acetone-tetrahydrofuran mobile phase (8). Supelco's reversed-phase Supelcosil LC-8 and LC-18 columns, using acetone-acetonitrile (63.6:36.4) mobile phase, are also reported to effect triglyceride separations (9).

Derivatized Fatty Acid Analysis by HPLC
Scholfield (10) reported the use of a C_{18}/Corasil column and aqueous acetonitrile mobile phase for separation of fatty acid methyl esters (FAMEs) by unsaturation

TABLE 3.4
Selected AOCS Official Test Methods for Soap and Soap Raw Materials

Test/AOCS Method for:	Fats and Oils	Fatty Acid Blends
Acid value	Cd 3d-63	Te 1a-64
Free fatty acid	Ca 5a-40	
Insoluble impurities	Ca 3a-46	
Iodine value	Cd 1-25	Tg 1a-64
Moisture/volatiles, 130°C	Ca 2c-25	
Colorimetric color, Wesson method	Cc 13b-45	
Colorimetric color, Lovibond method	Cc 13e-92	
Photometric color	Cc 13c-50	Td 2a-64
Alcoholic saponification color	Cc 13g-94*	
Refined and bleached color	Cc 8d-55	
R&B; alcoholic saponification color	Cc 13f-94*	
Saponification value	Cd 3-25	Tl 1a-64
Titer	Cc 14-59	Tr 1a-64
Unsaponifiable matter	Ca 6a-40	
Water, Karl Fischer method	Ca 2e-84	Tb 2-64

Test/AOCS Method for:	Soap and Soap Product	Soap with Detergent
Acid value	Da 14-48	
Anhydrous soap content	Da 8-48	Db 6-48
Chlorides	Da 9-48	Cb 7-48
Free fatty acid/free alkali	Da 4a-48	Db 3-48
Glycerin	Da 23-56	
Iodine value	Da 15-48	
Moisture and volatiles	Da 2a-48	Db 1-48
Saponification value	Da 16-48	Db 8-48
Titer	Da 13-48	

*AOCS Recommended Practice

and chain length. Work by Warthen (11) achieved analytical separation of FAMEs, including *cis* and *trans* isomers, using a μ-Bondapak C_{18} column and methanol-water mobile phase.

Work was reported by Jordi (12) for the separation of the phenacyl and naphthacyl derivatives of fatty acids using a μ-Bondapak column and the fatty acid analysis column (Waters) with an acetonitrile-water gradient. HPLC of geometric isomers of the fatty acids of coconut oil and other seed oils has also been reported by Wood et al. (13). This method also utilized the phenacyl and naphthacyl derivatives chromatographed on a C_{18} reversed-phase column using an acetonitrile-water gradient.

Free Fatty Acid Analysis by HPLC
Analysis of free fatty acid, without derivatization, by HPLC has been described by King et al. (14). In this method, a free fatty acid column (Waters) was used with various combinations of a ternary mobile phase consisting of tetrahydrofuran, acetonitrile, and water at a reduced pH, and employing a refractive index detector (Waters, Model 401 Differential Refractometer). Using this system, the identification

and semiquantification of fatty acid mixtures derived from industrial oils and alkyd resins are accomplished in about 15 min.

Additional work has been reported by George (15) about quantitation of fatty acids from triglycerides and soap without derivatization. The triglycerides are first saponified and then acidulated to free the fatty acids. The fatty acids are then dissolved in methanol or tetrahydrofuran-methanol solution for injection. The mobile phase consists of 45% acetonitrile, 20% tetrahydrofuran, 34.5% water, and 0.5% glacial acetic acid at a flow rate of 1.1 mL/min on a Waters fatty acid stainless steel column. Work was also done using radial compressed stainless steel columns. Soaps are dissolved in methanol and then injected directly into the HPLC, where they are acidulated on the column by the acidic mobile phase. A refractive index detector is used in conjunction with the isocratic reverse-phase chromatography.

Free Fatty Acid Analysis by Packed-Column GC

Gas chromatography of fatty acids containing carbon numbers from C_6 to C_{24} is of particular interest to the soap industry because these are the principal components of the fats and oils normally used as raw materials for soap making. Much work has been reported in the literature utilizing packed columns for both free fatty acids and derivatized fatty acids. Ottenstein et al. (16) reported the use of highly polar stationary phases for the separation of C_{14} to C_{20} free fatty acids. Supelco's literature (17) describes the separation of whole milk free fatty acids from $C_{4:0}$ to $C_{18:3}$. Williams et al. (18) describe the analysis of free fatty acids in vegetable oils by pyrolytic methylation with trimethylphenylammonium hydroxide upon injection of the sample.

Derivatized Fatty Acid Analysis by Packed-Column GC

Because of the high polarity of the carboxylic acid group and its tendency to form hydrogen bonds, fatty acids are most often derivatized to the less polar and more volatile methyl esters for best qualitative and quantitative results. Both Perkins et al. (19) and Ottenstein et al. (20) used FAMEs with packed-column technology to determine *trans* unsaturation. Chapman (21) described a procedure for preparing methyl esters of free fatty acids in vegetable oils using methylurea for analysis on packed columns. Polyester and cyanosilicone packed columns, along with suggested GC conditions, are discussed in Supelco's literature (22). Six techniques of preparing methyl esters are described by Jamieson et al. (23). The official AOCS method for FAME preparation (24) uses boron trifluoride-methanol reagent. The official AOCS method for FAME analysis by GC (25) also utilizes packed-column technology.

Derivatized Fatty Acid Analysis by Capillary GC

Capillary GC on wall-coated open tubular fused silica columns has greatly enhanced the accuracy and speed of analysis due to the development of technology to attach very polar stationary phases to highly deactivated fused silica. Slover (26) used capillary glass columns coated with SP2340 for quantitation of FAMEs. The columns were used for extended periods and with up to 1,900 samples analyzed on a single column during an 11-month period with little deterioration of the column. Separation included resolution of *cis* and *trans* isomers. Capillary GC analysis of *cis* and *trans* FAMEs is also described in Supelco's literature (27) in which most of the C_{18} isomers are resolved. In this method, the FAMEs are injected into a GC equipped with a glass split injection port (split ratio 100:1), SP2560 column (Supelco, #2-4056) 100 m × 0.25 mm i.d., 0.20-mm film, and a flame ionization detector (FID). Using an oven temperature of 175°C and helium as the carrier gas, excellent resolution of all the C_{12} through C_{22} FAMEs is obtained including partial separation of most of the $C_{18:1}$ *cis* and *trans* positional isomers, but run times can exceed 60 min.

Lanza et al. (28) have reported the use of short glass capillary columns coated with SP2340 for the rapid analysis of fatty acids that provides some resolution of geometric isomers. Sampugna et al. (29) also report the rapid analysis of *trans* fatty acids using an SP2340 coated glass capillary column.

Glycerin Content of Soap by HPLC

Glycerin determination in soap base has been accomplished using HPLC as described by George et al. (30). A carbohydrate analysis column (Waters) is used with an acetonitrile-water (92.5:7.5) mobile phase, 1.0 mL/min flow rate, with a differential refractometer (Waters, Model 401). Sample analysis time is about 30 min including sample preparation time under the described conditions.

Triclocarban and Triclosan in Soap by HPLC

George et al. (31) have also developed an HPLC method for the determination of triclocarban and triclosan. This method uses a radially compressed C_{18} column (Waters, Radial Pak A). The sample preparation involves using C_{18} solid phase extraction cartridges to remove most of the soap and fragrance components from the sample. The mobile phase for the separation consists of tetrahydrofuran/water mixture (58:42) at a flow rate of 2 mL/min with a UV detector at 280 nm. Sample preparation time is reported to be 15 min with a 15-min analysis time.

Chelating Agents in Soap by HPLC

Identification and quantification of aminocarboxylate chelating agents used in bar soaps such as ethylenediaminetetraacetic acid (EDTA), N-hydroxyethylenediaminetriacetic acid (HEDTA), and diethylenetriamine pentaacetic acid (DTPA) by HPLC is described by Goldstein et al. (32). This method uses cupric sulfate to precipitate the soap and form a water-soluble aminocarboxylate-copper complex. The copper complex is then isolated and chromatographed on an anion exchange column (Wescan, Anion R Analytical) with a mobile phase consisting of 0.003M sulfuric acid at a flow rate of 1.5 mL/min. A UV detector was used at a wavelength of 254 nm. Under these conditions, the retention times for the copper complexes were reported to be 3.5 min for HEDTA, 6.2 min for DTPA, and 6.9 min for EDTA.

BHT in Soap by Capillary GC

A method for the determination of 2,6-di-tert-butyl-methylphenol (BHT) in soap products is offered by Goldstein et al. (33). The authors present a method that can quantitate BHT in fragranced bar soap without employing the cumbersome standard addition method that has been previously reported (34). After blending the sample with dimethylformamide and adding 2,4-di-*tert*-butylphenol (DTBP) as an internal standard, *bis*-trimethylsilyltrifluoroacetamide (BSTFA) is added to a filtered aliquot to convert the BHT and DTBP to their silyl derivatives. The sample is then introduced into a GC equipped with a glass split injection port (split ratio 200:1), methyl silicone column (Hewlett-Packard, #19091-60010) 12 m × 0.2 mm i.d., and FID. Using helium as the carrier gas, the column temperature is held at 100°C for 2 min and then programmed up to 144°C at 2°C/min. The program rate is then changed to 30°C/min up to 240°C and held for 5 min. Under these conditions, the reported retention times of the silylated derivatives of DTBP and BHT were 15.5 min and 22.4 min, respectively. This method can also be applied to neat soaps, pellets, and fatty acids.

Evaluation Methods for Soap Base and Finished Bars

This section will review several methods that are useful in evaluating various quality attributes, performance properties, and product stability characteristics of both soap base and finished bar soap. Included are methods for color evaluation, translucency measurement, lather volume measurement, hand lather performance, bar smear by water absorption, dry cracking evaluation, wet cracking evaluation, bar feel during washdown, bar hardness measurement, accelerated aging test methods, and quantification of foreign matter.

Color Evaluation of Soap and Soap Products

Visual Color Comparisons

Bar soap color evaluation has traditionally been performed by visual comparison of fresh product against various standards such as paper or plastic color chips and retained standard product samples. Visual color comparisons are prone to being subjective and can be affected by individual differences in color perception and background lighting conditions. All physical standards, including both color chips and retained product standards, must be carefully handled and protected to avoid fading. All such standards will change with time and must be periodically updated.

For visual color comparisons, a controlled lighting environment is a necessity. The use of a standardized light source and working environment, such as that which can be provided with a Pantone Color Viewing Light, can be helpful in providing such standardized conditions. The Pantone Color Viewing Light is available from Pantone, Inc., of Moonachie, New Jersey. The Pantone Color Viewing Light provides three different light sources including artificial daylight, fluorescent store light, and incandescent home light, which can be used independently or in combination. The interior of the cabinet has a matte neutral gray finish that minimizes reflection and glare.

Soap Color by Hunter Reflectance Color Measurement

The instrumental approach to color evaluation can provide many advantages over visual comparison techniques. The instrument in common use by many soap manufacturers is the Hunter Tristimulus Colorimeter, such as the Model D25A-9, which is manufactured, sold, and serviced by Hunter Associates Laboratory, Inc., of Reston, Virginia. Among the advantages of the instrumental approach are the objective basis for color comparison to standard, a defined target that is not subject to deterioration, and quantifiable data that can be retained indefinitely. Variables such as light source, background lighting, and differences in individual color perception are eliminated. Also, the nature of the instrumental approach easily lends itself to interfactory and intercompany color control.

The Hunter Opponent Color Scale (35) expresses color in terms of three values: *L* expresses lightness and ranges from 0 to 100 units; *a* expresses redness when positive, gray when zero, and greenness when negative; and *b* expresses yellowness when positive, gray when zero, and blueness when negative. The Hunter *L, a, b* Color Solid is illustrated in Fig. 3.3.

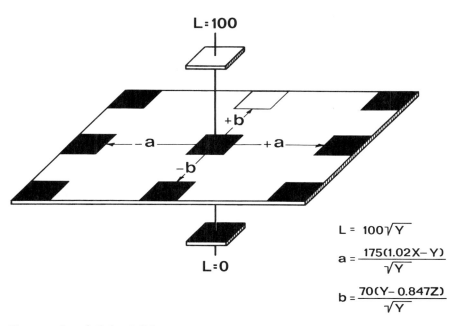

$$L = 100\sqrt{Y}$$

$$a = \frac{175(1.02X - Y)}{\sqrt{Y}}$$

$$b = \frac{70(Y - 0.847Z)}{\sqrt{Y}}$$

Fig. 3.3. *L, a, b* Color Solid.

The *L*, *a*, and *b* values are determined from fresh product samples that have been cut to a planar surface. The values obtained for each sample are then compared with previously established target values. The differences between sample and target *L*, *a*, and *b* values can then be interpreted individually, or they can be resolved into the total color difference (36) or delta-*E* (ΔE) value that serves as a very useful tool for color control at the production level. The total color difference or ΔE is expressed by the following equation:

$$\Delta E = \sqrt{\Delta L^2 + \Delta a^2 + \Delta b^2}$$

This Pythagorean equation resolves the three component differences, ΔL, Δa, and Δb, into the direct difference, in Hunter units, between the target and the sample. The ΔE serves as a very useful tool for routine quality control of bar soap color. For example, typical ΔE limits might allow the following:

- Product with a ΔE below 2.0 units to be packed without qualification
- Product with a ΔE between 2.0 and 4.0 units to be packed with ongoing effort to bring the product back below 2.0 units
- Product with a ΔE above 4.0 units not to be packed

Neat soap whiteness can also be measured using the Hunter colorimeter. The color can be reported in terms of *L*, *a*, *b* values, or can be expressed in terms of the Whiteness Index (37). The perception of whiteness, in terms of the Hunter Opponent Color Scale values, is favored by a higher *L* value (lightness) and a more negative *b* value (higher degree of blueness). The Hunter Whiteness Index is expressed as $WI = L - 3b$. It has been reported that differences of $1\frac{1}{2}$ units of Hunter Whiteness Index can be consistently perceived by the human eye (38).

In performing a color evaluation on neat soap, it is necessary to wait until the neat soap has thoroughly cooled to room temperature before measuring the whiteness. The sample needs to be cut to a flat surface before the Hunter readings are performed. It is recommended that the sample be read six times, turning the sample about 60 deg for each reading. The average of the six readings should then be reported.

In a similar manner, the color of soap to be commercially made from fatty acid blends can be reliably estimated by quantitatively preparing a sample of sodium soap in the laboratory. For an 80:20 tallow:coconut oil blend, for example, 130 g of fatty acid would be melted and weighed. A solution of sodium hydroxide would be prepared in a 600-mL beaker. The fatty acid would be added slowly with constant mixing to the caustic solution until the soap is formed. Soap would then be packed into a 125-mm Petri dish and let stand for about an hour to cool before covering. The soap would then set until the following day when the color would be evaluated. If the color is to be measured with the Hunter colorimeter, six readings should be taken and averaged. This procedure is not very practical as a routine quality control procedure but can be very useful in qualifying a vendor's material.

Soap Translucency by Hunter Opacity Measurement.

This technique has proven to be useful in quantifying the translucency of bar soap using the Hunter Tristimulus Colorimeter. The method is an adaptation of the "contrast ratio method" for measuring opacity (39). The procedure for performing the opacity reading is outlined in the Hunter instruction manual that accompanies those instrument models, such as the D25A-9, that are designed to perform this measurement.

The application of this procedure for measuring the translucency of bar soap involves preparing the soap sample for reading by cutting a flat slice to uniform thickness of ¼ inch. Then the standard routine for opacity measurement, as outlined in the Hunter manual, is performed. This procedure involves calibrating the instrument, followed by two simple operations with the meter. In the first step, the translucent slab is backed by the uncalibrated white tile to achieve maximum reflectance through the sample, followed by the second step, in which the sample is backed up by the black tile for minimum reflectance through the sample. The processor in the instrument then converts this data into an opacity value. The opacity value provides an inverse indicator of translucency. A reasonable scale of values for interpreting these results might rate an opacity value below 10% as excellent, between 10% and 25% as good, between 25% and 40% as fair, between 40% and 50% as poor, and greater than 50% as essentially nontranslucent product. Action steps in the plant can then be set for these various grades.

Lather Volume Determination

Two convenient methods are presented here that can be employed to quantify foam volume generated by soap and detergent bars.

The lather volume of a sample can be determined, using a conventional sudsing test (40), by dissolving a specified amount (1 g is a good starting point) of the product in 500 mL of deionized water at 20°C. A 50-mL portion of this solution is then placed into a 250-mL glass stoppered graduated cylinder. Next the cylinder is stoppered and shaken by inversion, rotating the cylinder about its midpoint without translational motion for 1 min at such a rate that 30 inversions are completed. The cylinder is then placed in the upright position, the stopper removed, and after 5 s the net volume of the foam is read and reported. The stability of the lather can also be evaluated by noting the net lather volume at a specified interval of time, for example, every 5 min, after the initial reading. After some experience, correlation of the results from this procedure can be made with in-use performance to provide a meaningful basis for interpreting data obtained by this method in the future.

Another useful method for evaluating lathering performance is based on the volume of test product solution required to generate a fixed volume of lather (Valley Products Company Analytical Method #604). A buret is selected on its ability to deliver 40 mL of test solution in 45 to 55 s with the stopcock wide open into a 50-mL graduated cylinder. The graduated cylinder walls are wetted with deionized water before each sample trial, and both the buret and cylinder are thoroughly rinsed between each trial. One g of soap (or up to 5 g for syndet bars), is finely divided, weighed, and dissolved in 100 mL of deionized water in a 250-mL beaker at room temperature, 22° to 25°C. The contents of the beaker are gently stirred on a magnetic stirrer for a minimum of 15 min or until the product is uniformly suspended or dissolved. The test solution is then carefully transferred to the buret without excessive agitation. Note the initial buret reading and adjust the buret so that the tip is 12 ± 0.5 cm directly above the lip of the graduated cylinder. Without delay, which might result in settling of the sample, open the stopcock wide until the *liquid plus foam height* fills the graduated cylinder to the 50.0 mL mark. Report the total volume, in milliliters, delivered from the buret as the "lather volume." Note the inverse relationship here; the smaller the volume of solution used to achieve the volume of foam, the better.

Additional procedures of interest are reported in the ASTM methods (41) and by Becher (42).

Hand Lather Evaluation

You are left to your own devices in setting up hand lathering methods. The hand lather test consists of rubbing or rotating a test bar 5 to 10 times in clean, wet hands using an individually devised technique. Lather grades can be described on a 1 to 10 scale or can be compared with standard bars. The benefit of hand lather evaluation comes with the qualitative evaluations that can be made regarding lather creaminess, consistency, quickness, stability, and afterfeel.

Bar Smear Evaluation by Water Absorption

As a soap bar absorbs water and tends to partially dissolve or form a gel, there is an increase in negative consumer reaction due to the appearance and feel presented by this condition. Because the bar smear is generally proportional to the amount of water absorbed by the bar, an objective evaluation of this tendency can be made by soaking bars for a specified period of time and then measuring the water absorption (Valley Products Company Analytical Method #605).

The bar weight, to the nearest 0.1 g, is determined and recorded. A small eye hook is then screwed into one end panel and the bar is reweighed; this weight is recorded as the dry weight. The bar is then suspended for 2 h in a large beaker of deionized water at room temperature in such a way that the bar is not in contact at any point with the sides or bottom of the beaker. After 2 h of soaking, the bar is carefully removed and suspended to drain for 15 min. The bar, including the hook, is reweighed in a tared container, and the wet weight is recorded. The percent weight increase is calculated:

$$\% \text{ Weight increase} = \frac{100 \, (\text{Wet weight} - \text{Dry weight})}{\text{Bar weight}}$$

From experience, the percent weight increase can be correlated with the degree of smear or slushing that will occur in use with a given brand. Typical results are generally in the range of 5 to 15% weight gain from water absorption for ordinary soap bars. Bars that absorb more water tend to slush and smear to a greater degree and conversely.

Dry Cracking of Bar Soap

Dry cracks are fissures that show when a bar is sectioned longitudinally into slices. These cracks are generally caused by air that is entrained in the soap at final extrusion resulting from poor vacuum or inefficient plodding.

The bars being evaluated should be at room temperature, should not have been exposed to any chilled condition, and should ideally be 4 to 8 h old. Each bar should be cut into several thin slices approximately a quarter of an inch thick. The panel slices should be discarded. The remaining slices should then be visually examined for signs of excessive interior cracks and fissures. The bars may be evaluated based on past experience or on a grading scale developed as a standard (Valley Products Company Analytical Method #603).

Wet Cracking of Bar Soap

From our experience, a number of factors can contribute to the development of cracks in bar soap during repeated usage. Among these factors are bar design, degree of distortion of the blank during stamping, method of stamping, composition of soap, composition and level of fragrance oil and other additives, and the efficiency of the soap finishing system.

To evaluate wet cracking characteristics of finished bar soap the following procedure can be used (Valley Products Company Analytical Method #602). The test bars are stored at 70°F and allowed to equilibrate; this can vary in time from 2 to 6 h depending on initial temperature. The bars are planed to a flat surface on one panel area by removing one-quarter to one-half inch of the surface. The bars are then placed in a specially designed tray, cut side up, and supported by pins that are attached to the inside bottom of the tray. The bars should have at least half an inch clearance from one another and from the surface of the tray. Next, 70°F deionized water is carefully poured into the pan so that it is about half an inch above the upper cut surface of the bars. The pan is then covered with a specially designed fitted lid. The tray of bars is then carefully placed into a 70°F incubator box for a period of 30 min.

At the end of the soaking period, the tray is removed from the incubator. The water is carefully drained off by tilting the tray. Any remaining water should be carefully wiped away from the inside surface of the tray without disturbing the bars. After the tray has been emptied of water and dried, the cover is placed back over the tray in such a way that it is supported above the rim of the tray by about 1 in. The tray is then placed back into the incubator for 16 to 24 h.

After the incubation period, the bars are removed from the tray and visually evaluated for wet cracks. The bars can be graded for wet cracking performance either based on the technician's experience or by comparison with a set of photographic standards to make the evaluation more objective.

Additional techniques for evaluating wet cracking are discussed by Pavlichko (43) and Osteroth (44).

Alcohol Immersion Test for Soap Flow Line Structure

This method is intended to reveal the flow line structure of soap bars. The bar is planed to a smooth surface, immersed in deionized water for 1 h at 25°C, followed by immersion in anhydrous methanol for 1 h. After the period of immersion, the bar is then dried for approximately 12 h, after which the bar is examined for flow line structure and cracks.

Bar Feel During Washdown

This test is conducted by washing bars by hand in water. A grade is assigned to each bar based on the defects that become apparent while washing the bar. The procedure is intended to uncover bar feel characteristics that the consumer would experience during use when the bar is in contact with the skin. Any defects noted would likely be caused by insoluble, semisoluble, or slower-dissolving material that would not be intended to be present in the bar. Varying types of bar feel characteristics may be caused by overdried or underworked soap, certain soap additives, foreign material, or marked differences in moisture content from one portion of the bar to another. The manifestation of the defects can range from barely detectable to extreme and may be evident as discrete hard specks or as generalized sandiness or roughness.

The procedure (Valley Products Company Analytical Method #601) involves initially preparing a basin of tap water that has been carefully regulated to a standard temperature, for example, at 80°F. The test bars should first be washed with both hands for about 1 min each to generally remove any rough edges or relief from the design of the bar. Next each bar is evaluated by rotating it in one hand for 10 to 15 s. Both the washing and evaluation should be done with the bar completely submerged. During the evaluation, any sandy or rough feel should be noted as well as the presence of any individual hard specks.

The bar feel can be evaluated and scored against a grading scale of your own design, such as that shown in Table 3.5. Obviously, such a scale could be expanded with more intermediate grades, and the descriptions of bar feel and speck count could be adjusted to meet individual company needs. Also, the specific washdown temperature should be defined to meet the desired quality objective.

Using this grading scale for example, bars with grade 1 that are perfectly smooth and grade 2 that are slightly sandy feeling could be packed without qualification. Product with grade 3, moderately sandy, might be packed only if correction

TABLE 3.5
Bar Feel Washdown Grading Scale

Rating	Bar Feel	Hard Specks
1	Perfectly smooth	No specks
2	Slightly sandy feel	1–3 specks
3	Moderately sandy feel	4–6 specks
4	Pronounced sandy feel	7–12 specks
5	Extreme sandy feel	13–20 specks

is immediately forthcoming. Product with a grade of 4 or 5 would not be packed, but would be diverted to recycle if otherwise appropriate.

Another approach in evaluating the bar feel during washdown is to gradually increase or decrease the water temperature and to report the point where the bar makes the transition between acceptably smooth and sandy feel. This transition point can then be reported as the washdown temperature. Predetermined ranges of temperature can be established for purposes of process control.

It should be noted that an individual's sensitivity to bar feel may deteriorate after several trials. Therefore, if large numbers of bars are to be evaluated, they should be grouped in lots of a dozen or so bars with an interval of up to 30 min between sessions. Also, the causes of poor bar feel performance can sometimes be partially determined by repeating the washdown procedure at approximate 10° incremental increases in temperature. For example, traces of overdried or under-worked soap, which are noticeable at 80°F, will likely no longer be evident at 90°F or 100°F. This is in contrast to certain soap additives that may not be significantly affected in terms of solubility or rate of solution by the increased temperature.

Bar Hardness

As a process control method, bar hardness can be easily and objectively measured using a Precision Universal Penetrometer to measure the penetration value of the soap under defined conditions. This device is manufactured by Precision Scientific of Bellwood, Illinois, and is available through most scientific equipment distributors. In addition to the penetrometer, the standard ASTM "fat cone" (Precision's item number K19900) should also be specified.

The penetration test should be performed on soap bars that are fresh from the finishing line. In performing this test, the soap bar is placed on the table directly beneath the penetrometer cone. The penetrometer cone is adjusted until the point of the cone is just touching the bar surface. The penetration cone should be fitted with a 200-g weight on top of the cone shaft. The release lever is pressed, held for 10 s, and released. The cone arm is carefully raised, the bar is moved so that the cone is just above a new point on the bar, and the procedure is repeated for a second time. Immediately, the bar is moved and the process is repeated for a third time in rapid succession. The top shaft button is held down to obtain a dial reading for penetrometer depth divisions of 0.1 mm. The reading obtained is the cumulative total of the three trials.

Another approach to measuring bar hardness utilizes a specially constructed Model LC Mullen Burst Strength Tester from the B.F. Perkins Company of Chicopee, Massachusetts. This device, used in the paper industry, essentially measures the pressure at which either the bar structure collapses or the bar no longer deforms under the applied pressure.

The Mullen test should also be performed on fresh bars taken directly from the finishing line. The soap bar is placed on the table of the tester and is centered over the rubber diaphragm. The clamp is tightened carefully until good contact is made. The pressure gauge is adjusted to zero. The wheel that controls the plunger is turned at a rate of about 1 revolution per second. The wheel is turned continuously until either the bar ruptures or the pressure remains constant. The gauge is read and the reading is reported. Also, measure and report the soap temperature at the time of the test.

In both approaches above, experience from initial use will help to develop meaningful guidelines to interpreting future data obtained for a given product. Acceptable values from one product to another may vary due to inherent differences related to product formulation.

Accelerated Oven Aging Evaluation

There is no truly reliable substitute for actual shelf testing over real time in determining the stability of soap and soap products. However, accelerated oven aging is commonly used to gain at least preliminary information about stability of soap base, functional additives, colorants, and fragrances in soap products. In many instances the manifestations of accelerated oven aging procedures exaggerate the deterioration of the product far beyond that which would ever be experienced over long periods of real time under typical shelf life conditions. The harshness of accelerated oven testing might best be thought of as an indicator for potential product degradation that would not occur for extremely long periods of time, if ever, under normal conditions.

The most valid use of rapid aging tests occurs when one has had the opportunity to correlate accelerated aging results, under defined conditions, with ambient real-time aging characteristics. This correlation generally would need to be done on a brand-by-brand basis to have high confidence in the results. With this comparative data, one would then be in position to use accelerated aging tests to evaluate product on an ongoing basis if required. It can also be used with reasonable confidence for initial stability studies when relatively minor reformulation of a brand is undertaken.

For fragrance stability, the soap bar should be placed in a clean jar fitted with a lid. The sample should then placed in an evenly heated, well-regulated oven at a set temperature, perhaps 110°F, for a prescribed period of time, perhaps one week. The sample is then compared with a duplicate bar that has been well protected in foil and refrigerated. For general bar soap products, rapid aging tests may be carried out by placing individual bars in tightly sealed jars and placing them in the oven for one week at 140°F. If necessary, two bars of the same formulation may be placed in the same jar. At the end of the test period, remove the jar from the oven and allow it to cool to room temperature before opening. Remove the soap and evaluate the color of the aged sample. Examine all stamped surfaces for discolored spots or other signs of bad aging characteristics.

If Hunter colorimeter values are to be read as part of the rapid aging procedure, the bars should be surfaced before the oven aging procedure. If more than one bar is to be evaluated for color, bars of like formula may be placed in the same jar for oven aging, but do not allow the cut surfaces to touch each other or the sides of the jar.

Foreign Particulate Matter in Soap

A convenient method for quantifying the presence of any particulate foreign matter in soap base or finished bars utilizes a Model J Bulk Tank Sediment Tester (used in the dairy industry) which is available from the Clark Dairy Equipment Company of Greenwood, Indiana.

The procedure involves finely dividing 100 g of sample and dissolving in 1 to 2 L of hot deionized water on a steam bath with agitation. The resulting soap solution is then filtered through a specially designed 3.5-cm filter disk that is designed to be used with the sediment tester device. The sediment tester is mounted on a large filter flask that is connected to an aspirator for suction. The particulate matter trapped on the filter disk can then be visually compared with a standard set of photographs to provide a grade for the sample.

Acknowledgements

Copies of Valley Products Company Analytical Methods cited in this chapter may be obtained by addressing a written request to the author at Valley Products Company, 384 East Brooks Road, Memphis, TN 38109, USA.

References

1. Walker, R.C., ed., *Official Methods and Recommended Practices of the American Oil Chemists' Society,* 4th edn., American Oil Chemists' Society, Champaign, Illinois, 1994.
2. *Annual Book of ASTM Standards,* vol. 15.04, American Society for Testing and Materials, Philadelphia, Pennsylvania, 1994.
3. Grompone, M.A., *J. Am Oil Chem. Soc. 61:* 788 (1984).
4. *Fatty Acid Data Book,* Unichema Chemicals, Inc., 2nd edn., Chicago, Illinois, 1987, pp. 4–5.
5. *Better Rendering,* The Procter & Gamble Company, Cincinnati, Ohio, 1967, p.17.
6. Arnold, J.G., and R.A. Ford, *The Chemist's Companion: A Handbook of Practical Data, Techniques, and References,* Wiley, New York, 1972, p. 488.
7. Plattner, R.D., G.F. Spencer, and R. Kleiman, *J. Am. Oil Chem. Soc. 54:* 511 (1977).
8. *Waters Sourcebook for Chromatography Columns and Supplies,* Waters Division of Millipore Corp., Milford, Pennsylvania, 1986, p. 43.
9. *One-Step Triglyceride Separation,* HPLC Bulletin 787B, Supelco, Inc., Bellefonte, Pennsylvania, 1980.
10. Scholfield, C.R., *J. Am. Oil Chem. Soc. 52:* 36 (1975).
11. Warthen, J.D., *J. Am. Oil Chem. Soc. 52:* 151 (1975).
12. Jordi, H.C., *J. Liq. Chromatogr. 1:* 215 (1978).
13. Wood, R., and T. Lee, *J. Chromatogr. 254:* 237 (1983).
14. King, J.W., E.C. Adams, and B.A. Bidlingmeyer, *J. Liq. Chromatogr. 5:* 275 (1982).
15. George, E.D., *J. Am. Oil Chem. Soc. 71:* 789 (1994).
16. Ottenstein, D.M., and W.R. Supina, *J. Chromatogr. 91:* 119 (1974).
17. *Fatty Acid Analysis,* GC Bulletin 727J, Supelco, Inc., Bellefonte, Pennsylvania, 1975.
18. Williams, M.G., and J. MacGee, *J. Am. Oil Chem. Soc. 60:* 1507 (1983).
19. Perkins, E.G., T.P. McCarthy, M.A. O'Brien, and F.A. Kummerow, *J. Am. Oil Chem. Soc. 54:* 279 (1977).
20. Ottenstein, D.M., L.A. Wittings, G. Walker, V. Mahadevan, and N. Pelick, *J. Am. Oil Chem. Soc. 54:* 207 (1977).
21. Chapman, G.W., *J. Am. Oil Chem. Soc. 56:* 77 (1979).

22. *Fatty Acid Methyl Esters,* GC Bulletin 746G, Supelco, Inc., Bellefonte, Pennsylvania, 1975.

23. Jamieson, G.R., and E.H. Reid, *J. Chromatogr. 17:* 230 (1965).

24. Walker, R.C., ed., *Official Methods and Recommended Practices of the American Oil Chemists' Society,* 4th edn., American Oil Chemists' Society, Champaign, Illinois, Method Ce 2-66, (1994).

25. *Ibid.,* Method Ce 1-62.

26. Slover, H.T., and E. Lanza, *J. Am. Oil Chem. Soc. 56:* 933 (1979).

27. *Capillary Analyses of Positional Cis/Trans Fatty Acid Methyl Ester Isomers,* GC Bulletin 822, Supelco, Inc., Bellefonte, Pennsylvania, 1985.

28. Lanza, E., J. Zyren, and H.T. Slover, *J. Agric. Food Chem. 28:* 1182 (1980).

29. Sampugna, J., L.A. Pallansch, M.G. Enig, and M. Keeney, *J. Chromatogr. 249:* 245(1982).

30. George, E.D., and J.A. Acquaro, *J. Liq. Chromatogr. 5:* 927 (1982).

31. George, E.D., E.J. Hillier, and S. Krishnan, *J. Am. Oil Chem. Soc. 57:* 131 (1980).

32. Goldstein M.M., and W.P. Lok, *J. Am. Oil Chem. Soc. 65:* 1350 (1988).

33. Goldstein, M.M., K. Molever, and W.P. Lok, *J. Am. Oil Chem. Soc. 59:* 579 (1982).

34. Sedea, L., and G. Toninelli, *J. Chromatogr. Sci. 19:* 290 (1981).

35. Hunter, R.S., and R.W. Harold, *The Measurement of Appearance,* 2nd edn., Wiley, New York, 1987, pp. 173–174.

36. *Ibid.,* pp. 174–175.

37. *Ibid.,* pp. 206–207.

38. Appleby, D.B., and K.A. Halloran, in *Soap Technology for the 1990's,* edited by L. Spitz, American Oil Chemists' Society, Champaign, Illinois, 1990, p. 104.

39. Hunter, R.S., and R.W. Harold, *The Measurement of Appearance,* 2nd edn., Wiley, New York, 1987, p. 90.

40. *Federal Test Method Standard No. 536A,* Part 2001.2, "Sudsing Test," General Services Administration, Washington, D.C., 1975.

41. *Annual Book of ASTM Standards,* vol. 15.04, Method D 1173-53, "Foaming Properties of Surface-Active Agents," American Society for Testing and Materials, Philadelphia, Pennsylvania, 1994.

42. Becher, P., and R.E. Compa, *J. Am. Oil Chem. Soc. 34:* 53 (1957).

43. Pavlichko, J.P., A.M. Fleischner, and A. Seldner, *Cosmet. Technol. 3(6):* 40 (1981).

44. Osteroth, D., *Soap Cosmet. Chem. Spec. 53(4):* 29 (1977).

Appendix

Regulatory Considerations for Soap Products in the United States

The following sections discuss the status of "soap" relative to cosmetics and over-the-counter (OTC) drugs, and some of the evolution of the so-called "soap exemption" as it relates to cosmetic regulation. For detailed guidance in preparing or reviewing product labeling, refer to the *CTFA Labeling Manual* (1), a publication of the Cosmetic, Toiletry and Fragrance Association. This manual contains a review of specific requirements for labeling of product based on the *Food, Drug & Cosmetic Act* (FD&C Act) (2) and the *Fair Packaging & Labeling Act* (FPLA) (3). Numerous examples are discussed, accompanied by several illustrations designed to clarify many of the requirements. The manual also includes an article of special interest to those in the soap industry entitled *Federal Regulation of Soap Products,* which addresses the "soap exemption," "cosmetic soap," "liquid soap," and pertinent statutory and regulatory considerations.

"Exempted Soap" Status

The legal definition of *soap* is more narrow than the common definition because courts and regulatory agencies tend to restrict the meaning of words that exempt products from compliance with health-related statutes and regulations. The FD&C Act was enacted in 1938 to include cosmetics, but with an exemption for soap. The burden to demonstrate exempted soap status rests with the manufacturer and distributor. The following discussion will review this more narrow definition of soap based on current Food and Drug Administration (FDA) regulations and will provide perspective for judging the soap-*vs.*-cosmetic issues that must be considered with respect to labeling and other regulatory matters.

The Food, Drug & Cosmetic Act's Soap Exemption

The statutory exemption for soap appears in the FD&C Act, which went into effect in 1938. The definition of *cosmetic* (4) is as follows:

> The term "cosmetic" means (1) articles intended to be rubbed, poured, sprinkled, or sprayed on, introduced into, or otherwise applied to the human body or any part thereof, for cleansing, beautifying, promoting attractiveness, or altering the appearance, and (2) articles intended for use as a component of any such articles; except that such term shall not include soap.

The evolution of this soap exemption is important in understanding its scope and limitations in practice. In 1935 Senate Bill No. 5 in the 74th Congress was introduced with an exemption for "ordinary toilet and household soaps" (5). This was later amended in Senate action to read "soaps or household cleansers for which no medicinal or curative qualities are claimed by manufacturers or retailers in labels or advertisements" (6). The bill's sponsor, Senator Copeland of New York, was a physician who was concerned with the safety of drugs and cosmetics. Copeland said on the Senate floor, "The purpose of the exemption is to make it clear that those soaps which the Senator in his household and I in mine use are not affected by the bill" (7).

Senate Bill No. 5 was rewritten in 1936 into a limited inclusion stating that the term *cosmetic* "will include soaps only when medicinal or curative qualities are claimed therefor" (8). Finally, in 1937 after House inaction, the Senate Bill No. 5 was reintroduced into the new Congress with the current language that "cosmetic . . . shall not include soap."

This evolutionary process, then, helps add meaning to what Congress intended the law to mean with respect to the rather vague "soap exemption." The state of activity in soap making in 1938 becomes important because it establishes what might have been in common use in soap generally found in households prior to 1936. For example, if it can be demonstrated that soaps existed in that era containing additives such as cocoa butter, cold cream, pumice, or lanolin, it is likely that soaps containing such ingredients today would still be excluded from the category of cosmetic products. Conversely, inclusion of ingredients developed or introduced into soap since 1936 would probably fall outside of the intentions of the soap exemption and would be considered as cosmetics.

The Administrative Definition of Soap

As an exemption, the *soap* term is narrowly defined. Current FDA regulations (9), in effect since September 26, 1958, define soap very narrowly:

Detergent substances, other than soap, intended for use in cleansing the body.

(a) In its definition of the term "cosmetic," the Federal Food, Drug, and Cosmetic Act specifically excludes soap. The term "soap" is nowhere defined in the act. In administering the act, the Food and Drug Administration interprets the term "soap" to apply only to articles that meet the following conditions:

   (1) The bulk of the nonvolatile matter in the product consists of an alkali salt of fatty acids and the detergent properties of the article are due to the alkali-fatty acid compounds; and

   (2) The product is labeled, sold, and represented only as soap.

(b) Products intended for cleansing the human body and which are not "soap" as set out in paragraph (a) of this section are "cosmetics," and accordingly they are subject to the requirements of the act and the regulations thereunder. For example, such a product in bar form is subject to the requirement, among others, that it shall bear a label containing an accurate statement of the weight of the bar in avoirdupois pounds and ounces, this statement to be prominently and conspicuously displayed so as to be likely to be read under the customary conditions of purchase and use.

The Court Definition of "Soap"

Two court cases, both involving shampoo, are relevant to soap issues. The FDA brought a seizure action against La Maur Egg & Lanolin Shampoo in Iowa, alleging that it was a cosmetic. The FDA argued that the manufacturer's claim of soap exemption coverage should be rejected because the statutory term *soap* did not include shampoos. The court agreed and ruled for the FDA.

In 1971 the FDA brought a seizure action against Beacon Castile Shampoo, which was manufactured by a Chicago private-label manufacturer for Topco, a large chain operation. The shampoo was seized in Cleveland and the private label manufacturer, Consolidated Royal Chemical Company, went to court to claim the goods. Consolidated asserted that the soap exemption covered the product and that the FDA lacked jurisdiction. The court imposed a burden of proof on the manufacturer to demonstrate that the product was within the intended meaning of the soap exemption and, thus, exempt from regulation as a cosmetic.

The court applied the FDA definition of *soap* to examine the Beacon Castile product. The nonvolatile matter consisted of potassium oleate, which all agreed is a soap, and the quantity of it did comprise the bulk of the nonvolatile matter. However, the court then went beyond FDA's bulk criterion to a minor ingredient, 1-1.5% of volume, which was Neutronyx 600. Expert testimony was heard on this synthetic surfactant additive. The court found that this additive was a nonionic synthetic detergent. The claimant's experts argued that the additive functioned only as a soap scum dispersant and used Ivory Soap as an example of basic soap in their demonstration. The court held that the detergent properties of the Beacon product were not due exclusively to the alkali-fatty acid compounds, and the product had not been proven to fall within the "soap exemption." The seizure order was upheld.

The manufacturer and distributor must be able to prove by preponderance of the evidence that (1) additives in the soap product do not have separate detergent properties apart from those properties of the soap base; (2) the product is labeled, sold, and represented only as soap; and (3) the bulk of the nonvolatile matter consists of soap. Absence of any one element would remove the product from consideration for exemption from the cosmetic requirements if the case were to be presented to a court.

Cosmetic Soaps

For products that (a) contain synthetic detergents, (b) contain ingredients other than soap that have in themselves significant detergent properties above and beyond properties attributable to the soap base, or (c) make representations of beautification, deodorancy, or cleansing beyond those attributable to soap ingredients alone, the FDA may take the position that cosmetic regulations apply including cosmetic ingredient labeling and use of approved cosmetic colorants. Addition of deleterious ingredients to them could trigger FDA penalties; manufacture, packing, or storage under unsanitary conditions is also subject to FDA action. Labels must bear, in addition to the ingredient declaration, accurate statements of contents and the name and place of business of the manufacturer or distributor. Voluntary cosmetic registration and reporting programs with the FDA should also be considered, but mandatory registration of neither manufacturers and distributors nor products currently exists for cosmetics.

Drug Soaps

Likewise, for products that make "medicinal" or "medicated" claims or that claim disease-treatment effect, or that claim to affect the structure of the skin, there is a strong possibility that FDA would insist on drug compliance. The cosmetic consequences just described may also apply to this type of drug product. Owners and operators of drug establishments are required to register their firms and to list their drug products with the FDA (10). In addition, current good manufacturing practices (GMPs) (11) for drugs should be followed and labels should state the established name of the active ingredient(s). The over-the-counter (OTC) monographs may in the future require warnings or limit acceptable claims for these products' labels.

Federal Agency Jurisdiction over Soap Products

Food and Drug Administration

All cosmetic soaps, drug soaps, and combinations thereof, come under the jurisdiction of the FDA. The FD&C Act, and all pursuant regulations that are applicable to a given category of product, must be satisfied, including formulation, conditions of manufacturing, documentation, and labeling. The FDA is also responsible for enforcement of the provisions of the FPLA for these categories of product. However, it should be noted that the FPLA applies only to products that are sold at retail level. Pursuant to the FD&C Act, certain labeling requirements such as product identity, net contents statement, name and place of business of manufacturer or distributor, and the listing of active ingredient (for drug products) are required on the immediate labeling of all such products regardless of manner of distribution. Additional requirements, such as cosmetic ingredient declaration, and conditions such as the placement and type size used, must be accommodated on the outermost labeling for those products which come under FPLA. It is interesting to note that the listing of cosmetic ingredients is required on the label of a cosmetic product *only when the item is intended for retail sale* because the cosmetic ingredient labeling regulations (12) were issued pursuant to the FPLA and not the FD&C Act.

Environmental Protection Agency

Any soap product that makes claims such as a "disinfectant" household product or as a "flea-killing" pet product would come under the domain of the U.S. Environmental Protection Agency (EPA) and must comply with the provisions of the *Federal Insecticide, Fungicide, and Rodenticide Act* (FIFRA) (13).

Federal Trade Commission

Ordinary soaps, which do not come under control of the FD&C Act or FIFRA, are also required to comply with the provisions of the FPLA when involved in retail trade. The FPLA for these products is enforced by the Federal Trade Commission. No ingredient labeling requirement currently exists for these "exempted soaps."

Consumer Product Safety Commission

In addition, "exempted soaps" are subject to regulation by the U.S. Consumer Product Safety Commission (CPSC) under the Consumer Product Safety Act (14) and the Federal Hazardous Substances Act (15).

Other Regulatory Concerns for the Soap Industry

Among other federal laws that impact on the soap industry are the following:

- The Clean Water Act (16)
- The Comprehensive Environmental Response, Compensation, and Liability Act (CERCLA or "Superfund") (17)
- The Resource Conservation and Recovery Act (RCRA) (18)
- The Toxic Substances Control Act (TSCA) (19)
- The Hazardous Materials Transportation Act (20)
- The Occupational Safety and Health Act (OSHA) Workers' Right-to-Know Rules (21)
- The Emergency Planning and Community Right-to-Know Act (EPCRA) (22)

These laws all have some potential for requiring a certain level of corporate compliance for any given company, the magnitude of which is somewhat dependent on the size and scope of the company's business. In some cases there are also related state and local compliance requirements with which to contend.

In the case of the last two sets of regulations just listed, specific documentation, training, and reporting procedures are required for hazardous chemicals as defined by the regulations. For the "workers' right-to-know" standard, material safety data sheets (MSDSs) must be available to the employees for any hazardous substances in their work area, a written hazard communication program must exist at each manufacturing site, hazardous materials must be labeled with specific hazard information or codes and certain other information, and training must be provided to the employees regarding the provisions of this law. For the "community right-to-know" regulations, specific annual facility reporting of inventory and location information for certain hazardous chemicals to state and local authorities is required among other provisions of the law.

References

1. *CTFA Labeling Manual,* 5th edn., The Cosmetic, Toiletry and Fragrance Association, Inc., Washington, D.C., 1990.
2. The Federal Food, Drug and Cosmetic Act, *21* United States Code (USC), Sec. 301 *et seq.*
3. The Fair Packaging and Labeling Act, *15* USC, Sec. 1451 *et seq.*
4. *21* USC, Sec. 201(i).
5. S. Rept. 361, 74th Congress, 1st Session (1935).
6. *79 Cong. Rec.,* 4905 (1935).
7. *79 Cong. Rec.,* 4845 (1935).
8. *80 Cong. Rec.,* 10230 (1936).
9. *21* CFR, Sec. 701.20.
10. *21* CFR, Sec. 207.20.
11. *21* CFR, Sec. 210 and 211.
12. *21* CFR, Sec. 701.3.
13. Federal Insecticide, Fungicide and Rodenticide Act, *7* USC, Sec. 136, *et seq.*
14. Consumer Product Safety Act, *15* USC, Sec. 2051 *et seq.*
15. Federal Hazardous Substances Act, *15* USC, Sec. 1261 *et seq.*
16. *33* USC, Sec. 1251 *et seq.*
17. *42* USC, Sec. 9605 *et seq.*
18. *42* USC, Sec. 6901 *et seq.*
19. *15* USC, Sec. 2601 *et seq.*
20. *49* USC, Sec. 1801 *et seq.*
21. *29* CFR, Sec. 1910.1200.
22. *40* CFR, Sec. 300 *et seq.*

Chapter 4

Kettle Saponification, Concurrent and Countercurrent Systems, and Computer Modeling

Joseph A. Serdakowski[a] and Edmund D. George[b]

[a]AutoSoft, East Greenwich, Rhode Island, U.S.A.

[b]Original Bradford Soap Works, West Warwick, Rhode Island, U.S.A.

This text is not designed to be a self-contained primer on the production of kettle soap via the full boil kettle process. It is designed to demonstrate an original method of doing this, utilizing the computer to achieve a high degree of accuracy in process control.

Historical Information

There have been many publications (Davidsohn, Spitz, Woolatt, Wigner, and so on) on the topic of full boil kettle processing. It is beyond the scope of this text to report on all of them. Obviously, more traditional approaches to soap making have been utilized successfully for many years. The appendix to this chapter, graciously submitted by Thomas Wood at Valley Products Company in Memphis, Tennessee, is an excellent illustration of a successful implementation of traditional soap-making wisdom.

Definitions and Terminology

The symbols in the curly brackets { } will represent the shorthand notation used in the algebra. *Processing steps* will be represented by sequential numbers spanning 0 to k + 1, with 0 being the loading and k being the number of washes. The processing step will be represented as subscripts when applicable. *Ingredients* are the materials that are either added to or removed from the kettle. The ingredients will be represented as subscripts when applicable. They include the following:

$\{f_1, f_2, \ldots, f_i\}$ Fats and oils (total number = i)

$\{a_{i+1}, a_{i+2}, \ldots, a_{i+j}\}$ Fatty acids (total number = j)

$\{c\}$ Caustic—50% solution of NaOH in H_2O

$\{b\}$ Brine—saturated solution of NaCl in H_2O

$\{l_0, l_1, l_2, \ldots, l_k\}$ Lyes generated by process steps (k = number of washes; k = 0 is the spent lye for glycerol recovery)—solutions of glycerol, NaCl, NaOH, and H_2O

Spent lye—a by-product of the kettle process that is high (> 15%) in glycerol and low (< 0.5%) in NaOH

Wash lye—a lye that is generated and consumed by the kettle process

$\{y_0, y_1, y_2, \dots, y_{k-1}\}$ Lyes added to process steps

$\{u_0, u_1, u_2, \dots, u_k\}$ Curd (k = number of washes; k = 0 is the curd resulting from loading)— an intermediate remaining after lye removal

$\{n\}$ Neat —the finished product of the kettle soap process

$\{r\}$ Seat (or nigre)—remains in the kettle after neat soap removal

$\{w\}$ Water—the liquid phase of H_2O

$\{t\}$ Steam —the vapor phase of H_2O

Components are the chemical compounds present in the ingredients. The components will be represented as superscripts when applicable. They include the following:

$\{s\}$ Soap

$\{\omega\}$ H_2O

$\{g\}$ Glycerol

$\{d\}$ Sodium chloride (NaCl)

$\{h\}$ Sodium hydroxide (NaOH)

Physical properties are quantitative characteristics of the components and/or ingredients. They include the following:

$\{M\}$ Mass (lb)

$\{X^\chi\}$ Mass fraction of component χ (lb/lb)

$\{W\}$ Molecular weight (lb/lb-moles)

$\{T\}$ Temperature (°F)

$\{\rho\}$ Density (g/cc)

$\{P\}$ Heat capacity (BTU/lb°F)

$\{\Gamma\}$ Heat of reaction (BTU/lb)

Miscellaneous Parameters

$\{\overline{T}\}$ Reaction temperature of kettle (220°F)

$\{D\}$ Day of the year

$\{E_0, E_1, E_2, \dots, E_{k+1}\}$ Electrolyte settling ratio (where k + 1 is the finish step)–the ratio of the different electrolytes as they settle through different phases, specifically, (NaCl)/(NaOH)

$\{\delta\}$ Separation efficiency—the fraction of the available lye that separates from the curd phase

$\{G\}$ Glycerol concentration factor—a measure of glycerol's preference to concentrate in the lye phase during phase separation

Cooling constants determine how fast the kettle cools. These values are site specific and are a function of the kettle geometry, insulation, and environment. They include the following:

{T_∞} Equilibrium temperature (°F)

{D_T} Half-life days

Evaporation constants determine how fast the kettle loses water due to evaporation. These values are site specific and are a function of the kettle geometry, insulation, and environment. They include the following:

{ω_∞} H_2O loss at infinite time (°F)

{$D\omega$} Half-life (days)

Conservation Equations

The principle that matter and energy can be neither created nor destroyed can be used in the analysis of the kettle process. We apply this in three distinct ways: conservation of mass, conservation of mass of each component, and conservation of energy.

Kinetics

The saponification reaction is not spontaneous. As described in Woollatt (p. 154), "the reaction with neutral fats . . . does not start readily. It is autocatalytic, that is catalyzed by the product of the reaction, soap. Hence, the reaction rate accelerates greatly until most of the fat is reacted, when it slows down again." The secret to successful computer simulation is to keep things as simple as possible, but not too simple. The reaction time is much less than the batch time. One simplifying assumption we can make is that everything happens instantly.

Phase Diagram Theory

The kettle soap process has five components and, strictly speaking, a five-component phase diagram is required to represent it. This is too complicated. We simplify the diagram into a three-component system. The components are soap; total electrolyte, which is a linear combination of the sodium chloride and the sodium hydroxide present; and solvent, which is a linear combination of glycerol and water. See Fig. 4.1.

The component list is then simplified to include the following:

{s} Soap

{v} Solvent

{e} Electrolyte

with the linear combinations defined as follows:

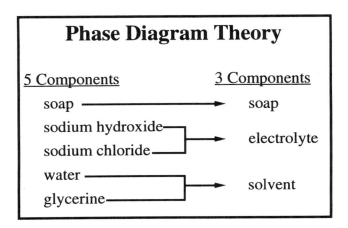

Fig. 4.1. Simplification of phase diagram from 5 to 3 components.

$$M^e = z^d \times M^d + z^h \times M^h \qquad\qquad (4.1)$$

$$M^v = M^\omega + M^g - (1 - z^d) \times M^d - (1 - z^h) \times M^h \qquad (4.2)$$

Here z is defined to be the *graining efficiency*, a traditional soapmaking term, which is a measure of how much of the particular electrolyte will have to be added to move the resultant mixture a certain distance in the x direction on the phase diagram. Other electrolytes can also be used, as described in Spitz (p. 119). The z factors are normalized such that $z^d = 1$. Wigner reports that $z^h = 1.25$; however, this value is not universally accepted. We will use Wigner's value in a later example. Equation (4.1) determines the total amount of electrolyte present. Equation (4.2) determines the total amount of solvent present. The final two terms in Eq. (4.2) are necessary to ensure that the conservation of mass components are maintained.

Figure 4.2 is a phase diagram of a typical 80/20 tallow/coco soap as illustrated in Woollatt (p. 153).

Also note the inclusion of several x-axes. The values for the x (electrolyte) axis depend on the chain length distribution of the soap. The graining index data presented in Spitz (p. 118) allow us to determine the phase diagrams for a number of different soaps. The relative graining indexes are aligned and the electrolyte is scaled in proportion to yield the phase diagrams for all listed soaps. As a first-order approximation, the x-axis is scaled in proportion to the graining index. For example, the coordinates of the point of intersection of the D and Q regions occur at 6.3% electrolyte for the 80% tallow/10% coco soap illustrated in Fig 4.2. This soap has a

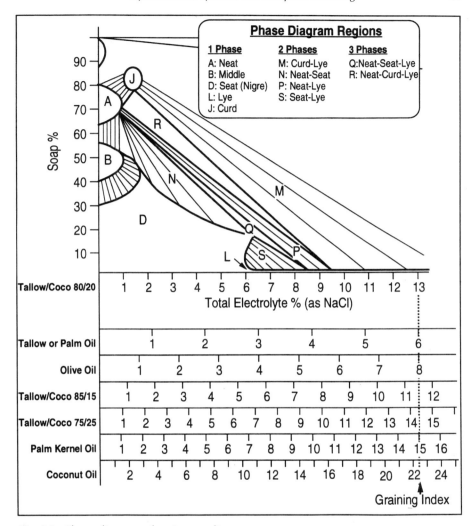

Fig. 4.2. Phase diagram of various sodium soaps.

graining index of 13. A pure coconut oil soap with a graining index of 22.5 will have the point of intersection of the *D* and *Q* regions at $(22.5/13) \times 6.3\% = 10.9\%$. In this fashion, phase diagrams for soaps of all chain length distributions can be determined.

The labeling is the same as Woollatt and is described in Table 4.1.

The phase diagram is further approximated for computerization. Only the two phase regions M and N are required for modeling. Both regions are approximated by

TABLE 4.1
Description of Phase Diagram Regions

Region	Number of Phases	Description
A	1	Neat soap
B	1	Middle soap
D	1	Nigre
J	1	Curd
L	1	Lye
M	2	Curd—lye
N	2	Neat—nigre
P	2	Neat—lye
Q	3	Neat—nigre—lye
R	3	Curd—neat—lye

straight-edged quadrilaterals, that is, linear approximations, which have proven to be sufficient. Higher-order approximations (quadratic) have been tested. The higher-order approximations complicate the mathematics but do not provide any improvement to the model. A specific linearized phase diagram will be discussed later.

Kettle Soap Boiling—General Discussion

Before the mathematics is detailed further, the kettle soap process will be discussed in general terms. More detailed discussions of this process are available in the references already cited.

Step 1—Loading

Typically a kettle of 20,000 to 200,000 lb. capacity is used. The nigre often remains in the kettle from the prior batch. The nigre is brought to a boil by the introduction of live steam into the bottom of the kettle through a specialized nozzle called a rosebud (because of its appearance) and through a series of open steam coils.

Precise amounts of fats, oils, and/or fatty acids are combined with caustic, brine, and water. In the case of the countercurrent process, recycled lyes from the first wash of a prior kettle are also added. The materials are added such that the rate of saponification is maximized.

Since *spent* lye is the desired output of this kettle, the electrolyte or x-axis of the phase diagram should be composed of only NaCl, with only enough NaOH added to the kettle to saponify the fats, oils, and fatty acids. This poses a problem for the soapmaker because high excess levels of NaOH drive the saponification reaction to completion; however, there should be no excess and perhaps even a slight deficit of NaOH, at the conclusion of the loading process to ensure formation of a *spent* lye.

We now turn our attention to identifying the region of maximum saponification on the phase diagram. It is slightly lower in electrolyte than region M (the two-phase curd-lye region). The exact location of this region is subject to some debate. Most published phase diagrams illustrate three distinct regions, those being M, P, and R; however, it is our experience that for all practical matters those regions are indistinguishable during the production process. That being the case, the point of maximum saponification will occur in region Q or perhaps even region N.

The actual *location* of this point of maximum saponification with regard to regions R, P, Q, N, and so on is inconsequential when one's principle priority is optimizing production. An experienced soapmaker inherently knows this region by the appearance of the kettle contents. To identify this crucial point in the soapmaking process, simply sample the kettle at the point when the experienced soapmaker *knows* that the kettle "looks" best. Between 10 and 12 kettles should provide more than enough data to define this point for the fat and oil blend being used. Once the first fat and oil blend has been identified, other similar blends can be extrapolated using the relative graining index method just outlined.

The percent soap, or *y*-axis, has a limited working range, because levels in excess of 55% soap result in a mixture that is too viscous to permit good agitation using only live steam, and levels below 40% result in excessive amounts of spent lye, reduced kettle capacity, and low glycerol concentrations in the spent lye.

Remember that the loading starts not with an empty kettle but with a nigre, which should have a composition on the border between region D (the one-phase nigre region) and region N (the two-phase neat-nigre region). The loading should proceed to bring the partial contents of the kettle to the saponification point as soon as possible, and then keep the kettle composition at the saponification point for the remainder of the loading process.

Step 2—Graining

After all of the fats, oils, and/or fatty acids have been saponified, the kettle needs to be positioned on the phase diagram at a point that will result in an unstable mixture of curd and lye. This area is in region M (the two-phase curd-lye region). Only a limited area in region M will effect good separation of the lye from the curd, this area being just over the border from region R. Complicated interactive forces at the molecular level exceed gravitational forces; thus the lye and curd do not completely separate. The percentage of total separation is the *separation efficiency*, in which 83% seems to be a realistic maximum for industrial kettle soap processes. Movement away from this border results in an "overgraining" condition in which even though the lye and the curd are two distinct phases, quite visible to the naked eye, they do not separate. In these cases, separation efficiencies can drop below 50%, yielding a process that cannot be economically viable because the resulting curd will not be high enough in soap percentage to allow for effective fitting.

Movement from region R to region M is done by the addition of brine. In theory, this could also be done by the addition of rock salt if the amount of generated lye needs to be minimized, or by the addition of NaOH if the presence of excess NaOH in the (now not) spent lye is acceptable. This process is called graining the kettle because the kettle's appearance changes from being smooth to being very grainy. There are a number of traditional soapmakers checks that can be made to ensure that the proper grain has been achieved. These tests are discussed in the traditional references outlined.

Note again that the exact point of "best" settling is known by the experienced soapmaker. Sampling a small number of kettles will define this point and allow the computer to bring the soapmaker to this point on a routine basis.

The efficiency of kettle agitation can be enhanced by installation of a *recirculation pipe*, as depicted in Fig. 4.3.

This recirculation pipe allows lye that accumulates on the bottom of the kettle to flow up the pipe and be disbursed on the top of the kettle. This process allows for more rapid saponification and full consumption of the NaOH. The recirculated lye can be sampled and tested for both free alkali and salt levels. Once the desired levels are achieved, then the graining process is considered complete. The desired levels are determined from the phase diagram by constructing a *tie line* that passes through the graining point. The intersection of this tie line with region L (the one-phase lye region) determines the electrolyte concentration. The absence of free alkali in the recirculated lye sample indicates that the saponification reaction is complete.

Step 3—Settling and Spent Lye Removal

The kettle is allowed to settle, which results in an accumulation of lye at the bottom of the kettle. The composition of this lye is predicted by the use of the tie line, as just described. The total quantity of lye is determined by the ratio calculation standard to all phase diagrams. The available lye is determined by multiplying the available lye by the separation efficiency, remembering that 17% or more of the available lye cannot be removed without the aid of increased gravitational forces (as with a centrifuge).

A properly grained kettle can have lye removal occur almost immediately. This immediate removal of lye does not come without a price, however. The solubility of soap in the lye is a partial function of the lye temperature. Lye removed immediately upon graining will have temperatures in excess of 220°F and will carry with it in excess of 1% soap. Upon storage and subsequent cooling, this soap will precipitate out of the lye and float to the surface, eventually creating a solid mass inside the lye storage tanks. This soap can be added back to subsequent kettles but requires management to ensure that lye storage capacity is not clogged with precipitated soap. Kettles allowed to settle for longer time periods will yield cooler lyes and less precipitated soap problems.

Countercurrent processing will net glycerol concentration in the lye in excess of 15%. Concurrent processing (that is, lack of countercurrent processing) will yield spent lyes with less than 12% glycerol.

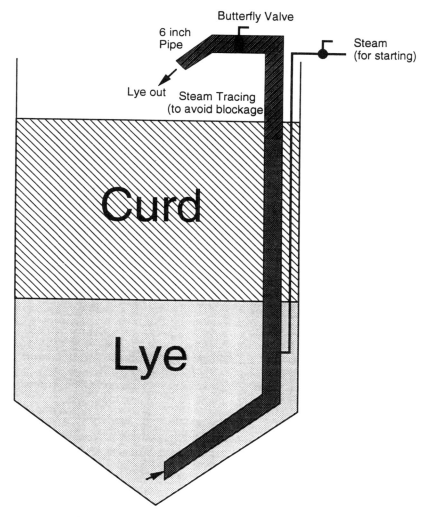

Fig. 4.3. Kettle recirculation schematic.

The exact concentration of glycerol in the lye can be calculated once the mechanisms at work are considered. The solvent in the simplified phase diagram consists of glycerol and water. Upon completion of the graining process, there is a definite ratio of glycerol to water in the solvent. One of three mechanisms can occur: the ratio of glycerol to water can increase in the solvent-rich or lye phase relative to the entire mass of the kettle, the ratio can stay the same, or the ratio can decrease. The first mathematical models of the system assumed that the ratio was constant throughout. Comparing actual results to the model netted slightly higher glycerol concentrations in the lye than predicted. A glycerol concentration (or "fudge") factor was defined. A value of 1.1 matched the model to the actual results, meaning that glycerol had a slightly higher tendency to migrate into the solvent-rich or lye phase in preference to the water.

Step 4—Kettle Washing

Washing is the process of adding additional amounts of caustic, brine, and water to a settled curd. Remember that the settle curd is located in region M close to region J (the single-phase curd region). Washing moves the kettle composition down the tie line toward region L. Again, the same constraints apply with regard to overgraining the kettle.

The washing is performed for several reasons: to improve the color and odor of the soap, to reduce the concentration of glycerol in the soap, and to control the ratio of free alkali to salt. The loading and graining steps require the generation of a *spent* lye, in which the electrolyte is composed purely of NaCl. Because the spent lye is removed from the kettle process and used as the feedstock for a glycerol evaporator, it is important to minimize the free alkali content to reduce the treatment costs associated with glycerol recovery. Thus the free alkali to salt ratio is effectively *zero*. Such high levels of salt, if carried through to the neat soap, will produce such an increase in salt that subsequent processing and bar pressing will be severely compromised, if not impossible. By washing the kettle with precise amounts of caustic and brine, the free alkali to salt ratio can be shifted to provide a soap base with superior handling characteristics.

The washing is performed in such a way as to bring the kettle to a point of instability (as just described in the graining step). Again, the recirculation pipe is utilized to effect better mixing and to allow sampling of the lye. Once the recirculated wash lye has achieved the desired free alkali and salt concentration, the washing is complete.

Step 5—Settling and Wash Lye Removal

The kettle is allowed to settle, which results in an accumulation of lye at the bottom of the kettle. Again, lye removal can proceed almost immediately if the kettle has been properly grained. This lye is stored and is used during the loading of subsequent kettles. A properly designed kettle process will yield an amount of lye from a wash to match the amount of lye to be recycled back into the prior step of the subsequent kettle. The computer model can greatly simplify this problem, as will be demonstrated later.

At the conclusion of lye removal the kettle is in region M, with the best location being as close to region J as possible, meaning most of the lye being removed.

Steps 4 and 5 can be repeated as many times as necessary to achieve the proper color, odor, glycerol concentration, and free alkali to salt ratio. Diminishing marginal returns occur after two well-defined washes.

Step 6—Finishing or Fitting the Kettle

It is this step that traditional soapmakers appear to hold as most mysterious and difficult. However, a properly designed kettle soap process will result in consistent finishes. At the start of this step, the kettle has been drained of all available wash lye and the desired free alkali to salt ratio has been achieved. The kettle is in region M close to region J. Water is used to finish the kettle. Addition of water to the kettle moves the kettle's composition directly toward the origin on the phase diagram. The kettle passes through region R (the three-phase curd-nigre-lye region) and into region P (the two-phase neat-lye region). It would be nice if effective settling were possible in region P, because this would yield only neat soap and lye; however, this is not observed, probably because of the insufficient gravitational forces generated on Earth (one has to wonder if future generations of soapmakers will ply their trade on Jupiter to take advantage of increased gravity there). Further addition of water will move the kettle's composition into region Q (the three-phase neat-nigre-lye region). Good separation can be found in this region; however, the addition of the *nigre-lye* phase increases the variability of the process and complicates processing. Best fitting occurs just over the border into region N (the two-phase neat-nigre region). Here gravity can just overcome the molecular-level forces and permit the neat soap and nigre to separate. Addition of excessive water will result in relatively large amounts of nigre and subsequently smaller kettle yields. A minimum of 8 h will be required before the neat soap can be removed from the kettle, and longer times, if available, will provide a more consistent product.

Secondary H_2O Considerations

To accurately calculate the process just defined, high levels of accuracy and precision are required because the critical areas of maximum settling are relatively small. Traditional methods may overlook the following contributions to kettle soap H_2O.

Steam used for agitation and heating condenses into the kettle mass. This amount must be calculated by using the temperatures, heat capacities, and heat of reactions of the various ingredients. Evaporation occurs during the settling process and must be considered. Finally, the kettle cools during settling and requires condensed steam to reheat.

Countercurrent Illustration

The countercurrent nature of this process is now illustrated with a four-kettle system in Fig. 4.4. The processing steps are listed across the top: load, first wash, second wash, third wash, and finish. Loading of every kettle generates a neutralized or spent lye, which is then sent to glycerol recovery. The wash lye removed from the first wash in kettle 1 goes into the loading of kettle 2. The wash lye removed from the second wash of kettle 1 goes into the first wash of kettle 2. The wash lye removed from the third wash in kettle 1 goes into the second wash of kettle 2. Kettles 2, 3, and 4 follow the same pattern. The seat, or nigre, generated during the

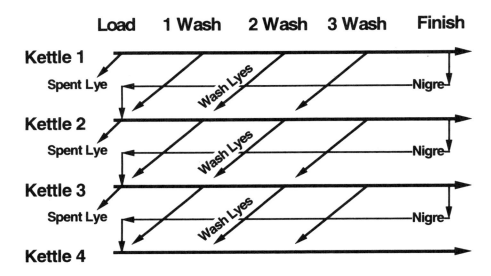

Fig. 4.4. Countercurrent soapmaking process summary.

finishing and the fitting of the first kettle is used for the loading of the second kettle. The other seats are handled in the identical fashion. The countercurrent flow become evident. The lines representing the production of soap go from left to right, and the lines representing the flow of glycerol go from right to left.

Variations on a Theme

There are many possible variations to the process just outlined. The process defined by Thomas Wood in the appendix to this chapter adds the coconut oil during the first wash. Other options include graining the nigre "off-line" and reintroducing this concentrated and washed nigre into a washing step, thus loading on an empty kettle. One can also hold back a relatively large amount of NaOH during the loading step, thus saponifying only a fraction (say 85%) of the fats during the first step of the process. These variations and others all have their features and benefits. However, all deal with the same phase diagram and the same concepts of graining and settling, and thus all can be calculated in the same fashion.

Kettle Soap Boiling—Detailed Discussion

In the following sections, a step-by-step detailed calculation of a kettle will be demonstrated. The phase diagram by Woolatt will be used for the basis of the calculation. Experienced soapmakers will note that some of the values resulting from this phase diagram will not match values achieved in real-life kettles. Hints are given at the end as to how to "massage" the published phase diagram to achieve predictions that match actual kettle performance. Unless otherwise defined, the physical properties used are listed in Table 4.2.

Loading the Kettle

The kettle will be loaded with 100,000 lb of fat at an 85/15 tallow-coco blend. Loading target of 50% soap and 3.75% total electrolyte is determined based on the criteria just discussed. This point, located in region R, is represented by the circle in Fig. 4.5. Note that this figure is a linearized version of the phase diagram depicted in Fig. 4.2 with only regions M and N fully defined. Also note that the x-axis in Fig. 4.5 is the 85/15 tallow-coco axis depicted in Fig. 4.2. Because spent lye is required, the electrolyte will be totally NaCl.

Loading Targets

$$X_0^s = 50.00\% \tag{4.3}$$

$$X_0^d = 3.75\% \tag{4.4}$$

$$X_0^h = 0.00\% \tag{4.5}$$

Stochiometry—Fats and Oils

The chemistry for the saponification reaction is

$$1 \text{ (Fat or oil)} + 3 \text{ (NaOH)} = 3 \text{ (Soap)} + 1 \text{ (Glycerol)}$$

which allows us to write

$$X_{f_n}^s = \frac{W_{f_n} + 3 \times W_h - W_g}{W_{f_n}} \tag{4.6}$$

TABLE 4.2

Physical Properties of Ingredients

	T	ρ	C	P	X^s	X^ω	X^g	X^e	X^h	W
f_1	140	0.882	0.477	37.76	1.033[a]		0.107[a]		−0.140[a]	860
f_2	130	0.892	0.467	47.03	1.042[a]		0.139[a]		−0.181[a]	662
a_3	140	0.830	0.477	37.76	1.078[a]	0.063[a]			−0.140[a]	284
c	94[a]	1.526	0.782	22.50		0.504			0.496	
b	75[a]	1.206	0.990	0		0.735		0.265		
l	140	1.120	0.962	0	0.010					
u		0.970	0.576	0						
n		0.970	0.576	0						
r		1.050	0.736	0						
w	75[a]	1.000	1.000	0		1.000				
t				1005						

[a]Calculated values.

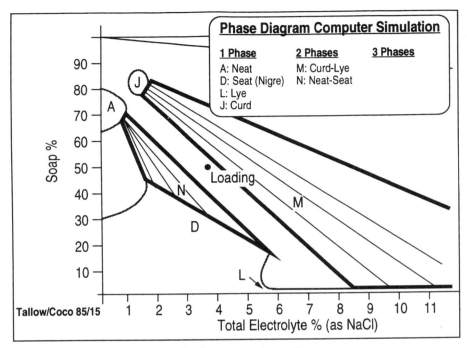

Fig. 4.5. Linear approximation of sodium soap phase diagram.

$$X_{f_n}^g = \frac{W_g}{W_{f_n}} \tag{4.7}$$

$$X_{f_n}^h = \frac{3 \times W_h}{W_{f_n}} \tag{4.8}$$

where

$$W_h = 40 \tag{4.9}$$

$$W_g = 92 \tag{4.10}$$

Stochiometry—Fatty Acids
The chemistry for the neutralization reaction is

$$1 \text{ (Fatty acid)} + 1 \text{ (NaOH)} = 1 \text{ (Soap)} + 1 \text{ (Water)}$$

which allows us to write

$$X_{a_n}^s = \frac{W_{a_n} + W_h - W_w}{W_{a_n}} \tag{4.11}$$

$$X_{a_n}^w = \frac{W_w}{W_{a_n}} \tag{4.12}$$

$$X_{a_n}^h = \frac{3 \times W_h}{W_{a_n}} \tag{4.13}$$

where

$$W_w = 18 \tag{4.14}$$

Pounds of Anhydrous Soap Produced
Calculate a weighted average:

$$X^s = 0.85 \times X_{f_1}^s + 0.15 \times X_{f_2}^s = 103.44\% \tag{4.15}$$

Total pounds of anhydrous soap = M^s:

$$M^s = X^s \times 100{,}000 = 103{,}435 \tag{4.16}$$

Pounds of tallow soap = M_o^{s1}

$$M_o^{s1} = M_o^s \times 0.85 = 87{,}920 \tag{4.17}$$

Pounds of coco soap = M_o^{s2}

$$M_o^{s2} = M_o^s \times 0.15 = 15{,}515 \tag{4.18}$$

Note that we are defining a ratio of 85/15 soap blend, not 85/15 fat blend; there is a difference between the two.

Total kettle mass:

$$M_o = M_o^s/X_o^s = 206{,}870 \tag{4.19}$$

NaCl loading:

$$M_o^d = X_o^d \times M_o = 7758 \tag{4.20}$$

Nigre Analysis
There is a nigre remaining in the kettle. This nigre is 20,000 lb and has an analysis of 25% soap, 1.5% NaOH, and 3.0% NaCl at an 80/20 tallow-coco blend. The glycerol content is estimated to be 4%. The exact glycerol content is not important for determining the loading quantities but can be used to determine the glycerol balance in the kettle for calculation of glycerol content in the spent lye.

$$M_r = 20{,}000 \tag{4.21}$$

$$X_r^s = 25.00\% \tag{4.22}$$

$$X_r^h = 1.50\% \tag{4.23}$$

$$X_r^d = 3.00\% \tag{4.24}$$

$$X_r^g \approx 4.00\% \tag{4.25}$$

Pounds of tallow soap = M_r^{s1}:

$$M_r^{s1} = M_r \times X_r^s \times 0.80 = 4{,}000 \tag{4.26}$$

Pounds of coco soap = M_r^{s2}:

$$M_r^{s2} = M_r \times X_r^s \times 0.20 = 1{,}000 \tag{4.27}$$

Pounds of NaCl = M_r^d:

$$M_r^d = M_r \times X_r^d = 600 \tag{4.28}$$

Pounds of NaOH = M_r^h:

$$M_r^h = M_r \times X_r^h = 300 \tag{4.29}$$

Amount of Lye to Add

The maximum amount of spent lye is to added to the kettle. The recycled wash lye is 1% tallow soap, 7.32% NaCl, 3.66% NaOH, and 11% glycerol; it is estimated as follows:

$$X_{y_0}^s = 0.010 \tag{4.30}$$
$$X_{y_0}^h = 0.0366 \tag{4.31}$$
$$X_{y_0}^d = 0.0732 \tag{4.32}$$
$$X_{y_0}^g \approx 0.11 \tag{4.33}$$

For the first iteration, assume that all of the added NaCl is to come from the lye. Added salt is

$$M_{y_0}^d = M_0^d - M_r^d = 7158 \tag{4.34}$$

Total lye added is

$$M_{y_0} = M_{y_0}^d / X_{y_0}^d = 97{,}782 \tag{4.35}$$

This will add soap,

$$M_{y_0}^s = M_{y_0} \times X_{y_0}^s = 978 \tag{4.36}$$

NaOH,

$$M_{y_0}^h = M_{y_0} \times X_{y_0}^h = 3{,}579 \tag{4.37}$$

and solvent:

$$M_{y_0}^v = M_{y_0} \times (1 - X_{y_0}^s - z^h \times X_{y_0}^h - zd \times X_{y_0}^d) = 85{,}172 \tag{4.38}$$

One can see that this is an impossibly large number; because the lye is predominantly solvent (water plus glycerol), in order to obtain all of the NaCl from the lye, too much water must be added simultaneously. This problem can be solved through a variety of numerical methods including matrix inversions, iterative schemes, and methods commonly employed in operations research. Wagner is a good source for these methods. It is not the intent of this paper to review such methods. It is important to note, however, that with sophisticated computer methods large amount of lyes can be consumed in the loading process, so that the concentration of glycerol in the spent lyes can be maximized.

Instead of maximizing the amount of lye to be consumed, we instead assume a fixed amount for the purpose of this illustration. Assume 40,000 lb of lye is to be consumed:

$$M_{y_o} = 40,000 \tag{4.39}$$

$$M_{y_o}^{s1} = 400 \tag{4.40}$$

$$M_{y_o}^{s2} = 0 \tag{4.41}$$

$$M_{y_o}^{d} = 2,928 \tag{4.42}$$

$$M_{y_o}^{h} = 1,462 \tag{4.43}$$

$$M_{y_o}^{g} = 4,400 \tag{4.44}$$

$$M_{y_o}^{\omega} = M_{y_o} - M_{y_o}^{s} - M_{y_o}^{d} - M_{y_o}^{h} - M_{y_o}^{g} = 30,808 \tag{4.45}$$

$$M_{y_o}^{e} = M_{y_o}^{d} \times z^{d} + M_{y_o}^{h} \times z^{h} = 4,758 \tag{4.46}$$

$$M_{y_o}^{v} = M_{y_o} - M_{y_o}^{s} - M_{y_o}^{e} = 34,842 \tag{4.47}$$

Note that

$$M_{y_o}^{v} \neq M_{y_o}^{g} + M_{y_o}^{\omega} \tag{4.48}$$

This is due to the graining efficiency correction and the application of the conservation of mass equation.

The balance of the added salt must be derived from the addition of brine to the kettle. This remaining salt is

$$M_{b_o}^{d} = M_{o}^{d} - M_{r}^{d} - M_{y_o}^{d} = 4,230 \tag{4.49}$$

The amount of brine to be added is

$$M_{b_o} = M_{b_o}^{d}/X_{b_o}^{d} = 15,961 \tag{4.50}$$

Tallow and coco must be added to the kettle. The amount of tallow soap to be derived from new tallow is

$$M_{f_1}^{s} = M_{o}^{s1} - M_{y_o}^{s1} - M_{r}^{s1} = 83,520 \tag{4.51}$$

and from coco,

$$M_{f_2}^{s} = M_{o}^{s2} - M_{y_o}^{s2} - M_{r}^{s2} = 14,515 \tag{4.52}$$

So the amount of tallow and coco to be added is

$$M_{f_1} = M_{f_1}^{s}/X_{f_1}^{s} = 80,852 \tag{4.53}$$

$$M_{f_2} = M_{f_2}^{s}/X_{f_2}^{s} = 13,930 \tag{4.54}$$

Caustic must be added to saponify the fats and oils. First determine the amount of NaOH needed in the caustic. We sum the amount required by the fats and oils and subtract the amount of NaOH already present in the lye and nigre:

$$M_{c_o}^{h} = (\sum_{1}^{i} M_{f_i} \times X_{f_i}^{h}) - M_{r}^{h} - M_{y_o}^{h} = 12,077 \tag{4.55}$$

The amount of added caustic is

$$M_{c_o} = M_{c_o}^{h}/X_{c_o}^{h} = 24,348 \tag{4.56}$$

We can now also calculate how much glycerol will be present in the kettle:

$$M_{o}^{g} = (\sum_{1}^{i} M_{f_i} \times X_{f_i}^{g}) + M_{r} \times X_{r}^{g} + M_{y_o} \times X_{y_o}^{g} = 15,787 \tag{4.57}$$

The final ingredient is water, the liquid phase of H_2O. H_2O is part of the nigre, lye, brine, and caustic. H_2O also condenses as a result of agitation and heating with open coils.

The total solvent (H_2O + glycerol - graining efficiency adjustment) present in the kettle can be determined from the loading targets:

$$X_o^v = 1 - X_o^s - X_o^d = 46.25\% \tag{4.58}$$

$$M_o^v = M_o \times X_o^v = 95{,}677 \tag{4.59}$$

The glycerol component of the total solvent has already been determined, allowing us to calculate the total amount of H_2O in the kettle:

$$M_o^\omega = M_o^v - M_o^g + (1 - z^d) \times M_o^d + (1 - z^h) \times M_o^h = 79{,}890 \tag{4.60}$$

Finally, additional water is added to achieve the correct amount of H_2O in the kettle. We now determine that number.

H_2O content of nigre:

$$M_r^\omega = M_r \times (1 - X_r^s - X_r^h - X_r^d - X_r^g) = 13{,}300 \tag{4.61}$$

H_2O content of lye:

$$M_{y_o}^\omega = M_{y_o} \times (1 - X_{y_o}^s - X_{y_o}^h - X_{y_o}^d - X_{y_o}^g) = 30{,}808 \tag{4.62}$$

H_2O content of brine:

$$M_{b_o}^\omega = M_{b_o} \times (1 - X_{b_o}^s - X_{b_o}^h - X_{b_o}^d - X_{b_o}^g) = 11{,}731 \tag{4.63}$$

H_2O content of caustic:

$$M_{c_o}^\omega = M_{c_o} \times (1 - X_{c_o}^s - X_{c_o}^h - X_{c_o}^d - X_{c_o}^g) = 12{,}271 \tag{4.64}$$

The amount of H_2O that condenses in the kettle as steam must now be determined. First the amount of energy required to bring the ingredients to the kettle reaction temperature is calculated. This is a function of the temperatures of the incoming ingredients. The nigre is assumed to be at the reaction temperature before sampling because the nigre needed to be brought to a boil to obtain a homogeneous sample. The other ingredients are below the reaction temperature and require condensed steam to heat.

Also, the heat of reaction/solution reduces the steam loading and must be considered. This number is quite substantial for the saponification reaction.

The temperatures of the brine, water, and caustic have seasonal fluctuations, reaching a high on August 9 of each year in West Warwick, Rhode Island. We know

$$T(D) = \frac{(T_{max} - T_{min}) \times \cos\left(\frac{2\pi (D - D_{max})}{365.25}\right) + T_{max} + T_{min}}{2} \tag{4.65}$$

The parameters to Eq. (65) in use at Bradford Soap are listed in Table 4.3.

TABLE 4.3
Raw Materials Seasonal Temperature Fluctuations

	T_{min}	T_{max}	D_{max}
c	64	94	Aug 9
b	45	75	Aug 9
w	45	75	Aug 9

The amount of steam required to heat component α to the reaction temperature $\overline{T}$ is

$$M_{t\alpha} = M_\alpha \times \frac{(\overline{T} - T_\alpha) \times P_\alpha - \Gamma_\alpha}{\Gamma_\tau} \tag{4.66}$$

where

$$\overline{T} = 220°F \tag{4.67}$$

Let us assume that the kettle is being loaded on August 9.

Pounds of steam to heat ingredients exclusive of water = $M_{t_o}{}'$

$$
\begin{aligned}
M_{t_o}{}' \quad &= \{M_{f_1} \times [(\overline{T} - T_{f_1}) \times P_{f_1} - \Gamma_{f_1}] &\{\text{Tallow}\} \\
&+ M_{f_2} \times [(\overline{T} - T_{f_2}) \times P_{f_2} - \Gamma_{f_2}] &\{\text{Coconut oil}\} \\
&+ M_{c_o} \times [(\overline{T} - T_c) \times P_c - \Gamma_c] &\{\text{Caustic}\} \\
&+ M_{y_o} \times (\overline{T} - T_{l_1}) \times P_{l_1} &\{\text{Recycled lye}\} \\
&+ M_{b_o} \times (\overline{T} - T_b) \times P_b \} &\{\text{Brine}\} \\
&\div \Gamma_t
\end{aligned}
$$

$$M_{t_o}{}' \quad = 7,147 \tag{4.68}$$

For the same kettle 6 months later (note the increase),

$$M_{t_o}{}' \quad = 8,187 \tag{4.69}$$

So on August 9 the total amount of H_2O in the kettle that is derived from the ingredients exclusive of water is

$$M_o^{\omega'} \quad = + M_r^\omega + M_{y_o}^\omega + M_{b_o}^\omega + M_{c_o}^\omega + M_{t_o}{}' = 75,258 \tag{4.70}$$

with the prime (') indicating exclusive of water.

So the amount of H_2O derived from water is (factoring the required steam to heat water to the reaction temperature)

$$M_{w_o} = \frac{\Gamma_t \times (M_o^\omega - M_o^{\omega'})}{\Gamma_t + (\overline{T} - T_w) \times P_w} = 4,048 \tag{4.71}$$

with the total steam condensing as

$$M_{t_o} = M_{w_o} \times (\overline{T} - T_w) \times P_w \div \Gamma_t + M_{t_o}{}' = 7,732 \tag{4.72}$$

So, in summary, the following ingredients are added:

Nigre	$M_r =$	20,000	(4.73)
Tallow	$M_{f_1} =$	80,851	(4.74)
Coco	$M_{f_2} =$	13,930	(4.75)
Lye	$M_{y_0} =$	40,000	(4.76)
Brine	$M_{b_0} =$	15,961	(4.77)
Caustic	$M_{c_0} =$	24,347	(4.78)
Water	$M_{w_0} =$	4,048	(4.79)
Steam	$M_{t_0} =$	7,732	(4.80)
Total	$M_0 =$	206,869	(4.81)

which agrees with the initial calculation.

Rates of Addition

As discussed earlier, the success of the kettle is a strong function of maintaining the proper point on the phase diagram to ensure maximum saponification. The loading target as defined is this maximum point of saponification. However, at the start of the loading process the kettle's composition is also the nigre's composition. In the simplified case of preblended fats, the first stage of loading is to move the kettle to the point of maximum saponification.

Fluctuating Fat Ratios

Often a preblend tank is not available, forcing the soapmaker to add the fats either serially or sequentially. In the serial case, the tallow/coco ratio varies as the kettle's ingredients are changed. To maximize saponification in this case, one has to "hit a moving target" because the maximum saponification point is moving. Using the various x-axes in the phase diagram of Fig 4.2, one can use a computer to predict this optimum point as a function of tallow/coco ratio; however, that calculation is too involved to discuss here. This problem is illustrated with Fig 4.6, where a 85/15 tallow/coco soap is being loaded on an 85/15 nigre.

At the onset of loading the coconut oil constitutes 15% of the total soap, as one would expect. If the coconut oil is loaded first, at the conclusion of the loading of the coconut oil the soap blend in the kettle is almost 70% coconut oil. Of course, the phase diagram of this blend is significantly different from the end product and will have a maximum saponification rate point far different from the end product as well. This sequential loading is the norm at Bradford, and the computer adjusts the intermediate loading figures to ensure that saponification is maximized throughout the loading process.

Percentage of Coconut Oil Soap During Kettle Loading of an 85/15 Tallow/Coco Soap

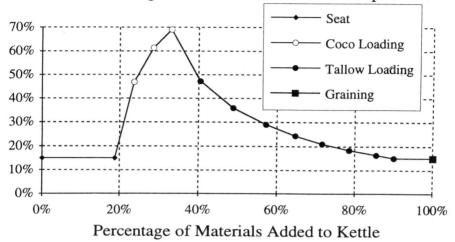

Percentage of Materials Added to Kettle

Fig. 4.6. Illustration of fluctuating fat ratios during kettle leading.

Fluctuating Electrolyte Ratios

The need to produce spent lye will result in no excess NaOH present when loading is completed. However, excess NaOH during the loading process increases the saponification rate. The computer ensures that there is excess NaOH during intermediate stages of the loading process. However, the computer must also reduce the NaCl concentration proportionately to ensure that the proper total electrolyte ratio is present. Figure 4.7 demonstrates how the excess NaOH or free alkali varies in time for an actual kettle loading sequence.

We will proceed with the assumption of a preblend tank for the fats and oils. No effort is made in this simplified example to provide free alkali during the intermediate stages of loading. This problem is left to the interested reader.

It is best if the optimum point is achieved as rapidly as possible. The nigre's composition is as follows:

$$M_r = 20,000 \tag{4.21}$$

$$X_r^s = 25.00\% \tag{4.22}$$

$$X_r^h = 1.50\% \tag{4.23}$$

$$X_r^d = 3.00\% \tag{4.24}$$

$$X_r^g \approx 4.00\% \tag{4.25}$$

$$X_r^\omega = 1 - X_r^s - X_r^h - X_r^d - X_r^g \approx 66.50\% \tag{4.82}$$

as we want to achieve:

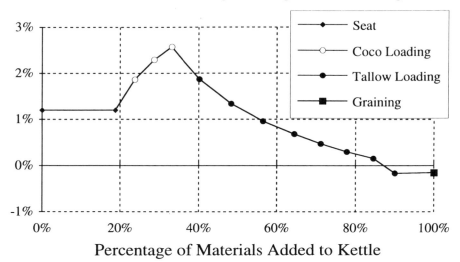

Fig. 4.7. Illustration of fluctuating free alkali during kettle leading.

$$M_{intermediate} = M_{int.} = Minimum$$
$$X_o^s = 50.00\% \qquad\qquad (4.3)$$
$$X_o^d = 3.75\% \qquad\qquad (4.4)$$
$$X_o^h = 0.00\% \qquad\qquad (4.5)$$

Without proceeding with the calculations step by step, let it suffice to say that again, advanced numerical techniques are required to minimize Mint.. It is beyond the scope of this paper to delve into these numerical methods. Note, however, that the limiting component is again the water content. The solution to this problem is as follows:

$$M_{int.} = 51,520 \qquad\qquad (4.83)$$
$$M_{f1\ int.} = 17,121 \qquad\qquad (4.84)$$
$$M_{f2\ int.} = 2,950 \qquad\qquad (4.85)$$
$$M_{c\ int.} = 5,304 \qquad\qquad (4.86)$$
$$M_{b\ int.} = 5,026 \qquad\qquad (4.87)$$
$$M_{t\ int.} = 1,119 \qquad\qquad (4.88)$$

These ingredients should be added in constant proportions until these quantities are loaded. At that point, the kettle is at the optimum saponification point and will be maintained there if the remaining ingredients are added at a fixed ratio.

This order of addition cannot be blindly followed with the expectation of successful loading. It is still paramount that the soapmaker, every step of the way, confirm that indeed the saponification reaction is proceeding as planned. There is currently no model that will accurately remove the responsibility from the soapmaker of ensuring proper agitation of the kettle. This is achieved only by a diligent soapmaker who is watching the kettle to ensure that proper mixing is occurring. Experienced soapmakers can perform this task quite well with a minimum of adjustments to the steam valves, thus allowing them to perform other tasks in close proximity to the kettle and thus improving their productivity. However, the soapmaker must continue to monitor the kettle to ensure that the reaction is proceeding as planned.

Graining the Kettle

Once the ingredients have been added, the kettle must be grained by moving its composition from region R into region M. This calculation can be accomplished through isoparametric mapping of the linearized phase diagram.

Tie Line Calculations

The relevant parts of the phase diagram have been approximated by quadrilaterals. We now need to determine the compositions and quantities of the two phases that comprise each point enclosed in the quadrilateral. An isoparametric mapping is performed to ease calculations (see Fig. 4.8). The nature and development of isoparametric mappings are discussed in detail in the books by Fried and Zeinkevic.

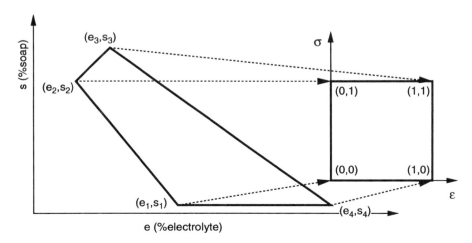

Fig. 4.8. Isoparametric mapping of curd-lye region.

The phase diagram of the lye-curd phase can be approximated by a quadrilateral defined by the following points:

$$(e_i, s_i) \ i = 1 \text{ to } 4 \tag{4.89}$$

We define the mapping such that

$$(e_1, s_1) = (8.3\%, 1.0\%) \Leftrightarrow (\varepsilon_1, \sigma_1) = (0,0) \tag{4.90}$$
$$(e_2, s_2) = (1.2\%, 78.5\%) \Leftrightarrow (\varepsilon_2, \sigma_2) = (0,1) \tag{4.91}$$
$$(e_3, s_3) = (1.6\%, 84.5\%) \Leftrightarrow (\varepsilon_3, \sigma_3) = (1,1) \tag{4.92}$$
$$(e_4, s_4) = (13.0\%, 1.0\%) \Leftrightarrow (\varepsilon_4, \sigma_4) = (1,0) \tag{4.93}$$

In this transformed coordinate system the coordinate of the kettle (e^*, s^*) is transformed to the coordinates $(\varepsilon^*, \sigma^*)$. The following relationships then apply:

- σ^* is the fraction of the total kettle mass that is curd.
- $(\varepsilon^*, 0)$ defines the composition of the lye phase in $(\underline{\varepsilon}, \underline{\sigma})$ coordinates.
- $(\varepsilon^*, 1)$ defines the composition of the curd phase in $(\underline{\varepsilon}, \underline{\sigma})$ coordinates.

We now need a convenient way to convert the point (e^*, s^*) to the $(\underline{\varepsilon}, \underline{\sigma})$ coordinate system, and the points $(\varepsilon^*, 0)$ and $(\varepsilon^*, 1)$ to the $(\underline{e}, \underline{s})$ coordinate system. See Fig. 4.9:

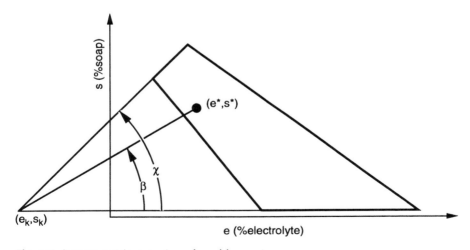

Fig. 4.9. Isoparametric mapping of curd-lye region.

The point of intersection (e_k, s_k) is defined by

$$e_k = \frac{(s_2 e_3 - s_3 e_2)(e_4 - e_1) - (s_4 e_1 - s_1 e_4)(e_2 - e_3)}{(s_2 - s_3)(e_4 - e_1) - (s_4 - s_1)(e_2 - e_3)} \qquad (4.94)$$

$$s_k = \frac{(s_2 e_3 - s_3 e_2)(s_4 - s_1) - (s_4 e_1 - s_1 e_4)(s_2 - s_3)}{(s_2 - s_3)(e_4 - e_1) - (s_4 - s_1)(e_2 - e_3)} \qquad (4.95)$$

The angle χ between these two lines is

$$\chi = \cot^{-1}\left[\frac{(e_2 - e_3)(e_4 - e_1) + (s_2 - s_3)(s_4 - s_1)}{(s_2 - s_3)(e_4 - e_1) - (s_4 - s_1)(e_2 - e_3)}\right] \qquad (4.96)$$

The angle β between line from intersect point of boundary lines to (e^*, s^*) and line defined by (e_1, s_1) and (e_4, s_4) is

$$\beta = \cot^{-1}\left[\frac{(e^* - e_i)(e_4 - e_1) + (s^* - s_i)(s_4 - s_1)}{(s^* - s_i)(e_4 - e_1) - (s_4 - s_1)(e^* - e_i)}\right] \qquad (4.97)$$

so the value e^* in the $(\underline{\varepsilon}, \underline{\sigma})$ coordinate system is defined by

$$\varepsilon^* = \frac{\beta}{\chi} \qquad (4.98)$$

the value for σ^* is determined in a similar fashion.

We know that the range of optimum settling is in region M, just over the border. This corresponds to the line $\varepsilon = \varepsilon^\dagger = 0.1$ in $(\underline{\varepsilon}, \underline{\sigma})$ coordinates. This line is defined by two points $(e_1^\dagger, s_1^\dagger)$, $(e_2^\dagger, s_2^\dagger)$ in $(\underline{e}, \underline{s})$ coordinates, which are calculated in the following fashion:

$$e_1^\dagger = e^\dagger \times e_3 + (1 - e^\dagger) \times e_2 = 1.56\% \qquad (4.99)$$

$$s_1^\dagger = e^\dagger \times s_3 + (1 - e^\dagger) \times s_2 = 83.9\% \qquad (4.100)$$

$$e_2^\dagger = e^\dagger \times e_4 + (1 - e^\dagger) \times e_1 = 8.77\% \qquad (4.101)$$

$$s_2^\dagger = e^\dagger \times s_4 + (1 - e^\dagger) \times s_1 = 1.00\% \qquad (4.102)$$

To grain the kettle properly, we want to add brine until the composition of the kettle intersects the line defined by $(e_1^\dagger, s_1^\dagger)$, $(e_2^\dagger, s_2^\dagger)$. Again, this numerical method is beyond the scope of this text. The addition of brine introduces NaCl, but also introduces H_2O from two sources, the brine and the steam required to heat the brine to the reaction temperature. For every pound of brine introduced into the kettle on August 9, 0.265 lb of NaCl are added, 0.735 lb of H_2O are added from the brine, and 0.143 lb of steam condense.

Intersection of the desired tie line will occur with the addition of 9,579 lb of brine. 1,368 lb of steam will condense heating the brine, yielding a kettle composition of 47.487% soap and 4.727% salt.

$$M_{b \text{ graining}} = 9,579 \tag{4.103}$$

$$M_{t \text{ graining}} = 1,368 \tag{4.104}$$

$$X^s_{\text{graining}} = 47.487\% \tag{4.105}$$

$$X^e_{\text{graining}} = 4.727\% \tag{4.106}$$

Again, blind addition of this amount of brine must be accompanied with careful monitoring of the kettle agitation and steam level. The recirculation pipe is started at this stage and the quality of the lye accumulating on the bottom of the kettle can be monitored.

Settling and Spent Lye Removal

A properly loaded and grained kettle can be drawn almost immediately. The separation efficiency, previously defined, is used to determine the kettle composition. Experience dictates that

$$\delta = 83\% \tag{4.107}$$

Therefore the composition of the curd after lye drawing is calculated to be

$$X^e_{u_o} = \delta \times e^\dagger_1 + (1 - \delta) \times e^\dagger_2 \ = 2.79\% \tag{4.108}$$

$$X^s_{u_o} = \delta \times s^\dagger_1 + (1 - \delta) \times s^\dagger_2 \ = 69.81\% \tag{4.109}$$

$$X^v_{u_o} = 1 - X^e_{u_o} - X^s_{u_o} \qquad = 27.40\% \tag{4.110}$$

Note that Eq. (110) appears to be too low in concentration. At first glance it appears that the value for δ is too high. However, the error comes from using the published phase diagrams. This problem will be discussed later.

The lye composition is

$$X^e_{l_o} = e^\dagger_2 \ = 8.77\% \tag{4.111}$$

$$X^s_{l_o} = s^\dagger_2 \ = 1.00\% \tag{4.112}$$

Note that the model is only as good as the phase diagram. In practice, the phase diagram coordinates (e_i, s_i) [i = 1 to 4] may have to be "massaged" in order to have the predicted results match reality. This often occurs because the fats and oils blend in use at a particular site may have a phase diagram somewhat different from the published data (or the published data could be in error).

The total amount of lye available is

$$M^\infty_{l_o} = (M_o + M_b \text{ graining}) \times \frac{e^\dagger_1 - X^e_{\text{graining}}}{e^\dagger_1 - e^\dagger_2} \tag{4.113}$$

$$M^\infty_{l_o} = 95,674 \tag{4.114}$$

Here the ∞ represents the total lye available. The actual lye removed is

$$M_{l_0} = 70,657 \tag{4.115}$$

This amount of lye removed will net the correct values of $X_{u_0}^e$ and $X_{u_0}^s$ just listed. Again, this calculation is beyond the scope of this text.

We now turn our attention to calculating how much glycerol is present in the spent lye. After the kettle is grained, we know that

$$M_{graining} = 217,817 \tag{4.116}$$
$$M_{graining}^g = M_o^g = 15,787 \tag{4.117}$$
$$M_{graining}^\omega = M_o^v - M_o^g + M_{b\ graining} \times X_b^\omega + M_{t\ graining} \tag{4.118}$$
$$M_{graining}^\omega = 88,299 \tag{4.119}$$

We make a first-order approximation that the ratio of glycerol to H_2O is constant during phase separation. As a result of empirical results, we have developed a correction factor, the glycerol concentration factor $\{G\}$, which adjusts the results to reflect the fact that glycerol preferentially concentrates in the lye phase. We have found that

$$G_0 = 1.1 \tag{4.120}$$

We therefore determine that

$$X_{l_0}^g = \frac{M_{graining}^g}{M_{graining}^g + M_{graining}^\omega} \times X_{l_0}^v \times G = 15.05\% \tag{4.121}$$
$$M_{l_0}^g = X_{l_0}^g \times M_{l_0} = 10,637 \tag{4.122}$$
$$M_{u_0}^g = M_{graining}^g - M_{l_0}^g = 5,151 \tag{4.123}$$

Kettle Washing

This entails the addition of brine, water, caustic, and recycled lye. The goal is to move along the desired tie line until the desired amount of lye is produced. One can also influence the relative ratio of NaCl to NaOH as well as reduce the concentration of glycerol in the soap. For the sake of this example, we will perform a final wash on the kettle and not include the calculations associated with recycling lye into the wash. These calculations follow the same pattern as just illustrated for adding lye to the loading step.

We first decide on a washing target. If we wanted to generate as much wash lye as spent lye, we could simply repeat the graining target. Instead, we desire a smaller amount of lye and will wash to a target of 55% soap.

$$X_1^s = 55.00\% \tag{4.124}$$

We must determine the intersection of this point and the desired tie line:

$$X_1^e = e_1^\dagger + (e_2^\dagger - e_1^\dagger) \times (s_1^\dagger - X_1^s) / (s_1^\dagger - s_2^\dagger) = 4.07\% \tag{4.125}$$

For illustration purposes only, we will demonstrate how to reduce the NaCl in the soap curd by performing a caustic wash. Recall that after the removal of the spent lye, there is no free NaOH in the kettle. If this were to continue, the resulting neat soap would be very high in NaCl and display poor handling in the rice and pressing stages. Caustic is introduced during the washing to replace some of the NaCl. In reality, both brine and caustic would be used during washes, because some NaCl is needed for proper handling and to avoid excessive levels of NaOH in the neat soap.

Again, precise amounts of caustic and water are added to achieve the desired washing point. Recall that steam again is required to heat both the caustic and water. The kettle has cooled since the graining stage, so steam will again be needed to bring the kettle to the reaction temperature. Also, some H_2O has evaporated since the graining was completed.

Evaporation Losses

The model is a simple exponential:
$$M_{\omega\ loss} = M_{\omega}^{\infty} \times \{1 - \exp\ [\ln(0.5) \times (\Delta D)/D_{\omega}]\} \tag{4.126}$$

where

$M_{\omega\ loss}$ = total H_2O loss (lb)

M_{ω}^{∞} = H_2O loss after infinite time; for a kettle of the size in this example, a value of 2,500 is about right. This is strongly dependent on the nature and size of the kettles in use.

ΔD = number of days the kettle has been evaporating

D_{ω} = half-life; we have found a value of 14 days to give good results.

Note that the actual mechanism is much more complex than this simple model. There is a period of relatively rapid evaporation due to convective flow in which the kettle is still very hot. The kettle then passes through a stage in which the surface begins to crust over, after which the evaporation rate slows. The relative magnitude of this loss is small, so this simple model suffices. For our example with a 16-h "quiet" period,
$$M_{\omega\ loss} = 81 \tag{4.127}$$

Kettle Cooling

The model is a simple exponential:
$$T_1 = \overline{T} - (\overline{T} - T^{\infty}) \times \{1 - \exp\ [\ln(0.5) \times (\Delta D)/D_T]\} \tag{4.128}$$

T_1 = temperature at the end of the settling period

T^{∞} = temperature after infinite time; for a kettle of the size in this example, a value of 100 is about right. This is strongly dependent on the nature and size of the kettles in use.

ΔD = number of days the kettle has been evaporating

D_ω = half-life; we have found a value of 7 days to give good results.

Note that the actual mechanism is much more complex than this simple model. The relative magnitude of this loss is small, so this simple model suffices. For our example with a 16-h "quiet" period,

$$T_1 = \overline{T} - (\overline{T} - T^\infty) \times \{1 - \exp[\ln(0.5) \times (\Delta D)/D_T]\} = 212 \qquad (4.129)$$

Preparing the Kettle for Washing

The kettle has cooled and lost water to evaporation. We must now reheat the kettle. The kettle composition evolves as listed in Table 4.4. Now caustic and water must be added to bring the kettle to the correct washing point. 5,660 lb of caustic, 28,809 lb of water, and 4,585 lb of steam are required to achieve the washing targets. Again, soapmaker diligence is required to ensure that the washing progresses smoothly.

Wash Lye Settling and Removal

The kettle has been washed to the point (0.10,0.65) in (e,s) coordinates. Therefore 35% of the kettle is lye.

The kettle is washed on the same tie line as it was grained after loading. Therefore the composition of the curd after lye drawing is calculated to be

$$X^e_{u_1} = \delta \times e^\dagger_1 + (1 - \delta) \times e^\dagger_2 = 2.79\% \qquad (4.130)$$
$$X^e_{u_2} = \delta \times s^\dagger_1 + (1 - \delta) \times s^\dagger_2 = 69.81\% \qquad (4.131)$$

Note that this is identical in (e,s) coordinates to the graining.

However, in this case the electrolyte is composed of both NaCl and NaOH. The electrolyte settling ratio {E} helps predict the relative ratio of NaCl to NaOH. For lye/curd separation of 85/15 tallow/coco soaps, it is essentially unity and can be ignored. We can then calculate

$$X^d_{u_1} = X^e_{u_1} \times M^d_1 / (M^h_1 \times z^h + M^d_1 \times z^d) = 1.50\% \qquad (4.132)$$
$$X^h_{u_1} = X^e_{u_1} \times M^h_1 / (M^h_1 \times z^h + M^d_1 \times z^d) = 1.03\% \qquad (4.133)$$

TABLE 4.4
Kettle Composition During Washing

Component	After Lye Removal	Before Reheating	After Reheating	Washing
Soap	102,728	102,728	102,728	102,728
H_2O	35,182	35,101	35,747	71,993
NaOH	0	0	0	2,807
NaCl	4,099	4,099	4,099	4,099
Glycerol	5,150	5,150	5,150	5,150
Total	147,160	147,079	147,726	186,779

The lye composition is

$$X_{l_1}^e = e_2^{\dagger} = 8.77\% \tag{4.134}$$

$$X_{l_1}^s = s_2^{\dagger} = 1.00\% \tag{4.135}$$

$$X_{l_1}^d = X_{l_1}^e \times M_1^d \: / \: (M_1^h \times z^h + M_1^d \times z^d) = 4.73\% \tag{4.136}$$

$$X_{l_1}^h = X_{l_1}^e \times M_1^h \: / \: (M_1^h \times z^h + M_1^d \times z^d) = 3.24\% \tag{4.137}$$

Again, soapmaker diligence and the use of the recirculation pipe will ensure that the kettle is within specifications.

The calculation to project the amount and composition of the lye removed from the kettle proceeds as just described. The results are listed in Table 4.5.

Model Dynamics

Note that the quantity of lye removed, 40,194 lb, is essentially the same as the quantity recycled into the loading stage (40,000). Also recall that it is possible to add additional amounts of lye during the loading stage, although 85,000 lb proved to be too much. This is important for a robust dynamic process, because various products may produce different amount of wash lyes for recycling, and the process should be capable of consuming more lye than it produces to avoid unwanted accumulation of wash lyes. Also note that the NaOH and NaCl levels in this wash lye are different from the levels in the recycled lye used during the loading. Because the calculation scheme can adjust for varying concentrations of different components, the correct targets are always achieved. This adjustment is also true for the quantity of lye removed. Every kettle will not yield exactly 33,905 lb of wash lye. Even a smooth process could expect ±10% variation in this number due to kettle settling over a weekend, insufficient mixing by the operator, and other unknown variables. Again, the model is dynamic and corrects itself as the batch progresses via operator feedback. The exact mechanics of the implementation will be discussed in a later section.

TABLE 4.5
Kettle Composition During Wash Lye Removal

Component	Washing	Wash Lye Removed	After Lye Removed	Reheated
soap	102,728	402	102,327	102,327
	55.00%	1.00%	69.81%	69.50%
H_2O	71,993	33,905	38,088	38,733
	38.54%	84.35%	25.98%	26.31%
NaOH	2,807	1,301	1,507	1,507
	1.50%	3.24%	1.03%	1.02%
NaCl	4,099	1,899	2,200	2,200
	2.19%	4.73%	1.50%	1.49%
Glycerol	5,150	2,687	2,463	2,463
	2.76%	6.69%	1.68%	1.67%
Total	186,779	40,194	146,585	147,229

Kettle Finishing or Fitting

The aspect of kettle production that has been traditionally shrouded in mystery is the easiest part of the process to model. Region N on the phase diagram, the neat-nigre 2 phase region, is approximated by a quadrilateral in the same fashion as region M.

$$(e_i, s_i) \; i = 5 \text{ to } 8 \tag{4.138}$$

We define the mapping such that

$$(e_5, s_5) = (1.50\%, \; 38.0\%) \Leftrightarrow (\varepsilon_5, \sigma_5) = (0,0) \tag{4.139}$$

$$(e_6, s_6) = (0.50\%, \; 66.0\%) \Leftrightarrow (\varepsilon_6, \sigma_6) = (0,1) \tag{4.140}$$

$$(e_7, s_7) = (0.63\%, \; 67.0\%) \Leftrightarrow (\varepsilon_7, \sigma_7) = (1,1) \tag{4.141}$$

$$(e_8, s_8) = (5.50\%, \; 17.0\%) \Leftrightarrow (\varepsilon_8, \sigma_8) = (1,0) \tag{4.142}$$

In this transformed coordinate system, the coordinate of the kettle (e^{**}, s^{**}) is transformed to the coordinates $(\varepsilon^{**}, \sigma^{**})$. The following relationships then apply:

σ^{**} is the fraction of the total kettle mass that is neat soap.

$(\varepsilon^{**}, 0)$ defines the composition of the nigre phase in $(\underline{\varepsilon}, \underline{\sigma})$ coordinates.

$(\varepsilon^{**}, 1)$ defines the composition of the neat phase in $(\underline{\varepsilon}, \underline{\sigma})$ coordinates.

The point of best finishing or fitting is just over the border from region Q into region R. Define the target to be the tie line $\varepsilon^{\dagger\dagger} = 0.95$ in $(\underline{\varepsilon}, \underline{\sigma})$ coordinates. This line is defined by two points $(e_3^\dagger, s_3^\dagger)$, $(e_4^\dagger, s_4^\dagger)$ in $(\underline{e}, \underline{s})$ coordinates, which are calculated in the following fashion:

$$e_3^\dagger = e^{\dagger\dagger} \times e_7 + (1 - e^{\dagger\dagger}) \times e_6 = 0.62\% \tag{4.143}$$

$$s_3^\dagger = e^{\dagger\dagger} \times s_7 + (1 - e^{\dagger\dagger}) \times s_6 = 66.95\% \tag{4.144}$$

$$e_4^\dagger = e^{\dagger\dagger} \times e_8 + (1 - e^{\dagger\dagger}) \times e_5 = 5.30\% \tag{4.145}$$

$$s_4^\dagger = e^{\dagger\dagger} \times s_8 + (1 - e^{\dagger\dagger}) \times s_5 = 18.05\% \tag{4.146}$$

To finish the kettle properly, we want to add water until the composition of the kettle intersects the line defined by $(e_3^\dagger, s_3^\dagger)$, $(e_4^\dagger, s_4^\dagger)$. Again, this numerical method is beyond the scope of this text. The addition of water introduces H_2O from two sources, the water and the steam required to heat the water to the reaction temperature. 43,840 lb of water properly finish this kettle. A total of 50,165 lb of H_2O are added due to steam condensation. The finish proceeds as outlined in Table 4.6. For the finish step, we know that

$$E_{k+1} = 0.98 \; [\text{electrolyte settling ratio}] \tag{4.147}$$

$$G_{k+1} = 1.0 \; [\text{glycerol concentration factor}] \tag{4.148}$$

Given these values for the electrolyte settling ratio and the separation efficiency, we can predict the composition and quantity of the neat and nigre phases. However, we

TABLE 4.6
Kettle Composition During Finish

Component	Ready to Finish	Finish	Neat Soap	Nigre
Soap	102,327	102,327	91,316	11,010
	69.50%	51.84%	66.95%	18.05%
H_2O	38,733	88,898	43,114	45,784
	26.31%	45.04%	31.61%	75.06%
NaOH	1,507	1,507	321	1,186
	1.02%	0.76%	0.24%	1.94%
NaCl	2,200	2,200	449	1,751
	1.49%	1.11%	0.33%	2.87%
Glycerol	2,463	2,463	1,195	1,269
	1.67%	1.25%	0.88%	2.08%
Total	147,229	197,395	136,395	61,000

are again plagued with the problem of incomplete separation of the two phases. Unlike the curd/lye separation, in which the process is modeled by the lower phase incompletely falling out of the upper phase, the neat/nigre separation is modeled by the upper phase incompletely rising out of the lower phase. The separation efficiency for this stage is

$$\delta_{k+1} = 0.95 \text{ [separation efficiency]} \tag{4.149}$$

Therefore the neat soap that is removed from the kettle has the composition of pure neat soap, but the nigre still has some neat soap contained in it. This level gives a kettle yield of 85%, which means that 85% of the soap in the kettle is removed as neat soap. This is outlined in Table 4.7. In this case the NaCl content of the neat soap is very low (0.33%) and the remaining electrolyte is supplemented by a high level of NaOH (0.24%). This level is much too high for a toilet soap, so post-finishing is required.

TABLE 4.7
Composition of Kettle at Conclusion of Process

Component	Neat Soap	Left in Kettle
Soap	86,751	15,576
	66.95%	22.97%
H_2O	40,958	47,940
	31.61%	7 0.69%
NaOH	305	1,202
	0.24%	1.77%
NaCl	427	1,774
	0.33%	2.62%
Glycerol	1,135	1,328
	0.88%	1.96%
Total	129,575	67,820

Post-Finishing

This process mimics the final stage of a continuous saponifier. Here fatty acid is injected into the neat soap stream just before flash drying. As a rule, it is better to use a lower-chain-length fatty acid than a higher chain length (coconut oil vs. tallow) because less fatty acid will be required to neutralize the excess NaOH. Simple injection of the fatty acid into the neat soap stream with a positive displacement metering pump is sufficient to neutralize the NaOH because the neutralization reaction proceeds spontaneously.

Further Comments on the Example

An experienced soapmaker will review the calculations and scoff at the value of some of the numbers, most notably a 26.31% H_2O content of the curd before finishing. We have not achieved this value in our kettles (and we seriously doubt that anyone else has, either). The values for the four corners of the lye/curd and neat/nigre regions were taken from Woolatt. We chose to publish the model as such to make it easier for the reader to understand the origin of the coordinates. The working model at Bradford Soap uses refined values that match actual results. Those numbers have not been published here because they are refined for Bradford's proprietary fat blends and are dynamic in nature to account for other fluctuations. Also, a detailed explanation of the dynamics of the Bradford mechanism for choosing the values in use is beyond the scope of this text. One can only speculate that Bradford's implementation of the published phase diagrams have somehow missed the mark because they do not mirror the actual processing.

The published phase diagrams used in the example given portray a significant neat/curd/lye region (R). Our experience indicates that this region appears to be significantly narrower than indicated on the phase diagram, with the regions A and J behaving more like a continuum than two distinct phases. The effect of this behavior is to modify the computer approximation for region M to include regions R and P. This places the pre-finish curd much closer to region N, and thus significantly less water is required to move the kettle into region N. We have also found the locus of regions M, P, and R to be higher in electrolyte than indicated the published phase diagrams, thus decreasing the slope of the region M tie lines. The observations just given are a direct result of modifying the model to fit observed kettle performance and are not a result of laboratory experiments to quantify the phase diagrams for specific fat blends.

When building a model for a particular fat blend, one can optimize the coordinates of the regions by comparison with actual data. We have found that one or two kettles will accurately predict the tie lines for all subsequent kettles of the same formulation. Also, once several formulations have been done, predicting new formulations is straightforward.

The values for all other parameters are the actual values used in the Bradford model and are consistent with prior published results.

Implementation at Bradford Soap

Our development of this process began in 1983. At that time Bradford had recognized that its kettle soap process had reached a plateau with regard to quality and performance, and a new approach was needed to achieve further improvements. The first step in improving the process was to develop a measurement system to gauge the existing process and the effects of subsequent process modifications.

Performance Indexing

The Performance Index was developed in order to quantify soap room performance during the implementation stage of the computer model. We continue to use the Performance Index with a time-weighted averaging scheme to smooth out any short-term fluctuations.

The Performance Index is measured on a scale from 0 to 100; a score of 100 was a perfect mark and a 0 meant that everything was wrong. Before the Performance Index, people would focus on one particular attribute of the kettle but could not put the entire kettle in perspective. So the trick was to come up with a combination of kettle parameters that were accurately measured and properly weighed to result in one number to describe the kettle accurately. One number results that defines kettle quality and performance.

The mathematics of the Performance Index were introduced in HAPPI (January 1987). We include it here for completeness:

$$\prod_{m=1}^{M} \left\{ \sum_{n=1}^{N} \left[1 - \left(\frac{X_n - A_n}{C_n - A_n} \right)^{Pn} \right] H(C_n - X_n) \, H(X_n - D_n) \, B_n \right\} E_m^{F_m} \tag{4.150}$$

where

N = the number of power-law parameters

M = the number of exponential parameters

X_n = the value of parameter n

A_n = the value of parameter n that yields maximum performance points

B_n = the number of performance points assigned to parameter n

C_n = the minimum value of parameter n for which any performance points are allowed

D_n = the maximum value of parameter n for which any performance points are allowed (for this special case, $D_n = 2 \times A_n - C_n$)

P_n = the power of the power-law fitted parameter n; for this simplified model, p is restricted to a positive even number (2, 4, 6, . . .)

E_m = the penalty for one occurrence or parameter m ($0 \le E \le 1$)

F_m = the total number of occurrences of parameter n

$H(y)$ = the Heavyside step function:

for $y > 0$, $H(y) = 1$; $y \le 0$, $H(y) = 0$

TABLE 4.8
Performance Index Example—Parameters

		N = 2				
n	Parameter	a_n	B_n	C_n	D_n	P_n
1	$H_2O\%$	30	60	29	31	2
2	Yield %	85	40	80	90	8

	M = 2	
m	Parameter	Em
1	Unscheduled Openings	0.8
2	Equipment Problems	0.85

We require additional features to generate our Performance Index, but the expression just given is sufficient to show the utility of this approach.

For example, assume that we are concerned with two power-law parameters (moisture of the kettle soap and yield) and two exponential parameters (the number of unscheduled openings and equipment problems). We then have the parameters shown in Table 4.8. Because P_1 where n = 1 is a relatively small number (= 2), a kettle in the middle of the desired range ($\Omega 30$) will receive maximum moisture performance points, but a kettle on the edge of the range (≈ 30.9) will receive few performance points. Because P_2 is relatively large (= 8), a kettle on the edge of the range (≈ 81) will still receive close to the maximum number of performance points for the parameter (40).

We examine three cases: an excellent, an acceptable, and a poor kettle; the results are shown in Table 4.9.

Decaying Memory Averaging

The Performance Index gave Bradford a technique for gauging individual kettles. Decaying memory averaging gives Bradford a technique to measure performance

TABLE 4.9
Performance Index Example—Three Kettles

		Excellent	Acceptable	Poor
X_1	Moisture %	30.1	29.4	30.9
	Performance Points	59.4	38.4	11.4
X_2	Yield %	84.9	89.0	79.0
	Performance Points	40.0	33.3	0.0
F1	Unscheduled Openings	0	0	1
E_1F_1	Multiplier	1	1	0.8
F_2	Equipment Problems	0	1	2
E_2F_2	Multiplier	1	0.85	0.7225
	Performance Index	99.4	60.9	6.6

trends with time. This mathematical technique was first discussed in Cosmetics and Toiletries (August 1993). It is presented here for completeness.

The decaying memory average is defined by

$$X_n = \frac{x_n + (\alpha - 1) X_{n-1}}{\alpha}$$

(4.151)

The decaying memory deviation is defined as follows:

$$S_n = \sqrt{\frac{(x_n - X_n)^2 + (\alpha - 1) S_{n-1}^2}{\alpha}}$$

(4.152)

The decaying memory fraction out of specification is defined by

$$\Phi_n = \frac{\phi_n + (\alpha - 1) \Phi_{n-1}}{\alpha}$$

(4.153)

where

$$\phi_n = 0 \text{ if } x_n \text{ is within specification}$$
$$\phi_n = 1 \text{ if } x_n \text{ is out of specification}$$

Terms are defined as follows:

x_n = the nth or current observation of a process

X_n = the decaying memory average at the current time

X_{n-1} = the most recent prior decaying memory average

S_n = the decaying memory deviation at the current time

S_{n-1} = the most recent prior decaying memory deviation

F_n = the decaying memory fraction out of specification at the current time

F_{n-1} = the most recent prior decaying memory fraction out of specification

α = the relaxation parameter [$\alpha > 1$]. This is a measure of how far in the past we wish to consider. If $\alpha = 1$, then only the current information x_n is considered when evaluating the process. As a gets larger, we weigh more heavily the information from the past.

The choice of α is process dependent. The use of $\alpha = 5$ allows the most recent data point to contribute 20% of the total to the weighted average, and the fifth most recent data point contributes 8.192% to the weighted average. The nth most recent data point contributes $[(\alpha - 1)^{n-1}]/\alpha^n$ to the weighted average.

Startup.

Average:	define $X_1 = x_1$	(4.154)
Deviation:	define $S_2 = |x_1 - x_2| \div 2$	(4.155)
Fraction out of specification:	define $\Phi_1 = 0$	(4.156)

Implementation.

For most applications we shall use two different values for α. Use $\alpha = 5$ to measure recent performance (approximately the last 20 data points). Use $\alpha = 50$ to measure long-term performance (approximately the last 200 data points). The relative contributions of the data points are presented in graphical fashion in Fig. 4.10.

KSPS at Bradford

The implementation at Bradford is called the Kettle Soap Process Simulator (KSPS). The current version is written in Microsoft® Excel, a macro-driven spreadsheet. The spreadsheet format offers rapid program development. The hardware platform of choice is the Macintosh computer, which is still easier to use and requires less maintenance than its better known Windows™/DOS competition.

The soapmaker sits downs in front of the computer whenever a new batch of soap is to be made and fills in blanks in the spreadsheet, answering questions about the batch. The computer calculates quantities and rates of additions and provides a written copy for the soapmaker, who then takes the paperwork to the kettle for easy reference. The computer makes various assumptions about the batch based on an extensive database and the modeling techniques just outlined. The soapmaker can override many of the results of these assumptions. These overrides typically occur when the soapmaker's analysis disagrees with the computer's predictions of the analysis. The computer uses this corrected information in subsequent calculations.

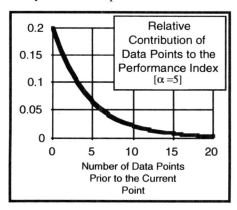 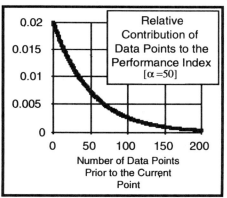

Fig. 4.10. Relative contribution of data points for various levels of α.

Item	computer generated	soapmaker input required	soapmaker override optional
Batch Information			
Batch Number:	x		x
Kettle:		x	
Soap Type:		x	
Lbs of Fat (if not standard):	x		x
Loading			
Date & Time Started:	x		x
Soapmaker:		x	
Load Lye Amount (0-1):	x		x
Load Lye Free:	x		x
Load Lye Salt:	x		x
Water Meter:		x	
Nigre Inches Down:		x	
Nigre Soap Type:		x	
Nigre Water:	x		x
Nigre Free:	x		x
Nigre Salt:	x		x
Fat:	x		x
Oil:	x		x
Fatty Acid:	x		x
Caustic:	x		x
Lye:	x		x
Brine:	x		x
Water:	x		x
Recycled Lye Free:	x	x	
Recycled Lye Salt:	x	x	
Date & Time Completed:		x	
Spent Lye Drawing			
Date & Time Started:	x	x	
Soapmaker:		x	
Inches Start:		x	
Inches Stop:		x	
To Tank [1-6]:		x	
Free:	x	x	
Salt:	x	x	
Wash			
Date & Time Started:	x		x
Soapmaker:		x	
Caustic:	x		x
Brine:	x		x
Water:	x		x
Recycled Lye Free:	x	x	
Recycled Lye Salt:	x	x	
Date & Time Completed:	x	x	

Item	computer generated	soapmaker input required	soapmaker override optional
Wash Lye Drawing			
Date & Time Started:	x		x
Soapmaker:		x	
Inches Start:		x	
Inches Stop:		x	
To Tank [1-6]:		x	
Free:	x	x	
Salt:	x	x	
Finish			
Date & Time Started:	x		x
Soapmaker:		x	
Caustic:	x	x	
Brine:	x	x	
Water:	x	x	
Finish Curd Water:	x	x	
Finish Curd Free:	x	x	
Finish Curd Salt:	x	x	
Predicted Free:	x		
Predicted Salt:	x		
Date & Time Completed:	x	x	
Nigre Pumping			
Date & Time Started:	x		x
Soapmaker:		x	
Inches Start:		x	
Inches Stop:		x	
To Kettle:		x	
Neat Pumping			
Date & Time Started:	x		x
Soapmaker:		x	
Inches Start:		x	
Inches Stop:		x	
To Tank:		x	
Water:		x	
Free:		x	
Salt:		x	
Glycerine:			x

Fig. 4.11. KSPS soapmaker inputs.

Figure 4.11 lists all of the soapmaker inputs, both mandatory and optional. Figure 4.12 is an actual screen shot of the first section of the KSPS computer screen. Figure 4.13 illustrates the printed output used by the soapmaker at the kettle.

Note the rows of intermediate loading targets. As the soapmaker progresses in the kettle loading, the maximum saponification rate is ensured if the soapmaker maintains the proper balance of ingredients as listed on the output sheet. The soap-

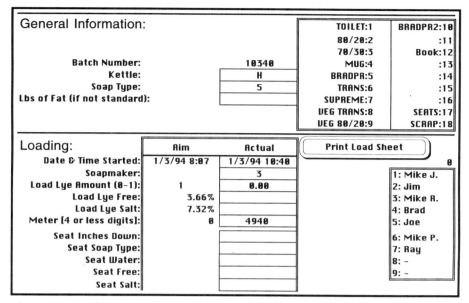

Fig. 4.12. KSPS human interface - input.

Kettle Soap Process Simulator - Loading Sheet

			Loading Targets:		Seat	
Batch Number:	10340					
Kettle:	H		% Soap:	49.27%	40.00%	
Soap Type:	5	BRADPR	% Free:	-0.15%	1.10%	
Version:	17.01		% Salt:	3.60%	2.20%	
Date & Time Started:	1/3/94 10:40		Fats Load:	70,000 lbs	160.0	:Inches Down
Soapmaker:	3: Mike R.				TOILET	:Seat Type
Load Lye Amount (0-1):	0.9				11,527 lbs	:Seat Weight
Load Lye Free:	3.66%		Comments:			
Load Lye Salt:	7.32%					
Meter [4 or less digits]:	4940					

Loading Quantities:

Fat (pounds)	Oil (pounds)	Fatty Acid (pounds)	Caustic (gallons)	Brine (cubic feet)	Lye (pounds)	Water (cubic feet)	Meter
	3,700		147	0	1,719	8	4,948
	7,400		293	0	3,437	16	4,956
	11,096		440		5,154	24	4,964
6,700			547	3	8,266	36	4,979
13,400			654	6	11,378	48	4,994
20,100			761	9	14,490	60	5,009
26,800			868	12	17,602	72	5,024
33,500			976	12	20,714	87	5,038
40,200			1,083	12	23,826	101	5,053
46,900			1,190	12	26,938	110	5,061
52,807			1,266	12	28,529	110	5,061

Fig. 4.13. KSPS human interface - output.

maker records the actual quantities on the output sheet as each milestone is passed. In this fashion a permanent record is made of each kettle and can be checked if necessary to diagnose any problems with the kettle at a later stage.

Figure 4.14 illustrates a small part of the spreadsheet which is accessed only by the process engineer. Here liberal use of color allows the engineer to quickly recognize which cells are parameters that can be varied and which cells are calculated values. This spreadsheet is password protected, and key calculations are locked to avoid unwanted changes. Another feature of the KSPS is the ease in which the process engineer can modify the process and observe long-term trends. The KSPS has an "iteration switch" that when activated calculates the steady-state performance of the kettle soap process. The lyes removed from a wash are automatically added to the prior wash, and the steady-state concentrations of all of the steps are determined. For example, the process engineer can vary the loading targets, and the modified volume and composition of the spent lye will be determined along with the new wash quantities and every component of every step. This has greatly assisted Bradford engineers in having tight control of the process. Most notably, NaCl levels in some translucent products have been experimentally varied and tightly controlled over the years to maximize translucency in the rice.

Results

Before and After

We now review the results of computer modeling using the 0 to 100 Performance Index discussed earlier. These results, displayed in Figs. 4.15 and 4.16, were originally presented in the HAPPI paper already mentioned.

Before computer modeling was implemented, Bradford's performance was acceptable. Over 30% of kettles were 80 or greater. They had another block of kettles that were between 70 and 80. Almost half of their kettles fell in the 60 to 70 range. Bradford was making a good soap most of the time, but some minor problems were being experienced. Some kettles fell in the 50–60 range. Almost 20% of Bradford's kettles were causing significant problems, being less than 50. About a year after the KSPS was implemented, over half of the kettles were in 80+ range, which equates to production rates and yields exactly or in excess of schedules. Another 25% of the kettles were between 70 and 80, experiencing only minor problems. Poorer performers still existed but were significantly reduced. Since 1987 the Performance Index definitions have been upgraded and expanded several times, so there is no easy way to compare today's performance with the results in 1987.

Variation

To demonstrate results over the last 12 years, we can focus on one parameter, that being moisture of the neat soap when it is removed from the kettle. Because that get has changed over the years, we focus instead on batch-to-batch variability of neat soap moisture. See Fig. 4.17. The horizontal axis is time, and the vertical axis

PHYSICAL PROPERTIES:	Density	Soap %	Soap Blend %	NaOH %	Glycerine %	Water %	NaCl %	Temp. °F	Specific Heat BTU/LB°F	Heat of Reaction BTU/LB
Tallow:	0.882	103.26	84.88	-13.95	10.70			140	0.477	37.76
Coconut Oil:	0.892	104.23	15.12	-18.13	13.90			130	0.467	47.03
Stearic Acid:	0.830	107.75		-14.08		6.34		150	0.477	37.76
Fat Blend:	0.883	103.40		-14.58	11.18					
Caustic:	1.526			49.61		50.39		65	0.782	22.50
Brine:	1.206					73.50	26.50	46	0.990	
Water:	1.000							46	1.000	
Load Lye:	1.120	1.00		3.66	10.27	77.75	7.32	140	0.962	
208 MW FA:	0.892	110.58		-19.23		8.65				
In-process Curd	0.970			Iter.Switch				211.178	0.576	
Finish Curd:	0.97			FALSE					0.576	
Seat:	1.050	22.70		1.20	4.50	68.00	3.60		0.736	
Steam:										1005

Loading Material Flow	Soap	Free	Salt	Water	Glycerine	Total	Electro	Sigma	Lambda
Actual Loading Targets:	49.22%	-0.03%	4.10%	41.51%	5.20%		4.10%		
Pounds Loading:	43,249 lbs	(27) lbs	3,606 lbs	36,469 lbs	4,566 lbs	87,862 lbs			
Evaporation Loss:				18 lbs					
At Time of drawing:	43,249 lbs	(27) lbs	3,606 lbs	36,451 lbs	4,566 lbs	87,844 lbs			
Kettle Composition:	49.23%	-0.03%	4.10%	41.50%	5.20%		4.10%	28.51%	20.06%
Predicted Lye Composition:	1.00%	0.05%	13.25%	75.20%	10.49%		13.31%	100.00%	20.06%
Actual Lye Composition:	1.00%	0.05%	13.39%	75.06%	10.49%		13.47%		
Pure Curd Composition:	68.47%						0.43%		
Total Available Lye:						25,041 lbs			
Predicted Lye removed:						19,532 lbs			

Fig. 4.14. KSPS human interface - process engineer.

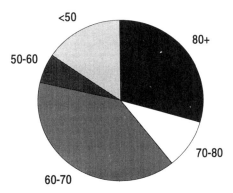

Fig. 4.15. Performance index before
KSPS implementation.

spans 0%–2%. The vertical axis is a measure of the standard deviation of the neat
soap moisture. Decaying memory statistics are employed in this analysis as
described earlier.

A typical neat soap moisture might be 30%–31%. Before the KSPS was conceived the standard deviation of neat soap moisture hovered around 1.6%. Early
versions of the KSPS solved many scheduling problems and raised the Performance
Index dramatically but did not affect the batch-to-batch variation of neat soap
moisture. Starting in 1988, refinements to the kettle soap process and the KSPS
mathematics resulted in "sawtooth" changes in neat soap moisture variation, with
each spike representing a change in the process model, and the gradual learning
curve and improvement of the process performance. Today's version is very stable,
and offers a level of variation approximate one-half the level of variation
experienced 10 years ago.

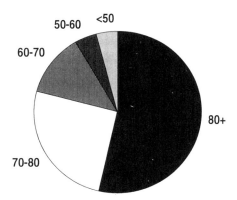

Fig. 4.16. Performance index before
KSPS implementation.

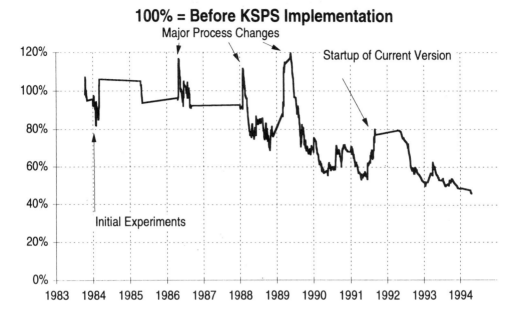

Fig. 4.17. Level of variation in kettle soap process.

Kettle Fitting

We now illustrate the level of control over the kettle fitting process offered by computer modeling. Figure 4.18 represents data on 921 production kettles collected over several years.

Each kettle was sampled at the time the soapmaker stopped the finishing process. Lab analysis was performed to determine the kettle's position on the phase diagram. These data represent many different fat and oil blends. The data was normalized for an 85/15 tallow/coco fat blend using the graining index normalization as defined in the phase diagram section. The data is plotted on a highly magnified area of the phase diagram, spanning 56%–62% soap and 1.2%–3.0% electrolyte. As can be seen, the locus of the points are in region N just over the border of region Q. The points are in a reasonably tight cluster, but note that there is still some batch-to-batch variability.

What is more enlightening is a contour plot of the Performance Index of the kettles. (Figure 4.19) Each kettle was assigned a simplified performance index value based on only the NaOH, NaCl, and H_2O concentrations, with low concentrations of all three measurements being preferred.

The contour graph clearly indicates that the best kettles result in region N, just over the border of region Q, with soap levels of 55%–60%. If a kettle's composition

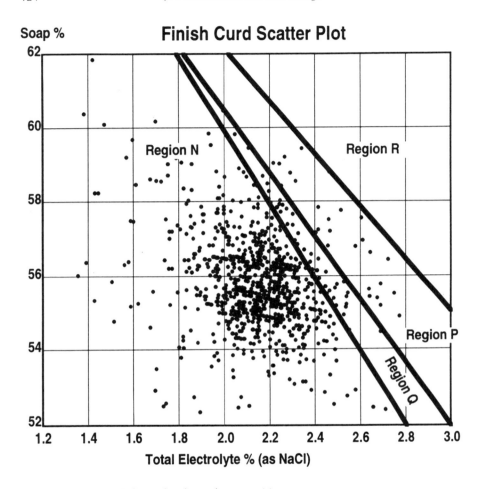

Fig. 4.18. Historical data—finish curd compositions.

is in this area of the phase diagram, then the neat and nigre will separate quickly and the resulting neat soap will meet or exceed all specifications.

Benefits

Bradford has received many benefits to computer modeling of their soap process. They include the following:

- Training the soapmaker in as little as 4 weeks
- Batch times as short as 24 h
- Quick and automatic adaptation to changing raw material blends

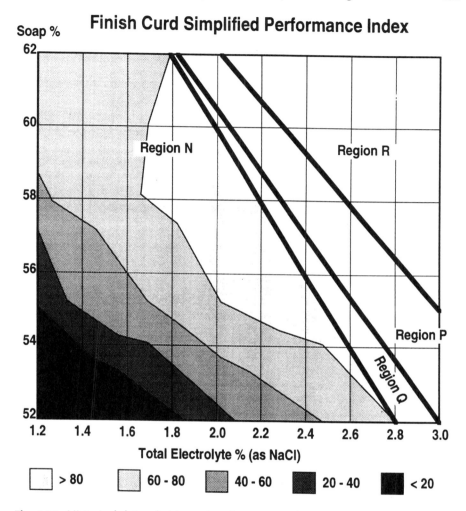

Fig. 4.19. Historical data—finish curd performance index.

- Control of neutralized, or spent, lye composition and volume
- Every kettle is thoroughly documented.
- The computer system ties into the inventory control system.
- Management summary reports summarize kettle performance.
- The process engineers control the process better in less time.
- New products can be designed using the computer, with batchmaking being simulated to optimize settling, glycerol concentration in spent lyes, and neat soap composition and yields.

- The growing database tracks long-term tends.
- The process engineers can identify cause and effect of raw material changes.
- The process engineers can tweak the process to improve kettle attributes, including color, size of nigre, size of lyes, and glycerol content of neutralized lyes.

In short, Bradford achieved better process control.

Future Plans

Recent work at Bradford has focused on computer network data acquisition and control. Although this technology has not yet been applied to the kettle soap process, recent technology advances now make it cost-effective to consider automating the opening and closing of valves and the regulating of pumps associated with kettle soap production.

Acknowledgments

We would like to thank the American Oil Chemists' Society for the opportunity to present our approach to kettle soap making, Luis Spitz for his support and assistance, Thomas Wood for his information on soapmaking as presented in the appendix to this chapter, and the senior management at Bradford Soap for allowing us the time and resources to compile this document.

References

Traditional soapmaking
1. Davidsohn, J., Better, E.J., and Davidsohn, A., *Soap Manufacture*, Vol. 1, Interscience (1953), pp. 309–374.
2. Palmqvist, Fredrik, *Soap Technology, Basic and Physical Chemistry*, Zander & Ingestrom AB, Stockholm, Sweden.
3. Spitz, *Soap Technology for the 1990's*, AOCS, Champaign, Illinois (1991), pp. 107–123.
4. unknown, "The Fitting Operation in Soap Making," SPC, February 1979.
5. Vincent, J.M., and Skoulios, A., "Study of Kettle Wax in the Sodium Palmitate-Water-Electrolyte System at 90C," *J. Am. Oil Chem. Soc. 40*: (1963).
6. Wigner, J.H., *Soap Manufacture: The Chemical Processes*, E. and F.N. Spon. (1940).
7. Wood, Thomas, personal communication, 1994.
8. Woollatt, Edgar, *The Manufacture of Soap, Other Detergents and Glycerine*, John Wiley & Sons, New York (1985), pp. 152–192.

Soapmaking at Bradford Soap
9. Davis, A.W., "Soap Manufacturing: Simulation, Data Acqusition, and Cost Management," *SciTech Journal*, Nov. 1993, p. 24.
10. George, E., and Serdakowski, J., "Computer Modeling in the Full Boil Soap Making Process," *HAPPI*, Jan. 1987, p. 34.
11. George, E., and Serdakowski, J., "Correlation of Fat and Oil Quality with Soap Base Color," *Cosmetics and Toiletries 108*:79 (1993).

Digital computer and numerical methods

12. Fried, Isaac, *Numerical Solution of Differential Equations*, Academic Press, New York (1979), pp. 4–53.

13. Serdakowski, J., "AppleTalk Data Acquisition and Control," *SciTech Journal*, November 1994, p. 30.

14. Wagner, H.M., *Principles of Operations Research*, Prentice Hall, Englewood Cliffs, New Jersey (1969), pp. 95–125.

15. Zienkiewicz, O.C., *The Finite Element Method*, McGraw-Hill, Maidenhead, England (1977), p. 178.

Appendix[1]

Traditional Countercurrent Kettle Soap-Making Process

The following presentation details a typical application of the traditional countercurrent kettle soap-making process in a modern soap manufacturing plant. Fundamentally, the process involves reacting triglycerides, such as tallow and coconut oil, with a strong base, usually sodium hydroxide, to form the metallic salt of the fatty acids contained in the fats and oils. By combining fats and oils with sodium hydroxide solution, the fatty acids in the fat combine with the sodium in the sodium hydroxide to form soap, and the combined glycerin in the triglycerides is released as free glycerin. The countercurrent kettle process serves as an efficient means to make soap and to separate the products and residual materials of the process.

The countercurrent kettle process, as described below, consists of five stages: lye reduction, saponification wash (or killing), strong wash, brine wash (or pickle wash), and fitting (or finish). Each stage consists of two parts: the active blending and boiling step and the passive settling step in which the phases separate. At the completion of each stage, the seat lye is pulled off before the next stage begins; the name associated with the lye changes from stage to stage, reflecting the name of the wash where it originated. The seat lye from each stage of the process from a given kettle is used in another kettle that is one step behind in the process except for the lye seat from the initial lye reduction stage that is delivered to the glycerin recovery operation. Refer to the countercurrent kettle process flowchart in Fig. 4A.1 for an overview of the movements of materials. Keep in mind that a given batch of soap remains in the same kettle until completion of the process, but that the lyes move from kettle to kettle backward through the system to maximize glycerin removal from the soap and to minimize the Na_2O content of the spent lye from each batch.

Soap boiling is typically done in large kettles that usually range in total capacity from 150,000 to 350,000 lb water weight in total capacity. The countercurrent process as it is described here is based on kettles with a total capacity of about 180,000 lb. Most kettle soap operations are done with the fat charge being 25% to 33% of the total kettle capacity, depending on the size and availability of kettle seat, to allow for adequate expansion that is needed during the boiling operations. Soap kettles are typically round tank constructions with cone bottoms. Each kettle should be fitted with both a closed steam coil for heating and an open steam coil for heating and agitation. The closed coil is used whenever needed during the process to avoid introducing excess water from the steam condensate. The open coil is used when vigorous agitation is needed such as on the brine wash and the fitting.

[1]The information presented in this appendix was provided by Thomas E. Wood, Valley Products Co., Memphis, Tennessee, USA.

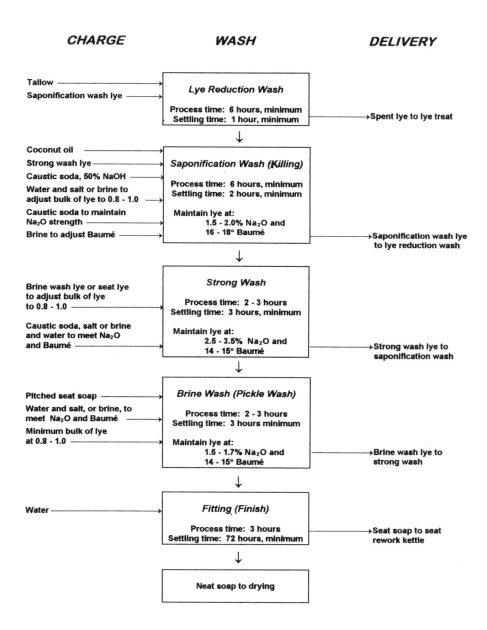

Fig. 4A.1. Countercurrent kettle process, 80:20 tallow:coconut oil charge.

Fats and oils, caustic soda, process lyes, water, brine, and steam are all controlled from a bank of valves on the operating level of the kettle house at the top of the soap kettles. The stocking lines enter each kettle at the top. Steam from the open coil enters each kettle near the bottom where the coil is placed. Process lyes and seat soap are generally removed through a line at the base of the cone bottom. Neat soap is typically removed through a swing line that enters on the side wall of the kettle midway up.

Kettle Charge

The following discussion of kettle charge for the countercurrent process assumes a total water weight capacity of 180,000 lb for the kettle. For larger or smaller kettles, proportionally larger or smaller fat charges would be appropriate. Ordinarily, it would be expected that pitched seat soap would be available, and thus the reduced charge would normally be used.

Reduced Charge

Use the reduced charge indicated as follows for kettles of 80% tallow and 20% coconut oil when pitched seat soap from a previously processed kettle will be available. The reduced charge amounts are as follows: 36,000 lb of tallow and 9,000 lb of coconut oil.

Full Charge

Use the full charge indicated as follows for kettles of 80% tallow and 20% coconut oil only when no pitched seat soap from a previously processed kettle seat will be available. The full charge amounts are as follows: 48,000 lb of tallow and 12,000 lb of coconut oil.

Lye Reduction Wash

The lye reduction wash is a minor step in the countercurrent kettle process that is performed before lye treatment step of glycerin recovery. Since the concentration of alkali in the saponification lye is relatively low, the quantity of soap produced in this stage is minor. This stage makes use of the valuable caustic soda remaining in the saponification wash lye and avoids excessive use of mineral acid to neutralize the lye before the glycerin recovery. The process control parameters for the lye reduction wash are given in Table 4A.1. Depending on scheduling requirements, the lye reduction need not be performed on each kettle at the time it is charged. Several saponification lye seats may be accumulated and reduced on a single kettle charge, allowing other kettles to proceed directly with the saponification wash at the time of receiving the fat charge.

TABLE 4A.1
Lye Reduction Wash Process Control Parameters

Operating Parameters	Limits
Free Na_2O of lye after boil	0.3%, maximum
Boiling time	4 h, approximately
Settling time	1 h, minimum
Bulk of lye	Variable

Throughout this presentation, the term Na_2O is used when referring to the alkaline strength of the lyes. Na_2O is a fictitious entity that may be thought of as the active portion of sodium hydroxide, anhydrous NaOH, or the metallic oxide of sodium that would be the precursor to NaOH.

Procedure
Pump all the tallow charge to the kettle. Next pump the saponification lye to the kettle. Circulate the lye and agitate with steam from the open steam coil until the lye Na_2O is reduced to the specified limit as shown by titration of lye sample with standardized sulfuric acid to the phenolphthalein end point. Shut off the steam, stop the circulation of lye, and settle the contents of the kettle for 1 h minimum. Finally, pump the spent lye to the lye treatment storage tank when recovering glycerin.

Notes
Free Na_2O remaining in the spent lye water after the lye reduction should be as low as practical, preferably not over 0.30%.

When edible tallow is used, the lye reduction may be difficult to accomplish because very little free fatty acid is present to start the reduction. The process could be expedited by the presence of coconut oil in the kettle, but this must be avoided because coconut oil soap that would be formed would be significantly more soluble in the lye. The dissolved soap would be discharged with the lye water, resulting in higher oil loss and increased oil and grease in the waste water. For this reason, coconut oil should never be pumped into the kettle before the lye reduction is completed.

Boiling time will vary depending on the amount and strength of the saponification lye, the amount of fat stock, and the free fatty acid level of the fat stock. The bulk of lye may be varied depending on the availability of saponification wash lye. The *bulk of lye* is discussed in more detail in the discussion of the saponification wash below.

Saponification Wash

The saponification wash is the main soap-making step of the kettle process. The operating parameters for the saponification wash are given in Table 4A.2. In preparation for the saponification wash, the coconut oil is delivered to the kettle where the tallow and the small amount of soap formed during the lye reduction are

TABLE 4A.2
Saponification Wash Process Control Parameters

Operating Parameters	Limits
Free Na$_2$O of lye after boil	1.5%–2.0%
Specific gravity of lye, 60°F	16°–18° Baumé
Boiling time	6 h, approximately
Settling time	2 h, minimum
Bulk of lye	0.8–1.0 lye to 1.0 fat

already present. The lye used in this stage of the process is strong wash lye originating from another kettle that is one step further along in the process. Since this is the wash where most of the triglycerides are converted to soap, it is in this wash that most of the fresh caustic soda is introduced to the batch. The saponification lye from this wash is the lye that will be used in the lye reduction stage of another kettle that is one step behind in the process.

Procedure
Pump the coconut oil charge into kettle. Pump in approximately 10,000 lb of strong wash lye and boil with the open coil until the *catch* is made. Add fresh caustic soda in small increments until thick soap is made. Add additional strong wash lye, water, or brine until a weight proportion of approximately 0.8–1.0 lye to 1.0 fat charge is reached. This ratio of lye to fat charge is referred to as the *bulk of lye*. Begin lye circulation and continue boiling with the open coil until all the fat is saponified. Take a lye sample every half hour, and titrate to determine the Na$_2$O level. If required, more caustic soda should be added to maintain Na$_2$O level at about 2%. When the percentage Na$_2$O remains nearly constant over a half-hour period with no caustic addition, saponification is complete. Adjust the specific gravity of the lye, as expressed in Baumé units, to specified limits with salt or brine. Shut off the steam and allow the kettle to settle for at least 2 h. Pump the saponification lye to lye storage or directly to another kettle for use in the lye reduction wash.

Notes
An excess of caustic must be maintained to complete saponification. This excess should not be more than 2% Na$_2$O at the end of saponification.

   The Baumé of the lye must be controlled to specified limits. If the Baumé is too low and a very soft grain is obtained, a proper settling will not be obtained at the end of the specified settling time, and the lye will contain large amounts of dissolved soap. In addition, the lower the Baumé of the lye, the more lye will be held up in the soap layer and, consequently, more glycerin will be retained in the soap. On the other hand, if the Baumé is so high that a very hard grain is formed, additional boiling time will be required at later stages to soften the grain to convert the kettle to neat soap. At the specified Baumé, 2-h settling time is sufficient to drop out most of the lye, with a minimum amount of lye held up in the soap.

The specified range for the bulk of lye ensures the most economical glycerin removal from the neat soap; the bulk of lye may be adjusted to the best economical advantage when recovering glycerin depending on the current glycerin market.

Strong Wash

The strong wash serves both to saponify any traces of neutral fat remaining in the batch after the saponification wash and to remove glycerin from the soap. The operating parameters for the strong wash are indicated in Table 4A.3. The lye used in this stage of the process is brine wash lye originating from the brine wash of another kettle that is further along in the process; lye from seat soap rework kettles may also be used on the strong wash if available. The strong wash lye from this stage is the lye that will be used in the saponification wash of another kettle that is one step behind in the process.

Procedure
Pump enough brine wash lye and/or seat lye into the kettle to meet the bulk of lye requirement. Caustic, salt or brine, and water are added as necessary to meet the Baumé and free Na_2O requirements. Boil the lye through the soap for 2–3 h. Settle the kettle for at least 3 h and pump the strong wash lye off to a kettle on the saponification wash or to lye storage for later use.

Notes
The strong wash is made from the lye seat of a brine wash of a previous kettle. Since brine washes are low in Na_2O and Baumé, caustic soda and salt or brine are added to bring the strength up so the soap will be sufficiently grained to drop out the maximum amount of lye. The caustic soda added during this wash will be consumed when the lye seat is used in the saponification step of another kettle.

The boiling need not be of long duration, but it must be vigorous enough to get the contents of the kettle thoroughly mixed. Since the electrolyte content is relatively high during the strong wash, the rate of settling will be rapid.

TABLE 4A.3
Strong Wash Process Control Parameters

Operating Parameters	Limits
Free Na_2O in lye after boil	2.5%–3.5%, approximately
Specific gravity of lye, 60°F	16°–18° Baumé
Boiling time	2–3 h, approximately
Settling time	3 h, minimum
Bulk of lye	0.8–1.0 lye to 1.0 fat

Brine Wash

The brine wash serves to reduce the level of free alkali, to adjust the salt level before the fitting, and to further remove glycerin from the soap. The operating parameters for the brine wash are detailed in Table 4A.4. Unlike the other washes, no lye is introduced to the batch during the brine wash. Instead, water and salt or brine are used to maintain the proper operating conditions for this stage. It is also recommended that any reworked seat soap be incorporated into the batch during the brine wash; if scheduling requires it, seat soap can be blended back during the strong wash instead.

Procedure
Begin boiling the kettle with carefully regulated steam from the open coil. Add pitched kettle seat soap from another kettle if available at this time. While keeping the soap boiling, add water and salt or brine to establish the specified limits for Baumé, free Na_2O, and bulk of lye. Continue boiling vigorously to ensure intimate mixing of soap and lye; this boiling with live steam should continue for 30–45 min. Shut off the steam through the open coil and continue simmering the soap with steam through the closed coil. In the absence of a closed coil, throttle down the open steam to just maintain gentle simmering action. Drop in a sample bucket, using a chain, down to the lye level, or use a sample cock on the bottom of the kettle, if available, to withdraw a lye sample to test for Na_2O and Baumé. If a sample bucket is used, wait about 15 min before withdrawing the sample bucket to ensure that a representative sample is obtained. Adjust the lye, as necessary, with water and salt or brine, and thoroughly boil the soap before another sample is taken. Boil thoroughly after each adjustment and for at least 30 min after a sample is found to be within limits, expanding the soap to near the top of the kettle. Shut off the kettle and settle for at least 4 h before pumping off the lye seat. Pump the brine wash lye off to storage or to a kettle on the strong wash.

Notes
The brine wash prepares the soap for the fitting, and a good fitting cannot be done without a proper brine wash. It is therefore imperative that the Baumé and Na_2O levels be controlled within close standard limits.

TABLE 4A.4
Brine Wash Process Control Parameters

Operating Parameters	Limits
Free Na_2O of lye after boil	1.5%–1.7%
Specific gravity of lye, 60°F	14°–15° Baumé
Boiling time	2–3 h, approximately
Settling time	4 h, minimum
Bulk of lye	0.8–1.0 lye to 1.0 fat

The glycerin level in the finished soap is also established on the last wash; the level of glycerin remaining in the finished soap is dependent on the level of glycerin in the lye seat. This is because the glycerin present in the kettle is soluble in the water of the soap as well as in the water of the lye, and is distributed in the same ratio as the water in the two layers. In practice, the soap usually contains one-third of the amount of glycerin in the lye beneath it. It is apparent from the foregoing that increasing the glycerin level in the lye, either by using glycerin-bearing brine or by lowering the bulk of lye, will result in a proportionate increase of glycerin in the finished soap. For this reason, it is important to maintain the bulk of lye as specified and not use brine containing glycerin on the last wash. The lye should be made with water and salt or brine only, and any additions must be well mixed into the soap by vigorous boiling of the kettle. This proper result cannot be achieved without bringing the kettle to a boil for the specified length of time.

The settling period for the brine wash is relatively long because, with a low electrolyte content, the grain is soft and lye drops out much more slowly than on the higher-electrolyte washes.

Fitting

The fitting is the final active step of the kettle process after the batch has been saponified and washed. The fitting is necessary to ensure removal of any impurities from the soap and to smooth the soap by removing the grain. The fitting is achieved using water to dilute the electrolyte content in the soap. As the electrolyte content decreases, the soap becomes more soluble and begins to dissolve in the water. This condition results in the formation of two different soap phases that coexist in the kettle: the upper neat soap phase that consists of the smooth soap with low electrolyte content, and the lower seat soap that is a mixture of soap, water, and electrolyte. The neat soap is the portion that will be delivered to the soap drying operation; the seat soap is the portion that contains most of the dirt and impurities that will be reworked in the kettle operation. Care must be taken on the fitting to avoid the use of excessive amounts of water since this will result in the formation of an undesirable gummy emulsion known as middle soap.

Procedure
Turn the steam on in the open coil and start boiling the soap. The steam must be carefully controlled to maintain gentle boiling action. While boiling, add a slow stream of water, in increments, making sure that the water is boiled into the soap after each addition. Continue adding water as long as the soap takes up the water and until the proper consistency is attained as indicated by the grain of the soap. Continue boiling and expand to near the top of the kettle, and then perform the trowel test. The soap should run off the trowel in a clear, smooth sheet, leaving half of the trowel dry. When the soap fitting is achieved, shut the steam off and settle the kettle for a minimum of 72 h before delivery.

Notes

Fitting a kettle of soap is a critical operation and requires considerable experience by the soap boiler since there are no analytical methods to determine the proper conditions in the kettle. The desired two-phase system of neat soap and seat soap is between two other possible phase combinations, and the range of water content to bring about a proper fitting is narrow.

The fitting is started by bringing the kettle to a boil with open steam and then adding water to reduce the electrolyte content. As water is added, it will at first dissolve some of the electrolyte but will not dissolve the soap. As additional water is added, the strength of the electrolyte is reduced until a point is reached when soap will begin to dissolve in the water. The critical point in this procedure is to determine when this point has been reached. If too little water is added, leaving the electrolyte content high, the soap produced will be grainy in character, with high salt and caustic, and will be unsuitable for further processing. On the other hand, an excessive amount of water will produce a gummy emulsion that will not settle out the seat soap and will leave the soap unpumpable.

When the correct amount of water has been added, the soap has cooked thoroughly, and the mass has expanded to near the top of the kettle, the trowel test will show the soap running off the trowel in a clean, smooth sheet, leaving half of the trowel dry. This is the best indication that neat soap has been formed.

Settling and Delivery of Neat Soap

Settling Time

Since seat soap will have an adverse effect on the quality and subsequent processing operations of the neat soap, it is important that uniform settling time be used and that settling be done in well-insulated kettles. These conditions will ensure that seat soap has dropped out uniformly and has not been dispersed into the neat soap by convection within the kettle.

Neat Soap Delivery

Proper delivery is important to avoid mixing seat soap back into the neat soap. Neat soap can be delivered by use of a swing pipe that allows the neat soap to be removed from the top of the kettle until seat soap is approached. When using this method, it is important that no seat soap be pumped off with the neat soap as the seat soap layer is approached. An alternative method is to first remove the seat soap from the bottom of the kettle before delivering the neat soap, but care must be taken not to agitate the seat soap that can result in contamination of the neat soap layer above.

Seat Soap Grain Out and Pitch

Seat soap from the countercurrent kettle process can be accumulated from several kettles in a rework kettle until a sufficient volume has accumulated to allow for economical processing. Seat soap can be grained out and pitched with the resulting neat soap delivered for use, in its entirety, as a lower-grade soap stock. As an alternative, especially when higher-grade feed stocks have been used, the reworked seat soap can be blended back to high-quality soap kettles during the brine wash stage of processing.

Seat Soap Grain Out
The seat soap is grained out by adding salt or brine to achieve approximately 15° Baumé lye while heating with steam from the open coil. The soap is expanded as in a brine wash. When the wash is completed, the heat is turned off and the kettle is allowed to settle for a minimum of 2 h. The seat lye is pumped off for use in strong washes.

Seat Soap Pitch
The seat soap pitch is done using heat from the closed steam coil. Add sufficient water to achieve medium grain as in a fitting. Continue to heat until the soap has expanded and a smooth, quick-moving appearance has been achieved. Settle for a minimum of 2 h after turning off steam. Pump the clean pitched seat soap to a kettle on the brine wash. The dark seat soap remaining from the grain out and pitch operations should be accumulated in a separate kettle for rework as a lower grade of soap stock.

Operating Tests for Kettle House

Specific Gravity of Soap Lyes

The specific gravity of kettle soap lyes is a direct indicator of the total electrolyte content of the lye. The total electrolyte consists of the sum of the Na_2O and NaCl present. The specific gravity is traditionally expressed in Baumé units. To perform this test, pour a sufficient sample to allow the hydrometer to float into a clean, dry jar or cylinder inclined about 30 deg from vertical. Lower the hydrometer into the liquid, keeping the stem above the final immersion point. Introduce a thermometer alongside the hydrometer and stir at all levels to constant temperature. Allow 1 to 2 min for the hydrometer to come to temperature equilibrium with the liquid. Note the temperature and remove the thermometer. Lower the hydrometer carefully until it floats freely. Read the hydrometer and record the Baumé as read. Remove the hydrometer and rinse it immediately with water.

For each degree above 60°F, add 0.0328° to the Baumé reading for correction to the standard temperature of 60°F. For convenience, use the correction chart for this conversion shown in Table 4A.5.

Free Na₂O in Lyes

The free Na_2O in soap lyes can be determined by titration of a sample of the lye with standardized sulfuric acid to the phenolphthalein end point. Pipet 10 mL of

TABLE 4A.5
Temperature Correction to 60°F for Baumé Readings of Kettle Soap Lyes

°Baumé	100°F	105°F	110°F	115°F	120°F	125°F	130°F	135°F	140°F	145°F	150°F	155°F	160°F	165°F	170°F	175°F	180°F	185°F	190°F	195°F	200°F
8.0	9.3	9.5	9.6	9.8	10.0	10.1	10.3	10.5	10.6	10.8	11.0	11.1	11.3	11.4	11.6	11.8	11.9	12.1	12.3	12.4	12.6
8.5	9.8	10.0	10.1	10.3	10.5	10.6	10.8	11.0	11.1	11.3	11.5	11.6	11.8	11.9	12.1	12.3	12.4	12.6	12.8	12.9	13.1
9.0	10.3	10.5	10.6	10.8	11.0	11.1	11.3	11.5	11.6	11.8	12.0	12.1	12.3	12.4	12.6	12.8	12.9	13.1	13.3	13.4	13.6
9.5	10.8	11.0	11.1	11.3	11.5	11.6	11.8	12.0	12.1	12.3	12.5	12.6	12.8	12.9	13.1	13.3	13.4	13.6	13.8	13.9	14.1
10.0	11.3	11.5	11.6	11.8	12.0	12.1	12.3	12.5	12.6	12.8	13.0	13.1	13.3	13.4	13.6	13.8	13.9	14.1	14.3	14.4	14.6
10.5	11.8	12.0	12.1	12.3	12.5	12.6	12.8	13.0	13.1	13.3	13.5	13.6	13.8	13.9	14.1	14.3	14.4	14.6	14.8	14.9	15.1
11.0	12.3	12.5	12.6	12.8	13.0	13.1	13.3	13.5	13.6	13.8	14.0	14.1	14.3	14.4	14.6	14.8	14.9	15.1	15.3	15.4	15.6
11.5	12.8	13.0	13.1	13.3	13.5	13.6	13.8	14.0	14.1	14.3	14.5	14.6	14.8	14.9	15.1	15.3	15.4	15.6	15.8	15.9	16.1
12.0	13.3	13.5	13.6	13.8	14.0	14.1	14.3	14.5	14.6	14.8	15.0	15.1	15.3	15.4	15.6	15.8	15.9	16.1	16.3	16.4	16.6
12.5	13.8	14.0	14.1	14.3	14.5	14.6	14.8	15.0	15.1	15.3	15.5	15.6	15.8	15.9	16.1	16.3	16.4	16.6	16.8	16.9	17.1
13.0	14.3	14.5	14.6	14.8	15.0	15.1	15.3	15.5	15.6	15.8	16.0	16.1	16.3	16.4	16.6	16.8	16.9	17.1	17.3	17.4	17.6
13.5	14.8	15.0	15.1	15.3	15.5	15.6	15.8	16.0	16.1	16.3	16.5	16.6	16.8	16.9	17.1	17.3	17.4	17.6	17.8	17.9	18.1
14.0	15.3	15.5	15.6	15.8	16.0	16.1	16.3	16.5	16.6	16.8	17.0	17.1	17.3	17.4	17.6	17.8	17.9	18.1	18.3	18.4	18.6
14.5	15.8	16.0	16.1	16.3	16.5	16.6	16.8	17.0	17.1	17.3	17.5	17.6	17.8	17.9	18.1	18.3	18.4	18.6	18.8	18.9	19.1
15.0	16.3	16.5	16.6	16.8	17.0	17.1	17.3	17.5	17.6	17.8	18.0	18.1	18.3	18.4	18.6	18.8	18.9	19.1	19.3	19.4	19.6
15.5	16.8	17.0	17.1	17.3	17.5	17.6	17.8	18.0	18.1	18.3	18.5	18.6	18.8	18.9	19.1	19.3	19.4	19.6	19.8	19.9	20.1
16.0	17.3	17.5	17.6	17.8	18.0	18.1	18.3	18.5	18.6	18.8	19.0	19.1	19.3	19.4	19.6	19.8	19.9	20.1	20.3	20.4	20.6
16.5	17.8	18.0	18.1	18.3	18.5	18.6	18.8	19.0	19.1	19.3	19.5	19.6	19.8	19.9	20.1	20.3	20.4	20.6	20.8	20.9	21.1
17.0	18.3	18.5	18.6	18.8	19.0	19.1	19.3	19.5	19.6	19.8	20.0	20.1	20.3	20.4	20.6	20.8	20.9	21.1	21.3	21.4	21.6
17.5	18.8	19.0	19.1	19.3	19.5	19.6	19.8	20.0	20.1	20.3	20.5	20.6	20.8	20.9	21.1	21.3	21.4	21.6	21.8	21.9	22.1
18.0	19.3	19.5	19.6	19.8	20.0	20.1	20.3	20.5	20.6	20.8	21.0	21.1	21.3	21.4	21.6	21.8	21.9	22.1	22.3	22.4	22.6
18.5	19.8	20.0	20.1	20.3	20.5	20.6	20.8	21.0	21.1	21.3	21.5	21.6	21.8	21.9	22.1	22.3	22.4	22.6	22.8	22.9	23.1
19.0	20.3	20.5	20.6	20.8	21.0	21.1	21.3	21.5	21.6	21.8	22.0	22.1	22.3	22.4	22.6	22.8	22.9	23.1	23.3	23.4	23.6
19.5	20.8	21.0	21.1	21.3	21.5	21.6	21.8	22.0	22.1	22.3	22.5	22.6	22.8	22.9	23.1	23.3	23.4	23.6	23.8	23.9	24.1
20.0	21.3	21.5	21.6	21.8	22.0	22.1	22.3	22.5	22.6	22.8	23.0	23.1	23.3	23.4	23.6	23.8	23.9	24.1	24.3	24.4	24.6
20.5	21.8	22.0	22.1	22.3	22.5	22.6	22.8	23.0	23.1	23.3	23.5	23.6	23.8	23.9	24.1	24.3	24.4	24.6	24.8	24.9	25.1
21.0	22.3	22.5	22.6	22.8	23.0	23.1	23.3	23.5	23.6	23.8	24.0	24.1	24.3	24.4	24.6	24.8	24.9	25.1	25.3	25.4	25.6
21.5	22.8	23.0	23.1	23.3	23.5	23.6	23.8	24.0	24.1	24.3	24.5	24.6	24.8	24.9	25.1	25.3	25.4	25.6	25.8	25.9	26.1
22.0	23.3	23.5	23.6	23.8	24.0	24.1	24.3	24.5	24.6	24.8	25.0	25.1	25.3	25.4	25.6	25.8	25.9	26.1	26.3	26.4	26.6
22.5	23.8	24.0	24.1	24.3	24.5	24.6	24.8	25.0	25.1	25.3	25.5	25.6	25.8	25.9	26.1	26.3	26.4	26.6	26.8	26.9	27.1

clear lye into about 100 mL of distilled water in a 300-mL Erlenmeyer flask. Add several drops of phenolphthalein indicator solution and titrate with 0.38N sulfuric acid solution to the end point where the last pink color disappears. Approach this end point very slowly to avoid overtitration. Note the titration reading in milliliters of 0.38N sulfuric acid required. Calculate the percentage free Na_2O by multiplying the milliliters of titrant required by 0.10.

Chapter 5

Continuous Saponification and Neutralization Processes

Clóvis F. Villela[a], and Elemér A. L. Surányi[b]

[a]PPE Engenharia S/C Ltda., Campinas, Brazil

[b]Sage Associados A. C. & P. Ltda., São Paulo, Brazil

Introduction

This paper discusses the basic principles of modern continuous saponification of neutral fats and neutralization processes of fatty acids. It begins with an analysis of Wigner's model, a fundamental cornerstone of the theory of modern soapmaking. Definitions and terminology can be found in the Appendix.

All commercially available saponification processes for neutral fats follow one of two basic routes of saponification, or a combination of the two systems. The two processes are: the Stirred Tank Reactor (STR) process and the Concentrated Caustic Pre-Emulsified (CCPE) process. In this paper the two systems are described side by side, with their similarities and differences noted. The description outlines sound guidelines for choosing the best process for a particular application and how to further optimize it. Some important details of the equipment involved are also provided. A typical neutralization process for fatty acids is described.

During the first forty years of the 20th century, soap was boiled and washed in kettles, in batch operations, in spite of the increasing scale of production spurred by demand. Boiling and fitting a kettle of soap was the work of the soapmakers, professional craftsmen with long years of practical experience and a feel for the operation. Expressions such as "killing change" for the first boil, "strengthening change" for the second boil, "fitting" and "pitching" for the second wash, "niger soap," "neat soap," and many more, were popular among soapers in those days. Those were also the times when a soapmaker would taste the strength of the lye with the tip of his tongue and tell the result with a precision that would rival that of titration. He also could judge visually the graining of the soap as "open" or "closed" and decide on the best graining for a particular stage of the process. He would take a sample of the soap on a trowel and let it run off by tilting the trowel at an angle: the exact angle!

Continuous soapmaking processes began to make their appearance only around the mid-1940s (1). Some of these processes were designed around continuous high pressure hydrolysis of oils and fats, producing fatty acids. The latter, after separation from the sweet water containing the glycerin, were neutralized with caustic soda solutions. The process patents granted to Procter & Gamble Mfg. Co. and Victor Mills (2) was similar in nature. The continuous saponification of neutral fats with strong caustic lyes under high pressure and flashing the products into a vacuum chamber was patented by Clayton and Thurman (3). Sharples Specialty Co. devised and patented a multistage low pressure continuous process for saponification of neutral fats with strong caustic lyes, with centrifugal separation of soap and lyes between each stage (4). In the fourth and last stage, the soap was washed and separated into neat soap and niger soap, again in a centrifuge.

The following years and decades saw the gradual emergence of many semi-continuous and continuous processes aimed at improving the washing yields and reducing the processing time between feeding the fats and pumping off the neat soap. All of the large soapmakers had their proprietary approaches, and a fierce competition fueled the search for improvements. Some engineering companies also offered their processes, adding to the sequence of technological advancements. A number of years passed before specialists realized that a simple but powerful model had been discovered and proposed by J. H. Wigner in 1940 (5). This model is so important for modern soapmaking that an entire section will be devoted to it in this paper. The application of this model to the several steps of the soapmaking process allows one to treat them as normal Unit Operations of the chemical process industry. This is crucial for evaluating and adopting equipment, as well as control instrumentation for modern continuous soapmaking processes.

At present, there are several commercially available continuous soapmaking processes with fairly similar designs. The main differences between the processes occur in the saponification stage. There are only two basic processing alternatives for the continuous saponification stage. The commercial processes use one or the other design approach, or a combination of elements of these, sometimes complemented with minor proprietary adaptations.

The choice of the saponification stage is very important. Once selected, it will determine the saponification reactor, the processing equipment and the operation sequences that must be adopted in the remaining steps. The importance of the selection is further underlined by the fact that, at present, almost all commercially available installations employ washing columns and centrifuges for soap washing and lye separation, arranged in suitable ways. The washing and lye separation steps must be in harmony with the soapmaking stage that they follow.

Wigner's Model

Soapmaking, in its rudimentary form, is a simple process and has been practiced for centuries. The intricacies of soapmaking appear during the separation and purification of the reaction products. This is done by "adjusting" the crude reaction mix and washing it with a watery solution of electrolytes. The adjustments for the washing and settling procedures developed into a highly specialized craft, before the underlying principles were understood. Earlier in this century, J.H. Wigner (5) derived an empirical model based on observations of the electrolyte contents of the soap and lye phases in equilibrium.

Using Wigner's model, it is possible to calculate the complete material balance of the saponification process, including the electrolytes and glycerol. The word "glycerol" will be used consistently for the chemical species 1,2,3-propanetriol, and "glycerin" for the more or less pure commercial product consisting mainly of glycerol. This model allows a scientific quantitative approach to all of the stages of soapmaking that depend on the interaction and equilibrium of components in separate phases. Because of the quantitative nature of the results, it is possible to apply them with success to the process engineering and dimensioning calculations of the equipment used in modern soap manufacturing. Modern soapmaking, with the aid of Wigner's model, can be transformed into a sequence of properly defined Unit Operations, and no longer depends on specialized craftsmanship.

Wigner's model (see Fig. 5.1) states the following:

1. Soap curds consist of a soap hydrate, containing 66% Total Fatty Matter (TFM) mixed with free solution. The 66% TFM soap hydrate can never be separated as a discrete phase, and soap curds will always contains some free solution.

2. Electrolytes (salt and caustic soda) are present in the free solution and in the separated lye, but not in the hydration water of the soap hydrate. The electrolyte contents of the free solution and of the separated lye are identical.

3. The sum of the hydration water plus free solution is the Lye-in-Soap. Glycerol is present in all of the aqueous phases at the same concentration, including in the water of hydration.

What does Wigner's model tell us about soap? It gives us a simple model of the soap structure. It includes the statement that, given the presence of enough water and electrolytes, soap will separate into two phases: a soap hydrate and a surrounding lye.

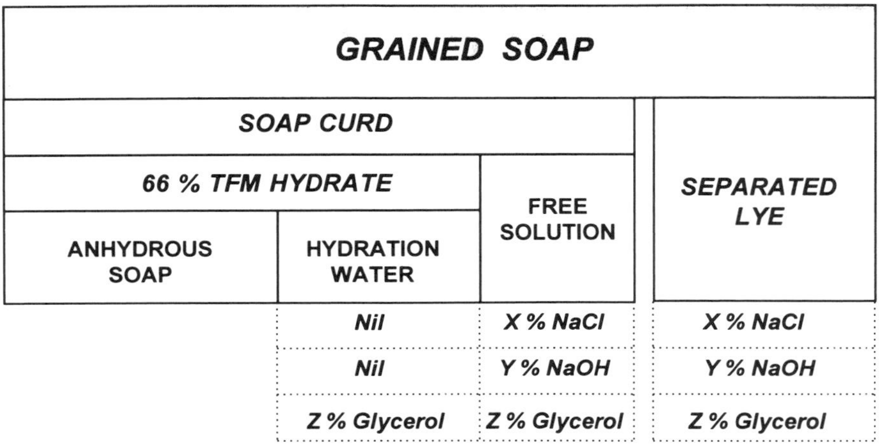

GRAINED SOAP			
SOAP CURD			
66 % TFM HYDRATE		**FREE SOLUTION**	**SEPARATED LYE**
ANHYDROUS SOAP	**HYDRATION WATER**		
	Nil	*X % NaCl*	*X % NaCl*
	Nil	*Y % NaOH*	*Y % NaOH*
	Z % Glycerol	*Z % Glycerol*	*Z % Glycerol*

Fig. 5.1. Graphic presentation of Wigner's model.

The soap hydrate is dry soap linked to water similar to the crystallization water in inorganic salts. When glycerol is present in the soap mass from which the soap curd separates out, the "hydration water" will be a mixture of water and glycerol. Thus, for all practical purposes, soap hydrate is a structure made up by dry soap "hydrated" with water and glycerol. The soap content of the soap hydrate is always equivalent to 66% TFM. The soap hydrate does not contain any electrolyte: salt and caustic soda are present only in the lye surrounding the soap hydrate.

It has been known for a long time that soap separates into phases in the presence of enough water and electrolytes, but the core of Wigner's discovery is that it always separates into a soap hydrate and a liquid phase. Now we know that a soap with more than 66% TFM will not separate, independently of how much electrolyte we add. Soap must be first diluted to less than 66% TFM to allow it to form the hydrate and throw out the excess water carrying all of the electrolytes. Soap has first to absorb water to form the hydrate. This is why there is now a practical ratio, which we will see later, called the "absorption ratio," used by engineers in design and production to define appropriate flow rates and composition of dilution and washing liquids in soap manufacturing.

It is not possible to separate soap hydrate from lye completely by settling or centrifuging: the soap obtained by these processes always contains less than 66% TFM. This is why part 1 of Wigner's model states that soap hydrate can not be separated as a totally independent phase; it will always carry with itself a certain amount of the surrounding lye. One could imagine this occluded lye as a sort of mortar joining bricks of soap hydrate. This occluded lye is in its nature and composition exactly equal to the lye that separates into the liquid phase. But because it always accompanies the soap hydrate, it is given a special name: "free solution". Free solution, thus, is lye that was not or could not be separated from the soap hydrate.

The hydration water (actually a water and glycerol mixture) plus the free solution, taken together receive the name of "lye-in-soap." Therefore, the Lye-in-Soap is made up hydration water, "bound" to the dry soap crystals, and free solution that is lye, "free" to circulate between the soap hydrate particles.

Apart from this enlightening picture of the soap structure, Wigner's model provides us with the quantitative relations between the elements of the structure. Soap hydrate always contains 66% TFM and does not contain any electrolytes. Glycerol is contained in the hydration water, in the free solution and in the separated lye: its concentration is the same in all of these three elements. Electrolytes, mostly salt and caustic soda, are present only in the free solution and in the separated lye: their concentrations are equal in these two elements. With these quantitative assumptions we can perfectly characterize soap and its components. The assumptions of this model are accurate because their application to process calculations has led, consistently, to correct results.

To illustrate the usefulness of Wigner's model two examples are presented. The first example is relatively simple and is presented step-by-step, showing every calculation and the reasoning behind it. From there we move into a fictitious process example, showing how we can calculate a complete material balance around some Unit Operations. Later, in other parts of this paper, we will return to Wigner's model, showing more calculated examples for the application of washing theory.

Example 1
Assume a centrifuge separating neat soap from lye with the following composition: Total Fatty Matter (TFM), 63.00%; Salt (NaCl), 0.55%; Caustic Soda (NaOH), 0.08%; and Glycerol, 0.20%. Applying Wigner's model, determine the composition of the separated lye.

Solution
Applying Wigner's Model, the composition of the separated lye can be calculated by Rule (a):

$$\text{Free solution} = 63\% \text{ TFM soap} - 66\% \text{ TFM soap}$$
$$= 100.0 - (100.0 \times 63\%/66\%) = 4.55 \text{ kg}$$

Rule (b):
The salt concentration in the neat soap of the example is 0.55%. From Wigner's model we know that all of this salt is contained in the free solution; thus the problem is to find the concentration of salt in the free solution using Rule (b):

$$\% \text{ Salt in free solution} = \% \text{ Salt in separated lye}$$
$$= [(100.0 \times 0.55\%)/4.55] \times 100 = 12.10\%$$

The caustic soda concentration in the neat soap of the example is 0.08%. From Wigner's model we know that of all this caustic soda is contained in the free solution, thus the problem is to find the concentration of caustic soda in the free solution again using Rule (b):

$$\% \text{ Caustic soda in free solution} = \% \text{ Caustic soda in separated lye}$$
$$= [(100.0 \times 0.08\%)/4.55] \times 100$$
$$= 1.76\%$$

Rule (c):
The lye-in-soap is the sum of hydration water plus free solution. Consequently, the remaining part is anhydrous soap. In other words, by Rule (c), the amount of the lye-in-soap is equal to the amount of soap minus the anhydrous soap.

$$\text{Lye-in-soap} = 63\% \text{ TFM soap} - \text{anhydrous soap}$$
$$= 100.0 - (100.0 \times 63\%) \times 1.09 = 100.0 - 68.67$$
$$= 31.33 \text{ kg of Lye-in-soap}$$

Note: the factor 1.09 above is the average ratio of pure anhydrous soap to fatty acids in the soap, the latter actually measured by TFM.

% Glycerol in the lye-in-soap = % glycerol in the separated lye

$$= [(100.0 \times 0.2\%)/31.33] \times 100$$

$$= 0.64\%$$

From the calculations above, the separated lye has the following composition: salt (NaCl), 12.10%; caustic soda (NaOH), 1.76%; glycerol, 0.64%; and (by difference) water, 85.50%.

In this example, we have gone through the calculations in detail. From the steps it is easy to derive the following formulae, which are much simpler to use.

$$C_L = \frac{C_S}{1 - (F_S/66)} \qquad S_L = \frac{S_S}{1 - (F_S/66)} \qquad G_L = \frac{G_S}{1 - (F_S \times 1.09/100)} \qquad (1)$$

where

C_L = concentration of caustic in the lye as percent
S_L = concentration of salt in the lye as percent
G_L = concentration of glycerol in the lye as percent
F_S = concentration of fat (TFM) in the soap as percent
C_S = concentration of caustic in the soap as percent
S_S = concentration of salt in the soap as percent
G_S = concentration of glycerol in the soap as percent

Thus, to know the concentration of glycerol in the lye, calculate:

$$G_L = \frac{G_S}{1 - (F_S \times 1.09/100)} = \frac{0.20}{1 - (1.09 \times 63/100)} = \frac{0.20}{1 - 0.6867} \cong 0.64 \quad (2)$$

In the second example, the material balance of a fictitious process will be examined.

Example 2

Assume that 1,200 kg/h of neutral fat is being saponified in a continuous high pressure reactor with 1% excess caustic soda. The saponification value of the fat is 216. The caustic soda is 48.5% wt/wt. At the outlet of the reactor, the soap is thoroughly mixed with lye of the same composition as in the previous example, so that TFM is brought down to 55%. This diluted soap is then separated in a centrifuge into soap of 63% TFM and lye. Work out a complete material balance, from inlet to outlet and show flows and compositions of each stream.

Solution

Molecular weight (MW) of the fat and oil mixture:

$$56,100/216 = 259.72$$

Caustic soda (100%) for the saponification:

$$(1,200 \times 40/259.72) \times 1.01 = 186.66 \text{ kg/h}$$

Caustic soda (48.5%) is:

$$186.66/(48.5/100) = 384.87 \text{ kg/h}$$

Glycerol generated:

$$(92/3) \times 1,200/259.72 = 141.69 \text{ kg/h}$$

Total soap stream from reactor:

$$1,200 + 384.87 = 1,584.87 \text{ kg/h}$$

of which water, added with caustic soda:

$$384.87 - 186.66 = 198.21 \text{ kg/h}$$

Excess caustic soda (100%):

$$186.66 \times (1 - 1/1.01) = 1.85 \text{ kg/h}$$

Glycerol, as calculated above:

$$141.69 \text{ kg/h}$$

Neutralized soap, by difference:

$$1,584.87 - (198.21 + 1.85 + 141.69) = 1,243.12 \text{ kg/h}$$

TFM in the soap:

$$1,243.12 \times [(259.72 - 92)/(3 + 17)]/[(259.72 - 92)/(3 + 40)]$$
$$= 1,136.85 \text{ kg/h}$$

Table 5.1 summarizes these findings.

To dilute the soap down to 55% TFM, we have to add:

$$1,584.87 \times [(71.73/55.0) - 1] = 482.09 \text{ kg/h of dilution lye.}$$

The diluted soap stream will be:

$$1,584.87 + 482.09 = 2,066.96 \text{ kg/h}$$

Table 5.2 summarizes the quantities and composition of the diluted soap mixture, using the composition of the lye to find each component's flow rate.

The separation in the centrifuge of the diluted soap will produce two streams. The soap stream will have a TFM content of 63%, and thus a flow rate of $1,136.85/(63.00/100) = 1,804.52$ kg/h. The flow rate of the separated lye will be the difference of the total flow into the centrifuge minus the soap stream leaving the centrifuge, i.e.:

$$2,066.96 - 1,804.52 = 262.44 \text{ kg/h}$$

TABLE 5.1
Material Balance of Soap Mixture

Material	Quantity (kg/h)	Composition (%)
Soap	1,243.12	78.44
Glycerol	141.69	8.94
Caustic soda	1.85	0.12
Water	198.21	12.50
Total	1,584.87	100.00
TFM in soap	1,136.85	71.73

TABLE 5.2
Quantities and Composition of Diluted Soap Mixture

Material	Soap (kg/h)	Dilution lye (kg/h)	Diluted soap (kg/h)	Composition (%)
Soap	1,243.12	0	1,243.12	60.14
Glycerol	141.69	3.09	144.78	7.00
Salt	0	58.33	58.33	2.82
Caustic soda	1.85	8.48	10.33	0.50
Water	198.21	412.19	610.40	29.54
Total	1,584.87	482.09	2,066.96	100.00
TFM in soap	1,136.85			55.00

Table 5.3 shows the flow rates and compositions of these two streams. The mass balance equations and Wigner's model equations were used to find the concentrations and flow rates of glycerol, salt and caustic soda, shown in Table 5.3. For each of these components, there is a simple system of two equations with two unknowns: the concentrations of the component in the separated soap and lye. The equation systems are as follows:

For glycerol

$$(G_L \times 262.44/100) + (G_S \times 1,804.52/100) = 141.69 \text{ from mass balance} \qquad (3)$$

$$G_L \times [1 - (63.00 \quad 1.09)/100] - G_S = 0 \text{ from Wigner's model} \qquad (4)$$

For salt

$$(S_L \times 262.44/100) + (S_S \times 1,804.52/100) = 58.33 \text{ from mass balance} \qquad (5)$$

$$S_L \times [1 - (63.00/66)] - S_S = 0 \text{ from Wigner's model} \qquad (6)$$

For caustic soda:

$$(C_L \times 262.44/100) + (C_S \times 1,804.52/100) \text{ from mass balance} \qquad (7)$$

$$C_L \times [1 - (63.00/66)] - C_S \text{ from Wigner's model.} \qquad (8)$$

Solving these systems of equations, we arrive at values shown in Table 5.3. Water was calculated by difference.

TABLE 5.3
Flow Rate and Composition of Centrifuged Soap and Lye

Material	Diluted soap (kg/h)	Centrifuged soap (kg/h)	(%)	Centrifuged lye (kg/h)	(%)
Soap	1,243.12	1,243.12	68.89	0	0
Glycerol	141.69	96.72	5.36	44.93	17.12
Salt	58.33	13.89	0.77	44.43	16.93
Caustic soda	10.33	2.53	0.14	7.87	3.00
Water	610.40	445.19	24.67	165.21	62.95
Total	2,066.96	1,804.52	100.00	262.44	100.00
TFM in soap	1,136.85	1,136.85	63.00		

Continuous Soapmaking Process

At present, almost all commercially available installations employ washing columns and centrifuges for soap washing and lye separation. Figure 5.2 illustrates the continuous soapmaking process. The main differences among the processes are in the saponification stage where there are the following two basic alternatives:

1. The Stirred Tank Reactor Process used by Mazzoni, LB, Binacchi, Meccàniche Moderne, Gianazza, and Weber & Seelander.
2. The Concentrated Caustic Pre-Emulsified Process previously used by Snurran and Monsavon (6) and presently used by PPE Engineering.

The choice of the saponification process stage is very important; once selected, it will uniquely determine the saponification reactor type and the processing sequence and equipment that must be adopted in the remaining steps. Two typical installations will be described to explain the theory and principles involved in soapmaking: the Mazzoni SCNC-C process which utilizes the Stirred Tank Reactor (STR) principle and the PPE Engineering process which uses the Concentrated Caustic Pre-Emulsified (CCPE) principle.

Saponification

The Saponification Reaction

Saponification is the alkaline hydrolysis of fats to yield glycerol and soap (7). Most fats and oils used to make soap are triglycerides, which are the condensation products of three fatty acid molecules and one glycerol molecule. In the saponification reaction, there is a sequential cleavage of the triglycerides, first to diglycerides, by splitting off one molecule of fatty acid, then to monoglycerides, by splitting off another molecule of fatty acid and finally to glycerol and the last molecule of fatty acid. The fatty acid molecules react with alkali to produce soap. The neutralized fatty acid molecules are removed from the hydrolysis reaction equilibrium, pulling the reaction mixture toward more hydrolysis. Figure 5.3 illustrates the saponification reaction.

Saponification is an exothermic reaction. The heat liberated by the hydrolysis of the three ester bonds in a triglyceride generates 18 cal/mol of fat. When the fatty acids are neutralized by the alkali, they generate a further 14 cal/mol of fatty acid. Thus, the total heat liberated during saponification is $18 + (3 \times 14) = 60$ cal/mol of fat. In modern saponification reactors, these values are used to calculate the energy balance around the reactor. This heat can be used to preheat the incoming raw materials streams, or it can be recycled for some other purpose.

Rate of Saponification

Because fats, oils and caustic soda are only slightly soluble in each other, the reaction mixture is heterogeneous. The size of the interface area between the reactants is the

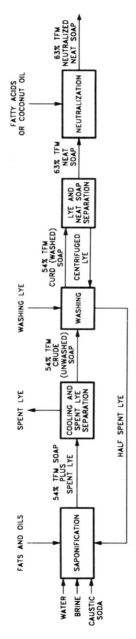

Fig. 5.2. Continuous soapmaking process block diagram.

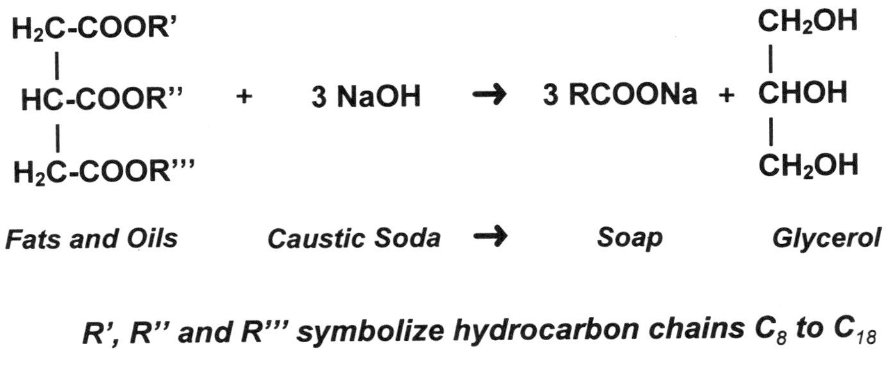

$$
\begin{array}{ccccc}
\text{H}_2\text{C-COOR'} & & & & \text{CH}_2\text{OH} \\
| & & & & | \\
\text{HC-COOR''} & + & 3\ \text{NaOH} & \rightarrow & 3\ \text{RCOONa} & + & \text{CHOH} \\
| & & & & | \\
\text{H}_2\text{C-COOR'''} & & & & \text{CH}_2\text{OH}
\end{array}
$$

Fats and Oils   Caustic Soda ➜   Soap   Glycerol

R', R'' and R''' symbolize hydrocarbon chains C_8 to C_{18}

Fig. 5.3. The saponification reaction.

most important controlling factor which sets the reaction rate. In the initial stage, the reaction rate is slow because the interface is small. The soap formed during the reaction accelerates this rate, because soap is a good emulsifier and promotes the increase of the interface area. The reaction rate decreases again close to its endpoint because the concentration of the reactants drops.

The process, illustrated in Figure 5.4 may be divided into three stages:

1. Slow Emulsification Stage (Induction Period): because the reactants form two immiscible liquid phases, the saponification reaction rate depends on the degree of emulsification obtained when caustic and fats are mixed. The reaction rate will slowly accelerate as emulsification increases due to the presence of soap formed.

2. Fast Autocatalytic Stage (Constant Rate Period): a period of uniform reaction rate (fairly rapid) during which the presence of the soap already formed maintains a large interface area and through reaction with caustic is removed from the equilibrium.

3. Slow Final Saponification Stage (Falling Rate Period): a slowing down period as the reaction proceeds toward completion. In this final stage of the reaction, the rate depends on the physical conditions and overall mixing efficiency to ensure that the last traces of fat undergo reaction.

The conclusion that we can draw from the two previous sections, concerning saponification is that an efficient saponification reactor has to provide a minimum number of operating conditions as follows (8):

1. A very intimate mixing of the main reactants to create the largest possible interface area. Heat may have to be added during the initial stage to promote this step.

2. Controlled temperature and pressure conditions that ensure the presence of water in which the alkali (usually caustic soda) can dissolve and react with the

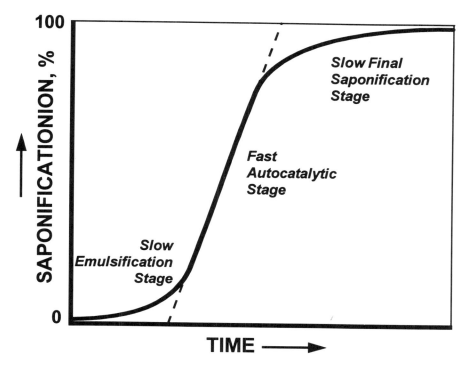

Fig. 5.4. Rate of saponification.

fatty acids generated during the hydrolysis. The soap that is formed in this reaction contributes to a good emulsification. Moreover, the fatty acids combining with the alkali are removed from the hydrolysis equilibrium and pull the reaction toward its completion.

3. Sufficient residence time and good mixing conditions to convert all fats completely to soap during the final slow saponification stage.

Saponification Processes

Stirred Tank Reactor Process (STR)—Mazzoni SCNC-C Process

This process, shown in Fig. 5.5, uses a reactor divided into four compartments in which measured and controlled streams of fats, caustic soda and dilution half-spent lye generated in the washing are injected into a stream of freshly formed soap in a recirculation loop which acts as a STR. The reactants become dissolved in the soap micelles or in the liquid crystals, and the reaction can be considered homogeneous. The reaction rate increases until all of the fat has been used up. The reactor is pressurized to 2 bar by a control valve on its outlet pipe and operates at a temperature

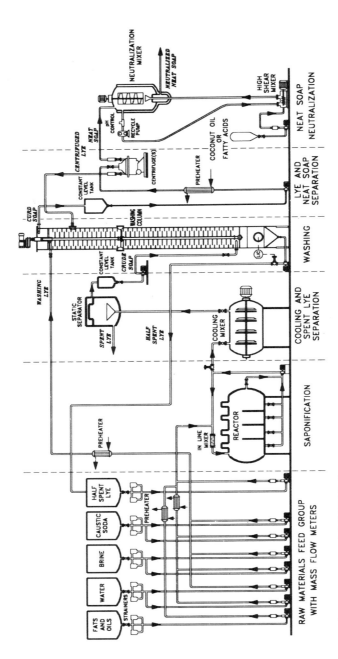

Fig. 5.5. Mazzoni SCNC-N plant.

of 120°C. As the reactor works with diluted caustic soda, a two-phase system is formed—soap at 54% TFM plus spent lye.

The stream leaving the reactor must be cooled and the spent lye separated from the soap before sending it to washing. Cooling occurs in a special type of heat exchanger with an internal rotating shaft with assembled disks. The disks have a double wall with water circulation supplied from a cooling tower. The water flow is adjusted and controlled to drop the temperature of the soap/spent lye mixture from 120–90°C. The cooled mixture is sent to a static separator where the spent lye is separated from the soap and sent to glycerin recovery. The soap drops from the top of the static separator into a buffer tank which feeds the washing column.

Concentrated Caustic Pre-Emulsified Process (CCPE)—PPE Engineering Process

This process, shown in Fig. 5.6, uses a combination of a high shear mixer and a small pressurized reactor, and works with high concentration caustic soda. Measured and controlled streams of fats and caustic soda at 36–37% NaOH diluted in line with spent lye are pumped into a high shear mixer followed by a pressurized reaction column. The high shear mixer is used to emulsify the caustic and the fat to increase the reactants' interface. A good emulsification shortens the induction period, and the pressurized reactor column provides sufficient residence time for the constant rate period. The column is sized to produce plug flow conditions. The internal pressure is controlled by a control valve on the reactor outlet. A soap with 67–68% TFM is produced and leaves the reactor at 118–120°C.

Spent lye extracted from the washing column, cooled to 60°C, is injected straight into the soap pipe at the reactor outlet. The mixture of the spent lye at 60°C with the soap at 120°C brings the temperature down to the required 90°C to feed the washing column. The amount of spent lye is used not only to cool the soap but also to complete the hydration of the saponified soap and to dilute it to 54% TFM.

Lye Absorption

The lye absorption process is illustrated in Fig. 5.7. Suppose we want to saponify 100 kg fat (95% fatty acids) with caustic soda at 50% NaOH. Let us assume that 14 kg NaOH at 100% are necessary to saponify the 100 kg fat.

The resultant TFM will be:

- caustic soda at 50% NaOH = $100 \times 14/50 = 28$ kg
- % TFM = $[(100 \times 0.95/100) + 28] \times 100 = 74.2\%$

According to Wigner's model, when this freshly formed soap meets the half-spent lye in the reactor, part of the half-spent lye is absorbed by the soap at 74.2% TFM to:

1. Complete the hydration to 66% TFM (no salt, no caustic, water only) or
2. Dilute the soap to 54% TFM.

The calculation of these two amounts show:

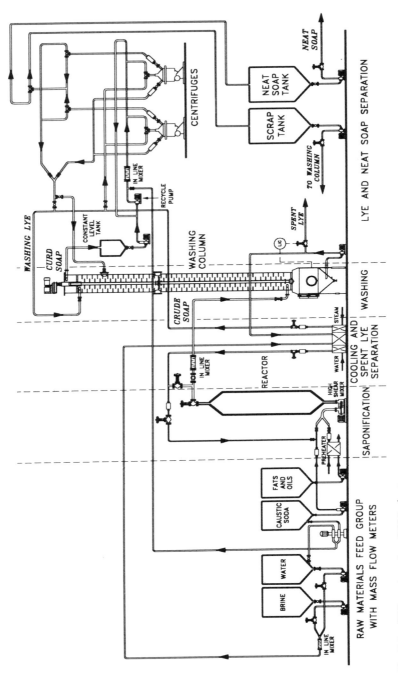

Fig. 5.6. PPE engineering CCPE plant.

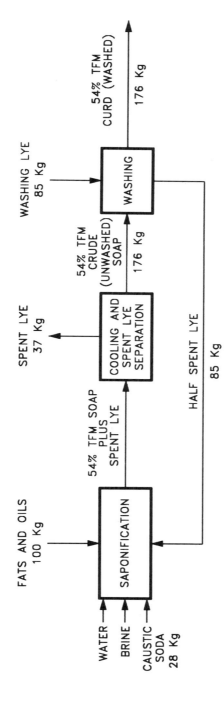

Fig. 5.7. Lye absorption process block diagram.

1. Water to complete hydration
   = Soap at 66% TFM − Soap at 74.2% TFM =
   = (100 × 95/66) − 128 = 15.93 kg

2. Half Spent Lye for dilution
   = Soap at 54% TFM − Soap at 66% TFM =
   = (100 × 95/54) − (100 × 95/66) = 31.98 kg

3. Total lye absorbed = 15.93 + 31.98 = 47.91 kg

The total lye absorbed related to fat is defined as:

Absorption Ratio = 47.91/100 = 0.48

Assume now that the soap above at 54% TFM (175.92 kg) is washed in the washing tower with 85 kg of washing lye. As the soap comes to the tower at 54% TFM and leaves the tower at the same 54% TFM, the amount of half-spent lye is equal the amount of washing lye (85 kg).

The amount of washing lye (or half-spent lye) related to fat is defined as:

Internal Ratio = 85/100 = 0.85

As shown above, this half-spent lye (85 kg) will be divided into two streams: the part absorbed by the soap = 47.9 kg, the difference 85 − 47.9 = 37.1 kg, the actual amount of spent lye that is sent to glycerin recovery.

The amount of spent lye (out of the system) related to fat is:

External Ratio = 37.1/100 = 0.37

Note that:

Internal Ratio = Absorption Ratio + External Ratio

The numerical values of these three ratios depend on the following:

The Internal Ratio: depends on the washing efficiency of the washing tower, (number of stages and the design), glycerol contents in the brine and economical ratios of recovered glycerol prices and lye evaporation steam costs.

The Absorption Ratio: depends on the total amount of water added—in any form except half-spent lye—during the saponification: as water, as brine or as live steam. Normally water should not be added; if it is, the absorption ratio decreases.

The External Ratio: depends on the values of the internal and the absorption ratios:

External Ratio = Internal Ratio − Absorption Ratio.

The control of these ratios is very important in the soapmaking process for the following reasons:

1. All of the glycerol generated in the reaction, plus glycerol that comes with brine, (less the glycerol that leaves the process with the washed soap) is incorporated into the spent lye.

2. The External Ratio determines the glycerol concentration in the spent lye: therefore if water is added to the saponification, the concentration of glycerol

in the spent lye decreases for an equal Internal Ratio (the Absorption Ratio decreases).

3. Maximizing the lye absorption in the soap, the amount of lye sent to the glycerin section is minimized. As a consequence, the glycerol content in spent lye increases.

4. On the contrary, absorbed lye decreases if water, brine or weak lyes are added to the saponification reactor. In this case, the glycerol content in the spent lye will be lower.

The only rational point for introduction of water is with the washing lye. If water is added at any other point (e.g., before the centrifuges), it should be brought back to the washing column as part of the washing lye to avoid spent lye dilution. The dilution of soap should only be made with high glycerol content lye.

The lye absorption also changes the electrolyte content in spent lye. Assume that the 85 kg of half-spent lye has a NaCl content of 10%. What will be the new salt content after lye absorption?

Assuming that the CCPE process is used:

- Salt weight in the lye = $85 \times 10\%$ = 8.5 kg.
- Water absorbed for hydration = 15.93 kg (as calculated).
- The 8.5 kg salt will be dissolved in $85 - 15.93$ kg lye.
- New salt content = $[8.5/(85 - 15.93)] \times 100$ = 12.3%.

The water absorbed by the concentrated soap for its hydration does not contain salt or caustic soda (Wigner's model) and thus the following occur:

1. In the CCPE process, the electrolyte concentration of the remaining spent lye (not absorbed) increases, causing some overgraining of the soap.

2. In the STR process, a large part of the total electrolyte in the half-spent lye is caustic soda, which is consumed in the reaction; because of this, salt has to be added to correct the graining. When the soap enters the washing column and meets the half-spent lye (high caustic), it becomes overgrained.

3. In both cases, the overgrained soap slightly reduces the washing efficiency, but this problem is unavoidable. The problem is minimized by adding extra washing stages in the washing column during the design of the unit.

In some saponification reactors, the process cannot handle overgrained soaps and the addition of water is unavoidable. The trade-off will be that the spent lye sent to glycerin recovery will be more diluted.

Soap Washing

Washing Theory

The objectives of soap washing are to remove all impurities soluble in the lye and to recover the glycerin liberated by the saponification reaction. The washing theory is based on the following assumptions:

1. Glycerol is totally soluble in washing lye (brine).

2. Glycerol present in soap is dissolved in the lye in the soap (Wigner's model).

3. When soap is mixed with washing lye, the glycerol moves from the "lye-in-soap" to the washing lye until the concentrations in both are equal.

4. When the mixture is allowed to settle, it separates into an upper layer of soap and a lower layer of washing lye; the glycerol concentration in the "lye-in-soap" is equal to that in the washing lye.

5. When washing is increased, more glycerol is extracted at the expense of more lye. This can continue until the cost of handling the evaporation of the additional spent lye is equal to the value of the extra glycerol recovered.

Washing lye with an electrolyte concentration (salt plus caustic soda) close to the limit of lye solubility is used, in which the glycerin is completely soluble, whereas the soap remains undissolved. Washing is best carried out in Rotating Disc Contactor (RDC) type washing columns.

Washing Column

Today all of the commercially offered saponification processes utilize the Rotating Disc Contactor (RDC) type washing column (9). RDC columns have been used for decades in many chemical industries prior to their application in the soap industry. Figure 5.8 illustrates a sample washing column.

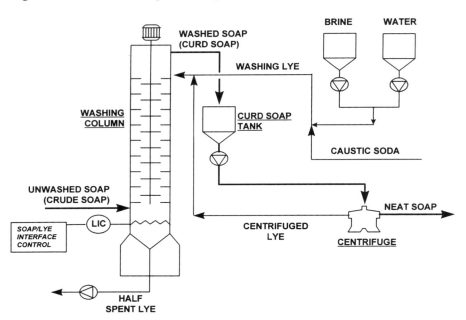

Fig. 5.8. Washing column.

The RDC Column is a mechanically agitated countercurrent extractor with a series of stator rings fixed inside a vertical shell and a vertical shaft with rotating discs mounted in it. Washing lye with a sodium chloride and caustic soda concentration, which must not be less than the limit lye concentration, enters the column below the washed soap outlet point, at a sufficient distance to allow separation of the lye. The soap/lye interface is maintained in the bottom separation zone. The crude soap (unwashed soap) enters the column above the interface level. Lye is collected at the washing column base which acts as a reservoir. The RDC design varies from 28 to 56 discs (2 columns of 28 discs in series). The best compromise between investment costs, size and efficiency indicates the use of washing columns of around 38 discs (40 washing stages) (Table 5.4).

One of the liquids is dispersed in the other, and the size of the droplets formed is controlled by the rotor speed. The lye in contact with the first disc is projected into the soap, forming small droplets. The lye droplets circulate within the compartment formed by the two stator rings and fall by gravity into the next lower compartment. In theory, the dispersed phase should be the less viscous phase, i.e., lye should be dispersed into the soap, and, in order to achieve this situation, the interface is maintained below the bottom disc.

At a given speed of rotation, a droplet of lye dispersed in soap will be smaller than a droplet of soap dispersed in lye, because lye is less viscous than soap; therefore at a given lye/fat ratio, the interfacial area will be greater in the former case. In practice, the discs cannot be perfectly flat and there will be some turbulence created in the washing column. This turbulence will be damped more easily if soap rather than lye is the continuous phase.

There are studies which assert that, in practice, the continuous phase is the soap phase. These studies, made in glass RDCs, would indicate that the interface control should be located at the top of the extraction column. Thus far the theoretical considerations prevail, and excellent control results are obtained by controlling the column operation from its bottom. These are practical results that stand against the theoretical studies described above. The discs should be perfectly flat and situated exactly between the stator rings to ensure that the motion of the discs creates no turbulence vertically and avoids "back-mixing" which would decrease the washing efficiency.

TABLE 5.4
Glycerol Content in Washing Lye at 0.44 Absorption Ratio

Glycerol in Washing Lye (%)	Washing stages			
	30 stages		40 stages	
	Internal Ratio	External Ratio	Internal Ratio	External Ratio
0.07–0.11	0.85	0.41	0.80	0.36
0.12–0.20	0.85	0.41	0.80	0.36
0.21–0.27	0.90	0.46	0.85	0.41
0.28–0.30	0.95	0.51	0.90	0.46

The main variables for the internal dimensioning of an RDC are the rotor diameter, the stator opening and the disc spacing. With increasing rotor diameter or increasing stator opening, the capacity increases but the washing efficiency decreases.

Extraction Efficiency

The glycerol content in washed soap should be between 0.1 and 0.2%. This is feasible only if the following conditions are met:

1. All of the operating conditions and the electrolyte contents of the washing lye have been adjusted for the specific fat charge formulation in use.
2. The glycerol in the washing lye is below 0.3% (equivalent to a maximum of 1.5% glycerol residue in the solid salt recovered in the glycerin evaporators) because the washed soap in the last washing stage is close to equilibrium with the washing lye.
3. The correct internal ratio is used, which in turn depends on the number of washing stages in the washing column and the glycerol concentration in the washing lye.

The glycerol content in the spent lye depends on the internal ratio (number of washing stages, glycerol in washing lye), the absorption ratio (amount of water added to the process, except with washing lye), the glycerol concentration in the recovered salt, which is part of the mass balance, the total amount of glycerol available for extraction in the fat charge formulation.

The glycerol level in spent lye should be consistent with the maximum of 0.2% in the washed soap. Economical considerations can change these specifications if the cost of spent lye evaporation is considered against the price of glycerin.

With an 80/20 (tallow/coconut) fat charge, a brine within the specifications and full lye absorption, the glycerol content achievable in spent lye as a function of the number of washing stages is summarized in Table 5.5.

Adjustment of the NaCl/NaOH Ratio

Concentrated electrolyte solution, close to the limit lye solubility, is used for washing. In such a lye, the glycerin is completely soluble, whereas soap remains undissolved. The total electrolyte concentration varies for different fat charge formulations, but for a given fat charge it remains constant. According to Wigner's

TABLE 5.5
Extraction Efficiency

Washing stages	30	40	(2 × 30 in Series) 60
Glycerol in Spent Lye (%)	22–23	28–29	31–32

model, the salt and caustic soda concentrations in the free solution of the washed soap are the same as those in the washing lye. For a given fat charge, the NaCl/NaOH ratio in washed soap is dependent on the washing lye composition. The neat soap—final product—should follow the specifications for NaCl and NaOH according to the needs of the user. To produce a neat soap with the required NaCl/NaOH ratio, two alternatives exist.

Mazzoni SCNC-N Process.   Adjust the NaCl content in the washing lye to give a salt content in the free solution of the washed soap equal to that required in the free solution of the specified neat soap. The total electrolyte content required in the washing lye is complemented with caustic soda. The neat soap will leave the centrifuges with excess caustic soda and must be neutralized in a separate soap neutralization section. The excess caustic soda present in the half-spent lye is used up inside the saponification reactor.

PPE CCPE-process.   Wash the soap with a lye high in salt and low in caustic soda and correct the NaCl/NaOH ratio by injecting calculated amounts of water and caustic soda in the soap stream, at the inlet of the centrifuges. The amounts are predicted by Wigner's model and are based on the adjustment of the NaCl/NaOH ratio of the free solution of the washed soap, before the centrifuges, to that required in the free solution of the specified neat soap. In this process it is not necessary to neutralize the neat soap to meet the specification.

  In both processes, the centrifuged lye returns continuously to the washing column, as part of the washing lye. The procedure to calculate the exact amounts of caustic soda and water that must be injected can be illustrated in the following example:

  Assume that 10 ton/h of 54% TFM washed soap (curd soap) leaves the RDC and that a sample of this washed soap separates a lye containing 10% NaCl and 2% NaOH. The soap should be adjusted to obtain a 63% TFM neat soap containing 0.4% NaCl and 0.1% NaOH. What are the quantities of 50% NaOH solution and water that must be injected to adjust the soap?

- 54% TFM soap = 10.0 ton/h
- 66% TFM hydrate = $10 \times 54/66 = 8.18$ ton/h
- Quantity of free solution = $10.0 - 8.18 = 1.82$ ton/h
- Salt in the free solution = $(1.82 \times 10.0)/100 = 0.182$ ton/h
- Caustic soda in the free solution = $(1.82 \times 2.0)/100 = 0.0364$ ton/h

  To adjust the soap it is necessary to inject caustic soda to bring the NaCl/NaOH ratio to its correct value. Water is added to reduce the overall concentration.

The required NaCl/NaOH ratio is:

$$NaCl/NaOH = 0.4/0.1 = 4$$

As calculated, the washed soap contains 0.182 ton/h of NaCl and 0.0364 ton/h NaOH.

To correct this, the following quantity of 50% strength caustic soda must be added:

$$(0.182/4) - 0.0364 = 0.0091 \times (100\%/50\%) = 0.0182 \text{ ton/h}$$

For 63% TFM neat soap, some water has to be added. 66% TFM soap hydrate:

$$(63\%/66\%) \times 100 = 95.45\%$$

Free solution:

$$100.0 - 95.45 = 4.55\%$$

NaCl in the neat soap:

0.4% (as required)

NaCl in the free solution:

$$(100 \times 0.4)/4.55 = 8.8\%$$

The free solution quantity is 1.82 ton/h and contains 10% NaCl. To dilute this to 8.8%, as calculated above, the lye quantity would be:

$$(1.82 \times 10)/8.8 = 2.07 \text{ ton/h}$$

The free solution in the 54% TFM washed soap is 1.82 ton/h, as calculated above. To this we added 0.0182 ton/h of 50% caustic soda. The sum of these two quantities is:

$$1.82 + 0.0182 = 1.838 \text{ ton/h}$$

The net amount of water to be added is the difference between the total quantity lye calculated above (ton/h), i.e., $2.07 - 1.838 = 0.232$ ton/h

Therefore, to adjust the washed soap in the example, 0.0182 ton/h of 50% caustic soda and 0.232 ton/h of water must be added.

Neat Soap and Lye Separation

For soaps, the standard centrifuge arrangement consisting of a disc stack, distributor and separating disc has been replaced by a wing insert (see Fig. 5.9). The washed soap flows through the feed inlet into the distribution space of the wing insert and flows up into the separation chamber through holes in the base of the wing insert.

The mixture is separated into a light phase (soap) and a heavy phase (washing lye). The light phase (soap) flows toward the center of the separation chamber and is discharged by a centripetal pump. The heavy phase (washing lye) flows together with the solids toward the bowl periphery, rises along the outer edge of the wing insert and is discharged by another centripetal pump. The regulation ring determines the position of the separation zone. The separated solids collect in the sediment-holding space and must be removed by hand when the bowl is at a complete standstill.

Installation Arrangements for Centrifuges

A recirculation loop in the washing lye line should be installed above the centrifuges, branching off the washing lye line before the washing column (10). Because the washing lye is in equilibrium with the soap in the centrifuge, the above piping and flow arrangement avoids the introduction of foreign materials during start-up and final cleaning of the centrifuges.

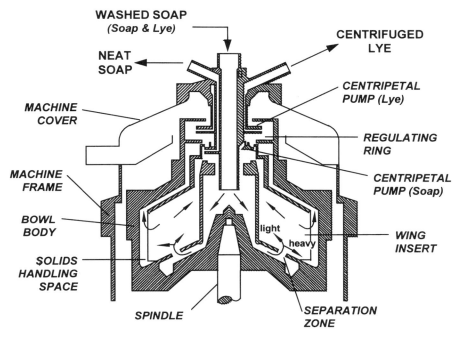

Fig. 5.9. Westfalia TA-140 centrifuge. Source: Westfalia Separator, AG.

The start-up and shut-down procedures done with hot washing lye (a recommended procedure!) prevent under- or overgrained soap from entering the centrifuge. In this way the soap will always be in contact only with washing lye, which has the correct electrolyte content. To avoid the contamination of neat soap with glycerol, the start-up and shut-down procedures should not be made with half-spent lye or any other liquid with high glycerol content. Soap flushed out during cleaning must always be sent back to the washing column.

With this flexible arrangement it is possible to stop or start one centrifuge in a bank of centrifuges working in parallel without disturbing the operation of the others. Individual streams to each centrifuge should be measured and controlled accurately and independently. A closed control loop in the soap inlet line helps to keep constant inlet pressures to each centrifuge, regardless of their number in a bank of parallel units. Cleaning of the centrifuges at regular time intervals is an absolute must. The cleaning frequency will depend on the quality of the raw materials used for saponification. In a continuous soapmaking process utilizing centrifuges it is essential to use bleached and filtered fats in the saponification to avoid problems with base odor, color of the soap base and the need for frequent cleaning of the centrifuges. In a continuous soapmaking process using centrifuges, there is no niger separation and any dirt present is retained in the centrifuge bowl.

Back pressures inside the centrifuges determine the position of the separation interface. The way to correctly control the interface position is to activate the back pressure inside the centrifuges. Large pressure drops upstream and downstream in the soap and lye lines must be avoided to prevent crippling the operation of the pressure and flow control loops for these fluids. If lay-out constraints exist that do not allow one to meet those requirements, the remedy is to install intermediary decoupling buffer tanks.

Neat Soap Neutralization

In the STR–type process, the soap leaves the centrifuges with a high caustic soda content (usually 0.3–0.2% NaOH) due to the adjustment method employed for the NaCl/NaOH ratio. The excess caustic should be neutralized to bring the soap to the neat soap specification set out by the user, usually in the range of 0.04–0.08% final NaOH. This neutralization of neat soap is done in a special mixer with a crutcher design, provided with a recirculation loop using a gear pump and a high shear mixer (turbodisperser). Fatty acids or coconut oil are injected into the turbodisperser to correct the free alkalinity. The process control is made by a pH instrument, located in the suction side of the recirculation pump.

Advantages of the Continuous Saponification Processes

The main advantages of the continuous saponification processes are the following:

1. The extra glycerol recovered (not left in the neat soap) and higher concentration of glycerol in the spent lye. In a batch kettle process with three countercurrent washing operations, the minimum glycerol content left in the neat soap is approximately 1.5% and the maximum glycerol content in the spent lye is 15%. Processing the same fat charge in a continuous saponification process, the glycerol content left in the neat soap is reduced to between 0.1 and 0.2%, and the minimum glycerol content in the spent lye increases to between 28 and 32%.

2. The reduced steam consumption in the soap making process (batch process, 0.4–0.5 ton of steam vs. continuous process, 0.1–0.15 ton), and for the lye evaporation and glycerin recovery operations (batch process, 15% glycerol in the lye vs. continuous process, 28–29%).

3. Smaller inventory of products in process. Batch processes require over 40 h of production materials, whereas continuous process require less than 4 h.

4. Better flexibility to switch soap bases/fat charges: because there is less material in process and a continuous sequence of operations, it is easier to switch between soap bases. This is an important factor in dealing with price changes and the availability of oils and fats in the market.

5. In continuous soapmaking, there is no degradation of the so-called "niger soaps," which accumulate in batch processes and have to be degraded and used

in cheaper soap bases after a certain number of fitting operation cycles have been carried out in the soap pans.

6. The continuous soapmaking processes allow easier achievement of a good and constant product quality because the operations are all based on instruments and do not have to rely on operator inputs.

7. The continuous soapmaking processes are easily equipped with computerized process management information systems, providing not only a more accurate process control but also valuable process accounting information. Automated process management is a must in the modern competitive manufacturing environment.

8. The continuous soapmaking processes are very compact, requiring much less area for their installation and thus leading to much lower building investment costs.

All of the above items contribute to a better cost effectiveness of the continuous soapmaking processes.

The main economical justification for the initial investment of installation comes from the relative values of the glycerin selling price, the cost of working capital, the cost of steam and the cost of manpower. The same economical justification factors also apply to the investments for the conversion or replacement of a batch type soapmaking plant by a continuous one. The smallest continuous saponification and neutralization processes offered are rated for 2 ton/h capacity. The most cost effective processes are plants with 3 ton/h capacity and above.

Fatty Acids Neutralization

The first large-scale continuous fatty acid neutralization process was developed by the Procter & Gamble Company. The process (see Fig. 5.10) started with the continuous counter-current splitting of fats, followed by the distillation of the crude fatty acids and neutralization of the distilled fatty acids with caustic to produce neat soap. The first complete facility of this kind was installed in Procter's Chicago plant in 1938.

$$RCOOH + NaOH \rightarrow RCOONa + H_2O$$

Fatty Acids   **Caustic Soda**   **Soap**   **Water**

$$2\,RCOOH + Na_2CO_3 \rightarrow 2\,RCOONa + H_2O + CO_2$$

Fatty Acids   **Soda Ash**   **Soap**   **Water**   **Carbon Dioxide**

R symbolizes hydrocarbon chains C_7 to C_{17}

Fig. 5.10. The neutralization reactions.

In 1956, Mazzoni developed the SC process (11) for fatty acid neutralization with caustic soda and, in 1960, the SCC process (12) in which preneutralization with soda ash is followed by neutralization with caustic soda. The success of these two processes based on was the invention of the special pH/millivolt system for the accurate adjustment and control of the final alkalinity of the soap.

Many factors enter into the decision to select the fatty acid route or the more widely practiced conventional triglyceride saponification as a process route. The fatty acid process is generally more sensitive to raw materials costs, availability and existing site constraints than differences in operating costs between the two process routes.

Mazzoni SC Neutralization Process with Caustic Soda

Measured and controlled streams of fatty acids and caustic soda enter the turbodisperser (high shear dynamic mixer) where they contact each other in a flow of preformed soap which is circulated by the recycle pump (see Fig. 5.11). Caustic soda is diluted in line, the mixture is homogenized in an in-line static mixer and the brine is fed into the caustic soda circuit before its entry into the turbodisperser. Minor additives such as chelating agents, glycerin and superfatting agent (if used) are dosed in the soap after the mixer. All of the flows are measured and controlled by mass flow meters which vary the speed of gear pumps. The exception is EDTA which is dosed by a dosing pump, due to the small amount required. From the turbodisperser, the soap flows into the mixer. The recirculated soap passes to the chamber of the pH control device where an electrode assembly is mounted. The potential mV is a function of the soap alkalinity. This potential is connected to an amplifier installed in the panel board from which a low impedance signal emerges. The signal is used to measure and record the soap alkalinity as well as for the automatic control of the fatty acids used to correct alkalinity. Control accuracy of the newest processes is better than ±0.01% NaOH. The neat soap is transferred to the holding mixer via a small in-line homogenizer in which the minor additives are dosed. The mixer is a typical crutcher type design and ensures vigorous mixing of the entire soap mass. The plant is fitted with a hot water circulation system heated by steam or electricity. In case of noncontinuous production this arrangement allows the plant to be full of soap during both the night and weekend stops. In this way, stopping and restarting operations are shortened considerably.

Instrumentation and Control

Proportioning Processes

The instrumentation and control in a continuous soapmaking process are based on the measurement and control of flows of different liquids (fats, fatty acids, caustic soda, brine, water, lyes, liquid soap and additives), temperatures, pressures, tank

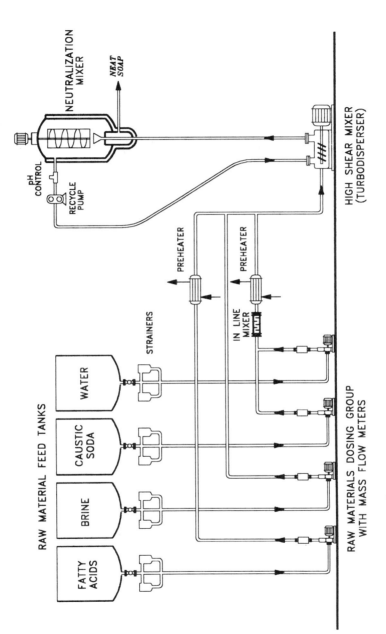

Fig. 5.11. Mazzoni SC plant.

levels and densities. The flows can be adjusted by a dosing pump or measured by suitable flow meters and controlled in different ways. Two of the most widely used alternatives for flow control are the following:

1. Flow measurement by a flow meter (mass, electromagnetic or orifice plate), which sends a signal to a system with a centrifugal pump and a control valve or to a system with a positive displacement pump (gear, trilobe or helical) driven by a variable speed motor. The control system by flow meter is superior to a dosing pump because such machines, though mechanically sound, are normally blind to the fluctuations of process variables. When a pumping head in a dosing pump is not working correctly or is not properly set, the operators cannot detect the fault until they can observe it directly or an analysis shows that there is something wrong. The correct functioning of a process based on dosing pumps depends on operator skill and attention to manual adjustment of the flows, which is frequently the cause of mistakes.

2. The modern processes employ computer control, and the best system is the one consisting of a mass flow meter/computer/inverter/positive displacement pump train, because it is very reliable and very quick to react to process changes.

The accuracy of the mass flow meter is very good (±0.15–0.2%), and the signal is directly in mass flow units, independent of viscosity, density or temperature. The mass flow meter can also actuate a control valve, keeping practically the same accuracy. This kind of loop is adequate for liquids such as brine, water and lye that can be pumped well with centrifugal pumps.

Programmable Logic Controller (PLC)/Personal Computer (PC) Control

The PLC/PC mass flow meter system (see Fig. 5.12) allows on-line, real-time process calculations and corrections to be made in the PC by downloading the set points of all mass flow meters from the PC to the PLC and uploading the measured field results to the PC. Process conditions are instantaneously detected and any actual situation outside the preset intervals can be counteracted or an emergency

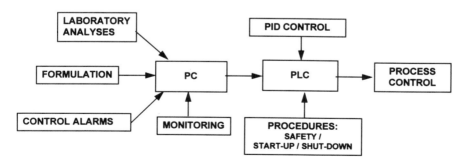

Fig. 5.12. PLC/PC control system.

action is triggered, putting the plant immediately in recirculation mode whenever the process cannot be readjusted. Practically every situation resulting in out-of-specification materials products can be avoided with this system.

The standard configuration of the mass flow meter has built-in devices that evaluate density and temperature of the fluids being measured. Knowledge of the values of these two variables, together with the process set points, allows the continuous correction of the caustic soda and brine concentrations. Furthermore, the PLC/PC mass flow meter system can provide a complete management information system covering items such as mass balance data together with gain and loss reports, process efficiency data complemented by glycerol extraction yields and quality control data with point values and trends over time.

Appendix 1

Definitions and Terminology

Saponification: Saponification is the chemical reaction between fats, oils and caustic soda to form soap (neat soap) and glycerin. It is also the alkaline hydrolysis of fats to yield glycerol and soap.

Neutralization of Fatty Acids: Neutralization of fatty acids is the chemical reaction between fatty acids and caustic soda to form neat soap.

Neutralization of Neat Soap: Neutralization of neat soap products by a continuous neutral fats saponification system refers to the addition of fatty acids or fats to the neat soap, leaving the centrifuge, to reduce its alkalinity.

Washing: Washing is the removal of glycerol from liquid soap by extraction with a brine and centrifuged lye mixture in an extraction column. This is done by addition of brine and caustic soda solution (the washing lye) to crude soap, to separate at the top of the washing column 54% TFM curd soap and from the bottom of the column half-spent lye.

Fitting: In the kettle process after settling, the curd soap will separate into a top layer of clean 61–63% TFM neat soap and a bottom of dirt carrying 15–33% TFM niger (seat). In the continuous saponification systems currently in use, there is no fitting stage: the neat soap is neutralized only to an optimum level.

Total Fatty Matter (TFM):   Total Fatty Matter (TFM) is a measure of the total content of fatty material (fatty acid) in a soap mixture.

Total Fatty Acid (TFA):   The Total Fatty Acid content is similar to TFM, but excludes any nonsaponifiable organic matter.

Grain: Grain is an indication of the electrolyte concentration of the soap and lye mixture. At high electrolyte content, the soap is "open" or "grained out," it will separate readily and generate a thin lye with almost no TFM. As the electrolyte level is reduced, the grain becomes finer and the TFM of the separated lye increases until, at a sufficiently low electrolyte level, no separation occurs, and the soap is designated as "closed."

Soap Curd: Soap curd refers to any soap phase that separates from a soap and lye mixture.

Crude Soap (Unwashed Soap):   The unwashed crude soap is the saponified mixture of fats, oils, caustic soda and brine which enters at the lower section of the washing column and which will be washed countercurrently with the washing lye fed to the top of the column.

Curd Soap (Washed Soap):   The washed curd soap is the 54% TFM soap which leaves the washing column and is fed to the centrifuge or leaves by some other means of settling or separating.

Neat Soap: Neat soap is the 63% TFM soap phase which leaves the centrifuge or separates from an aqueous layer.

Limit Lye Concentration (LLC):   Limit lye concentration (LLC) is the lye concentration (total electrolyte content—caustic soda and salt) at which soap will start to dissolve in the lye. LLC is also referred to as "limit lye," "limit lye solubility" or "graining index."

Washing Lye: The washing lye is a mixture of caustic soda, brine and water fed to the top of the washing column. To avoid introducing water into the saponification and washing systems, it usually is prepared by blending brine and centrifuged lye or brine, caustic soda and centrifuged lye. For good efficiency of glycerol removal during washing, the washing lye should have the lowest possible glycerol content. The total electrolyte concentration in the washing lye should not be less than the limit lye concentration.

Half-Spent Lye: The half-spent lye is the washing lye that leaves the bottom of the washing column, and is recycled to the saponification reactor for an extra washing stage.

Spent Lye: Spent lye is the glycerol-enriched lye which leaves the static separator after the saponification reactor or washing unit. It contains from 12 to 32% glycerol, 0.2 to 0.3% TFM and 0.1 to 0.2% caustic soda.

Centrifuged Lye (Niger Lye): Centrifuged lye is a lye with a small amount of glycerol separated from the washed soap in the centrifuge(s). The centrifuged lye is recycled to the upper section of the washing column. In a batch-type kettle process it is the "niger lye" that is separated in the fitting or settling step.

Internal Ratio: Internal ratio is the total amount of washing lye in relation to the amount of fats and oils.

Absorption Ratio: Absorption ratio is the amount of lye used up to complete soap hydration, plus the amount necessary to dilute the soap to 54% TFM, in relation to the amount of fats and oils.

External Ratio (Bulk of Lye): The external ratio, also referred to as the "bulk of lye," is the actual amount of spent lye sent to glycerol recovery in relation to the amount of fats and oils saponified.

Washing Ratio: The washing ratio relates the quantity of spent lye to the quantity of neat soap produced.

References

1. Safrin, L. (1944) New Soapmaking Processes, *Soap and Sanitary Chemicals 20,* pp. 25, 39.
2. U.S. Patent 2,159,397.
3. U.S. Patent 1,968,526, 2,249,676.
4. U.S. Patent 2,300,750.
5. Wigner, J.H. (1940) Soap Manufacture, *The Chemical Processes,* Chemical Publishing Co., New York.
6. Lachampt, F., and Perron, R. (1958) Monsavon Continuous Process for Soap Manufacture, in *Progress in the Chemistry of Fats and Other Lipids,* Vol. 5, Pergamon Press, London, pp. 32–50.
7. Palmquist, F.T.E. (1983) *Soap Technology—Basic and Physical Chemistry,* Zander & Ingerstrom AB, Stockholm, Sweden.
8. Palmquist, F.T.E., and Harris, B.M. (1964) Neutral Fat to Finished Soap Base in Twelve Minutes, in *Soap, Perfumery and Cosmetics 37,* 865–872.
9. Cusack, R.W., Fremeaux, P., and Glatz, D. (1991) Liquid-Liquid Extraction, *Chem. Eng.,* pp. 66–76 (2); pp. 132–138 (3).
10. Platt, G.R. (1970) Manufacture of Soap Bases by a Completely Continuous Process, *Soap, Perfumery and Cosmetics.*
11. G. Mazzoni S.p.A., Italian Patent 550,133 (1956).
12. G Mazzoni S.p.A., Italian Patent 667,905 (1964).

Chapter 6

Glycerine Recovery from Spent Lyes and Sweetwater*

Daniel D. Anderson and Dale Hedtke†

Crown Iron Works Company, Minneapolis, Minnesota, U.S.A.

Introduction and History

Definitions and Typical Reactions

Glycerine is used in a number of pharmaceutical, personal care items and industrial applications, and while a significant amount of synthetic product is produced from propylene, most of the commercial natural supply is made from vegetable oils and animal fats.

The terms used to describe the product vary and often lead to confusion. Although the terms "glycerine," "glycerin," and "glycerol" are often used interchangeably, subtle differences in their definitions do exist. The Soap and Detergent Association uses the following definitions:

1. *Glycerine* is the commonly used commercial name in the United States for products whose principal component is glycerol.

2. *Glycerin* refers to purified commercial products containing 95% or more of glycerol.

3. *Glycerol* is the chemical compound 1,2,3-propanetriol and the anhydrous content in a glycerine product or formulation (1).

Commercial glycerine is available in a number of grades, from chemically pure (C.P.) concentration, (>99.7%) to Dynamite grade (>98.7%). Glycerol is actually a trihydric alcohol $C_3H_5(OH)_3$, and while other polyols such as sorbitol, propylene glycol, and pentaerythritol compete with glycerine, demand for the product continues to grow.

Much of the naturally produced (as opposed to *synthetic*) glycerine is a by-product of saponification of fats using caustic soda. Saponification is used to produce soap product with the glycerine generated as a co-product.

This reaction is as follows:

H				H		
|				|		
H-C-OOC-R_1				H-C-OH		R_1COONa
H-C-OOC-R_2	+	3NaOH	→	H-C-OH	+	R_2COONa
H-C-OOC-R_3				H-C-OH		R_3COONa
|				|		
H				H		
Triglyceride		Caustic Soda		Glycerine		Soap

*Major portions of this article appear in the 5th Edition of *Bailey's Industrial Oils & Fat Products* published by John Wiley & Sons, Inc.
†Correspondence should be addressed to Dale Hedtke, Larex, Inc., 2852 Patton Rd., Roseville, MN, USA 55113.

After refusing the soap phase, the resulting by-product stream, called *spent soap lye (SSL)* is liberated. SSL typically contains 10–30% glycerine (depending on the production process), as well as contaminates such as salt (NaCl), water, and various organics usually known as MONG (Matter Organic Non-Glyceride). Crude glycerol specifications may also include a limit for trimethylene glycol (TMG, which indicates a degraded product), arsenic and sugars.

Another abundant source of glycerine arises from the hydrolysis of fats and oils as fatty acids are liberated.

This reaction is as follows:

```
      H                                      H
      |                                      |
  H-C-OOC-R₁                             H-C-OH            HOOC-R₁
  H-C-OOC-R₂        +      3HOH    →     H-C-OH     +      HOOC-R₂
  H-C-OOC-R₃                             H-C-OH            HOOC-R₃
      |                                      |
      H                                      H
  Triglyceride            Water           Glycerine        Fatty Acid
```

The aqueous phase of this reaction, commonly known as sweet water because of the characteristic sweet taste imparted by the glycerine, generally contains 12–18% glycerol, water, and unreacted fats and fatty acids. Depending upon the quality of the raw materials, sweet water may contain some nitrogenous bodies from proteinaceous materials in the fat, and in some instances, sulfuric acid or sulfates.

Glycerine can also be recovered from the by-products of methyl esters (from fats and acids) and fatty alkanolamides. The amount produced from these sources remains relatively small, but has increased significantly due to worldwide construction of fatty alcohol plants and sulfonation fieldstock requirements, especially abundant in Asia. While markets for biodiesel and other applications have not developed as anticipated, this technology bears investigation and monitoring. A change in a subsidy program could shift the economies of production, setting off a great expansion in this market. Not only would this affect the price of the fuelstock, but it would also dramatically increase the supply of glycerine. There are some other interesting sources of glycerine that will be discussed later in this presentation.

Brief History of Glycerine Processing

Glycerine was first identified in 1779 by K.W. Scheele, a Swedish chemist, who produced the substance by heating olive oil and litharge (lead monoxide). Later, in 1784, he observed that the same substance could be produced from other vegetable oils and animal fats such as lard and butter. He called this new substance "the sweet principle of fats" because of glycerine's characteristic sweet taste (2). In 1811, the Frenchman M.E. Chevreul, while studying "Sheele's Sweet" as it had come to be called, coined the modern name *glycerine* from the Greek word *glyceros* meaning sweet. After closely studying glycerine, he was awarded the first patent relating to its manufacture in 1823. Chevreul also distinguished himself by doing some important early research on fats and soaps. By 1836, the formula for glycerol had been determined by Pelouze, and finally, in 1883, Berthelot and Luce published the structural formula (3).

Nitroglycerine was discovered in 1847 by Ascanio Sobrero, a Frenchman. This compound is dangerously unstable, which limited its potential for commercial applications. In 1863, Alfred Nobel demonstrated nitroglycerine's explosive capabilities, and in 1866 he invented *dynamite*, which is still in use today. In 1875, he followed that discovery by the invention of *blasting gelatin*, a mixture of nitroglycerin and nitrocellulose. Nobel's commercial success and humanitarian efforts are well known, but equally important was the vital role his inventions played in advancing the Industrial Revolution. The demand for his explosive products helped to form a large and growing demand for glycerine.

As indicated earlier, glycerine is a co-product of other operations, and as such, the history of glycerine is closely related to the history of soap. One of the earliest commercial sources of glycerine was recovery from soap lyes, and soap lyes continue to be a common feedstock for glycerine plants today. In the early 1870s, the first U.S. patent "for the recovery of glycerine from soap lyes by distillation" was issued. The process was further developed by the Englishman Runcorn in 1883. In the decades following, the soap industry began recovering glycerine from this "waste" from their soap making operations on a relatively large scale, thus making glycerine a readily available commodity.

Another prominent historical root in the development of commercial glycerine is from the "sweet waters" of fat splitting, which initially came from the manufacture of stearine for candle making (2). The famous "Twitchell Process" for fat splitting was developed at the turn of the century. Twitchell's process for splitting fat used a catalyst and dilute sulfuric acid that produced an acceptable product. This was followed by high pressure "autoclave splitting" which relied on high pressure steam for hydrolysis of fat to produce a superior product. Today's modern fat splitting plants, using stainless steel vessels, represent the full development of this process. The high quality sweet water obtained from the Twitchell process was one of the factors allowing further refining of the product into the high purity grades of glycerine used today.

Processing Principles and Details

Sources of Glycerine

Natural glycerine is a co-product of certain processes producing products from animal or vegetable fat and oils. As indicated earlier, the primary sources include the following:

1. Fat splitting (hydrolysis) under high pressure and temperature to obtain fatty acids and *sweet water* glycerine.
2. Saponification of fats with caustic soda yielding spent soap lyes containing glycerine, water, salt (NaCl) and other impurities. Depending on the process for saponification and soap washing, the concentration of the glycerine varies greatly.

3. Transesterification of fats, typically with methanol, in the presence of sodium methoxide or other suitable catalyst, to produce methyl esters and glycerine. The chemical reaction for this process is as follows:

$$
\begin{array}{ccccccc}
\text{H} & & & & \text{H} & & \\
| & & & & | & & \\
\text{H-C-OOC-R}_1 & & & \text{NaOCH}_3 & \text{H-C-OH} & & \\
\text{H-C-OOC-R}_2 & + & \text{3CH}_3\text{OH} & \rightarrow & \text{H-C-OH} & + & \text{3RCOOCH}_3 \\
\text{H-C-OOC-R}_3 & & & & \text{H-C-OH} & & \\
| & & & & | & & \\
\text{H} & & & & \text{H} & & \\
\text{Triglyceride} & + & \text{Methanol} & \text{Catalyst} & \text{Glycerine} & + & \text{Methyl Ester}
\end{array}
$$

While the fundamental chemical reactions for these applications yield very high theoretical concentrations of glycerine, processing washing and purification steps dilute the glycerine to typical values as shown in Table 6.1.

4. Synthetic glycerine is produced (Dow Process) by chlorinating propylene to allyl chloride, converting this to dichlorohydrin, which is then converted to glycerine through the addition of small amounts of dilute NaOH and Na_2CO_3.

Treatment Processes

Before the aqueous phases generated from these applications can be processed into a high purity glycerine product, impurities in the stream must be removed. The basic intent of the initial purification (or treatment), processes is to remove these contaminants, and the basic process is as shown in Fig. 6.1. Of special emphasis is the removal of non-glycerol organic materials (measured as MONG) as completely as possible. If not removed, these impurities could cause operating problems in subsequent processing steps, which would result in a poor quality finished product.

Of primary importance when processing soap lyes is careful handling of the material prior to treatment. It is essential to keep the lye cool and, if possible, neutralized with fatty materials to reduce the formation of polyglycerol. Lyes which have not been properly stored are also subject to destruction of the glycerol by microorganisms. If permitted, this destructive process not only results in serious glycerine losses, but also produces gases and acids during decomposition. The quality of the fat feedstock is also an important consideration in the treatment process, because a poor quality product will contain significant quantities of trimethylene glycol (TMG). The gases produced by the bacteria cause trouble in SSL evaporation, and the TMG is dif-

TABLE 6.1 Typical Glycerine Concentrations in Aqueous Phase

Process	% Glycerol
Fat splitting	12–18
Kettle saponification	10–14
Continuous saponification	18–30
Transesterification	25–30

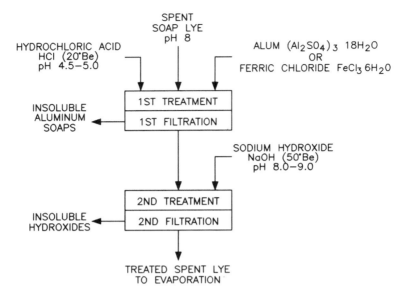

Fig. 6.1.   Soap lye treatment block diagram.

ficult to separate from the glycerine due to similar vapor pressure characteristics. Finally, excess impurities such as sulfates in the lye can act as nucleation sites for crystal growth in the evaporator, spawning the growth of a large number of small crystals and discouraging the growth of larger, more easily separated salt crystals.

A simplified process flow diagram is shown in Fig. 6.2. The treatment process generally consists of a settling step, a chemical reaction step to create an insoluble precipitate followed by a physical step of filtration and/or decanting, and a neutralization step followed by a final filtration. The ultimate success of the treatment process is measured by the reduction in MONG content of the crude glycerine produced.

Because they are drawn from the continuous saponification area or soap kettles, spent soap lyes consist primarily of glycerine, sodium chloride, and water. Small quantities of caustic soda, sodium sulfate, sodium carbonate, soap, fatty acids, and some albuminous and oleaginous matter are also present. To remove these impurities, the soap lye is treated according to the following four treatment steps.

First, a cooling and settling step is used to separate the soaps and fatty components. As the solution is cooled, fatty materials are less soluble and rise to the surface, where they can be removed by skimming. Other heavy materials will settle to the bottom of the tank and must be periodically removed. Minimization of phosphatides, traces of soap, and other emulsifying materials is essential to avoid poor separations. Generally, soap lyes are now commonly neutralized with fatty material before leaving the soapmaking department. In any case, this cooling and separation step is best done in the soap plant before the lyes are transferred to the glycerine treatment plant.

Second, the skimmed and settled lye is then transferred to the first treatment tank. It is common practice to heat the lye to 50–60°C at this point to facilitate the treatment process. In this first tank, the pH 9–12 SSL is acidulated to pH 4.6–4.8 with sulfuric or hydrochloric acid. Hydrochloric acid is recommended to avoid the formation of sulfates, but generally it is a higher cost reagent. In addition to interfering with proper crystal growth, these sulfates can build up in the recovered salt and interfere in the soap washing step. During this treatment step, fatty acids can be decanted from the top phase of the vessel and can be returned as a low grade soap. This acid treatment reaction is as follows:

$$NaOH + HCl \rightarrow NaCl + H_2O$$

Third, a coagulant is added next to cause the residual organic material to form insoluble precipitates. Generally, sufficient coagulant material is added to reduce the pH by 0.5. During coagulation, compressed air is introduced into the tank and is used for mixing and to promote oxidation. The coagulant is usually either aluminum sulfate (alum) or ferric chloride. If ferric chloride is used, sufficient aeration is required to oxidize the

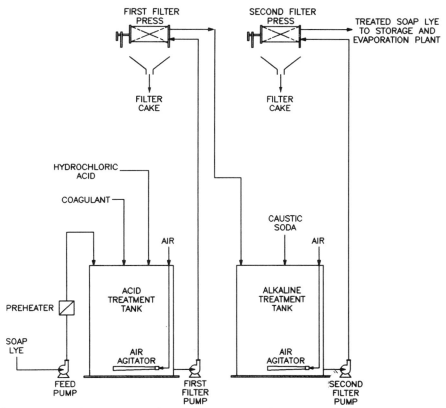

Fig. 6.2. Soap lye treatment plant (Crown Iron Works Company).

iron to its ferric state (4). While ferric chloride is more difficult to handle than aluminum sulfate, is strongly corrosive to iron, and is more expensive, it is usually preferred because alum introduces sulfates into the SSL. The precipitates of either aluminum hydroxide or ferric hydroxide will adsorb other impurities such as fatty acids and proteinaceous matter and allow relatively easy removal of these impurities from the lye.

About 0.25% by weight of ferric chloride ($FeCl_3 \cdot 6H_2O$) is normally used, with less required for high quality feedstocks. If alum is used, about 0.05–0.1% aluminum sulfate [$Al_2(SO_4)_3 \cdot 18H_2O$] is required. The completion of the first treatment step can be verified by filtration of a lye sample followed by addition of a small amount of dilute coagulant to the filtrate. Any resulting precipitate indicates that the reaction is incomplete (4,5).

The precipitation reaction is as follows:

$$Al_2(SO_4)_3 \cdot 18H_2O + 6RCOONa \rightarrow 2Al(RCOO)_3 + 3Na_2SO_4 + 18H_2O$$

After the chemical reaction has been completed, a physical filtration process is used to remove the precipitate. The contents of the first treatment tank are purified in any number of industrial filtration devices such as a plate and frame or pressure leaf filter. As the filter becomes full, the filter cake is discharged periodically for disposal. One problem experienced in filtration of these products is that the precipitate tends to block filter element pores, quickly reducing filtration capacity. To improve filtration rates, filter aid can be added.

Fourth, the filtrate from the first press flows to the second treatment tank, where caustic soda (50°Baume) is added to produce a slightly alkaline product. While a pH of 9 is preferred when ferric chloride is used, a slightly lower pH is appropriate when alum is used as the coagulant. Compressed air is again used for agitation and aeration, and, when the reaction is complete, the mixture is filtered through a second press. A smaller amount of precipitate is produced in this step and, therefore, the filter can be sized smaller. Filter aid may again be added to improve filtration rates. In addition, spent bleaching carbon containing residual glycerine from the distillation plant can be introduced at this point to recover residual glycerol.

The alkaline reaction is as follows:

$$Al_2(SO_4)_3 \cdot 18H_2O + 6NaOH \rightarrow 2Al(OH)_3 + 3Na_2SO_4 + 18H_2O$$

Treatment can also be carried out using a continuous or semi-continuous operation. In each case, accurate, in-line pH control is required to regulate the addition of the treatment chemicals. Without accurate control, the quality of the resulting glycerol can be affected by dehydration (low pH) or polymerization (high pH). Filtration can be accomplished on a semi-continuous basis by sizing the treatment tanks to allow surge capacity for filter clean-out intervals, or it can be a continuous operation using a vacuum drum filter.

Treatment of Sweet Water

Treatment (purification) of sweet waters is generally less involved than the treatment of soap lyes because salts are generally not present. The purpose for treating sweet waters is to remove any unreacted fats, eliminate any nitrogenous bodies, and prevent bacteriological growth in the glycerine. The process block diagram is shown in Fig. 6.3. Typically, the sweet water is settled for several hours to allow any fatty materials to rise to the surface, where they are skimmed off the top. It has been demonstrated that addition of small quantities (under 10 ppm) of a suitable poly-electrolyte can enhance the quality of the crude glycerine produced (4). Then, in the presence of air agitation, slaked lime [$Ca(OH)_2$] is added to neutralize the fatty acids still present. An excess of 0.23% as calcium oxide (CaO), is normally used, with the batch adjusted appropriately to obtain this excess (6). Filtration is normally used to enhance the separation. Sodium carbonate (Na_2CO_3) is sometimes used to remove the calcium excess as carbonate.

The sweet water treatment reactions are as follows:

$$Ca(OH)_2 + 2RCOOH \rightarrow Ca(RCOO)_2 + 2H_2O \quad \text{Lime Treatment}$$
$$Na_2CO_3 + Ca(OH)_2 \rightarrow CaCO_3 + 2NaOH \quad \text{Soda Ash Treatment}$$

Some processors treat with barium chloride and sulfuric acid, which may be more effective, but is also a more expensive process. Other processors report good results using ferric chloride or alum as described earlier for soap lyes.

It should be noted that saponification crude may produce a better crude product because the inorganic salts formed in soap lyes are not present. However, because of concerns related to the quality of the final distilled glycerine product, it is common to mix sweet waters and soap lyes with the mixture processed as a soap lye.

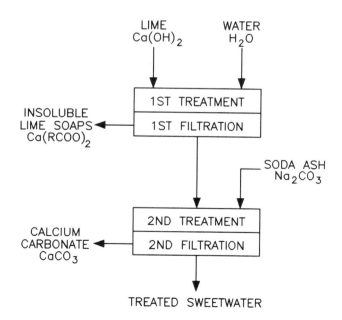

Fig. 6.3. Sweet water treatment block diagram.

Evaporation of Soap Lyes or Sweet Water

Following treatment, the soap lyes or sweet waters are concentrated to 80–88% glycerine in an evaporation process. This is an energy intensive process and, depending on the plant's capacity and energy costs, the evaporation plant may be equipped with a single-, double-, or even a triple-effect system. Economics play a large role in determining the plant configuration because generally, the greater the number of effects a plant is equipped with, the lower the energy costs will be. As a result, much of the design emphasis of an evaporation plant is placed on its evaporation economy, expressed in terms of kilograms of water evaporated per kilogram of steam supplied.

With soap lyes, a secondary consideration is the method for handling the salt that is "dropped" during the concentration process. While salt recovery is generally not a factor in a sweet water evaporation plant, salt will crystallize and must be removed from a lye evaporation facility. Because the salt contains substantial amounts of glycerol and water, both of which must be recovered prior to discharging, provisions for separating and handling these materials must be included in the plant design.

The amount of water to be evaporated is calculated from the amount of liquor and its glycerine concentration. Table 6.2 shows the concentrations that occur, starting with 1000 kg of dilute soap lye and concentrating to 80% crude glycerine; losses are ignored.

Thermal Economy

With emphasis on steam efficiency, several varieties of evaporation plants have been developed over the years. At atmospheric conditions, 1 kg of steam will evaporate 1 kg of water at its boiling point, but as the pressure is reduced in the evaporation vessel, the boiling point falls. The designs that have evolved over the years are designed to capitalize on this basic fact.

Early evaporation plants typically had a single effect natural circulation evaporator, as shown in Fig. 6.4. This type of unit relied upon a thermosyphon to circulate liquors, and in operation often processed a batch of soap lye in two steps. During the first cycle, a half-crude product (40–45% glycerol) was produced and stored. After a sufficient quantity of product was generated, the conditions were changed and the batch was processed to full crude (80%) strength. These evaporators are no longer manufactured, having been replaced by more efficient designs, but many of these units have been used over the years, and some are still in operation.

Unlike natural convection on which the calandria evaporator depended, forced circulation heating is used for most new evaporation plants for several reasons. Primarily, pumping the liquor allows the selection of a higher flow rate and velocity than can be achieved by natural convection, significantly improving the overall heat

TABLE 6.2   Example of Concentrations in Soap Lye and Crude Glycerine

	Soap Lye (%)	Weight (kg)	Crude glycerine (%)	Weight (kg)	Change
Glycerine	10	100	80	100	—
Salt	11	110	9	14	96 kg salt
Water	79	790	11	16	774 kg water

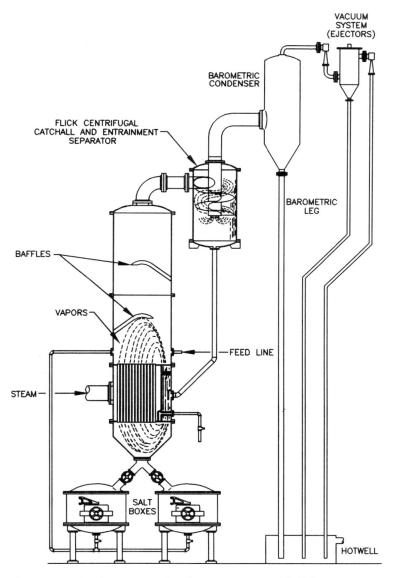

Fig. 6.4. Calandria type single-effect evaporator with flick separator.

transfer coefficient, thus reducing the surface area and/or temperature differential required for heating. Higher velocities also reduce fouling and scaling on the tubes. Drawbacks are the extra energy usage for the pump motor and, if the velocity is too high, problems with salt crystal growth. Material selection is also critical for conditions encountered in the systems as well.

Multiple-effect evaporators have all but eliminated the simple single-effect system. In addition to providing a continuous operation, multiple-effect evaporators allow a greater degree of flexibility in addition to substantial reductions in operating costs.

Multiple-effect evaporation is accomplished by joining two or more evaporators in series and using the latent heat in the vapor from one effect to heat the liquor in the subsequent effect. Live steam is added in the first effect only and provides the driving force for controlling the plant. The difference in the working pressure in each effect (each effect working at a lower pressure than the previous one) allows the use of lower pressure heating steam in the following effects. If the design permits, the pressure difference also allows transfer of liquor from one effect to the other.

As a general rule, it can be assumed that approximately 0.8 kg of water can be evaporated per kg of steam for each evaporator effect. Thus, one effect will evaporate 0.8 kg of water per kg of steam, two effects will evaporate 1.6 kg of water/kg steam, and three effects will evaporate 2.4 kg of water/kg of steam.

Multiple-effect evaporators are most often used in larger glycerine recovery plants (over 15–20 MTD), with the typical plant having a two-effect evaporator system. As plant capacities increase, with some feed rates approaching 350 MTD of treated product, triple-effect evaporators are becoming increasingly more attractive. The user must calculate and compare the savings in steam usage for operating additional effects against the additional equipment and maintenance costs, as well as the space and operational complexities. In the calculation, the user should consider not only the heating steam requirements, but also other utility demands such as the motive steam (or electricity) demand necessary for the vacuum system.

Even the basic design of multiple-effect evaporation plants has changed over the years. While early double-effect systems capitalized on steam economies, many concentrated the lye to a half-crude product, and then used a separate single-effect finisher to produce crude product. These plants used steam at slightly above atmospheric pressure, and the supply commonly included exhaust from steam pumps and small steam engines (4). Over the years, the designs have evolved to eliminate the need for the finisher and to recover greater amounts of thermal energy using innovations such as the lye/condensate economizer.

One of the most significant innovations came from development of thermal recompression. While electrically driven compressors are not particularly attractive for these applications, jet recompression is widely used when motive steam of at least 7 barg is available. One of the best descriptions of the thermocompressor used in a multiple-effect evaporation system (see Fig. 6.5) is presented by Swenson Process Equipment and is as follows (7):

> "To reduce energy consumption, water vapor from an evaporator is entrained and compressed with high pressure steam in a thermocompressor so it can be condensed in the evaporator heat exchanger. The resultant pressure is intermediate to that of the motive steam and the water vapor. A thermocompressor is similar to a steam-jet air ejector used to maintain vacuum in an evaporator.
>
> Only a portion of the vapor from an evaporator can be compressed in a thermocompressor with the remainder condensed in the next-effect heat exchanger or a condenser. A thermocompressor is normally used on a single-effect evaporator or on the first effect

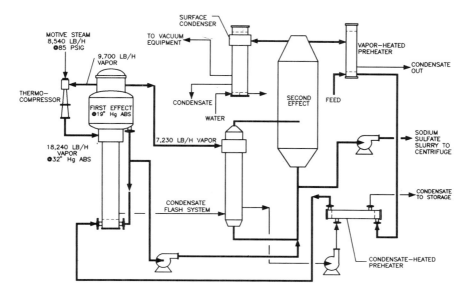

Fig. 6.5. Swenson double-effect thermal recompression evaporator.
Source: Swenson Process Equipment, Inc. (7).

of a double- or triple-effect evaporator to reduce energy consumption. As with mechanical recompression, thermal recompression is more applicable to low boiling-point rise liquids and low to moderate ΔTs in the heat exchanger to minimize the compression ratio. To illustrate the energy effectiveness of a thermocompressor, compare the steam usage for a double-effect evaporator with that of the double-effect evaporator with a thermocompressor. Motive steam at 85 psig is utilized to compress first-effect vapor from 19″ Hg to 32″ Hg absolute. The forward-feed, double-effect evaporator is used to concentrate a weak sodium sulfate solution. A rising-film evaporator is used for the first effect and a forced-circulation evaporator is required for the second effect where sodium sulfate crystals precipitate.

	Lb/hr of 85 psig steam	Lb evaporation/ lb of steam	Ratio of motive/ evaporated water
Double effect	12,800	1.94	0.515
Double effect with thermocompressor	8,540	2.90	0.345

The double effect evaporator with thermal recompression requires 33% less steam than the conventional double effect. In essence, the steam usage for the double effect with thermal recompression is comparable to that of a triple-effect evaporator. The main advantage of thermal recompression is improved steam economy for a moderate capital expenditure—less than that for an additional effect."

The operation of a triple-effect, continuous evaporator including vapor recompression is shown in the accompanying Fig. 6.6. Treated soap lye or sweet water is fed continuously and regeneratively heated with condensates coming from the heaters. Water is evaporated in the first evaporator chamber, which is kept at a pressure slightly above atmospheric, using heat provided by the thermocompressor. The thermocompressor takes part of the vapor from the first heater and recompresses it using the high pressure motive steam to increase its temperature and pressure. The remaining vapors liberated in the first effect heat the liquor circulated in the second effect heater. Vapors are flashed from the second effect and are condensed in the vacuum system. In the first evaporator, about 45–50% of the water is evaporated. The lye is pumped to the third effect which, in this system, is again heated with steam. In other designs, the final effect is under a greater vacuum, using the vapors from the second effect, providing a greater degree of thermal economy.

The glycerine product is concentrated to 80% for soap lyes or up to 88% for sweet water crude (containing no salt) and then pumped to crude glycerine storage.

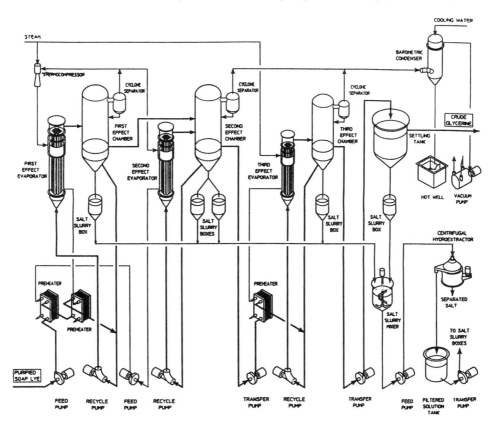

Fig. 6.6.   Mazzoni LB "CSL" triple-effect soap lye evaporator with thermocompressor. *Source:* Mazzoni LB, S.p.A.

Entrainment Separation

Although modern evaporation systems are designed with low vapor velocities, it is inevitable that some liquor is entrained in the vapors leaving the system. Entrainment separation is necessary to reduce losses of glycerine entrained in the vapor stream from each evaporator. An entrainment separator should be installed at the outlet of each evaporator body, and the glycerine recovered should be returned to its evaporator. The separator is usually of the cyclonic type which can be enhanced with internal impingement baffles (Flick type). Mesh- or baffle-type separators may be considered, but care must be taken to prevent any build-up on the internal components which would cause excessive pressure drop and ultimately plugging. Conductivity devices can be provided to measure the electrolyte present in a sample of condensed vapors. Because salt will be present only in the liquid phase, an indication of increased conductivity is an indication of carryover.

Salt Handling

Because the glycerine concentration increases during evaporation, salt crystals begin to form as the liquor saturates. These agglomerated salt crystals tend to sink to the cone bottom of the evaporator. This salt must be removed as soon as possible after forming, or the liquor will become unduly thick and potentially plug the system. Proper treatment and filtration of the soap lye is important to insure that there are not excessive amounts of impurities in the lye that can hamper the growth of the crystals. There are several methods used to recover the salt from the evaporators, and all of the methods address the need to separate the salt/glycerine/water slurry as effectively as practical. Subsequent washing is generally done with water, with the purpose of recovering glycerine that is removed with the salt. The wash liquor is subsequently returned to the treatment plant. Obviously, any water that is added must be re-evaporated, thus a balance must be achieved between glycerine loss and washing water used.

Salt Extractors

All salt removal systems start with the salt crystals (and some glycerine) being removed from the evaporator, and collected in a *salt receiver* or a *salt extractor*, depending on the configuration of the plant. If a batch-type salt receiver is used, the salt level is allowed to build to a certain level, determined by observation through a sight glass, or the level is estimated by using a timed fill system. When full, the receiver is isolated from the evaporator and the salt slurry is blown to a salt extractor, using compressed air or steam. Salt extractors are usually provided in pairs and used alternately. When emptied, the receiver is then placed back on line for receiving salt slurry. While the receiver is isolated from the evaporator, salt will continue to deposit in the lower section of the flash tank. Obviously the receiver must be placed back on line quickly to avoid a process upset. It is also important to refill the receivers with lye and evacuate the headspace to the evaporator before placing it back on line. Many serious process upsets have occurred by placing a receiver that is full of air on line. If salt receivers are not used, the salt is dropped directly into the salt extractor, as described above. Salt receivers are placed below each evaporator and under a salt settling tank.

Regardless of which option is selected, it is imperative that a quiet zone be provided for the salt to accumulate and drop.

The salt extractor is fitted with a screen to catch the salt and allow the liquor to pass out of the extractor, returning to the salt receiver from which it originated. Several batches of salt from the receivers are accumulated in the extractor before the salt is removed. When full, the extractor is isolated and the salt is washed to remove residual glycerine. The wash fluid can be soap lye for the first wash and water for subsequent washes. The wash water is returned to the treatment plant, or may be delivered to the first effect. Caution must be exercised to avoid a process upset when returning the liquor. The salt bed is finally steamed to provide a dry salt product. Residual glycerine in the extracted salt will be between 1 and 3%.

Some older plants are designed to remove the salt manually, by shoveling it out through a door in the salt extractor. Removal of dry salt can be a labor-intensive practice. The salt is usually returned to the soap plant, where it is dissolved into brine to be used again in the kettle soap-making operation.

A more efficient practice is to dissolve the salt within the salt extractor. There, after the final wash, water is introduced into the extractor, the salt is dissolved, and the brine is pumped back to the soap plant. This process can be automated if desired, reducing the labor requirements.

Centrifugal Separation

Centrifuges are the most effective and convenient devices to extract and wash the salt obtained in the evaporation process. The centrifuge used for this process can be either a continuous or a batch operation model. The continuous model, described below, typically offers higher capacity and efficiency, but at a higher cost. In either case, the salt slurry is fed to the centrifuge in a similar fashion as described earlier for feeding a salt extractor, but is typically delivered to a slurry tank prior to introduction to the centrifuge. Because a continuous flow of liquor is extracted from the flash vessels, the salt receiver batching is not required and the process is not subject to many of the potential upset problems inherent in a salt extractor system.

A batch centrifuge is equipped with a basket having a fine screen (or cloth bag) which catches the salt while the liquor is drained back to the evaporation plant. In operation, the centrifuge is fed with salt slurry from the receiver tank and is spun until the basket is full of salt. The salt is then washed with treated lye and/or water to remove residual glycerine. The salt can then be removed by mechanical means or dissolved, for reuse in the saponification plant. In the case of the cloth bag-type of centrifuge, the entire bag full of salt is lifted out of the basket while a new bag is installed. The salt is then dissolved in a separate tank, freeing up the centrifuge for additional salt extraction. Because the feed is delivered in intermittent bursts requiring a certain amount of judgmental skill and because the recovered salt must be handled manually, batch centrifuges require a relatively high amount of operator attention. For that reason, the modern trend is toward automatic continuous centrifuges.

The continuous centrifuge is fed with salt slurry from the evaporators. The cutaway diagram of a centrifuge shown in Fig. 6.7 shows the main components and operating principles for this type of unit. The salt slurry is continuously fed to the top inlet

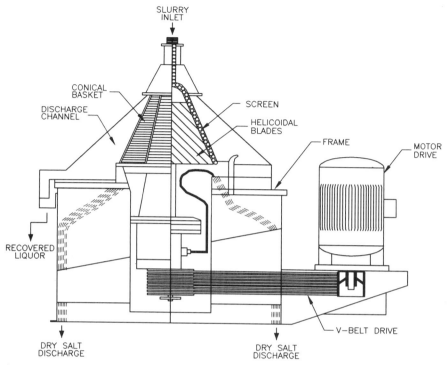

Fig. 6.7.   Guinard model "C" continuous salt extractor centrifuge diagram.
Source: Guinard centrifugation.

where it falls between the helicoidal blades of the extractor and the conical basket. The basket has a specially designed slotted screen installed over it to retain the salt crystals. The slurry is subjected to a very high centrifugal load due to spinning of the basket. The liquid passes through the screen and is collected in the discharge channel. The extractor rotates at a positive differential speed relative to the screen which pushes the salt that has settled on the screen toward the bottom of the screen where it is discharged out the bottom of the machine's frame. The salt can be continuously washed while it is on the screen. Some machines have more than one washing zone, allowing the use of different wash liquor concentrations for more efficient washing. Some designs include a means to separate the wash water from the recovered liquor. The efficiency of salt removal is dependent on crystal size, but a residual glycerol of less than 1% is very common while using less than $\frac{1}{3}$ kg wash water per kg of dry salt. For best results the crystals should be at least 100 μm in diameter.

Vacuum System

The vacuum system in the evaporation plant is used to condense the water evaporated, remove any noncondensables in the system, and provide the required vacuum of approximately 50 torr. The vapor contacting device is either a barometric condenser, utilizing a direct contact water spray and a barometric leg discharging into a hot well, or a water-cooled surface condenser. Despite a somewhat higher equipment cost (and a somewhat higher utility consumption), the surface condenser system offers several significant advantages over a barometric condenser: First, any glycerine in the vapor that is condensed is recovered and is not lost into the hot well water, and second, the surface condenser is cooled with clean water, typically circulated from a cooling tower. In a direct contact system, the spray water becomes contaminated, requiring a separate cooling tower system from the "clean" cooling water used elsewhere in the plant. The "dirty" cooling tower water must be periodically replaced to prevent excess build-up and fouling in the cooling system. A "dirty" tower will also require additional downtime and incur additional maintenance costs as well.

Following the condensing section, and depending on the utility situation at the plant site, either a liquid ring vacuum pump or two-stage steam jet with intercondenser is used. A booster ejector is generally not required to maintain the vacuum level required for evaporation.

Glycerine Refining

Purification of glycerine is normally accomplished using distillation with steam, under high vacuum, and at elevated temperatures. The boiling point of glycerine at atmospheric pressure is 290°C, and because glycerine will begin to polymerize at about 200°C, distillation must occur at low pressure. When distilling with steam, the partial pressure of the glycerine is reduced, while maintaining the same total pressure, as related by Dalton's law (assuming ideal gas behavior):

$$\frac{\text{Weight Glycerine Vapor}}{\text{Weight Water Vapor}} = \frac{\text{Partial Pressure Glycerine Vapor}}{\text{Partial Pressure Water Vapor}} \times \frac{\text{Molecular Weight Glycerine}}{\text{Molecular Weight Water}}$$

The details of glycerine refining are given in the description of the Crown Iron Works process (Fig. 6.8), which is representative of the continuous distillation plants utilized by many suppliers for sweet water or soap lye crude glycerine.

The crude glycerine is preheated regeneratively by hot distilled glycerine. The liquor then enters the still heater where it is further heated to about 165°C and circulated by means of a circulation pump. The circulated liquor is partially vaporized through the aid of vacuum (6 torr) and sparging steam in the flashing chamber. The vapor rises through an entrainment separating section and then enters the condensing section. Here the glycerine vapors are condensed in a layer of packing, wetted by recirculated, cooled, distilled glycerine.

The remaining vapors enter the salvaging condenser where remaining traces of glycerine are condensed and recovered as 80–90% substandard glycerine, which is sent to intermediate storage. The substandard glycerine is rerefined after sufficient quantities have been collected for a 2–3 d run every month. Typically the substandard glycerine is processed into a lower grade of glycerine, such as High Gravity or Dynamite Grade.

The residue in the bottom of the still is continuously discharged from the crude still to the foots still. The residue should be kept rich in glycerine (>25%) to improve handling and distillation characteristics. A small amount (0.5–1%) of phosphoric acid is added to keep the residue soft by lowering the pH to retard formation of

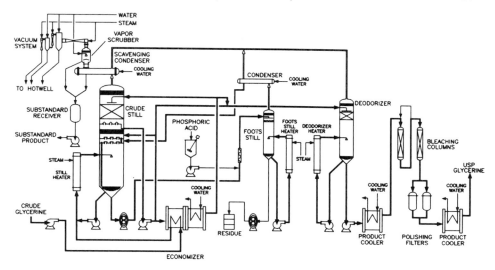

Fig. 6.8. Glycerine refining plant. *Source:* Crown Iron Works Company.

polyglycerines. In the foots still system, the product is reheated by recirculation through an external heater to about 175°C and partially vaporized, under vacuum and with about 25% stripping steam. Most of the vapors are condensed in the foot still's condenser and the condensate is returned to the crude still. Any remaining glycerine is recovered when the vapors pass through the salvaging condenser. The residue at the bottom of the foots still is discharged to a drum for disposal.

The distilled glycerine from the crude still is re-evaporated in the deodorizer at about 130–140°C, again in the presence of high vacuum and stripping steam and with external heating, to ensure optimal removal of odoriferous materials and residual moisture. The vapor passes through a packed section where the incoming feed condenses the vaporized glycerine. Proper reflux rates are required to ensure removal of close boiling impurities. Residual volatiles are condensed in the salvaging condenser.

The deodorized glycerine is cooled before flowing through columns of activated carbon, where color impurities and trace odoriferous materials are removed. The bleached glycerine is filtered to remove trace carbon particles, cooled further, and sent to storage.

As the description indicates, the operating temperature of the refining still is about 165° at a pressure of 5–6 torr. Although the vapor pressure of pure glycerine at this temperature range would indicate that a lower operating temperature may be possible, the boiling point of the crude feedstock is actually elevated due to dissolved substances. For example, soap lye crude, which contains a substantial amount of salt, will require an evaporation temperature 10–15° above the boiling point temperature of pure glycerol (4). Temperature control is very critical, because certain undesirable chemical reactions can occur in crude glycerine at distillation temperatures. Examples of these reactions are as follows:

1. Formation of nitrogen compounds by thermal breakdown of proteinaceous matter present in the crude glycerine (not removed in the treatment process). These, along

with volatile decomposition products, form impurities in the refined glycerine.

2. Formation of volatile glycerine esters by reaction with soaps (low MWs) by the following reaction:

$$C_3H_5(OH)_3 + R\text{–}COONa \rightarrow C_3H_5(OH)_2\text{–}OCOR + NaOH$$

3. Formation of polyglycerines which occur in the presence of NaOH.
4. Formation of acrolein (CH_2=CHCHO) which imparts objectionable odors to the final product.

Therefore, it is important to limit the time the glycerine is at high temperature, as well as the maximum temperature it is exposed to. As the time and temperature increase, additional polyglycerides are formed, which are normally removed as residue from the still. While these may not directly affect the quality of the distilled product (polyglycerols are essentially nonvolatile at these conditions), their formation does represent a loss of glycerol. The pH in the still is also an area of concern. While ester formation (as a balanced reaction) is suppressed by alkaline conditions, polyglycerol formation is enhanced in the presence of caustic soda. It is important, therefore, to keep the pH as low as possible while producing an acceptable distillate product.

The amount of total stripping steam for distillation is about 20% of the amount of glycerine processed. This amount is greater with poorer quality feed stocks. It is clear that it would be undesirable to produce crude glycerine at a higher concentration in the evaporation stage, because additional steam would be needed.

While some designers actually pass the crude product through a dehumidifier to remove residual moisture, other designers have elected to capitalize on this moisture present in the feedstock. Because the water in the incoming crude glycerine flashes to steam, providing a significant portion of the stripping steam requirements, this approach is pragmatic. In fact, it is postulated that residual moisture in the crude feedstock may help suppress polyglycerol formation, even though the water is rapidly vaporized (4).

Residue Recovery and Disposal

The residue that accumulates at the bottom of the still contains some glycerine, polymerized glycerine, aldehyde resins, organic products of decomposition, and salt, which must not be allowed to become too concentrated, because build-up of residue will cause quality and capacity problems in the plant.

If the residue is allowed to concentrate in the still, it displaces volume for incoming crude glycerine, reducing the still's capacity, and because residue from soap lye crude is similar in consistency to toffee, circulation of this material becomes increasingly difficult. At least three methods of removing the residue are used: First, a residue receiver vessel is placed below the still accumulating residue which is periodically discharged to the treated lye tank for reprocessing. Second, residue is removed continuously from the still and redistilled to recover the remaining glycerine as described earlier. The concentrated residue from the bottom of the foots still is discharged to a drum for disposal, typically to land fill as solid waste. Last, a scraped surface dryer is used to concentrate the residue, recovering nearly all of the glycerine and discharging the residue as a solid.

Scraped Surface (Wiped Film) Dryers

Wiped film dryers have successfully been applied for residue processing, and at least one system supplier has designed the total distillation plant around the wiped film dryer. Considering first the dryer for residue recovery, under proper conditions, the volatile fractions will be vaporized nearly instantaneously while a dry residue is discharged from the base of the evaporator. Figure 6.9 provides some of the specifics for the wiped film dryer.

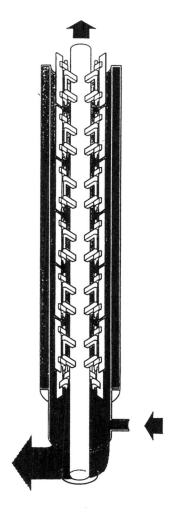

Fig. 6.9. Agitated thin film processor. *Source:* LCI Corp.

LCI Corporation's literature describes the evaporator function as follows:

"The process utilizes the counter current method to separate vapor flowing up through the thin film dryer from the liquid feed, which is flowing downward. The thin film dryer consists of two major assemblies: a heated body and a rotor. Preheated crude glycerine enters the dryer tangentially at the top, above the heated zone, and is distributed evenly over the inner circumference of the body wall by the rotor. Product spirals down the wall while bow waves developed by the rotor blades generate highly turbulent flow and optimum heat flux. Volatile components are rapidly evaporated. The heating temperature is selected for evaporation of the glycerine, which is condensed in a separate heat exchanger. The evaporation takes place under high vacuum. Non-condensables are removed by the vacuum system. The condensed glycerine is pumped from the system to an additional purification stage, if required. The non-volatile components, or residue, is discharged from the bottom of the dryer as a powder.

The rotor removes any encrusting solid or viscous material (usually rich in NaCl) from the heated surface, thus maintaining the original high heat transfer condition for long periods without shutting down for cleaning."

Advantages of the wiped film dryer for this application are obtained primarily due to the nearly instantaneous vaporization of the glycerol. The yield can be very high when drying residue, and because the residence time of the glycerol is very short (as is exposure to the high temperature heating surface), very little polyglycerol is formed. Ester formation is also quite low, as is the generation of other undesirable compounds. Waste disposal volume is reduced, and the consistency of the residue is changed from a sticky, tar-like substance to a dry powder. Steam stripping is not required with wiped film dryers, reducing utility consumption.

Wiped film dryers do have some special considerations that limit their use, however. As saponification crude lacks salt, scraped surface dryers are generally applied only to soap lye distillation plants. The equipment is rather capital and maintenance intensive, and does require careful control of the jacket heating. The MONG/salt ratio in the feedstock must be kept high, and upsets in the feedstock can cause the evaporator to become imbalanced.

While much of the applicability for wiped film evaporation focuses on residue recovery, the Buss process for glycerine distillation is a two-stage process using a wiped film dryer for evaporating all of the glycerol (see Fig. 6.10). First, the crude glycerine is concentrated by flashing the water under moderate vacuum. After this pre-evaporation, the glycerine is distilled in a scraped surface thin film heat distiller under high vacuum. The thin film process operates with high heat transfer coefficients, allowing moderate temperatures and short residence times to keep decomposition and polyglyceride formation low. The glycerine vapors are condensed in a three stage condensing system, where chemically pure, dynamite, and technical grades (A, B, and C) are collected. The yield of CP glycerine is typically about 60% with total glycerine recovery of 97%. One of the main advantages of scraped surface distillation is that salts and organics are concentrated and discharged as a free-flowing residue containing >2% glycerine, thus reducing glycerine losses associated with residue disposal.

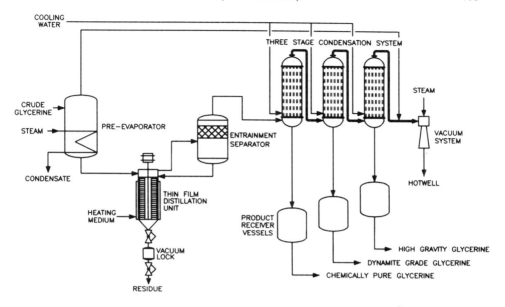

Fig. 6.10. Buss glycerine refining plant. *Source*: BUSS, A.G.

Ion Exchange

In addition to the purification processes listed, ion exchange resins can be used to purify glycerine in a process that circulates the crude glycerine through cation and anion resin columns. There the cations (Na^+, Ca^+, and Mg^{2+}) are exchanged for H^+ ions and the anions (Cl_- and SO_4^{2-}) are removed by the resin. The cation resin is regenerated with mineral acid (HCl or H_2SO_4), while the anion resin is regenerated with caustic soda. While the process is effective, it has not been widely used due to the costs involved with regenerating the resins and with waste disposal (8).

Vacuum System

The vacuum system for the distillation plant can be a common system for the still, deodorizer, and, if supplied, the foots still and/or the wiped film dryer. To achieve a minimum vacuum of 10–12 torr required in the distillation equipment, the vacuum system is usually designed for 6–8 torr, to allow for such events as system pressure drops or noncondensable loading. Most systems use a hybrid-type vacuum system, with a typical arrangement including a surface condenser combined with a barometric condenser, followed by a booster jet and multiple-stage vacuum system. A vacuum pump may also be used in place of the steam jets. The cooling and condensing water going to the hot well is normally uncontaminated because virtually all condensables are removed by the noncontacting surface condenser. Because a higher level of vacuum is used than that required for the evaporation plant, distillation vacuum systems are generally more complete and complex.

Storage and Stability

Dilute and crude glycerines contain certain amounts of suspended material (e.g., precipitates and salt) that tend to settle out during storage. Therefore, to avoid introducing these materials into the process when liquor is introduced, it is recommended that discharge nozzles be located above the lowest level of the tank and that the tank be emptied and cleaned periodically.

Dilute solutions of glycerine (<50%) are subject to fermentation, which reduces yields and introduces breakdown products that degrade the glycerine. Holding the glycerine above 70°C and/or at higher concentrations will help alleviate this problem. Naturally, this is a trade-off condition, because the higher the temperature (especially in an alkaline condition), the greater the formation of polyglycerols will be. Concentrated glycerine is difficult to pump at lower temperatures because of its high viscosity. Glycerine should be pumped at 40–50°C because lower temperatures make pumping difficult and higher temperatures can affect the product quality. If heating coils or steam tracing is used, it is important to use low pressure steam and provide sufficient agitation in the holding vessel so as not to overheat the glycerine and cause breakdown products.

Stainless steel or lined vessels are recommended to prevent the formation of color complexes, especially if moisture or residual fatty acids are present in a carbon steel tank. Because glycerine is hydroscopic, care should be taken to exclude moisture from the refined glycerine storage tank. Glycerine subjected to heat should not be stored in vessels containing copper or iron, because iron and copper salts will catalyze oxidation of glycerine under those conditions (9).

Odor and Color

As mentioned in other sections, color and odor problems can generally be avoided by using a high quality raw material, treating and storing crude glycerine properly, and avoiding high temperatures for an extended time period. Impurities in crude glycerine, especially MONG, affect the quality and quantity of distilled glycerine. If the MONG content [defined as 100 – (ash + glycerol + moisture) %] is high (3–5%), odor, taste, and color reversion problems may exist in the final product. Trimethylene glycol, which is present in MONG, can affect the color of glycerine and also lead to problems in storage. Acrolein, which has an odor that is distinctive and is easily detected in small quantities, may be formed in the presence of iron, iron salts, acid or neutral salts.

A properly designed deodorizer or "stripper" vessel, located after the distillation column and operating at high vacuum and with open or "sparging" steam, will remove close boiling impurities and many odor bodies. Remaining color bodies can be absorbed by passing the deodorized product through an activated carbon bed.

Glycerine left inside processing vessels, especially carbon columns, can degrade and develop hard to remove odors if left for an extended downtime period. Therefore, prior to shutting down the process for several days or longer, process equipment should be cleaned, with carbon columns blown down with air and washed with water. Fresh carbon (and ion exchange resins) can initially contain odor-causing impurities. This temporary condition can be limited if the bed is first washed with distilled glycerine that can be subsequently redistilled.

Plant Operating Considerations

Plant Layout and Construction Practices

Careful selection of materials of construction is important to ensure adequate life of equipment in the glycerine processing plant and a high quality finished product. This is especially true when recovering glycerine from soap lyes, because of the salt and free chlorides as described earlier, but is also true for certain sweet waters. The reader is encouraged to consult reference materials and manufacturers' data sheets regarding corrosion for the compounds and materials encountered in the glycerine plant.

Treatment Plant

When treating sweet water glycerine, ductile iron can generally be used for pumps, although lined ductile iron and stainless steel pumps can be used for longer life. The primary consideration is resistance to sulfuric acid and other reagents. More care must be used when selecting material for treating spent soap lye, because the treatment program is generally more involved (and more equipment is required) and a greater variety of potentially damaging reagents and conditions are encountered. Treatment vessels can be either lined carbon steel or stainless steel. FRP (fiberglass reinforced polyester) is also a good choice for both tanks and piping in the treatment plant, but care must be taken not to exceed the temperature range of the FRP material. From a mechanical standpoint, the system designer must consider the strength limitations of the FRP equipment (especially considering the operating temperatures). Regardless of the materials selected, care must be taken to prevent aggressive chemical reagents from attacking upstream components. In one facility, a failed check valve allowed hydrochloric acid to find its way back to the soap plant, attacking carbon steel piping materials on its way.

The open plate and frame filter presses typically used usually have cast iron frames with polypropylene plates that give excellent life. Filter cloths are usually constructed of synthetic materials such as polypropylene felt, rather than natural fibers such as cotton. Suitable tools must be used to clean cake from the cloths to avoid damage.

A plate and frame heat exchanger is sometimes used to preheat incoming soap lye, because treatment is more effective at an elevated temperature. The plates should be constructed of suitable chemically resistant materials, such as 316L stainless or titanium.

As the designer considers the plant layout, it is best to seal all tanks and vessels as completely as possible, and use a vent fan to draw vapors from the vessel head spaces. Because these vapors can be rather odoriferous, these should be discharged away from inhabited areas. The area should be designed for complete wash down and ideally have liner panel on the walls to protect structural components from corrosive attack. Provisions should be made to contain potential chemical spills, and safety showers/eye wash stations should be provided. Easy access should be provided to clean the pH probes and, in a continuous system, interfacial draw-off zones should be easily inspected and readily adjusted. A chemical resistant flooring system should be considered, as should isolation and skimming facilities for washdown water and spill containment. It is not uncommon for the floor in the treatment area to become slippery. It is wise to consider precautions against this condition to prevent serious operator injuries, especially in the areas in which reagents may be present.

Evaporation Plant

Material selection for the evaporation plant is one of the greatest challenges in the glycerine recovery plant, especially in the presence of salt. For evaporating sweet water, carbon steel vessels may be suitable, but better results (especially glycerine color) will be obtained using stainless steel. The corrosive nature of the brine/glycerine solution makes carbon steel plate and most grades of stainless steel a poor choice for evaporators processing soap lye. The heavy walled, cast iron calandria that have been used extensively for soap lye evaporation in the past have very long lives. However, economic considerations and modern use of high pressure steam (and resulting higher temperatures) have resulted in more severe operating conditions that make the older style equipment obsolete.

The material of choice for the past 20+ years for welded plate evaporators (and related components handling salt and/or brine) is copper nickel alloy. The two principal alloys to consider are 9010 (90% copper/10% nickel), which has very high corrosion resistance to the glycol/brine solution and 7030, which is very suitable but is more expensive.

An interesting class of materials that should be considered for evaporator service are the duplex stainless steel alloys. These are high performance alloys that have a "duplex" structure, in that the alloy contains both the austenitic and ferritic phases in the grain structure at the same time. These materials have shown high resistance to stress corrosion cracking in chloride-bearing environments and also possess generally high corrosion and erosion resistance. Two alloys that show particular promise for this application are Sandvik's SAF 2205 and SAF 2507. Traditionally, the fabricated cost for duplex stainless vessels has been 30–40% less than comparable 9010 cupro vessels. With limited availability of the stainless material coupled with recent increases in all stainless materials, this price differential is less attractive.

Other more expensive alloys such as Ferralium 225, Hastelloy, and Monel have been used in various applications. Mechanical designs must avoid high stress areas, especially for seam and joint areas.

Pumps for the evaporation plant must stand up under the hot and abrasive glycerine/brine solution. Cast stainless steels (CF8M alloy) are often used for this application, but a superior material to consider is a chromium-nickel-copper-molybdenum stainless steel known as CD-4m. Although CD-4m pumps are slightly more expensive, the long-term life of these materials more than offsets this premium. For the circulating lye applications, the low head/high flow duty mandates a centrifugal style pump and although magnetic drive pumps without seals are an interesting possibility, economics generally dictate selection of a mechanical seal pump. Proper selection of the mechanical seal is one of the greater challenges for the design engineer. If the fluid contains salt crystals, which are very abrasive, it is mandated that a water-flushed seal be used. The water must be cool and free of dirt and other impurities. In an least one installation, seal failure was a recurring problem that defied explanation. Upon inspection, it was found that iron and other materials in the water were depositing on the sealing faces and plugging the water channels. Both seal faces on the pump side should be of a hard material; using carbon for one of the seal halves has not proven successful. The reader is urged to contact seal manufacturers for their recommendations. Like the treatment plant, the evaporation facility should be built to withstand possible corrosion from salt solution.

Refining Plant

Materials for the refining plant can be selected on the basis of the impurity content of the glycerine being handled. While a saponification crude refinery may be made from a lower grade material, the crude distillation vessel or residue still for a spent lye should be fabricated from a high grade of stainless steel. The "L" grades generally exhibit slightly better corrosion resistance than their base versions. 316L stainless steel is recommended for stills and residue purification equipment handling soap lye crude glycerine, while 304 stainless steel is adequate for sweet water stills, vapor scrubbers, condensers, deodorizers, and bleaching equipment where salt is not present.

In some cases, even more corrosion resistant materials have been used; the "Ittner" refining stills in use at some Colgate-Palmolive plants were fabricated from pure nickel. The life expectancy of such equipment is exceptionally long. In some other instances, lower grade alternatives, even mild steel, have been supplied with sacrificial anodes of aluminum.

Pumps in the refining plant can be centrifugal as long as the temperature and viscosity of the product is satisfactory for the pump being used. For lower temperatures and for pumping residue, a positive displacement pump with a pressure relief loop (internal) should be used. Of special concern is the residue pump on the foots still. As a viscous, toffee-like material containing less than 35% glycerine is desired, pumping limitations alone may dictate the glycerine recovery from this material.

Instruments, Controls, and Computerization

The basic control parameters for glycerine plants have not changed much over the years, as the fundamental controls include pHs, levels, temperatures, and vacuums in the various portions of the plant. Most glycerine processes *can* be operated manually, and traditionally, manual operation of batch plants has been the norm. However, the development and refinement of digital single loop controllers and associated sensing and measuring equipment have increased the ease and accuracy of control, and make modern continuous operation feasible. One example of instrumentation allowing process development is the advent of in-line, reliable pH indicators coupled with highly accurate chemical metering pumps. With these, continuous treatment operations become feasible, reducing not only chemical consumption and waste, but also allowing a more consistent product to be produced. Another example is application of mass-type flow meters, which not only provide accurate process and inventory control, but also indicate instantaneous process yield and efficiency values.

The process designer must take care, however, to select instrumentation that can handle the sometimes severe conditions in the glycerine recovery plant. The crude glycerine in the evaporation and crude refining sections contains both dissolved and suspended salt, while the residue in the refining plant contains salt as well as other impurities which make the viscosity of the residue very high. Vane-type flow meters have been used successfully for measurement for difficult flows such as residue from the crude still. Rotometers often will be plugged by suspended solids and are recommended only for flows that are "clean" and will maintain a constant viscosity over time. Mass flow meters provide accurate measurement of flow, temperature, and density, but are expensive and add complexity to the plant.

To avoid potential galvanic reactions for wetted portions, the instrumentation should be constructed of the same metal as the host equipment. It is recommended that a diaphragm seal be installed to protect pressure gauges that are installed in fluid lines that contain suspended or entrained solids.

Many plants are now "computer controlled" through the use of Programmable Logic Controllers (PLCs) that handle the start-stop functions, sequencing, as well as process control PID loops. Systems utilize an operator interface with the PLC so that the user can observe the details of the process and control the operation often from a remote location. The interface can be anything from a simple I/O output connected to a hardwired annunciator and run/start/stop status lights, to the most modern graphical operator interface. In most cases, the PLC actually controls the process, giving and receiving instructions from a PC running the data acquisition software. In this manner, a loss in the PC function allows the process to continue to operate while a technician "forces inputs" to change process variables until the PC is back on line. Modern operator interface packages provide a host of display options, data recording and trending, and interfacing to marketing, accounting, inventory, and process management reports.

Product Specifications and Quality Assurance

Glycerine Grades
As indicated earlier, commercial glycerine is available in a variety of grades. Some manufacturers have developed proprietary products with specific functional properties for custom applications. Some of the more common standards used for standardized trading are as follows:

Crude glycerine (see Table 6.3) procured from the soap plant or from the splitter is available for the refiner, and should comply to one of the two listed specifications. The major distinction comes from the source process, because the soap lye crude obviously contains a significant amount of salt and has a lower glycerol content.

TABLE 6.3   Composition of Crude Glycerine

	Crude from soap lye (%)	Crude from sweet water (%)
Glycerol	80–84	84–88
NaCl	8.0–9.0	—
Ash	8.5–9.5	0.5–1.0
Alkalinity	0.5–1.5	—
Acidity	—	0.01–0.05
Water	5.0–9.0	5.5–9.0
MONG[a]	0.5–2.0	0.5–1.0
TMG[b]	0.1–0.5	0.1–0.5
Nitrogen	0.01–0.04	0.02–0.04

[a]Matter organic nonglycerol.
[b]Trimethylene glycol. *Source:* Radhakrishnan (6).

USP Glycerine (US Pharmacopoeia) or CP (chemically pure) glycerine standards have been established for the highest grades of glycerine, water-white in color, and with glycerine content of not less than 99%. This product conforms to standards given in *U.S. Pharmacopeia* (USP XXII, 1990) and is used by the food and pharmaceutical industries because of its high purity. Both natural and synthetic glycerines meet these specifications, with synthetic product usually of the highest purity (and cost).

High Gravity Glycerine is a commercial grade of glycerine, near white in color, and with high glycerine content (>98.7% with specific gravity minimum of 1.2595 at 25/25°C). It conforms to Federal Specification O-G-491C and the *Standard Specification for High-Gravity Glycerine, D-1257*, issued by the ASTM.

Dynamite Glycerine meets all High Gravity Specifications, except that a darker color is allowed.

Table 6.4 illustrates USP trading specifications and Table 6.5 provides those standards for BS (British Standard) glycerol.

Test Methods

The American Oil Chemists' Society (AOCS) and other groups such as the European Pharmacopeia have published standard methods for the sampling and testing of glycerine. The AOCS tests are the most widely used standards, some of which are published in *Official Methods and Recommended Practices of the American Oil Chemists' Society* (4th Edition, 1989). Included are procedures for sampling crude product and test methods for glycerine, including apparent specific gravity at 25/25°C, moisture, color, ash, alkalinity, salt, and organic residue. Refractive Index (RI) readings at standardized temperatures can be used to estimate glycerol concentration (10). The RI (as obtained by a refractometer) is a common field located process control device.

Color is determined using the APHA scale. APHA Color is determined using AOCS Official Method Ea 9-65 which measures "color by comparison with artificial empirical standards."

TABLE 6.4 USP Refined Glycerine Specifications

	99.7% USP	99.5% USP	99.0% USP	96.0% USP
Glycerol %, minimum	99.7	99.5	99.0	96.0
Specific gravity, 25°C/25°C	1.26092	1.26073	1.25945	1.25165
Color, APHA, Hazen scale, typical	12	12	12	12
Ash %, maximum	0.01	0.01	0.01	0.01
Chlorides, ppm, maximum	10	10	10	10
Sulfate, ppm, maximum	20	20	20	20
Arsenic, ppm, maximum	1.5	1.5	1.5	1.5
Heavy metals, ppm, maximum	5	5	5	5
Chlorinated compounds, ppm, maximum	30	30	30	30
Fatty acids and esters, maximum	1.0 mL of 0.5N NaOH	1.0 mL of 0.5N NaOH	1.0 mL of 0.5N NaOH	1.0 mL of 0.5N NaOH
Acidity or alkalinity	Neutral	Neutral	Neutral	Neutral

Source: Various commercially offered USP glycerine specifications.

TABLE 6.5 BS Refined Glycerine Standards

	Chemically Pure BS 2625	Dynamite grade BS2624	Technical grade BS2623
Glycerol %, minimum	99.0	99.0	99.0
Specific gravity, 20°C/20°C	1.261 to 1.264	1.261 to 1.264	1.261 to 1.264
Color, 5-1/4" Lovibond	5.0Y, 1.2R	5.0Y, 1.2R	
Sulfated ash %, maximum	0.01	0.01	0.01
Chlorides, ppm, maximum	0	100	100
Iron, ppm, maximum	—	—	2
Arsenic, ppm, maximum	2	—	—
Lead, ppm, maximum	1	—	—
Heavy metals, ppm, maximum	5	—	—
Saponification equivalent, MEQ/100 g	0.64	0.64	-Acidity or alkalinity
MEQ/100 g	0.064	0.32	0.32
Odor	Free from odor when tested by method BS5711		

Source: British Standards Institute—Standards of 1979 and reconfirmed in 1994 (11).

Losses

Losses are a natural part of any processing plant, and control and reduction of the losses will, in part, determine the economic effectiveness of the plant's operation. All areas of the glycerine recovery plant are potential sources for loss. In storage, fermentation and poor storage practices can cause serious losses of glycerine due to the formation of trimethylene glycol, gases and acids during decomposition. In the evaporation plant, glycerine can be carried out with the vapor during evaporation and lost in the vacuum system's condensing water. In the refining plant, carryover can also be a factor, as can production of an excess amount of substandard product, or excess residual glycerine in the foots residue. Some design features which can be incorporated to reduce loss include the following:

1. Installation of entrainment separators, or "catchalls," and mist eliminator pads at the vessel vapor draw-off points. Sight glasses in the separator and/or return lines are extremely valuable to monitor the function of recovery devices.

2. Control of the vapor velocity and maintenance of an adequate disengagement space at the top of the vessel. It is important to control the liquid level in order to maintain the required vapor space.

3. Installation of a surface condenser to recover any additional condensables that could escape into the vacuum system. Not only does the entrained glycerol represent a substantial loss, but every 1% of glycerol in the barometric condenser water increases the BOD loading by around 10,000 units.

Other sources of glycerine losses include the following:

1. Glycerine residual in the extracted salt from a soap lye evaporation plant. Proper washing of the salt is required to minimize this loss, and a continuous centrifuge may be justified on the economics of glycerol recovery alone.

2. The filter cake and skimmings from the treatment plant.

3. Spent bleaching carbon which contains glycerine of the highest purity produced in the plant. When the bed is taken off line, the column should be allowed to drain and then blown with compressed air or nitrogen to extract as much free glycerine as feasible. If the carbon has lost its effectiveness, the spent material can be added to the first treatment tank, where additional glycerine will be washed from the carbon. The carbon is separated in the filter press and disposed of with the rest of the filter cake. As indicated earlier, it is important to clean the bed prior to an extended shutdown, even if the carbon is fresh.

4. Improper operation resulting in improper temperatures and reflex rates in the distillation plant. If plant yield is down while substandard and/or waste treatment loading is increased, the crude still or the deodorizer are probably not being operated correctly.

5. Residue removed from the distillation process contains a high percentage of glycerine (typically 40–60%). Much of the glycerine can be recovered from this residue by a separate distillation step (such as the foots still).

6. Spillage from all sources. Careful plant design and operation are needed to minimize spills from overfilling tanks, equipment malfunctions, and other operational errors.

7. Operating the system infrequently, with alternating periods of runtime and downtime. For that reason, the user is advised to specify and operate the plant on a continuous basis that matches the capacity of the soap or saponification plant. Obviously, this must be balanced against the cost of staffing on a continuous basis. Frequent shutdowns and startups always involve a certain period of instability during which the plant is not operating at peak efficiency. Material can be lost from opening the equipment. Where possible, the liquor should be blown into the next process vessel or carefully drained into a bucket for recovery. Any product that is recovered that is not badly contaminated can be returned to a storage tank upstream in the process. If a "substandard" glycerine tank is

provided in the refining section, it can be equipped with an easily removable cover and provide a plant-wide recovery point. Care must be taken when reintroducing this material back into the process stream, because recovered materials can contain high amounts of impurities that can upset the system.

Waste Management

The three primary waste streams from a glycerine recovery plant are skimmings and filter cake from the treatment plant, contaminants in the vacuum system condensing water, and residue (foots) from the glycerine refining plant. Filter cake discharge is typically sent to solid land fill. The cooled concentrated residue from a "foots still" or solid powder from a wiped film evaporator will typically solidify and must be disposed of as required by the local environmental authority.

Contaminants in the vacuum system's condensing water supply represent a significant problem. Obviously, the glycerine represents a loss of product which, at prices approaching $1.00 per pound, represent a serious loss of income. In addition, these contaminants will foul the condensing water supply, which is typically recirculated through a "greasy" cooling tower. A certain amount of the contaminated tower water is discharged into a plant or municipal water waste treatment facility and is replaced with makeup water to maintain an acceptable water quality level. An important alternative to consider is the inclusion of a *closed-loop* system incorporating a surface condenser cooled with *clean water* recirculated from a cooling tower. The surface condenser, usually a horizontal shell and tube heat exchanger, replaces the barometric condenser in the vacuum system and allows any condensables to be recovered without direct contact between the glycerine and the cooling water. Naturally, the initial and operating costs of the noncontact condensing system are higher than for its direct contact counterpart, and an economic evaluation is appropriate for this decision.

Ion exchange plants theoretically do not produce any waste streams, but, in practice, large amounts of acid and alkali waste water are produced when the resin beds are regenerated. In addition, the resins themselves must be periodically replaced, thus causing a disposal problem with the exhausted resins.

Usage, Applications, and Economics

Uses of Glycerine

The number of uses of glycerine is truly phenomenal. Depending on the publication surveyed, up to 1700 uses have been identified with the number of new applications growing. Its wide range of applications is in part related to a few of its key properties:
1. It is a natural product, is non-toxic, and is Generally Recognized As Safe (GRAS) for human consumption.
2. It is an excellent humectant, emulsifier, and plasticizer.

3. It is compatible with a wide variety of materials and mixes well.
4. It possesses antioxidant properties.

Below is a brief overview of some of the more prominent or interesting uses for glycerine (1,3,9).

Antifreeze—Properties include low freezing point, non-toxicity, and excellent compatibility with water and anticorrosion products. If ethylene glycol became highly regulated, the demand for glycerine for this application alone could greatly affect the glycerine market.

Cosmetics—Glycerine is widely used for cosmetic applications, acting as a bonding agent, emollient, humectant, lubricant, and solvent. It is present in skin creams and lotions, shampoos and hair conditioners, soaps and detergents.

Dental Creams—Up to 50% of typical dental creams are glycerine, which is used as humectant, to assure good dispersion, and to serve as a vehicle for dyes and flavorings. According to Joel Houston of Colin A. Houston and Associates, "Toothpaste and tobacco are examples of two (products) that won't switch to something like sorbitol. Toothpastes can be very complex systems" (1). Glycerol produces a certain taste and mouthfeel, and use of a substitute product may affect the consumer's perception of the product quality. It should be noted that certain new formulations, such as those containing baking soda, use high amounts of glycerine and are gaining popularity.

Explosives—A large amount of glycerine is consumed in the manufacture of nitroglycerine-based explosives.

Food and Beverages—Glycerine is used in a wide variety of food applications, where it serves as a solvent, carrier, emulsifier, conditioner, freeze preventer, and coating. It is used in wine, liqueurs, chewing gum base, confectioneries, and chewy bars. There are kosher uses, as well as kosher producers of glycerine.

Pharmaceuticals—Glycerine is used in several different types of pharmaceutical preparations including salves and dressings, antibiotic preparations, capsules and suppositories, as well as a vasodilator for angina pectoris (nitroglycerine).

Chemical Industries—Several important classes of resins are based on glycerine, including ester gums, phthalic acid and maleic acid resins, polyurethanes and epoxies. Urethane foams and cellophane wrap also consume great amounts of glycerine.

Tobacco—Large amounts of glycerine are consumed for use as a humectant, softening agent, and a flavor retainer.

Economics

The consumption of glycerine has remained strong over the years and trends indicate that this is not likely to end soon. While glycerine is a mature market product, the Soap & Detergent Association is one agency involved in developing new uses for the product. Edward T. Sauer, global business manager for glycerine and fatty acid chemicals at Procter & Gamble, presented the following ideas (during a March 1995 talk at Iowa State University) for potential new uses for glycerine (1):

Application	Advantage
Low fat cookies	Sweetness and water-carrying ability
Smokeless tobacco	Tobacco flavor with less secondhand smoke
Improved fabric softener	Environmentally friendly
Sports drink	Improved hydration during exercise
Deicing fluids	Environmentally friendly
Drilling fluids	Environmentally friendly

Glycerol has had a history characterized by price fluctuations. In 1995, the market price of natural 99.5% USP Glycerine in the United States was over $1.00 per pound, which is up considerably from 40 to 45 cents per pound in 1991. As the prices have fluctuated, making budgeting difficult at best for glycerine users, alternate polyols (discussed earlier in this report) have been substituted in some products. Currently, all viable alternatives to glycerine are less expensive, but because glycerine consumers are reluctant to change formulations and such, demand has been increasing. The US demand for glycerine is approximately 420 million pounds per year and has increased by 33% in the last three years alone. China and India alone have been importing around 12 million pounds annually whereas as recently as three years ago, these countries purchased virtually no glycerine. Clearly, the demand for glycerine is expanding to the point of creating a supply deficit.

Synthetic glycerine is produced in the US by only a single supplier with a total of only three producers worldwide. The strong supply of natural glycerine and rising cost of feedstock (propylene) is likely to pressure producers to revise their production. Also, some epichlorohydrin, an intermediate product, can be diverted into other products of potentially higher profitability. Finally, synthetic glycerine contains small quantities of chlorinated hydrocarbons, which has caused some concern in the health care industry.

Future Considerations

Technology

Future technologies that appear most likely to effect the production of natural glycerine center around the demand for other products derived from the triglyceride molecule. Currently, the availability of glycerine is largely determined by the demand for soaps, fatty acids, and detergent alcohols. A prime example demonstrating the tie of glycerine's availability to demand for other products is with the production of biodiesel. Because its production has not been increasing as earlier forecasted, glycerine supply has not grown as predicted. Should this situation change, and biodiesel become a significant fuel replacement, large amounts of high quality glycerine will become available. This alone could actually result in an oversupply of glycerine. Technological developments may make glycerine recovery from nontraditional sources a reality. Some technologies that show promise for use in glycerine recovery include ultrafiltration, reverse osmosis, and ion exchange. As these processes become more effective and efficient, and as economic conditions change, some of these processes may come into common use. Because much of this potentially recoverable product is currently lost in solid and liquid waste streams, economic purification and concentration of this material could result in favorable environmental considerations as well.

An interesting source of glycerine is from purification of ethanol fermentation stillbottoms by chromatographic separation (ADSEP™) coupled with an ion exchange process. In a proprietary process proposed by IWT/US Filter, raw stillage is first clarified by filtration through a membrane, concentrated (by evaporation), and then fed to their ADSEP™ chromatography process where the glycerine is separated. The glycerine is then further purified by ion exchange, followed by traditional concentration and distillation (12). Another interesting possibility is recovery of glycerol from the acid water stream of an edible oil refinery acidulation system. As tocopherol, sterol, and fatty acid fractionation and recovery from edible oil deodorizer distillate streams become more widespread, glycerol purification may be a potential revenue source. Clearly, technological developments will affect the availability of glycerol.

In addition to technological changes, the future economics of glycerine worldwide will change due to fundamental changes on the supply side (not even considering changes in the demand side). Animal fat is a significant raw material for glycerine production, both for fat splitting and saponification. Recent health trends emphasizing lower fat content in meats has caused animal producers to develop leaner meats, reducing the supply of animal fat. Other users of fat will add additional price pressures. In addition, the soap and detergent industry is expected to increase its usage of glycerine as a nonionic surfactant. An increasing number of bar soaps using more surfactants and less fatty acids from triglycerides will also have an impact on the supply of glycerol.

Summary

Glycerine is an important ingredient for many everyday consumer health products and for a host of industrial applications. Because natural glycerine is a by-product of other processes, its supply and demand curve is at times skewed by other forces. However, both the supply and demand are expected to grow annually by double-digit percentages for the foreseeable future. The market will continue to provide a dynamic environment for both the producer and supplier.

Acknowledgments

We would like to express our appreciation to the companies and individuals listed in the text who have provided graphic material used in this chapter and especially to the following organizations and individuals for their cooperation and assistance in preparing this presentation: Special mention is given to our good friend, Luis Spitz, of L. Spitz, Inc., for his tremendous effort in assisting us in this endeavor.

American Oil Chemists Society (AOCS), Champaign, IL
LCI Corporation, Process Division, Charlotte, NC
Guinard Centrifugation, Saint-Cloud, France
Robert Salazar, National Chem Industries, Houston, TX
Fernando Cardona, INCOSERI, Guatemala City, Guatemala
Hernan Paredes, Crown Iron Works Company, San Pedro Sula, Honduras
Buss, AG, Basel, Switzerland
Mazzoni LB S.p.A., Busto Arsizio, Italy

References

1. *INFORM 6*:1109 (1995).
2. Lamborn, L.L., *Modern Soaps, Candles, and Glycerine,* D. Van Nostrand Co, New York, 1906, p. 542.
3. Neuman, A.A., *Glycerine,* C.R.C. Press, Cleveland, 1968.
4. Woollatt, E., *The Manufacture of Soaps, Other Detergents, and Glycerine,* Ellis Horwood Ltd, Chichester, 1985.
5. Wurster, O.H., The Recovery of Crude Glycerine, *Oil & Soap 13*:246–253, 283–286, 1936.
6. Radhakrishnan, K.P., in *Soap Technology for the 1990s: Glycerine Processing from Spent Lyes and Sweet Water,* Spitz, L., ed., AOCS, Champaign, 1990, p. 128–153.
7. Swenson Process Equipment, Sales Bulletin SW-200R.
8. Jungermann, E., and N.O.V. Sonntag, *Glycerine: A Key Cosmetic Ingredient.*
9. *Glycerine: An Overview,* The Soap and Detergent Association, New York, 1990.
10. Miner, C.S. and N.N. Dalton, eds., *Glycerine,* Reinhold Publishing, New York, 1953, Chapter 6.
11. British Standards Institute—Standards of 1979 and reconfirmed in 1994.
12. Burris, B.D., *Recovery of Glycerine from Still Bottoms,* Illinois Water Treatment, 1987.

Chapter 7

Soap Drying Systems

Luis Spitz,[a] Piero Verde,[b,*] and Roberto Ferrari[b]

[a]L. Spitz, Inc., Skokie, Illinois, U.S.A.

[b]Mazzoni LB, S.p.A., Busto Arsizio, Italy

Soap Drying Systems

The objective of this chapter is to present an updated overview of toilet and laundry soap drying systems and equipment. Easy-to-interpret, comprehensive three-dimensional drawings are used to serve as practical guides for the selection of various drying processing systems and their components.

Today's soap drying systems are derived from the "vacuum spray drying" process patented by Mazzoni (1, 2) and introduced to the world market following World War II. The use of vacuum as a medium to simultaneously dry and cool the liquid neat soap is responsible for the beginning of the modernization of soap manufacturing. Plants were designed to convert liquid neat soap into dry soap pellets and could also produce continuously extruded laundry soap bars. The elimination of the laborious and costly laundry soap framing process was a revolutionary step in soap manufacturing.

Before the invention of vacuum spray drying, hot-air cabinet dryers were used; the neat soap was fed onto a steam-heated chill roll and then onto a steel belt of a long and large cabinet in which hot air completed the drying of the soap in flake form.

In 1955 tubular dryers (3) and in the mid-1960s plate dryers (4) and Miag's double expansion dryer (5) were patented. The Miag dryer consisted of two shell-and-tube heat exchangers, two atmospheric spray chambers, and a chill roll. A few tubular, plate, and double expansion dryers were installed, but over the years they did not gain acceptance.

Today vacuum spray dryers and combination atmospheric and vacuum spray dryers (6) are the most widely used systems worldwide. Their flexibility in handling many product formulations at different moisture levels, very simple operation, and minimum maintenance requirements have not been surpassed.

Toilet and Laundry Soap Drying

The principle of vacuum spray drying of liquid toilet soap and soap/synthetic (combo) products into a dry base in pellet form at various moisture levels are shown in Figs. 7.1 and 7.2. The figures illustrate how the vacuum spray drying process is affected and controlled by key variables, such as steam and soap pressure and temperatures, spray nozzle pressure, operating vacuum, dry soap temperature leaving the system, and the heating and evaporating phase distribution.

*Correspondence should be addressed to Piero Verde, Colgate-Palmolive S.p.A., Viale Palmolive 18, 00042 Anzio, Italy.

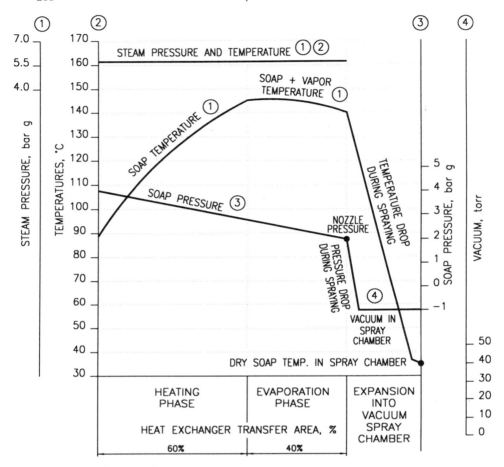

Fig. 7.1. Toilet soap drying.

Toilet Soap Vacuum Spray Dryer (Fig. 7.3)

The function of toilet soap vacuum spray dryers is to convert liquid neat soap into dry soap by removing moisture. Vacuum is used as a medium to obtain soap drying and cooling simultaneously.

Standard, superfatted, and translucent bases of different moisture levels can be produced with ease by only changing operating conditions. The moisture removal range is from 9% to 33%. The preheated soap and vapor mixture is sprayed into a vacuum chamber that has a rotating scraper assembly and a spray nozzle. As the vapors are flashed off because of the vacuum, the soap is dried to its final moisture. The rotating scrapers remove the dried and cooled soap, in flake form, from the chamber walls. The unevenly shaped dry soap flakes fall onto plodders to be pelletized. During spraying, soap fines are formed and are carried off with the high-

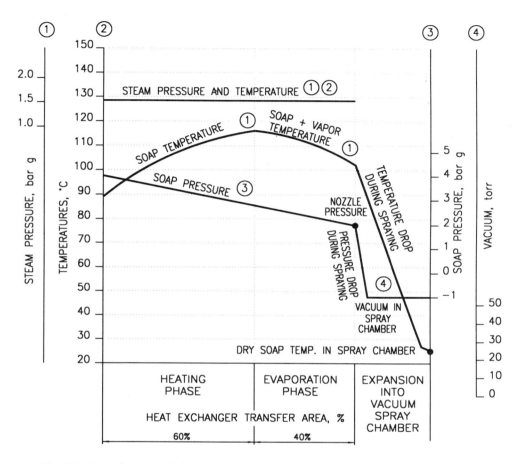

Fig. 7.2. Laundry soap drying.

velocity vapors into cyclone-type fines recovery systems before being condensed in a barometric or a surface condenser. A vacuum pump, steam ejector, or combination of both maintains the vacuum and removes the noncondensables.

Toilet Soap Combination Atmospheric/Vacuum Spray Dryer (Fig. 7.4)

The combination two-stage atmospheric and vacuum spray dryers are used for drying not only toilet soaps but also the high initial moisture content (40%–45%) combo soap/synthetic mixtures, which have to be dried to low (8%–9%) final moisture content.

The moisture removal range is from 8% to 50%. To remain in pumpable condition to the vacuum spray drying stage where the final moisture level is reached, the product in the atmospheric stage is flashed dried to 24%–25% moisture content.

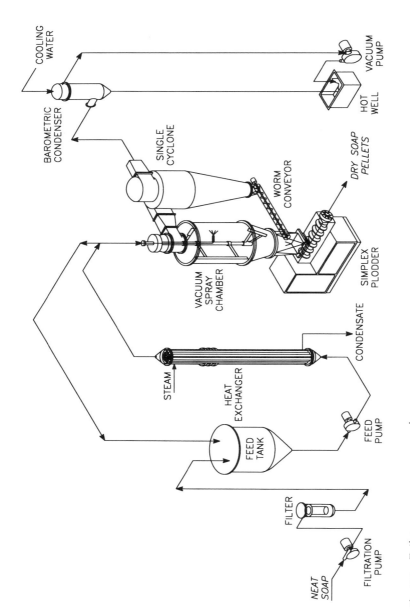

Fig. 7.3. Toilet soap vacuum spray dryer.

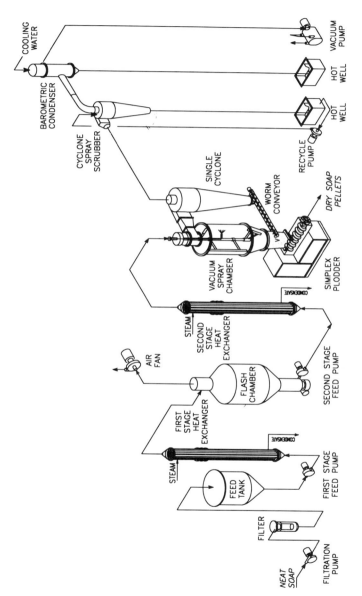

Fig. 7.4. Toilet soap atmospheric/vacuum spray dryer.

Laundry Soap Vacuum Spray Dryer (Fig. 7.5)

The plant shown in Fig. 7.5 represents a standard configuration for the production of laundry soaps directly in bar form. Pure or filled opaque and translucent laundry bars of different moisture levels are produced with these plants (Table 7.1). A crutcher must be used for the thorough mixing of additives and filling solutions into the neat soap before drying. The moisture removal range is from 21% to 33%.

Soap Flaking System (Fig. 7.6)

Chill rolls are designed to solidify molten materials into thin flakes of varied thickness. In the past, before the invention of vacuum spray drying, they were used with the band (also called apron) soap drying systems. Today, with the increased popularity of soap/synthetic (combo) and fully synthetic bars, the use of chill rolls has become relevant again for the production of the combo and syndet base materials. Chill roll design and performance data have not been published. The Buflovak soap flaking system with data shown in Fig. 7.6 is presented for the first time. A plodder is usually used to pelletize the flaked product, leaving the chill roll. Even after 90 years of experience in flaking numerous products, Buflovak still recommends laboratory testing for each product to accurately size a chill roll.

Products and Operating Conditions

Table 7.1 summarizes the operating parameters of vacuum spray dryers for the production of the most important toilet soap bases (in pellet form) and laundry soaps (in bar form). A plodder selection guide for the dryers is also included.

Drying System Components

Filtration Pumps and Filters

Continuous saponification systems produce clean neat soap, whereas soap made via a kettle, or semi-boiled process, usually has impurities. Filters protect the feed pump, heat exchanger, and spray nozzles from damage due to foreign bodies. Hollow disc type filtration pumps can handle impurities and still provide sufficient soap pumping pressures (3–4 bar g) to the dryer. Figure 7.7 illustrates a complete filtration section with pump, dual filters, and an automatic control system. Basket filters with drilled steel plate filtering element with or without screens are used in horizontal or vertical position.

Typical filtering elements are as follows:

- For toilet soap: drilled plate with 2 mm diameter holes, covered with 30–50 wire mesh screen
- For laundry soap: drilled plate with 0.5–0.8 mm holes
- A fast changeover standby filter is installed for continuous operations. Filter clogging is detected by a pressure differential transmitter, and the signal is used to stop the filtration pump or to switch over to the clean second filter. The use of bag filters is recommended when polyethylene contamination from fats occurs.

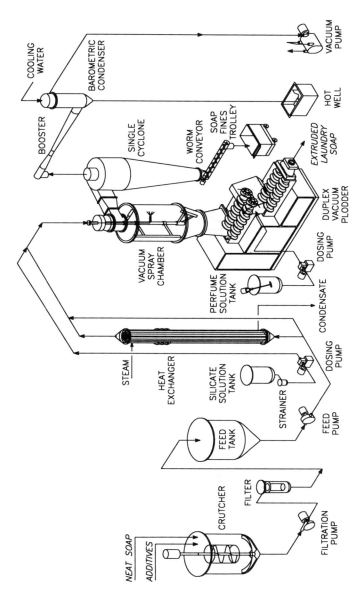

Fig. 7.5. Laundry soap vacuum spray dryer.

TABLE 7.1

Spray Dryers: Typical Products and Operating Data

Products	Toilet Soap Bases (Pellets)					Laundry Soap Bars				
	Standard	Superfatted	Translucent	Combo		Opaque Pure	Opaque Pure	Opaque Filled	Opaque Filled	Opaque Translucent
Total fatty acid content, % TFA	78–84	80–84	68–72	—		68–72	62–63	55	45–50	72
Moisture content, % H₂O	14–8	12–10	26–22	25–26	8–10	25–21	32–31	32	34–32	21
Operating Conditions				Atm. Stage	Vacuum Stage					
Heat exchanger steam pressure, bar g	5–6		0.5–1.0	1	4–5	0–2	0	0	0	3–4
Soap temperature at heat exchanger outlet, °C	140–150		105–115	110–115	120–130	110–115	85–90	85–90	85–90	120–125
Absolute pressure (vacuum), torr	30–40		20–25	Atm.	30–40	20–25	15–20	10–15	10–15	25–30

Plodder selection guide:

Toilet soap dryers: standard, superfatted, and combo pellets, one or two plodders. Dryer capacity determines if the plodder is a single or twin-worm.

Laundry soap dryers: pure bars, two single plodders or twin-worm plodder; filled bars, two twin-worm plodders; translucent bars, three plodders: first two are twin-worm; the last one (extrusion stage) is a single worm.

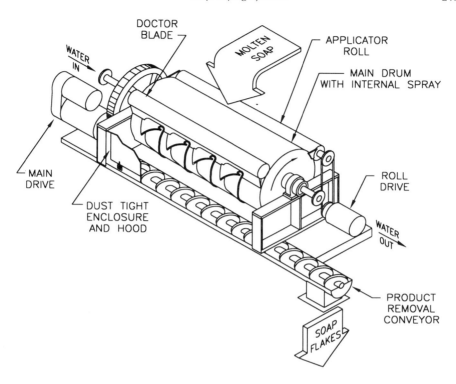

DRUM SIZE (meter)	CAPACITY (Kg/h)	POWER (HP)	COOLANT (m³/h)
1.5 x 3.6	3000–4200	12	68
1.5 x 2.8	2200–3200	9	50
1.2 x 2.4	1600–2200	7	45
1.2 x 1.2	800–1200	5	28
0.6 x 2.4	400–1000	3	31

TYPICAL OPERATING CONDITIONS:

FEED TEMPERATURE: 105–115°C COOLANT TEMPERATURE: −5 TO +15°C
FLAKE TEMPERATURE: 25–38°C DRUM SPEED: 3–7 RPM
FLAKE THICKNESS: 0.5–1.0 mm ROLL TEMPERATURE & SPEED: VARIABLE

Fig. 7.6. Buflovak soap flaking system. (Courtesy of Buffalo Technologies Corp.)

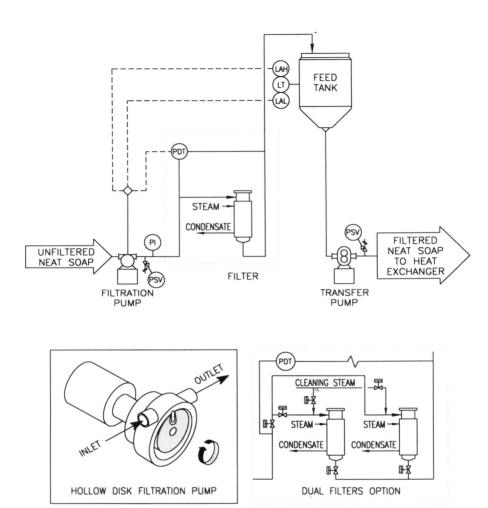

Fig. 7.7.   Filtering section: filtration pump automatic control.

Feed Pumps

Figure 7.8 illustrates the most widely used soap feed pumps. External and internal gear pumps provide excellent positive flow with nonpulsating discharge. Internal gear pumps offer the advantage of having only a single shaft and one mechanical seal or stuffing box. External gear pumps can be made with spur, helical, or fishbone gears.

Lobe pumps are also excellent for pumping *shear* sensitive fluids such as neat soap. They are available in bilobe or trilobe design. A particular feature of lobe pumps is that by removing the front cover, the rotors and seals can be accessed easily without disconnecting the process line.

SLIDING VANE

LOBE

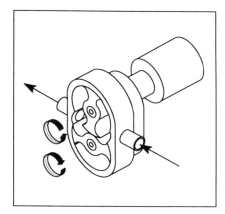

INTERNAL GEAR

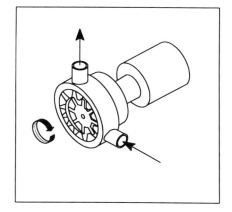

EXTERNAL GEAR

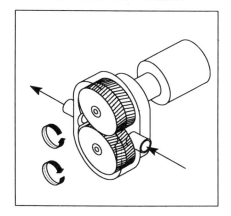

Fig. 7.8. Feed pumps.

For soap feed pump sizing and selection, note that the viscosity of 30%–32% moisture neat soap is in the range of 2000–3000 cp, and at 1 m/s pumping speed the viscosity is reduced to 50–100 cp.

Heat Exchangers (Fig. 7.9)

Shell-and-Tube

Single-pass shell-and-tube heat exchangers for soap applications are simple in design. A number of straight tubes are sealed between two perforated tubesheets. The tubes, plates, and cones are made of 304, 316, or 316L AISI stainless steel. To avoid the problem of stress corrosion cracking for soap formulations with higher

SHELL–AND–TUBE

PLATE–AND–FRAME

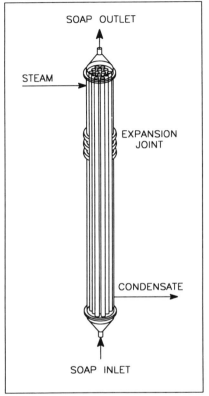

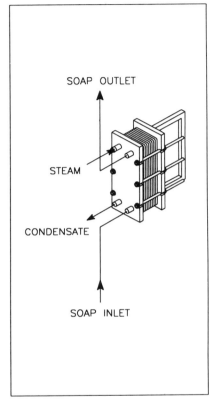

Fig. 7.9. Heat exchangers.

than 0.6%–0.7% salt content in the neat soap, high-nickel-content materials such as Alloy 825 (Unified Numbering System UNS-N08825) are recommended. Expansion joints should be used to prevent mechanical stresses. The optimum diameter of the tubes is 12 mm inside and 14 mm outside.

Shell-and-tube exchangers have the lowest capital cost per square meter of heat transfer area and can be installed in various configurations:

• A single vertically placed heat exchanger for any plant capacity
• Two equally sized heat exchangers in series
• Two different-size units, in series, to optimize preevaporation ratio and moisture control

Plate-and-Frame

Plate-and-frame (PAF) heat exchangers consist of a number of thin, corrugated metal plates and gaskets clamped together and fitted into a frame. The turbulence of the flowing material induced by the surface design of the plates results in three to five times higher heat transfer coefficient (*U;* see the appendix to this chapter). The limited use for soap drying of PAF exchangers has been due to the gasket materials and their replacement costs.

Today the use of EPDM (ethylene propylene diene monomer) gaskets designed for 200°C maximum temperature and 25 bar pressure has eliminated the previous problem that existed when less resistant gaskets were used. Drying of high moisture content, heat sensitive, soap/synthetic (combo) products requires high preevaporation rates in the atmospheric drying stage. PAFs provide for this application homogeneous two-phase flow with a higher steam rate. PAF heat exchangers cost more per square meter of heat transfer area than shell-and-tube units, but their high heat transfer rate, compact size, and use for special applications can make them cost-effective.

Vacuum Spray Chambers

Two types of vacuum spray chambers are used for soap drying plants (Fig. 7.10). The most widely used "rotating nozzle" type is offered by all of the manufacturers of vacuum spray drying plants. Weber & Seelander also offers the "rotating lower cone" (fixed nozzle) version. This is a more expensive solution and has found limited acceptance.

Soap Fines Recovery Systems

Four different cyclone-type recuperators are used to recover and recycle, or not to recycle, the soap fines produced in the vacuum spray chamber (Fig. 7.11).

Dual Cyclone

Two cyclones connected in series is the commonly used traditional system for most small and medium-size dryers. The soap fines recovered in the second cyclone are fed with a worm conveyor into the first cyclone from where another worm conveyor feeds it into the vacuum spray chamber. Both worm conveyors are fitted with bridge breakers.

Multicyclone

For large-capacity (over 3,000 kg/h) dryers, the multicyclone system is suggested due to its higher separation efficiency. A large single cyclone is connected to a group of four smaller cyclones. Worm conveyors feed the recovered fines into the vacuum spray chamber.

Single Cyclone

The latest development is the application of a single large-size cyclone, preferably in combination with a wet scrubber located downstream. The actual overall efficiency of this system is yet to be determined for various plant capacities drying different products and at various moisture levels.

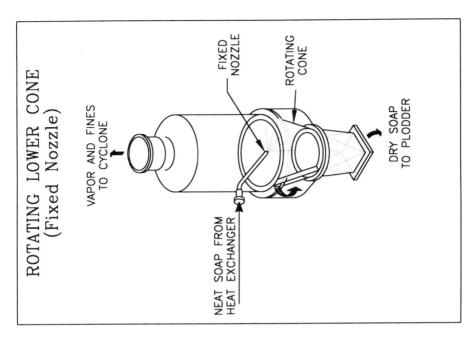

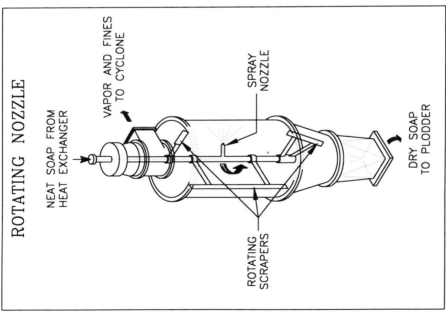

Fig. 7.10. Vacuum spray chambers.

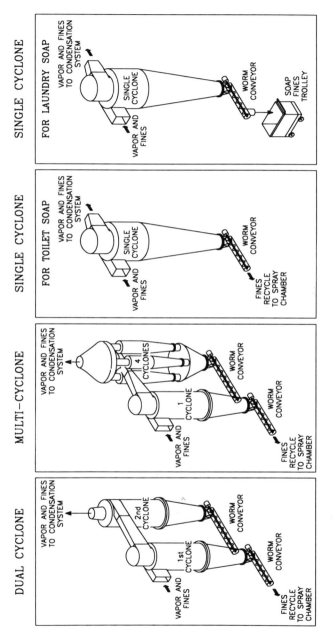

Fig. 7.11. Soap fines recovery systems.

If the performance matches and/or exceeds the more expensive and complex multicyclone system, the single cyclone could become a very attractive new standard for most soap dryers. In all cases the single cyclone is recommended for laundry soap applications that produce minimal or no soap fines. Figure 7.10 illustrates only soap fines not recycled into the spray chamber but rather discharged and collected outside for laundry soaps, but this application is also used for toilet soaps.

In this case the second single cyclone, or the four multicyclones, still feed continuously the fines into the first cyclone of each system from where the fines are discharged with the worm extruder conveyor. Because handling and reusing the fines are major problems, the noncontinuous recycle systems for toilet soaps drying are seldom used.

Scrubber

The installation of a centrifugal wet scrubber separator between a soap fines recovery group and a barometric or surface condenser captures about 50% of the fines passing through it. Approximately 0.3 m^3 of water per 1,000 kg/h of dried soap is sprayed on top of the scrubber. The water with the fines is collected in a separate hot well from where it is recycled to the scrubber. Figure 7.12 shows the scrubber installation and provides comparative scrubber performance data with a barometric and a surface condenser.

Vacuum-Producing Systems

The vacuum-producing systems for soap dryers consist of steam condensation equipment (barometric and surface condensers and cooling towers systems) and vacuum-producing equipment (vacuum pumps and steam jet ejectors).

Steam Condensation Equipment

The sole purpose of a condenser is to continuously convert the vapors generated during the drying process back into water. Two basic types of condensers are used: barometric and surface condensers.

Barometric Condensers
Barometric condensers are also called direct contact condensers because they condense the vapors by direct contact with the cooling water. The condensed vapors, the cooling water used for condensation, and the soap fines are discharged together into a hot well (Fig. 7.13). When the cooling water temperature is higher than 27–28°C, a steam booster must be used to compress the vapors at a residual pressure corresponding to 40–42°C.

Surface Condensers
Surface condensers are shell-and-tube heat exchangers. The vapors to be condensed flow inside the tubes and the cooling water on the shell side. The cooling water is

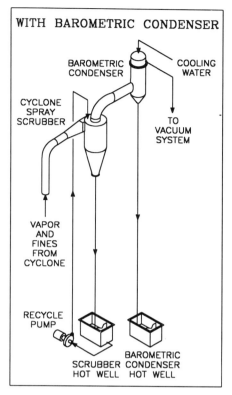

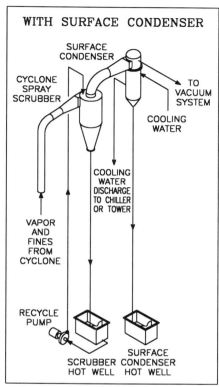

	BAROMETRIC CONDENSER		SURFACE CONDENSER		
	SCRUBBER DOWNLEG	B.C. DOWNLEG	SCRUBBER DOWNLEG	S.C. DOWNLEG	COOLING WATER
WATER FLOW (m³/h)	0.3	40	0.3	0.3	40
FINES	0.04%	2.8 ppm	0.04%	0.04%	—

BASIS: 1000 Kg/h OF 12% MOISTURE DRY SOAP
COOLING WATER ΔT = 4°C

Fig. 7.12. Scrubber.

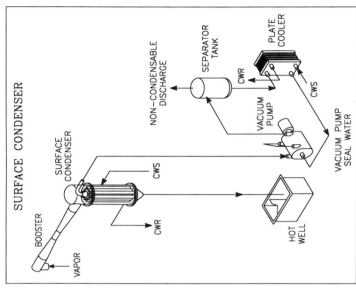

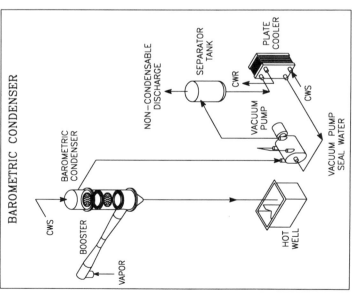

Fig. 7.13. Vapor condensation and vacuum formation.

not contaminated with the soap fines because they never come in contact with each other (Fig. 7.13). The condensed vapors that carry the final traces of soap fines are discharged into the hot well and the clean water, free from soap fines, is recycled. To limit the size and cost of a surface condenser, the cooling water temperature should not exceed 20°C. The cooling water can be efficiently used in closed circuit using a cooling tower in conjunction with a water chiller. Surface condensers can also be mounted at a lower height than barometric condensers using a collecting vessel under vacuum and a pump.

Vacuum-Producing Equipment

Vacuum Pumps

> Mechanical piston pumps were used extensively years ago with soap dryers. Now they are seldom applied due to their high cost.
>
> Today liquid ring pumps are the preferred choice. They are available on the market as a skid-mounted complete group with gas/liquid separators and plate coolers, or shell-and-tube heat exchangers to maintain the seal water temperature at 18°C maximum (Fig. 7.14).

Steam Jet Ejectors
Steam jet ejectors are very simple devices consisting of a nozzle, a mixing chamber, and a diffuser. Using motive steam, and according to the required operating vacuum level in the dryer, a single-stage or multistage ejector group is used (Fig. 7.15).

Cooling Tower Systems
Figure 7.16 shows two cooling tower system configurations for cooling the hot condensing water from the hot well and returning it to the barometric condenser.

Solid and Liquid Additive Systems

Soap formulations are becoming more complex, and various liquid and solid additives are added to enhance their performance characteristics. Figure 7.17 illustrates two specific additive systems. Other designs, dependent on the type and quantity of additives, their ease of mixing, and heat stability, are also available.

Plant Automation

Instrumentation

Electronic instrumentation is now standard for soap drying installations. The majority of new plants are computerized and existing ones are converted to computer control.

Besides computer control, a modern drying plant should include the following main instruments:

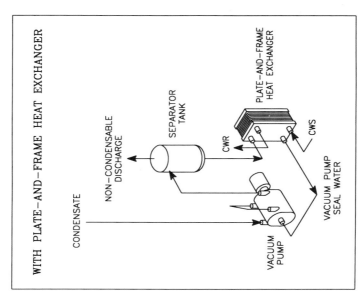

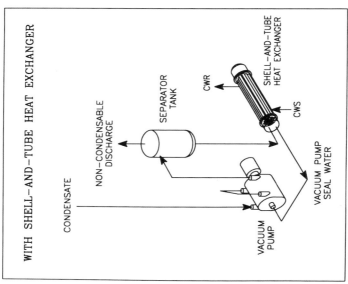

Fig. 7.14. Vacuum-forming systems.

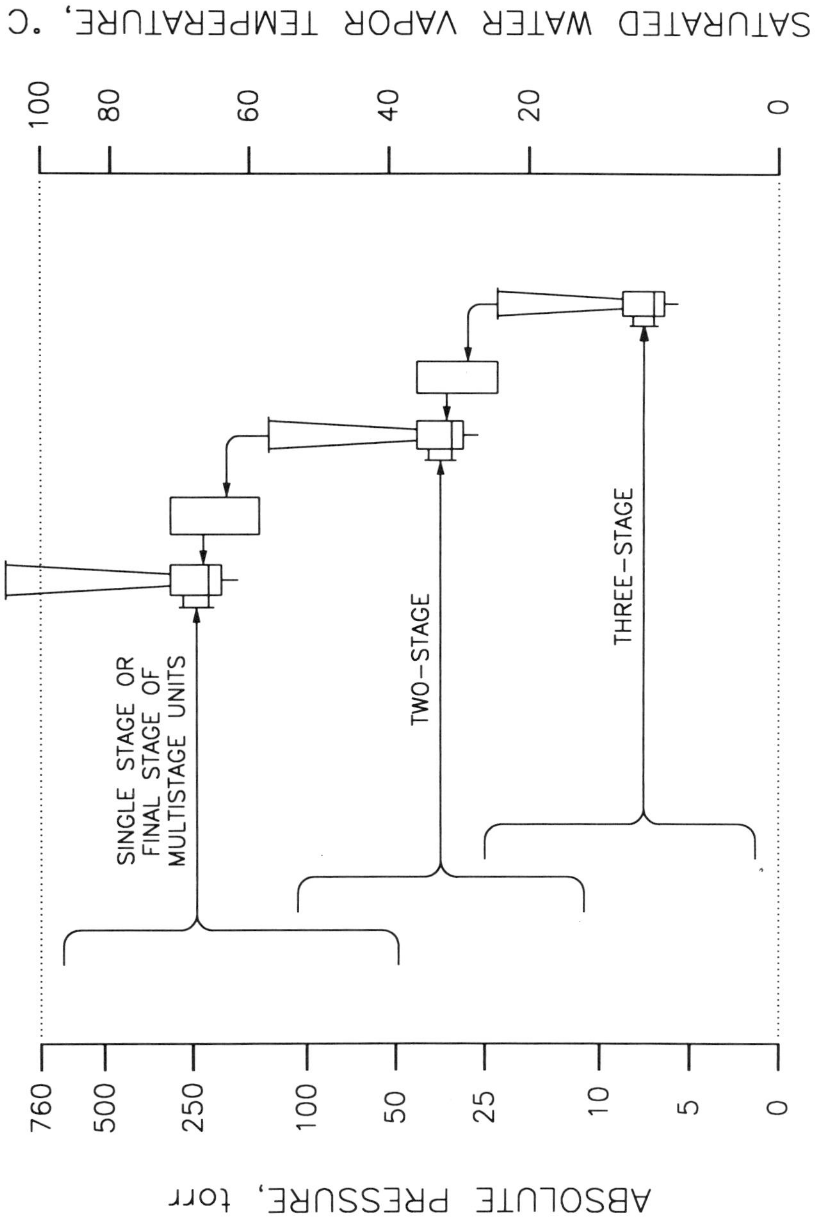

Fig. 7.15. Steam jet vacuum systems.

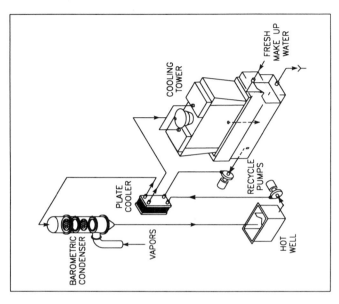

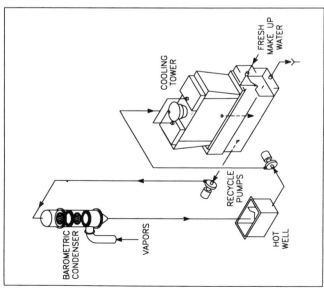

Fig. 7.16. Cooling tower systems.

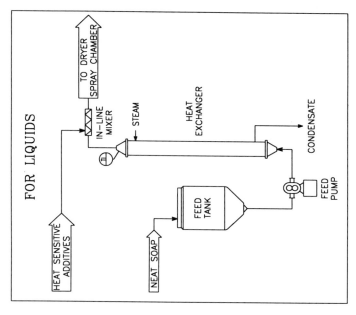

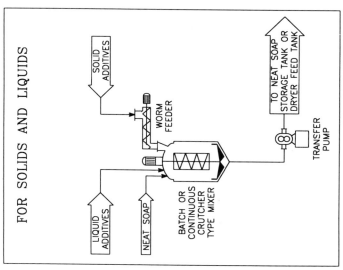

Fig. 7.17. Additive systems.

- Feed tank level control
- Soap flow control loop using a magnetic or mass flow meter operating via computer control systems (CCS) or single-loop controller for feed pump speeds
- Steam pressure control loop (independent for any single heat exchanger used)
- Vacuum chamber residual pressure (torr) transmitter
- Moisture control
- Vacuum control loop operating on booster or cooling water and air bleeding valve

Safety interlocks are used for all motors, man-doors, high soap level in chamber, and automatic valves. A soap recycle mode is used for efficient startup, shutdown, or emergencies, using automatic or a computerized set of valves that allow the operator to pump the soap from the heat exchanger back to the feed tank. Once the proper temperature is reached, the soap is fed to the spray chamber. Steam cleaning of all the soap pipes with automatic or computerized steam injection valves is part of the overall system. Figure 7.18 shows the flow diagram of a complete, fully instrumented, automated dryer with dual filters, booster with bypass, and moisture control system.

Moisture Control System

Final dry soap pellet moisture is the most important parameter to control for optimum dryer operation and subsequent bar soap formulation and efficient finishing.

Moisture Control System's MicroQuad 8000 instrument is illustrated in Fig. 7.19. The sensor has to view a uniform bed of pellets that must be placed on a short belt conveyor in front of the plodder. The sensor illuminates the pellets with near-infrared light measuring the amount of reflected light at a wavelength that water absorbs. The wetter the soap pellets, the less light is reflected. The processor converts the reflectance to percentage of moisture and outputs it to a digital display sending a 4-20 mA signal to a PID controller, which acts on a steam pressure automatic control valve to the heat exchanger. Minor changes in steam pressure give very precise control of the soap pellet moisture.

Computerization

Computer control systems (CCSs) are applied for soap dryer process control and alarms. CCSs based on hardware consisting in PC and PLC are favored because of lower investment costs and similar capability over more costly distributed control systems (DCSs), which are more suitable for large plants. Two main levels of computerization can be used:

- *Basic level:* CCS controls all process parameters by single-loop instruments, all plant motors, and a few automatic on-off valves.
- *High level:* CCS controls are all the on-off valves needed for startup/shutdown operated from the control station. All the procedures are automated.

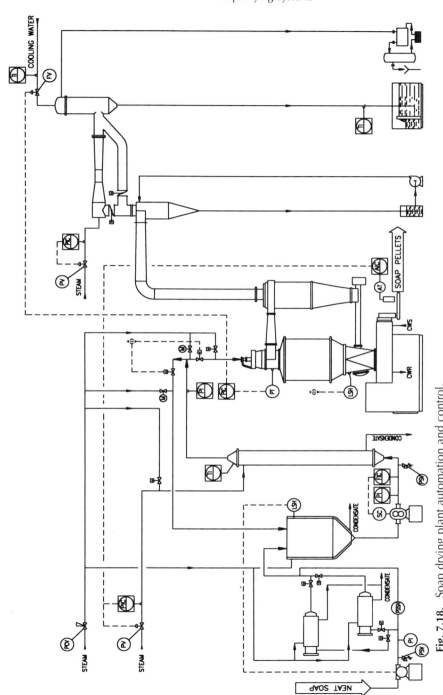

Fig. 7.18. Soap drying plant automation and control.

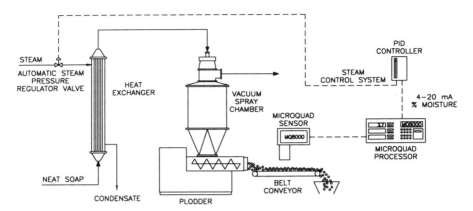

Fig. 7.19. Moisture control system. (Courtesy of Moisture Systems Corporation.)

CCS memorizes production recipes and operating conditions, totalizes neat soap and final product quantities, and records process parameter trends. It also performs emergency procedures and alarm detection.

Appendix

We have decided to include in this appendix the following key helpful items related to vacuum spray drying technical matters:

- Definitions and terminology relating to drying and to heat transfer
- Material balance calculations (Eq. 1) and graphs for liquid neat soap to dry toilet soap pellets (Fig. 7A.1) and laundry soap bar conversions (Fig. 7A.2)
- Formulas for calculating the amount of water evaporated (flashed off) under vacuum, preevaporated in the heat exchanger (Eqs. 2 and 3), and a graph illustrating toilet soap dried to different moisture levels (Fig. 7A.3)
- The well-known formulas for heat exchanger duty (Eqs. 4 and 5), and for heat transfer coefficient calculation for sizing a heat exchanger (Eq. 6).
- Overall material balances for toilet and "combo" product drying (Fig. 7A.4 and Fig. 7A.5)
- A guide to determine the actual operation vacuum (absolute pressure) and the relationship to the condensed water discharge temperature (downleg temperature) into the hot well in a drying system with a barometric or a surface condenser (Table 7A.1)
- Total fatty matter (TFM) and moisture content (%) calculation and table for toilet soaps (Table 7A.2)

Definitions and Terminology

Density: The density or specific weight of a fluid is its weight per unit of volume. The density of common neat soap base is 960 kg/m³.

Specific gravity: The specific gravity of a fluid is the ratio of its density to the density of water (dimensionless).

Vapor pressure: The vapor pressure of a liquid is the pressure (at a given temperature) at which a liquid will change to a vapor.

Viscosity (μ): The viscosity of a liquid is a measure of its tendency to resist a shearing force. Neat soap behaves as a non-Newtonian fluid exhibiting a nonlinear shear stress–shear rate behavior. Its viscosity decreases with increasing shear rate (velocity).

Specific heat (C_p): The specific heat of a substance is the amount of heat necessary to raise the temperature of a unit mass by one degree. The following formula is used to calculate the amount of heat gained (lost) by a given mass due to a change in temperature:

$$Q = m \bullet C_p \bullet \Delta T$$

where Q is the amount of heat (Kcal), m is mass (kg), C_p is the specific heat (Kcal/kg $\bullet$ °C) and ΔT is the change in temperature (°C). The specific heat is calculated according to the water content in the neat soap by the formula

$$C_p = 0.6 \times \% \text{ anhydrous soap} + 1 \times \% \text{ H}_2\text{O}$$

Thermal conductivity (k): Its value depends on the water content and its expressed by the formula

$$k = (0.58 \bullet W) \bullet 0.9 + (0.15 \bullet S) \bullet 0.9$$

where

$S = \%$ of soap and $W = \%$ of water. K is expressed in Kcal/m $\bullet$ °C.

Sensible heat: The sensitive heat (Kcal/kg) is the heat that produces a temperature rise in a fluid. For example, it takes 100 Kcal to name the temperature of 1 kg of water from 0°C to 100°C.

Latent heat of vaporization: The latent heat of vaporization (Kcal/kg) is the heat that produces a change of state without a change in temperature. For example, it requires 540 Kcal to convert 1 kg of water at 100°C to 1 kg of steam at 100°C.

Material Balances

In the soap industry, the production of soap dryers is referred to the plant output at a given moisture. The quantity of incoming liquid neat and resulting quantity of finished product at various moisture levels is given by Eq. (1), and it is illustrated for toilet and laundry soaps in Figs. 7A.1 and 7A.2.

$$G_{in} = G_{out} (100 - M_{out}) / (100 - M_{in}) \tag{1}$$

where

G_{in} = incoming neat soap quantity (kg)

G_{out} = outgoing dry soap (kg/h)

M_{in} = incoming neat soap moisture (%)

M_{out} = outgoing dry soap moisture (%)

The quantity of water evaporated by vacuum expansion is

$$W_v = G_{in} \bullet Cp_{he} \bullet (T_{he} - T_{vc}) / LH_v \tag{2}$$

where

W_v = quantity of water evaporated by vacuum expansion (kg/h)

LH_v = latent heat of water at vacuum chamber conditions (Kcal/kg)

$C_{p\,he}$ = specific heat of soap at heat exchanger outlet (Kcal/kg • °C)

T_{he} = temperature of soap at heat exchanger outlet (°C)

T_v = temperature of soap in vacuum chamber (°C)

The total amount of water to evaporate is

$$G_{total} = G_{in} = G_{out}$$

The water to be preevaporated in the heat exchanger is

$$W_{he} = G_{total} - W_v \tag{3}$$

The heat exchanger duty is the sum of the following two partial duties:

$$\text{Sensible heat} = G_{in} C_p (T_{he} - T_{in}) \tag{4}$$
$$\text{Latent heat} = W_{he} LH_{he} \tag{5}$$

where

LH_{he} = latent heat of water in the heat exchanger (Kcal/kg).

The amount of water evaporated (flashed) by expansion into the vacuum spray chamber and the quantity preevaporated in the heat exchanger, per Eqs. (2) and (3), is illustrated in Fig. 7A.3 for 31.5% moisture content neat soap and dry toilet soap pellets at different moisture levels.

The overall material balance for dry toilet soap pellets produced from heat soap with a single stage vacuum spray dryer is summarized in Fig. 7A.4. "Combo" pellets dried from a high moisture (45%) "combo" base with a combination atmospheric vacuum spray dryer are shown in Fig. 7A.5.

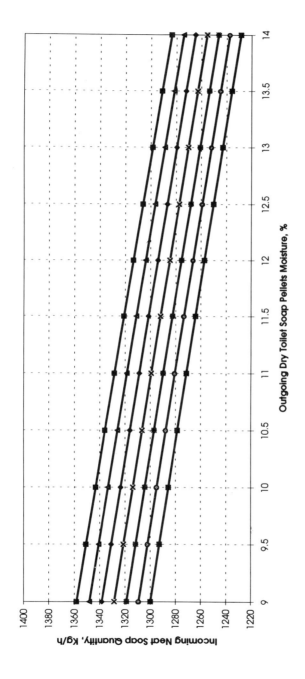

Fig. 7A.1. Toilet soap mass balance per 1,000 kg/h of dried product.

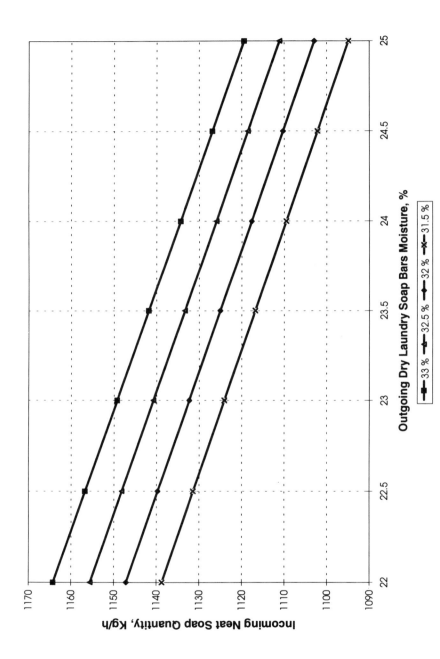

Fig. 7A.2.   Laundry soap mass balance per 1,000 kg/h of dried product.

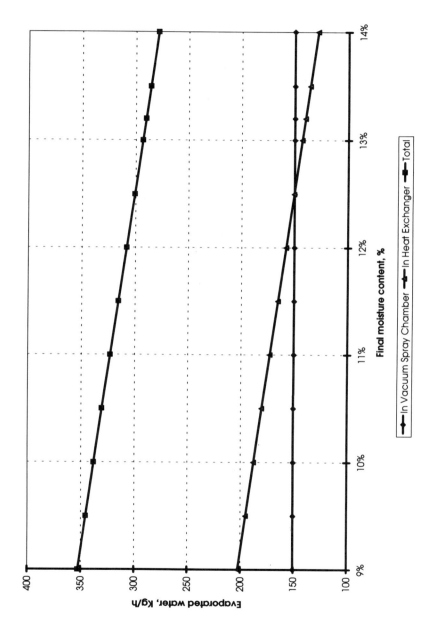

Fig. 7A.3. Evaporation distribution of toilet soap.

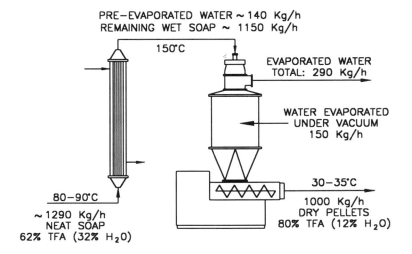

Fig. 7A.4. Vacuum spray dryer material balance for toilet soap drying.

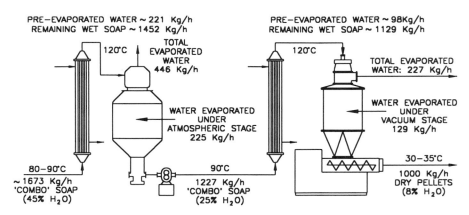

Fig. 7A.5. Atmospheric/vacuum spray dryer material balance for combo soap drying.

Soap behavior is pseudoplastic, and its viscosity range under shear is from 50 to 100 cp at working temperatures in the heat exchanger. The global heat transfer rate, tube side, is calculated by the following formulas:

Reynolds number: $Re = DVp/\mu$

Nusselt number: $Nu = hd/k$

Prandtl number: $Pr = Cp\mu/k$

Using the formula

$$(hd/k) = 1.86[(Re)(Pr)(d/l)]^{1/3} (\mu/\mu_p)^{0.14}$$

The value of *h* can be determined and substituted in

$$U = 1 \, (1/h + x/k_s + 1/h_v)$$

obtaining the overall heat transfer rate, where

D	=	tube diameter (mm)
V	=	velocity of soap flow (m/s)
p	=	density (kg/m^3)
μ, μ_p	=	viscosity (cp $\times$ 3.6)
d	=	equivalent diameter (mm)
k	=	thermal conductivity (Kcal/m • h • °C)
C_p	=	specific heat (Kcal/kg • °C)
l	=	tube length (m)
h	=	heat transfer coefficient (soap side) (Kcal/m^2 • h • °C)
h_v	=	heat transfer coefficient (steam side) (Kcal/m^2 • h • °C)
x	=	tube wall thickness (mm)
k_s	=	steel thermal conductivity (Kcal/m • h • °C)
U	=	overall heat transfer coefficient (Kcal/m^2 • h • °C)

The water evaporated by expansion under vacuum is normally not enough for industrial drying purposes except for the production of a few types of laundry soaps. The rest of the water must be preevaporated in the heat exchanger. Heat transfer areas utilized for vacuum dryer shell-and-tube heat exchangers are 20, 25, or 30 m^2/ton of dry soap; the size selected depends on the dry soap final moisture and the product type.

Laminar flow is assumed for calculating the required heat transfer area. This allows for adequate safety margin since turbulent flow occurs when water starts boiling (preevaporation) in the heat exchanger. Applying the equation $U = 1/(1/h + x/k_s + 1/h_v)$, the heat transfer coefficient (*U*) is in the range of 80–200 Kcal/m^2 • h • °C. The average value utilized is 150 Kcal/m^2 • h • °C.

Operating Vacuum (Absolute Pressure) and Condensate Temperature Using Barometric and Surface Condensers

With Barometric Condenser

For direct-contact barometric condensers, the condensate (cooling water plus condensed vapors) temperature in the downleg must be about 3°C below the saturated vapor pressure in the vacuum spray chamber.

Example:

Cooling water temperature into barometric condenser: 24°C

Condensate temperature into hotwell (downleg temperature): 30°C

Vapor pressure of water at 30°C (absolute pressure): 32 torr (from Table 7A.1)

Actual vacuum in the dryer at 30°C + 3°C = 33°C: 38 torr (from Table 7A.1)

With Surface Condenser

For noncontact surface condensers, the condensate must be about 2°C below the saturated vapor pressure in the vacuum spray chamber. A pressure drop of 3 torr through the unit also must be taken into account.

Example:

Cooling water temperature into surface condenser: 18°C

Cooling water temperature from surface condenser: 24°C

Condensate temperature into hotwell (downleg temperature): 30°C

Vapor pressure of water at 30°C (absolute pressure): 32 torr (from Table 7A.1)

Actual vacuum in the dryer at 30°C + 2°C = 32°C is 36 torr (from Table 7A.1) plus 3 torr due to pressure drop, or 36 + 3 torr: 39 torr.

Total Fatty Matter (TFM) and Moisture Content of Toilet Soaps

Assumptions

- 80/20 tallow/coco mixture with an average 216 acid value
- Average molecular weight (MW): 56,100/216 = 260
- 63% TFM neat soap contains: 0.40% glycerine, 0.05% NaOH, and 0.35% NaCl

Sample Calculation and Formulas

$$\text{Fatty acid} + \text{Caustic soda} \rightarrow \text{Soap} + \text{Water}$$
$$RCOOH + NaOH \rightarrow RCOONa + H_2O$$
$$260 + 40 \rightarrow 282 + 18$$

TABLE 7A.1
Vapor Pressure of Water

Temperature (°C)	Absolute Pressure (Torr)	Temperature (°C)	Absolute Pressure (Torr)	Temperature (°C)	Absolute Pressure (Torr)
10	9.2	21	18.6	31	33.7
11	9.8	22	19.5	32	35.7
12	10.5	23	21.1	33	37.7
13	11.2	24	22.4	34	39.9
14	12.0	25	23.8	35	42.2
15	12.8	26	25.2	36	44.6
16	13.6	27	26.7	37	47.1
17	14.5	28	28.4	38	49.7
18	15.5	29	30.5	39	52.1
19	16.5	30	31.8	40	55.3
20	17.5				

TABLE 7A.2
Total Fatty Matter and Moisture Content of Toilet Soaps

% TFM	% Anhydrous Soap	% Glycerine	% NaOH + NaCl	% Water
58	63.0	0.4	0.4	36.3
59	64.0	0.4	0.4	35.2
60	65.1	0.4	0.4	34.1
61	66.2	0.4	0.4	33.0
62	67.3	0.4	0.4	31.9
63	68.4	0.4	0.4	30.8
64	69.5	0.4	0.4	29.7
65	70.6	0.4	0.4	28.6
66	71.6	0.4	0.4	27.5
67	72.7	0.4	0.4	26.4
68	73.8	0.4	0.4	25.3
69	74.9	0.4	0.4	24.2
70	76.0	0.5	0.4	23.1
72	78.2	0.5	0.5	20.9
74	80.3	0.5	0.5	18.7
76	82.5	0.5	0.5	16.5
78	84.7	0.5	0.5	14.3
80	86.8	0.5	0.5	12.1
82	89.0	0.5	0.5	9.9
84	91.2	0.5	0.5	7.7

$$\text{Anhydrous soap yield} = (MW + 22)/22 = 282/260 = 1.085$$
$$\text{Anhydrous soap} = 1.085 \times \text{TFM}$$
$$\text{Soap moisture content } (\%) = 100 - (\% \text{ anhydrous soap}, \% \text{ glycerine},$$
$$\% \text{ NaOH}, \% \text{ NaCl, and so on})$$

Acknowledgments

We wish to thank Buffalo Technologies Corporation for preparing the Buflovak Soap Flaking System material and Moisture Systems Corporation for providing details on the Micro-Quad Infrared Moisture Analyzers.

References

1. G. Mazzoni, S.p.A., Italian Patent 386,583 (1940).
2. G. Mazzoni, S.p.A., USA Patent 2,945,819 (1951).
3. Bassett, G. H., U.S. Patent 2,710,057 (1955).
4. Palmason, E. H., U.S. Patent 3,673,380 (1963).
5. Miag GmbH, British Patent 1,063,715 (1964).
6. G. Mazzoni, S.p.A., Italian Patent 623,670 (1961).
7. Perry, R. H., *Perry's Chemical Engineers' Handbook,* 6th edn., McGrawHill, New York, 1984.
8. Waler, G., *Industrial Heat Exchangers: A Basic Guide,* 2nd edn., Hemisphere, New York, 1990.

9. Burley, J. R., Don't Overlook Compact Heat Exchangers, *Chem. Eng.,* August 1991, pp. 90–96.

10. Boyer, J., and G. Trumpfheller, Specification Tips to Maximize Heat Transfer, *Chem. Eng.,* May 1993, pp. 90–97.

11. Woollatt, E., *The Manufacture of Soaps, Other Detergents and Glycerine,* Ellis Horwood Limited, London, England, 1985.

12. Spitz, L., ed., *Soap Technology for the 1990's,* AOCS Press, Champaign, Illinois, 1990, pp. 154–173.

Chapter 8

Bar Soap Finishing

Luis Spitz

L. Spitz, Inc., Skokie, Illinois, U.S.A.

Bar Soap Finishing

Bar soap finishing is the final processing step for the production of standard, super-fatted, translucent, soap/synthetic (combo), and synthetic products into a solid bar (tablet) form. Bar *soap finishing* consists of *mixing* the main base with minor liquid and solid ingredients, followed by *refining* the fully formulated mixture into a uniform, homogeneous product, *extruding* into a continuous slug (billet) form, *cutting* into individual length slugs, and *stamping* into a final predetermined weight and shape bar (tablet) ready for packaging.

Mixing

There is no precise definition and no measuring criteria for mixing in the soap industry. Mixing of 1% to 5% of solid and liquid ingredients with the dry pelletized soap base in standard mixers can be called *macro mixing*. During macro mixing, the additives only coat the outer surface of the pellets. *Intensive* or *micro mixing* is achieved when the pellets are broken up to expose more surface area and to allow the ingredients to penetrate inside the pellets.

Refining

Refining is the work done on soap by the combined action of pressure and shear. The purpose of refining is threefold:

- To produce a fully homogeneous, uniform product
- To improve bar feel by eliminating low-solubility hard particles
- To enhance product lather, solubility, and firmness by affecting crystalline structure change

Refining is performed with plodders, roll mills, or both units used in combination.

Methods to Measure Degree of Refining

Washdown Temperature
To measure the degree of refining, soap feel (grittiness, sandiness, roughness) due to the presence of hard particles (specks), the popular "washdown temperature" test

is used. The bar is washed with both hands for 1 min in a sink with 30°C water. Once the bar surface is smooth and all protruding lettering and designs are washed away, the water temperature is decreased. Washdown temperature is the temperature at which the first hard specks can be detected. A smooth bar without any hard specks has a 22°C washdown temperature. The "slightly gritty" feel appears at 23°–24°C, "moderately gritty" at 25°–26°C, "gritty" feel is at 26°–27°C, and "very gritty" when the washdown temperature is 28°C.

Photoevaluation

A visual method can also be used for roughness evaluation. The bar is washed for 1 min in 20°C water and then left to dry. If the bar is held at an angle in front of a high-intensity light source and below eye level, the dry specks can be easily seen. A sample observation box fitted with the light source can be used to obtain better results. Using photographic standards, the bar can be graded as 0, 1, 2, 3, 4, or 0, 25, 50, 75, and 100. Zero grade represents a smooth product, and a number 4 or 100 refers to a very gritty bar.

Plodder and Roll Mill Refining

Figure 8.1 illustrates the basic method of how soap pellets are refined through a refining plodder and a roll mill. In a refining plodder, a refining screen is located inside a refining-pelletizing head group. The product forced through the screen by the forward motion of the plodder worm exits through a drilled pelletizing-refining head group in the form of 8 mm diameter—optimum diameter for refining—by about 12 mm, or other length compact pellets. The length of the refined pellets depends on the plodder output and the number of blades of the pelletizing/cutting knife. Through a roll mill the pellets are converted into flakes with a final flake thickness set by the gap setting between the last two rolls. As shown, an 8 mm × 12 mm pellet becomes a 0.15 mm × 12 mm × 603 mm flake.

Plodder Refining

Plodders perform the three main refining functions defined earlier only when they are fitted with a refining screen. The degree of refining depends on the size of the refining mesh size used. One *full stage of refining* is achieved with a plodder when a plodder is fitted with a 50 mesh refining screen, only partial refining is attained with 30, 20, and coarser mesh screens. This definition of a full refining stage is important for the comparison with roll mill refining and for the selection of the proper finishing line.

Refining Screens

The most widely used refining screen is the *square mesh wire* type. When selecting the desired mesh size for refining, it is very important to specify the wire diameter, width of opening, and percentage of open area. Screen suppliers offer different wire

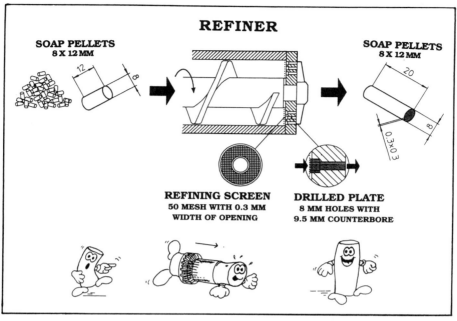

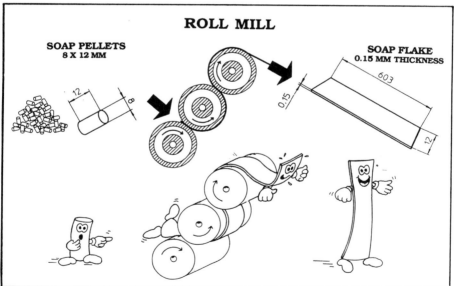

Fig. 8.1. Plodder and roll mill refining.

TABLE 8.1
Refining Screens

European Mesh Number[a]	Wire Diameter (mm)	Width of Opening (mm)	Open Area (%)	U.S. Mesh Number[b]	Wire Diameter		Width of Opening		Open Area (%)
					in.	mm	in.	mm	
9	1.00	2.09	45.4	8	0.041	1.04	0.084	2.13	45.2
					0.028	0.71	0.097	2.16	50.2
18	0.50	1.04	45.5	16	0.020	0.50	0.042	1.08	46.2
					0.017	0.43	0.038	0.98	48.3
22	0.45	0.78	39.9	20	0.018	0.46	0.032	0.81	45.3
					0.016	0.40	0.034	0.86	46.2
30	0.36	0.56	37.9	28	0.014	0.35	0.022	0.55	36.9
					0.013	0.33	0.020	0.51	37.1
55	0.20	0.30	36.3	50	0.008	0.20	0.012	0.30	36.0
					0.009	0.23	0.011	0.28	30.3

[a]European mesh number = number of openings per 25.4 mm.
[b]U.S. mesh number = number of openings per linear inch.

diameters for the same mesh sizes. Two of the common, commercially available diameter types are summarized with other important specifications in Table 8.1.

Roll Mill Refining

A roll mill's degree of refining depends on two factors: the amount of shearing action performed on the product as it passes at a certain clearance between the last two rolls, and the optimum speed differential between them.

The *clearance, gap setting,* or *nip* between the last two rolls establishes the film thickness. For most products, the gap should be 0.15 mm for maximum performance. The combination of gap setting and speed differential between the last two rolls determines not only the degree of refining but also control of product temperature. Unlike plodders, which during refining always increase the product temperature, roll mills are capable of maintaining and even reducing it.

Plodder and Roll Mill Comparison of Refining Stages

Refining with plodders, roll mills, or a combination of both machines has been and will continue to be a much debated subject. Since the selection of a specific finishing line should be based on the number of stages of refining required, a general comparison guide between plodders with various sizes of screens and roll mills with different gap settings is presented in Table 8.2.

Prerefining

Refining of aged, dry soap base before the addition of any minor liquid and solid additives is called *prerefining*. Lower final washdown temperature—higher degree of refining—can be achieved with a finishing line when the line is set up in a prerefining configuration rather than as a standard refining type layout. Figure 8.2

TABLE 8.2
Plodder and Roll Mill Comparison of Refining Stages

Number of Refining Stages	Plodder Refining Screen Mesh Screen	Roll Mill Top Roll Gap Setting (mm)
1 1/2	80	0.15
1	50	0.20
3/4	30	0.25
1/2	20	0.30
1/4	16	0.40

Note: when product passes through a plodder fitted with a 50 mesh screen, it is considered one stage of refining. The use of 80 mesh and finer screens is not recommended due to excessive temperature rise and reduced production rates.

illustrates the stage-by-stage washdown temperature (photoevaluation scale) changes for refining- and prerefining-type finishing lines with the same total number of refining plodders. Prerefining works only for products that have been aged at least 2 to 4 h. Prerefining plasticizes the base product and offers these main advantages:

- Improved refining (lower washdown temperature of the finished product)
- Better mixing of the liquid additives with the plasticized and partially refined base, facilitating the refining action of the subsequent machinery
- Easier processing of all hard products (syndet, translucent)

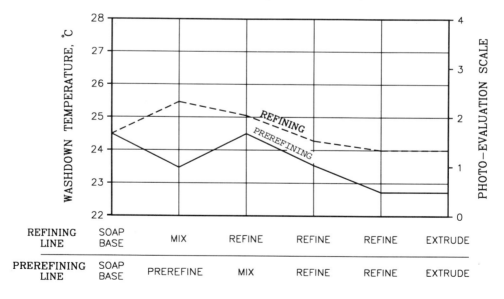

Fig. 8.2. Refining and prerefining.

Aging

If formulated soap base is stored and aged before final refining, extrusion, and stamping, line efficiency increases considerably. During aging, soap crystallization is completed and soap temperature is reduced. Most products should be aged at least 8 h, but it is advisable to determine experimentally the optimum aging time for each specific formulation.

Bar Soap Finishing Equipment

Mixers

Soap mixers are designed to mix soap and syndet pellets/flakes with all the liquid and solid ingredients that constitute the final product (Fig. 8.3).

Amalgamator with "Open Sigma Blades"
The most popular soap mixer, called an *amalgamator,* is a top-loading, bottom-discharge, nontilting unit with open sigma blades. The easy-to-clean, efficient open sigma design is derived from the sigma blade, which is the universal mixing blade in the chemical industry. These mixers do not break up the pellets; they only coat their outer surface with the additives.

Double-Arm Tilting Mixer with Sigma Blades
For the production of toilet and synthetic laundry bars, double-arm tilting mixers with two counterrotating sigma blades are used. The blades may be tangential or overlapping.

Paddle Mixer with Paddle Blades and Plow Mixer with Plow Blades
These units are also used in the soap industry, but they are less common than the amalgamators.

Intensive Mixers with Choppers
High-speed "choppers" of different designs can be fitted into paddle- and plow-type mixers. The high shear action of the choppers breaks up the pellets and allows for easier penetration and dispersion of the additives throughout the entire mixture. Intensive mixing for most products *is* equivalent to one stage of refining. In practice, this permits the elimination of one refining plodder from a finishing line, or alternatively by keeping the same number of refiners, a more refined and homogenized end product can be obtained.

Mixer-Kneaders
Mixer-kneaders consist of a double-arm mixer with two tangential sigma blades and a discharge screw with a pelletizing head built into the base of the mixer bowl. After the end of the mixing cycle, blade rotation is reversed to facilitate product

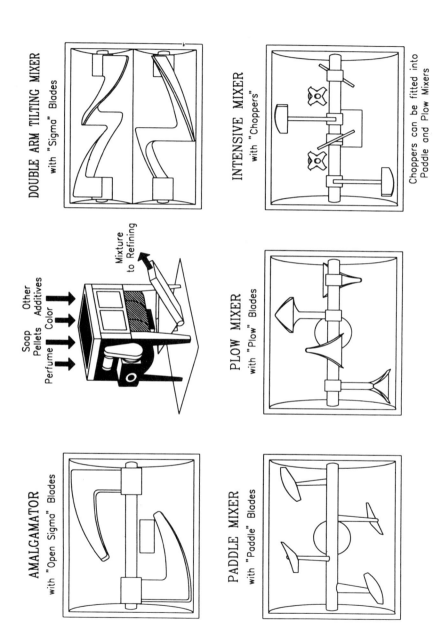

Fig. 8.3. Mixers.

discharge. Mixer-kneaders have not gained much acceptance in the soap industry because they offer no real advantage over the use of a batch amalgamator followed by a separate *simplex refiner*.

Continuous Processors

Continuous processors are continuous mixers that feature two parallel corotating agitators. As the agitators rotate, the blades move the ingredients forward and backward, providing intensive mixing and remixing in each section. The end can be fitted with a pelletizing group. Werner & Pfleiderer's ZSK Twin-Screw Extruder and Teledyne Readco's Twin-Screw continuous processor are two of the units used by a few soap producers. The high cost of these very efficient machines has limited their wider use.

All mixers can be provided with jackets for heating or cooling.

Roll Mills

Before the use of plodders as refiners, roll mills were used for refining and single-stage plodders for extrusion of the finished product. Even after the introduction of duplex vacuum plodders, refining was still performed by roll mills because the first stage of the duplex was designed to be used with a coarse screen to remove dirt instead of refining the product.

Roll Mill Contact Areas

Soap mills are available with three, four, and five rolls. The choice between them is still a much debated subject.

Roll configuration geometry is very important because it determines the total contact area available for refining and cooling. Figure 8.4 shows the contact areas for the most widely used three-, four-, and five-roll mills. It can be seen that there is very little contact area difference between three-roll mills with 490° configuration and four-roll mills with 515°.

For a three-roll mill, the clearance between the top roll and middle (fixed) roll should be 0.5 mm, and between the idle and the feed roll 0.20 mm. Typical roll speeds for three-roll mills are 20 to 40 to 240 rpm, respectively, an overall speed ratio of 1 to 12. The same total speed ratio is used for four- or five-roll mills. Because speed ratio is a very important variable for obtaining optimum refining, cooling, and output, in some cases separate adjustable speed drive motors are used for each roll.

Today, the tendency is to use three-roll mills due to the combined effect of the following factors:

- Contact areas are large enough to ensure proper product refining and cooling, as shown in Fig. 8.4.
- About 70% of the total power absorbed, that is, 70% of the heat input to the soap, takes place between the last two rolls of all roll mills.

THREE ROLL MILL

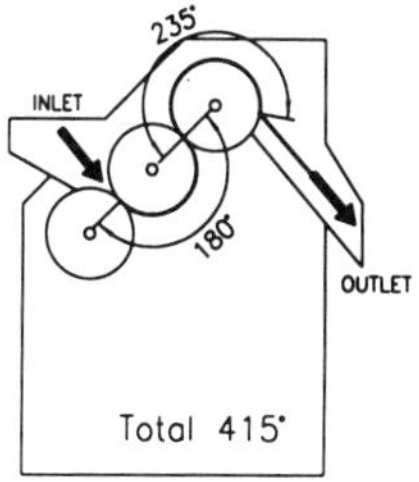

ROLL DIAMETER (mm)	ROLL LENGTH (mm)	CONTACT AREA (m²)
320	800	1.1
400	1000	1.7
400	1350	2.3
450	1500	2.9
450	1800	3.5

Total 490°

FIVE ROLL MILL

ROLL DIAMETER (mm)	ROLL LENGTH (mm)	CONTACT AREA (m²)
320	800	1.9
400	1000	2.9
400	1350	4.0
450	1500	5.0
450	1800	6.0

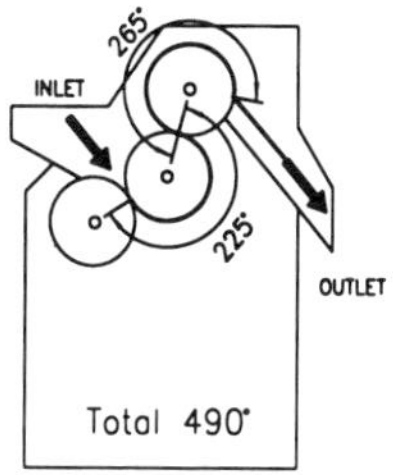

Total 850°

THREE ROLL MILL

ROLL DIAMETER (mm)	ROLL LENGTH (mm)	CONTACT AREA (m²)
320	800	0.9
400	1000	1.4
400	1350	1.9
450	1500	2.4
450	1800	2.9

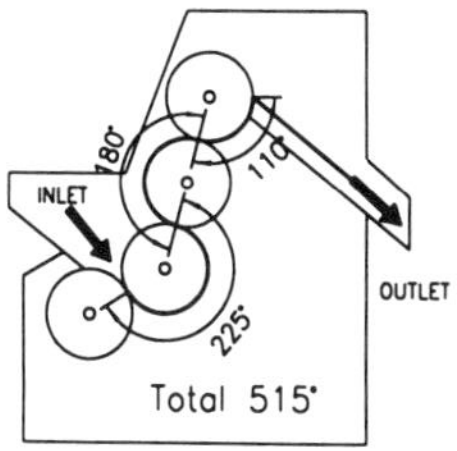

Total 415°

FOUR ROLL MILL

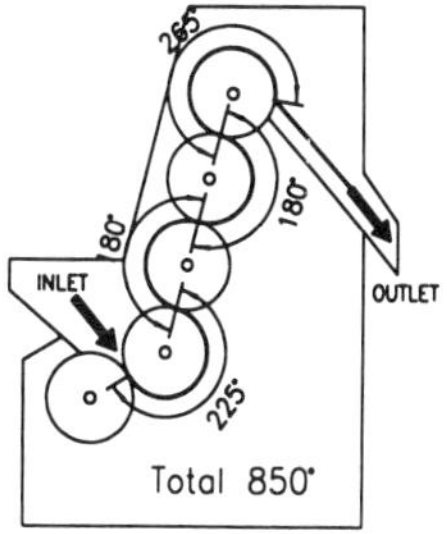

ROLL DIAMETER (mm)	ROLL LENGTH (mm)	CONTACT AREA (m²)
320	800	1.1
400	1000	1.8
400	1350	2.4
450	1500	3.0
450	1800	3.6

Total 515°

Fig. 8.4. Roll mill contact areas.

- Gap setting adjustment and control is easier than with the larger four- and five-roll mills.
- Three-roll mills costs less to purchase and to operate due to lower power and cooling water requirements.

Roll Mill Production Rates
Table 8.3 is a production rate guide for most of the commercially offered different size soap roll mills with 0.15 mm and 0.20 mm top gap setting.

Roll Design—Metallurgy
Standard-quality rolls are made of compound, chilled cast iron with 500–550 HB (Brinell) hardness. For better corrosion and wear, high-chromium cast iron rolls with 520–580 HB to a depth of chill of 15 to 20 mm is indicated. The cooling

TABLE 8.3
Roll Mill Production Rates

Roll Size (mm)	Roll Mill Models				
Length	800	1,000	1,350	1,500	1,800
Diameter	320	400	400	450	450
Top Roll Speed (rpm)	Production Rates (kg/h) (Top Roll Gap Setting @ 0.15 mm)				
180	1,011	1,580	2,133	2,667	3,200
190	1,068	1,668	2,252	2,815	3,378
200	1,124	1,756	2,370	2,963	3,556
210	1,180	1,844	2,489	3,111	3,733
220	1,236	1,931	2,607	3,259	3,911
230	1,292	2,019	2,726	3,407	4,089
240	1,348	2,107	2,844	3,556	4,267
250	1,405	2,195	2,963	3,704	4,444
260	1,461	2,283	3,081	3,852	4,622
270	1,517	2,370	3,200	4,000	4,800
Top Roll Speed (rpm)	Production Rates (kg/h) (Top Roll Gap Setting @ 0.20 mm)				
180	1,348	2,107	2,844	3,556	4,267
190	1,423	2,224	3,002	3,753	4,504
200	1,498	2,341	3,161	3,951	4,741
210	1,573	2,458	3,319	4,148	4,978
220	1,648	2,575	3,477	4,346	5,215
230	1,723	2,692	3,635	4,543	5,452
240	1,798	2,809	3,793	4,741	5,689
250	1,873	2,926	3,951	4,938	5,926
260	1,948	3,043	4,109	5,136	6,163
270	2,023	3,161	4,267	5,333	6,400

Notes: rates are based on 75% production efficiency; soap relative density (specific gravity) = 1.035

efficiency depends on the minimum thickness of the chilled white iron outer layer of the cast iron rolls, which has a thermal conductivity of 20 to 25 W/m°C; the roll's gray iron core is higher, ranging from 45 to 60 W/m°C.

Roll Cooling Systems

Water is sprayed onto the inside walls of the rolls in the spray system, which is the most widely used cooling system. The more efficient but much more expensive *peripheral* method has gained limited acceptance (Fig. 8.5). Roll mills, when properly set and cooled with chilled water sprayed onto the inside of the roll walls, can control the product temperature passing through the rolls. As described before for certain products and production rates, the temperature differential between the incoming and the milled discharged product can be maintained, or its increase minimized and in some instances even reduced.

Plodders

Classification

The soap industry uses three types of plodders available in single-worm and twin-worm versions: simplex refiners, duplex refiners, and duplex vacuum plodders.

A *simplex refiner* consists of one plodder designed to operate with a 50 mesh refining screen at a maximum pressure of 60 bar.

A *duplex refiner* consists of two simplex refiners mounted in tandem. Each plodder is fitted with a 50 mesh screen.

A *duplex vacuum plodder* consists of two plodders mounted in tandem and connected by a vacuum chamber.

The preliminary-stage plodder is exactly the same as a simplex refiner and can be used with a 50 mesh refining screen. In the final-stage plodder, the refined pellets are compacted and extruded as a continuous slug (billet) free from any entrapped air.

Twin-worm plodders exist with tangential (touching) counterrotating worms in a single barrel and nontangential (nontouching) counterrotating worms in two separate barrels. Tangential twin-worm plodders are recommended for processing sticky products and for high production capacity lines. Nontangential twin-worm plodders are used for the production of marbleized soaps and for high-speed lines.

It is very important to note that a plodder functions as a refiner only when it is fitted with a 50 mesh screen. When a 30, 28, or coarser screen is used, the refining degree diminishes, as shown earlier in Table 8.1.

Plodder Worms

Plodder worms perform three functions: refining, compression, and extrusion. Plodder worm design is still an art even in today's computerized world. The final

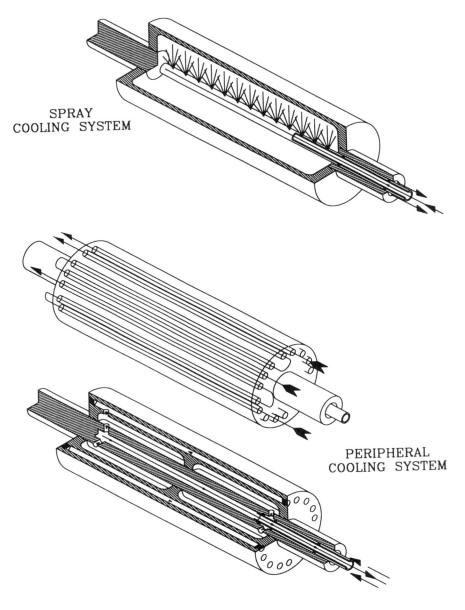

Fig. 8.5. Roll mill cooling. (Courtesy of Meccaniche Moderne).

designs depend on extensive testing. Figure 8.6 illustrates three main types of plodder worms.

- The *uniform style,* with constant root diameter and slightly decreasing pitch, is best suited for dry, nonsticky products.
- The *curved style*, with lesser decreasing pitch, slightly increasing root diameter, and large curved flights, is applicable for wet and sticky products.
- The *combination,* is a multipurpose worm designed to handle all types of products. As its name suggests, it is a combination of the uniform and the curved types.

Short and Long Lc/D Ratio Plodders

Figure 8.7 shows the definitions for plodder barrel length and diameter. Lc is the closed barrel section and D is the worm diameter. Please note that the Lc/D ratio is not the same as the L/D ratio, which is the total worm length L to the D worm diameter ratio.

Conventional plodders are designed with short 3-to-1 Lc/D ratios and operate with 12 to 15 rpm worm speeds. The longer 5-to-1 and 6-to-1 Lc/D ratio units function with up to 18 to 20 rpm worm speeds.

To increase production rates, plodders are sized with worm diameters that increase in 50 mm increments. Namely, 150, 200, 250, 300, 350, and 400 mm are the standard worms diameters of most plodders. In practice, the output of a short Lc/D ratio plodder is about equal to the production rate of a 50 mm smaller diameter long Lc/D ratio plodder. The longer barrel with greater heat transfer surface limits the increase of product temperature and also ensures proper refining. Up to now, for vacuum spray dryer plodders only the short Lc/D ratio single- and twin-worm plodders are utilized. For finishing lines, usage of the long Lc/D ratio plodders has been gaining considerable acceptance.

In 1984 Mazzoni introduced the Nova series of long Lc/D ratio plodders. In 1988 Binacchi entered with its Extenda models and TEMA in 1994 with the L series.

Instrumentation

Soap plodders can be fitted with the instruments shown in Fig. 8.7

- Pressure transducers (such as Dynisco's European Model IDA 372 or USA Model MG840-844): the pressure measured can be used as a control guide to indicate the need for screen changes.
- Temperature indicators can be placed into the worm supporting plate of each refining stage and also in the extrusion stage.
- Infrared noncontact thermometers can measure the continuously extruded slug temperature without coming in contact with it (such as Raytek, Inc., Thermalert IT Series).

L. Spitz

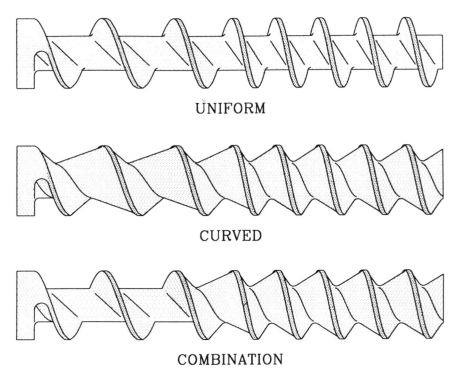

UNIFORM

CURVED

COMBINATION

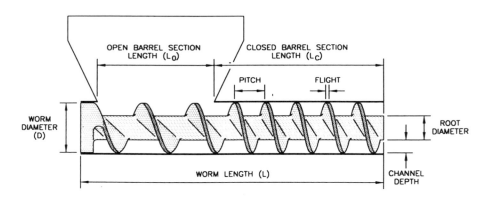

Fig. 8.6. Plodder worms.

REFINING GROUP

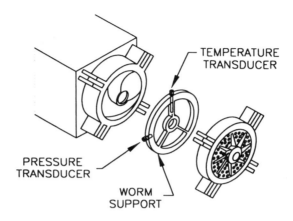

TEMPERATURE
TRANSDUCER

PRESSURE
TRANSDUCER

WORM
SUPPORT

EXTRUSION GROUP

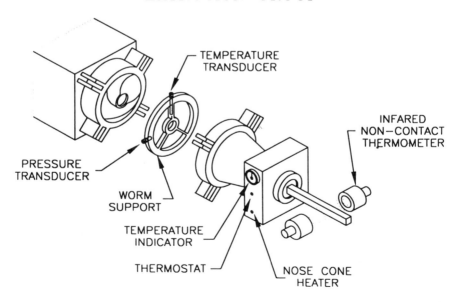

TEMPERATURE
TRANSDUCER

INFARED
NON−CONTACT
THERMOMETER

PRESSURE
TRANSDUCER

WORM
SUPPORT

TEMPERATURE
INDICATOR

THERMOSTAT

NOSE CONE
HEATER

Fig. 8.7. Plodder instrumentation.

Drive Motors

Fixed-Speed Motors and Clutches

Normally fixed-speed motors are used for all the plodder refining stages and also for roll mills. For plodders they are installed in combination with pneumatic clutches—mostly Eaton dynamatic Airflex types—to allow the plodder worm to stop and restart without having to stop and restart the motor itself. Most motor manufacturers now offer premium-efficiency motors that provide operating cost savings over standard motors.

Adjustable Speed Drive Motors

The extrusion stage of *duplex vacuum plodders* always requires the use of an adjustable speed drive (variable speed drive) motor.

Eddy current drives with a built-in eddy current clutch (coupling) for many years have been the most widely used drives for plodders. Since the clutch is an integral part of this drive, a separate pneumatic clutch is not required.

Brush type DC drives are used but to a much lesser degree because of their higher cost and higher maintenance requirements.

The present trend is to utilize the more economical and more efficient *adjustable-frequency AC drives (AFDs),* also called *variable-frequency drives (VFDs)*. The AFDs provide constant-torque operation from near zero to full speed and the power requirements are proportional to the operating speed. These two characteristics are of key importance for plodder drive and also for roll mills and soap press applications. The main features of eddy current, brush DC and AFD drives are listed in Table 8.4.

Modern microprocessor-based electronics corrected the past poor performance of adjustable speed drives relating to torque, speed range and the handling of wide dynamic load changes. Due to these new technologies, today there are many other adjustable speed drive systems. These are the *Flux, Vector, Reluctance, Brushless DC* and the latest *Direct Torque Control (DTC)* drives. *Hydraulic drives* for plodders are used only by Britannia from England.

It is common practice to use fixed-speed motors with pneumatic clutches for all the refining stages. Level controls with timers installed in the feed hoppers control the stop and restart cycles according to the product flow in each machine.

The application of AFDs instead of fixed-speed motors for all the refining stages and not only for the extrusion stage is especially advantageous for translucent, combo, and syndet bar production.

Employing AFDs also allows for speed coordination—synchronization—between all the machinery in a finishing line, minimizing the stop and start cycle time for each stage, resulting in lower product temperature rise and more uniform extrusion rates.

Cutters

Soap cutters are designed to cut the continuously extruded slug by the duplex vacuum plodder into individual slugs of predetermined length. Traditional soap presses

TABLE 8.4
Adjustable-Speed Drives

Drive System Characteristics	Drive Systems		
	Eddy Current	Standard DC	AC Adjustable Frequency[a]
Starting torque[b] (%)	100	150	180
Torque range[c] (rpm)	200–1,700	0–1,800	60–1,800
Speed range[d]	8:1	100:1	10:1
Speed regulation[e] (%)	0.6	0.1	2
Dynamic response[f] (mS)	200	80	400
Efficiency[g] (%)	50	80	90
Cost factor[h]	1.5	1.2	1.0
Service factor[i]	Low	High	Low
Motor type[j]	Standard AC	Brush type DC	Standard AC

Note: all ranges indicated are based on 1,800 rpm base motor speed.
[a]Performance based on present-generation IGBT (insulated-gate bipolar transistor) technology.
[b]Maximum torque produced for starting a load or "breakaway" torque.
[c]Range in RPM where the output produces rated torque of motor.
[d]Operational range of speed for continuous operation.
[e]Percentage of speed deviation from set speed.
[f]Response in milliseconds for the system to react and stabilize to change in load.
[g]Ratio of power produced to the electrical power provided at the system input.
[h]Relative cost of each system in relation to each other.
[i]Amount of maintenance anticipated.
[j]Type of motor required.
(Courtesy of ABB Drives, Inc.)

have to be fed with accurately cut single-length slugs. The new flashstamping presses in turn are fed with multiple-length slugs.

Mechanical Multiblade Cutters

There are two types of mechanical multiblade chain cutters:

- *Manual adjustable:* this is a fixed chain cutter. The cutting chain can be removed and each chain adjusted manually, link by link, within a certain length range.
- *Automatic adjustable:* the cutting length can be changed within the full range of the chain by simply turning a handwheel.

Electronic/Pneumatic Single-Blade and Multiblade Cutters

The first single-blade electro/pneumatic cutter was introduced by Meccaniche Moderne in 1987.

The TVG model is offered in single and twin versions. Roller printers, mainly for laundry soaps capable of printing on two or four sides of the bar, can be mounted on this cutter.

IMSA's VST-2 cutter was introduced in 1989 (Fig. 8.8). This cutter, like the MM unit, uses an encoder that measures the extrusion length (speed) and activates a pneumatically operated blade to cut both ways with a back-and-forth motion. SAS's Unicut model was introduced in 1993.

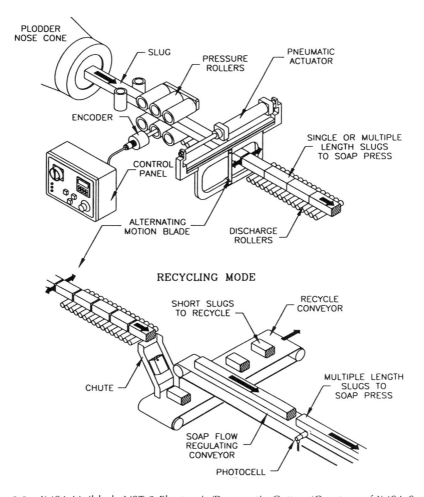

Fig. 8.8. IMSA Uniblade VST-2 Electronic/Pneumatic Cutter. (Courtesy of IMSA S.r.l.)

Electronic Single-Blade and Multiblade Cutters

TEMA's 1994 RC cutter uses rotating cutting blade(s) (Fig. 8.9). The latest electronic models are Mazzoni LB's TE (Figs. 8.10 and 8.11), Weber & Seelander's ARC, and Binacchi's MBE units.

The fully electronic design permits very accurate cuts at varying cutting speeds. Some of these cutters have three operating modes (Fig. 8.10). They can cut single-length slugs, multiple-length slugs, and short slugs to facilitate recycle (rework). In case long slugs are to be recycled, usually an expensive twin-worm plodder is required for the recycle stage.

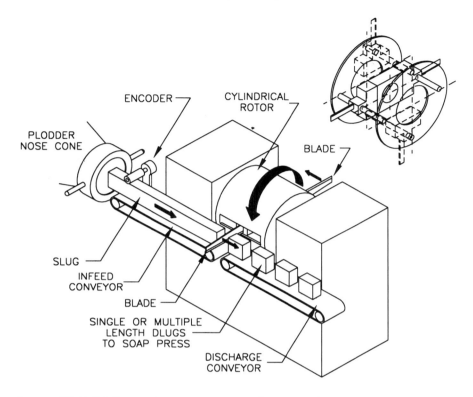

Fig. 8.9. TEMA RC Electronic Cutter. (Courtesy of TEMA Technologies & Manufacturing S.r.l.)

The automatic short slug cutting recycle mode is a very important practical feature that should become a standard for all of these new cutters. All of the presently offered electronic/pneumatic and electronic cutters are summarized in Table 8.5.

Presses

Soap Shapes

There are two basic soap shapes:

- *Banded* (with a side band): all soap shapes with vertical sides around their periphery
- *Bandless* (without a side band): all soap shapes with one parting line only, that is, without vertical sides around their circumference

All banded and bandless shapes can be further classified into four variations: *rectangular, round, oval,* and *irregular.*

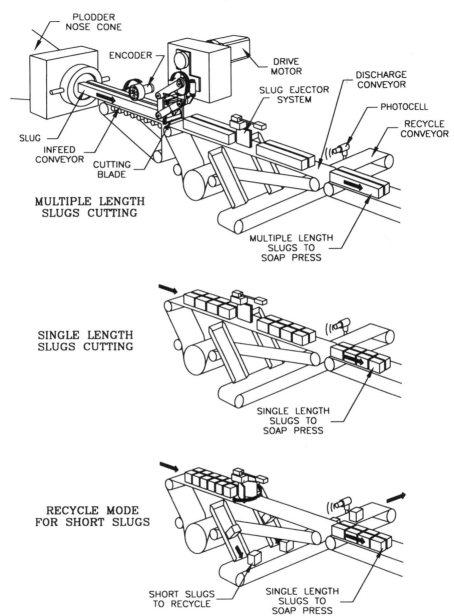

Fig. 8.10. Mazzoni TE Electronic Cutter. (Courtesy of Mazzoni LB, S.p.A.)

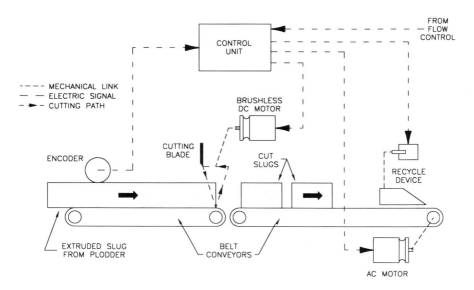

Fig. 8.11. Mazzoni TE Electronic Cutter Operating System. (Courtesy of Mazzoni LB, S.p.A.)

Soap Press Types

Soap presses can be classified into horizontal and vertical motion units. Both types are available today with the traditional die box system for stamping banded soaps only; disc die box flashstamping system for all banded, bandless, and irregular shapes; and the new flashstamping system using dies only, also for all shapes.

Stamping Systems

Die Box Stamping System for Banded Soaps

The horizontal motion presses were designed to stamp banded soaps with a four-station die box system. At the first station, the soap slug arriving on a belt conveyor from the cutter is pushed through a trimmer for accurate weight control into the cavity of the vertically mounted indexing die box. Rotating to the second station, the die and counterdie move into the die box shaping and embossing the soap with the internal surface of the die box discharged with a vacuum suction cup. The fourth station is used to clear any soap not discharged in the third station.

Disc Die Box Flashstamping System for All Shapes

As the dies come together and form the final shape of the bar and thus determine its weight, excess soap is formed around the periphery of the dies called flashing. Flashstamping always requires a slug 15% to 30% heavier in weight than the final stamped bar. The quantity depends on the bar shape. This system, introduced 20

TABLE 8.5
Electronic/Pneumatic and Electronic Cutters

Type and Make	Model	Number of Blades	Maximum Cutting Length (mm)	Maximum Cuts per Minute	Line Rate (kg/h) for Different Bar Weights (g)				
					80	100	150	200	250
Electronic/pneumatic									
Britannia	SS1	1	1,000	100	480	600	900	1,200	1,500
IMSA	VST2	1	1,000	250	1,200	1,500	2,250	3,000	3,750
MM	TGV	1	1,000	300	1,440	1,800	2,700	3,600	4,500
SAS	Unicut	1	1,000	150	720	900	1,350	1,800	2,250
Electronic									
Binacchi	MBE	1	600	300	1,440	1,800	2,700	3,600	4,500
Mazzoni LB	TE	1	1,000	640	3,070	3,840	5,760	7,680	9,600
TEMA	RC	1	600	160	770	960	1,440	1,920	2,400
		2	300	320	1,540	1,920	2,880	3,840	4,800
W&S	ARC	1	1,000	120	580	720	1,080	1,440	1,800
		3	333	360	1,730	2,160	3,240	4,320	5,400

Abbreviations: MM, Meccaniche Moderne; W&S, Weber and Seelander.

years ago, changed the soap industry by allowing 400-bars-per-minute stamping speeds of all bandless soaps with automatic flashing removal. These so-called universal presses used the disc die box method for bandless and irregular shape soap stamping, and the traditional die box was utilized for banded soaps.

Flashstamping System for All Shapes with Dies Only

The new flashstamping system does not require the use of any type of die box to stamp all banded, bandless, and irregular-shaped (specialty) products. The stamping is done only with dies and their corresponding counterdies.

New Generation of Flashstamping Soap Presses

Horizontal Motion Flashstamping Presses

Binacchi's model USN-500 horizontal motion flashstamping press, introduced in 1989, was the first dual mandril press designed to accommodate up to eight dies (four on each mandril) (Fig. 8.12), three dies for each stamped bar. One is mounted on a reciprocating die slide and the other two on a 180-deg rotating mandril. The deflashing and bar discharge operations are similar to those of the vertical motion presses. Britannia's Rotary Series, SAS Moldex, and Weber & Seelander's SPH presses are also of this type.

Vertical Motion Flashstamping Presses

Mazzoni LB's model STUR presses are illustrated in Figs. 8.13 and 8.14. The upper dies are mounted on a vertical motion reciprocating die slide and the lower dies on

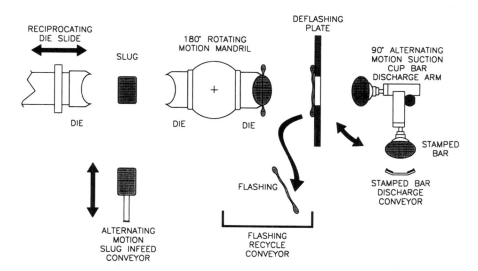

Fig. 8.12. Binacchi USN Soap Press Stamping System. (Courtesy of Binacchi & Co.)

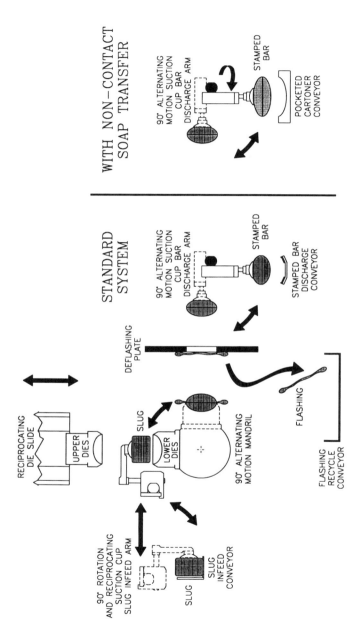

Fig. 8.13. Mazzoni STUR Soap Press Stamping System. (Courtesy of Mazzoni LB, S.p.A.)

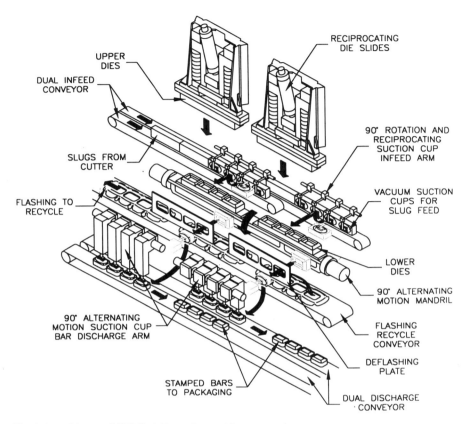

Fig. 8.14. Mazzoni STUR-8 Soap Press. (Courtesy of Mazzoni LB, S.p.A.)

a rotating 90-deg motion mandril. Soap slugs are fed to the mandril by a dual-motion vacuum suction cup (reciprocating horizontal motion combined with 90-deg rotation) slug infeed system. As the dies meet, the slugs are stamped and the mandril rotates, transferring the stamped bars and the flashing to the discharge position. During discharge by dual-motion suction cups, the bars pass through a deflashing plate that separates the bars from its flashing; then the cups rotate 90 deg and place the deflashed bars onto a discharge conveyor.

IMSA's VST-1 is a single mandril type soap press (Fig. 8.15). Meccaniche Moderne 8509 and 9107, SAS's Condorex (Fig. 8.16), and TEMA's Stemar Series belong to this category of vertical-motion flashstamping presses.

Table 8.6 summarizes all of the presently offered flashstamping presses. The support plates for single-mandril models are noted with one number. When the number is shown multiplied by two, the press is a dual-mandril (two separate mandrils) unit.

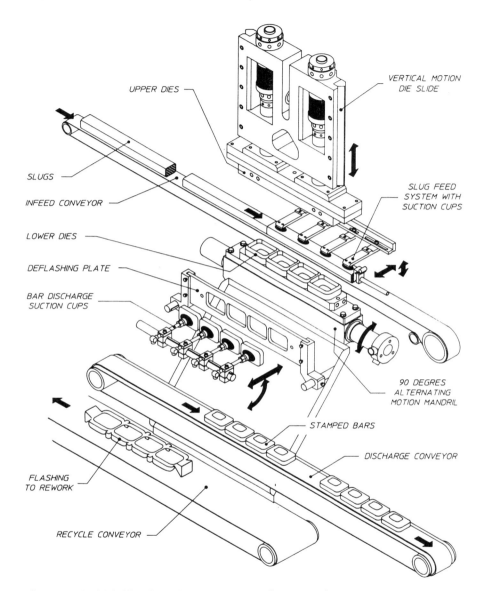

UPPER DIES

VERTICAL MOTION
DIE SLIDE

SLUGS

INFEED CONVEYOR

SLUG FEED
SYSTEM WITH
SUCTION CUPS

LOWER DIES

DEFLASHING PLATE

BAR DISCHARGE
SUCTION CUPS

90 DEGRES
ALTERNATING
MOTION MANDRIL

STAMPED BARS

DISCHARGE CONVEYOR

FLASHING
TO REWORK

RECYCLE CONVEYOR

Fig. 8.15. IMSA VST-1 Soap Press. (Courtesy of IMSA S.r.l.)

Hotel Soap Presses

Figure 8.17 illustrates the special vertical motion hotel soap presses that operate at
high speeds using only one set of dies. For banded soaps up to 300 bars per minute,
these presses utilize a traditional die box and, for bandless shape at speeds of 250

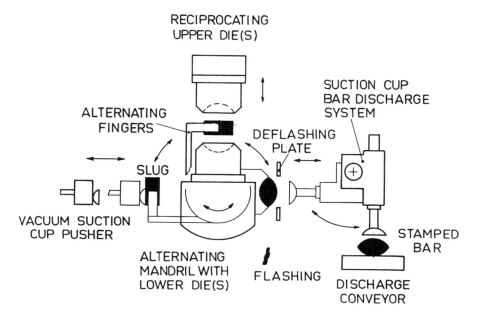

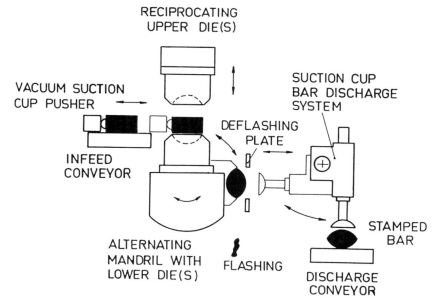

Fig. 8.16. SAS Condorex Soap Press. (Courtesy of SAS).

TABLE 8.6
Flashstamping Soap Presses

Make	Model	Number of Die Support Plates	Die Cavities				Stamping Strokes per Minute[a]	Stamping Speed (bars per min)		Mandril Rotation (degrees)
			Maximum Number for Each Support Plate		Total Number for All Support Plates			Bar Weight (g)		
			Bar Weight (g)		Bar Weight (g)			100	150	
			100	150	100	150				
Binacchi	USN-300	3	5	4	15	12	50	250	200	180
	USN-400	2 × 3	4	3	24	18	50	400	300	180
	USN-500	2 × 3	5	4	30	24	50	500	400	180
Britannia	Rotary-300	3	3	2	9	6	55	165	110	180
	Rotary-590	3	5	4	15	12	55	275	220	180
	Rotary-T/590	2 × 3	5	4	30	24	55	550	440	180
IMSA	VST-1.3	2	3	2	6	4	60	180	120	90
	VST-1.6	2	6	5	12	10	60	360	300	90
	VST-1.12	2 × 2	6	5	24	20	60	720	600	90
Mazzoni LB	STUR-4	2	4	3	8	6	75	300	225	90
	STUR-6	2	6	5	12	10	65	390	325	90
	STUR-8	2 × 2	5	4	20	16	60	600	480	90
MM	8509	2	3	2	6	4	35	105	70	60
	9107	2	4	3	8	6	65	260	195	60
SAS	Moldex	3	6	6	18	18	50	300	300	180
	Condor	5	2	2	10	10	90	180	180	90
	Condorex	2	3	3	6	6	60	180	180	90
TEMA	Stemar-201	2	3	3	6	6	100	300	300	60
	Stemar-401	2	5	4	10	8	70	350	280	90
	Stemar-801	2 × 2	5	4	20	16	60	600	480	90
W&S	SPH	3	5	4	15	12	60	300	240	180
	SPH-S	3	6	6	18	18	60	360	360	180
	SPH-D	2 × 3	6	5	36	30	60	720	600	180
	SPV	2	2	1	4	2	80	160	80	90

[a]Manufacturer's suggested stamping strokes per minute.
Abbreviations: MM, Meccaniche Moderne; W&S, Weber and Seelander.

BANDED SOAP STAMPING

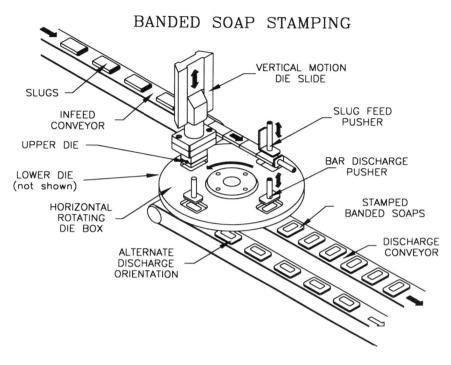

BANDLESS SOAP STAMPING

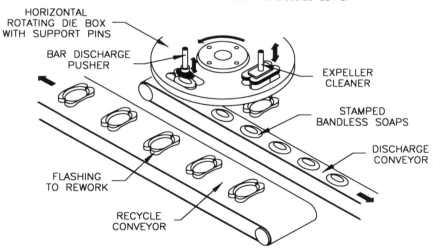

Fig. 8.17. Hotel soap presses.

bars per minute, a special die box with slug holding pins designed for flashstamping operating mode.

Figure 8.18 shows the operating sequence of TEMA's Stemar-201 vertical-motion new multidie flashstamping type press with 60-deg rotation mandril for hotel and regular-size soap application.

Die Material Specifications

Table 8.7 summarizes the most widely used materials, with detailed specifications for all items relating to dies, die boxes, and ejectors.

Direct Noncontact Bar Soap Transfers for Cartoner and Wrapper Interface

Binacchi Transfer

The Binacchi transfer to be applied with a bar soap cartoner was introduced in 1989 and broke the 300-cartons-per-minute speed barrier (Fig. 8.19). This was the first noncontact bar soap transfer system designed as an integral part of a soap press.

The transfer group first fitted on Binacchi's USN-500 press consists of an extra rotating mandril with two sets of suction cups. First the deflashed bars are twisted 90 deg clockwise and then placed on the mandril, which rotates 180 deg, and before discharging the bars it moves them laterally to match the cartoner bucket pitch. The cartoner must be provided with an extended chain to fit under the press discharge section.

In 1995 Binacchi introduced a similar transfer system for the direct interface with a new bar soap wrapper. (For more details, refer to the soaps and detergents packaging chapter in this book.)

Mazzoni LB and TEMA Transfers

Mazzoni LB's direct transfer unit DTU (Fig. 8.20) and TEMA's SDS system for cartoners have recently been introduced. Both systems use a single suction cup transfer group without an extra mandril. The stamped bars are lifted from the lower dies. After passing through the deflashing plate, they are turned 90 degrees downward followed by 90-degrees twisting from lengthwise position. Finally the bars move laterally to match the cartoner pitch and are placed accurately into the cartoner buckets. An electronic interface system couples the presses with the cartoner.

Water Chillers for Plodders and Roll Mills

Water chillers are designed to supply in closed circuit clean, chilled water to plodders and roll mills. Chilled water from approximately 8°C to 15°C circulates in the barrels of plodders and in the roll of roll mills. The exact operating temperature depends on the product and the output.

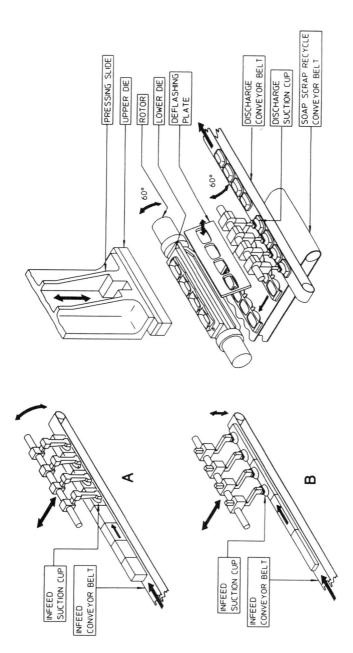

PRESSING SLIDE

UPPER DIE

ROTOR

LOWER DIE

DEFLASHING PLATE

60°

60°

DISCHARGE CONVEYOR BELT

DISCHARGE SUCTION CUP

SOAP SCRAP RECYCLE CONVEYOR BELT

A

B

INFEED SUCTION CUP

INFEED CONVEYOR BELT

INFEED SUCTION CUP

INFEED CONVEYOR BELT

Fig. 8.18. TEMA Stemar-201 Press for regular and hotel soaps. (Courtesy of TEMA Technologies & Manufacturing S.r.l.)

TABLE 8.7
Soap Die Set Material Specifications

Materials	ASTM and CDA Numbers	Al	Cu	Cr	Fe	Ni	Pb	Sn	Si	Zn	Other	TC	HDN
					Chemical Composition (%)							Properties	
Dies													
AMPCO 940 chromium copper	C18000		96.4	0.4		2.5			0.7			125.0	Rb 94
AMPCO 97 chromium copper	C18200		98.9	1.1								190.0	Rb 80
AMPCO 45 nickel aluminum bronze	B124	10.0	81.0		3.0	5.0					1.0	22.0	Rb 86
Muntz metal	B-171		61.0		0.1		0.7	0.2		38.0		71.0	Rb 65
Ejectors													
AMPCO 18 aluminum bronze	B150	10.5	85.5		3.5						0.5	34.0	Rb 84
bearing bronze	B550		83.0			1.0	7.0	6.5		2.5		34.0	Rb 72
Die boxes, die discs, trimmer plates, and mounting plates													
Naval bronze	B-21		62.0		0.1		0.2	1.0		36.7		67.0	Rb 82
Free cutting brass	B-16		63.0		0.4		3.7			32.9		67.0	Rb 78
304 stainless steel	A276			20.0	66.9	10.0			1.0		2.1	9.0	BHN 150
303 stainless steel	A582			19.0	66.9	10.0			1.0		3.1	9.0	BHN 160

Abbreviations: TC, thermal conductivity in BTU/ft2/hr/°F @ 68°F; HDN, hardness; Rb, Rockwell hardness b scale; BHN, Brinell hardness; ASTM, American Society for Testing and Materials (A and B numbers); CDA, Copper Development Association (C numbers).

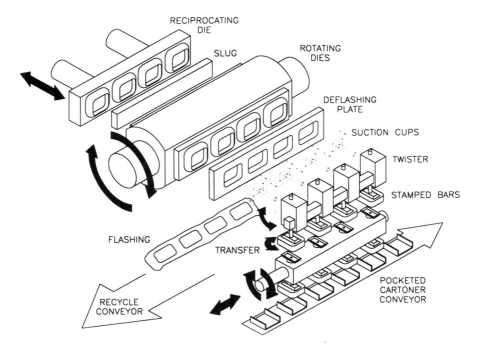

Fig. 8.19. Binacchi USN Press with noncontact soap transfer. (Courtesy of Binacchi & Co.)

The cost of nonrecycled cooling water, the water temperature limitation during certain seasons of the year, the elimination of periodic cleaning of the plodder barrels to remove hard water deposits, and the use of optimum chilled cooling water temperature in most cases justifies the initial investment and the operating cost of a chiller for one or more complete finishing lines.

Cooling units are rates in tons of refrigeration, which measures the cooling effect of 2,000 lb (a short ton) of ice melting in 24 h (288,000 heat units in 24 h, or 12,000 heat units per hour).

In the United States, one ton of cooling capacity (refrigeration capacity) of a water chiller is equal to 12,000 BTU/h (3,024 Kcal). The British equivalent is 14,256 BTU/h (3,592 Kcal/h). In Europe, the frigorie/h unit is also used, which amounts to 756 Kcal/h (3,000 BTU/h).

Chiller capacities in the United States are based on a flow rate of 2.4 gallons per minute per ton and a temperature drop of 10°F.

Chiller sizing is determined by the following formula:

$$\text{Tons of refrigeration} = \text{GPM} \times \Delta T / 24$$

where GPM is the water flow rate in gallons per minute and ΔT is the temperature differential in °F between the water leaving and entering the system.

L. Spitz

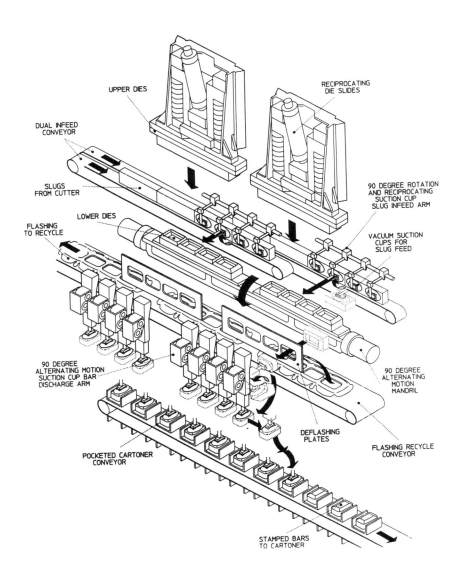

Fig. 8.20. Mazzoni LB STUR Press with noncontact soap transfer. (Courtesy of Mazzoni LB, S.p.A.)

The design capacities are also based on ambient air at 35°C for air-cooled chillers and 30°C condenser water temperature for water-cooled chillers. As a rule of thumb, a chiller's tonnage capacity is reduced by 2% for each 0.5°C below 10°C.

Low-Temperature Glycol/Water Chillers for Soap Press Dies

Low-temperature glycol chillers are designed to supply a glycol/water coolant solution to soap press dies. For soap press applications, they are designed and rated to supply the coolant at − 30°C (− 22°F) temperature.

The rating is expressed in tons of chilling capacity, but the actual load is temperature dependent. As the temperature rises from - 30°C to 0°C, the cooling capacity increases rather sharply. As an example, a unit rated for about 5,000 Kcal/h at - 30°C will provide 8,000 Kcal/h at − 20°C, 11,000 Kcal/h at − 10°C, and 17,000 Kcal/h at 0°C. The rating is also affected by the ambient temperature. At 28°C it is nominal but at 38°C it is about 12% lower.

The load requirement for a specific press must be provided by the soap press producer, and the exact rating of the lower temperature glycol chiller should be specified by the chiller manufacturer.

Temperature Control Units

Temperature control units are recommended for cooling and heating the water circulating in the barrel of each plodder and in each roll of a roll mill. Special low-temperature glycol control units are used for handling the glycol/water coolant solution circulating through the soap press dies.

Both of these independent self-contained units consist of a high-pressure centrifugal pump, an immersion heater, solenoid valve, and microprocessor-based controller, which provide the following:

- Improved heat transfer due to increased turbulence created by increased water and/or glycol/water solution flow
- The possibility to maintain water and/or glycol/water solution temperature within ± 0.5°C of the circulating fluid, which can be set to be different from the main chiller temperature

The first-soap related application of these units—commonly used in the plastics industry—was started by Application Engineering Corporation in the United States. Their trademark name, TurbuFlow, is now also used by the European soap equipment producers under the generic name of Turboflow units.

Water Cooling and Heating System for Plodders and Roll Mills

One central chiller with separate individual temperature controllers can be used to cool each plodder barrel with optimum water temperature (Fig. 8.21).

These closed-circuit individual temperature controllers are designed to cool and also to heat the water to different temperatures. The water heating capability will prevent plodder overloading when restarting after a prolonged shutdown and thus will protect against possible machine part damage. This application is essential for large-capacity lines and for all of the machinery producing synthetic, combo, and other quick-to-harden products regardless of size.

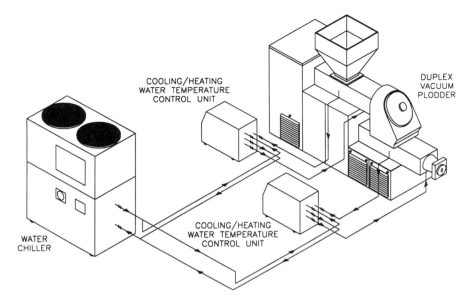

Fig. 8.21. Chilled water system for plodders.

Also, for roll mills, the dual cooling and heating feature is very important since for certain products the first roll is heated while the others must be cooled to different temperatures. The use of rotary joints is the best method for water distribution to and from each roll of a roll mill.

For best operating performance, each roll mill should have its own independent water temperature control system. A complete system for a three-roll mill consists of a central water chiller, which operates at full water flow requirements, and three separate temperature control units (Fig. 8.22).

Multizone Die Chiller System

Multizone die chilling increases line productivity by optimizing stamping efficiency. Stamped bar release from the dies of the reciprocating and rotating die groups of the horizontal flashstamping presses and the reciprocating and alternating motion die groups from the vertical-type units is best achieved with different temperatures for each group.

Figure 8.23 illustrates the setup for a central glycol/water chiller that operates at a predetermined temperature and two low-temperature control units that can be set to provide the coolant with different temperatures. The turbulent flow increases the heat transfer and also equalizes the temperature throughout each die group.

The use of low-temperature glycol/water chillers for chilling soap press dies have been common practice for years, but the utilization of additional independent low-temperature control units is relatively low.

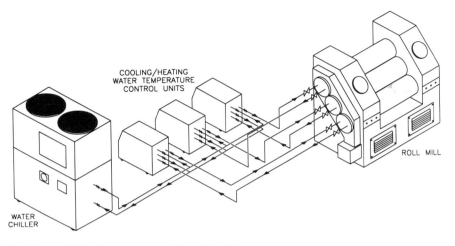

Fig. 8.22. Chilled water system for roll mills.

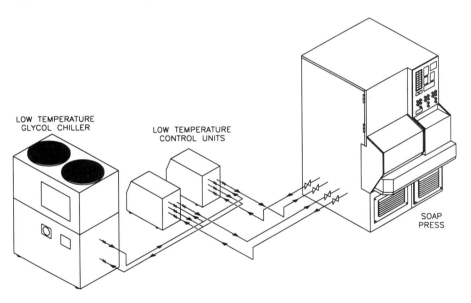

Fig. 8.23. Low-temperature glycol chiller system for soap press dies.

Bar Soap Finishing Lines

The selection of a finishing line depends on the following:

- Types of products to be produced
- Line operating speed

- Preference between an all plodder or a combination plodder and roll mill line
- Number of refining stages required
- Handling of recycle-based on total quantity and the location for its rework

Also, it is important to consider in the line selection the need for the following:

- Multifunction line layout
- Prerefining method for certain product
- Aging method

Figure 8.24 is a three-dimensional summary rendition of all the bar soap finishing processing steps and names of the machinery used for each operation.

All the standard and special bar soap finishing lines are illustrated in Fig. 8.25. Due to the large number of variations, enlarged portions of Fig. 8.25 are shown in Figs. 8.26, 8.27, and 8.28. The following description will help to clarify the two summary finishing lines illustrations in Figs. 8.24 and 8.25.

Standard Lines

The most widely used standard lines are the following:

- Line with three plodders: simplex refiner and a duplex vacuum plodder (two refining stages)
- Line with four plodders: duplex refiner and a duplex vacuum plodder (three refining stages)
- Line with three plodders and one roll mill: simplex refiner, roll mill, and a duplex vacuum plodder (three and one-half refining stages)
- Line with one roll mill and two plodders: roll mill and a duplex vacuum plodder (two and one-half refining stages)

Prerefining Lines

Any line can be made into a prerefining line by placing a simplex refiner or, alternatively, a roll mill before the mixer (Fig. 8.26).

Multifunction Lines

In the standard lines, the products have to pass from one machine to another without the possibility of bypassing one stage. For some products, using less mechanical work (less refining) is indicated (Fig. 8.27).

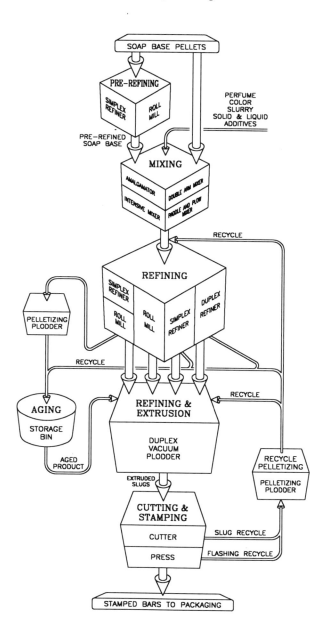

Fig. 8.24. Bar soap finishing lines.

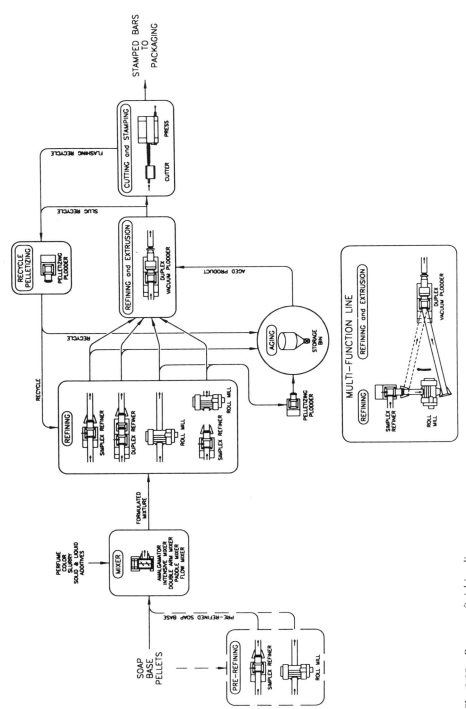

Fig. 8.25. Bar soap finishing lines.

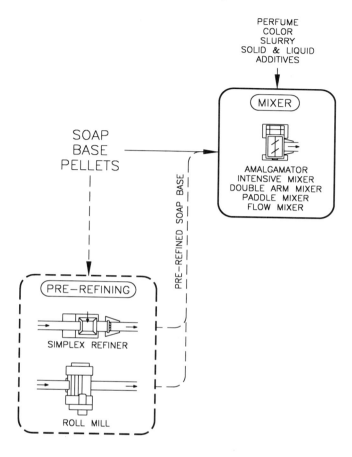

Fig. 8.26. Bar soap finishing lines—prerefining and mixing.

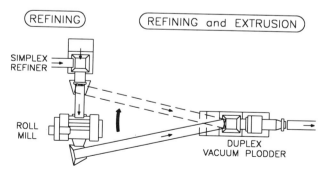

Fig. 8.27. Bar soap finishing lines—multifunction line.

As an example, a multifunction line with three plodders and one roll mill is shown. This is a dual-function line in which the roll mill can be bypassed if so desired. A similar line with four plodders would have a right-angle duplex refiner to allow bypassing the first-stage plodder when not needed.

Line with Refined and Aged Pellets

Storing refined soap in the middle of the finishing operation and aging will not only enhance the finished bar performance characteristics but will also increase overall line productivity due to lower processing temperatures (Fig. 8.28).

High-Speed Lines

For the production of 300 and more bars per minute, four types of plodder, cutter, and press combinations can be used (Fig. 8.29):

- One single-worm plodder with single extrusion from one nose cone
- One tangential twin-worm plodder with dual extrusion from one cone followed by one cutter and one press
- One tangential or nontangential twin-worm plodder with single extrusion from two separate nose cones, twin cutter and one press
- Two side-by-side single-worm plodders with single extrusion from two nose cones followed by two separate single cutters and one press

Computerization

Please refer to Chapter 17 of this book for an overview of computerization relating to bar soap finishing.

Future Challenges

Tomorrow's challenge is to minimize the total number of machinery required for each line and to allow the production of the new, more complex products more economically and with greater ease. The recent innovations in cutter, press, and bar soap transfer designs prove that there is new creativity and commitment in the industry to advance bar soap finishing technology. We are looking forward with great interest to the invention of exciting new machinery in the near future, especially in the intensive mixing and refining areas.

Acknowledgments

For the CAD/CAM-generated three-dimensional drawings and summary tables of finishing equipment in this chapter, I am indebted to many companies who offer soap finishing machinery. Their names are duly acknowledged on the respective illustrations—all prepared

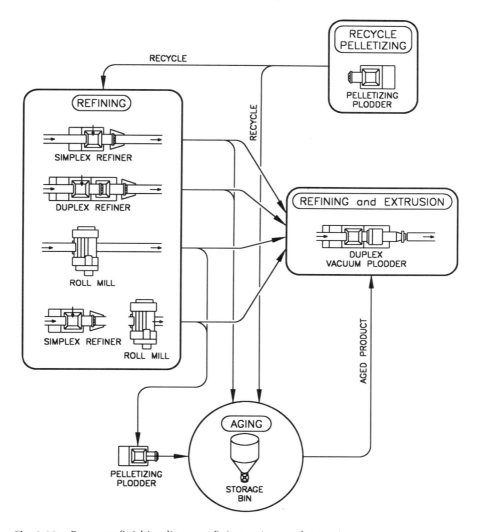

Fig. 8.28. Bar soap finishing lines—refining, aging, and extrusion.

for the first time—which they provided. Without this most valuable contribution, the chapter would be less visual and less comprehensive.

I wish to thank in particular Mazzoni LB, S.p.A. for their help in the preparation of several illustrations and for supplying technical material. The assistance of The Dial Corp. in various aspects of this presentation is greatly appreciated.

Finally, a special note of thanks to my old friend and associate, Fred Simonian, for his technical input and expert guidance. We both started our soap careers in 1961, when The Dial Corp., located in Chicago, was called Armour Grocery Products Company.

L. Spitz

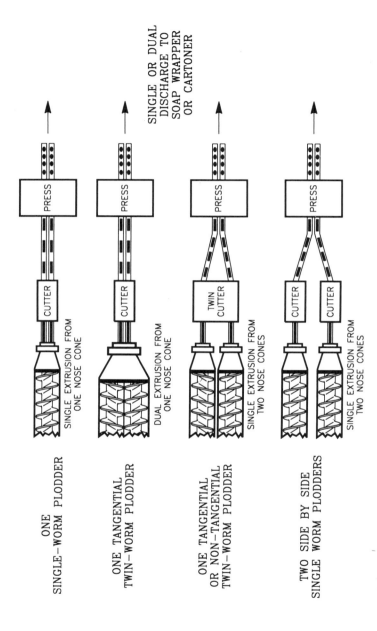

Fig. 8.29. High-speed finishing lines.

References

1. Spitz, L., ed., *Soap Technology for the 1990's,* AOCS Press, Champaign, Illinois, 1990, pp. 173–208.
2. Woollatt, E., *The Manufacture of Soaps, Other Detergents and Glycerine,* Ellis-Horwood Limited, London, England, 1985.
3. Hensen, F., ed., *Plastics Extrusion Technology,* Hanser-Gardner, Cincinnati, Ohio, 1988.
4. White, J.L., *Twin Screw Extrusion,* Hanser-Gardner, Cincinnati, Ohio, 1991.
5. McDonough, R.J., *Mixing for the Process Industries,* Van Nostrand Reinhold, New York, 1992.

Chapter 9

Surfactant Raw Materials: Classification, Synthesis, and Uses

K. Lee Matheson

VISTA Chemical Co./D.A.C. Industrie Chimiche S.p.A., Milan, Italy

In general, surfactants have two common structural features: a water-soluble head group and an oil-soluble hydrocarbon tail group. For many surfactants the hydrocarbon tail or chain makes up more than half the weight of the molecule (Fig. 9.1). The focus of this chapter is on the raw materials and processes used to produce the hydrocarbon chains that make up the tail groups.

Basic Raw Materials

With few exceptions, the hydrocarbon chains of large-volume commodity surfactants and many specialty surfactants are produced from one or more of the following raw materials:

- Natural fats and oils
- Petroleum
- Ethylene
- Propylene

Surfactant Intermediates from Fats and Oils

In natural fats and oils the hydrocarbon chains have already been formed in the raw material. Biological processes in living organisms synthesize long carbon chains in

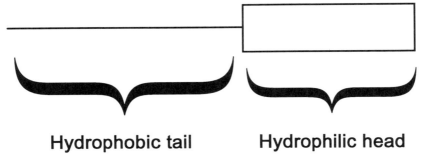

Hydrophobic tail   Hydrophilic head

Fig. 9.1. General surfactant structure.

the form of triglycerides. From plant and animal oils the triglycerides are separated and chemically converted into key surfactant intermediates. Coconut oil and palm kernel oil are preferred for the production of C12–C14 chain lengths. Animal fats and palm oil are preferred for the production of C16–C18 chain lengths (see Table 9.1).

Fatty Acids

In general, the ester linkage in the triglyceride molecules can be severed in two ways. In one process steam is used to hydrolyze the triglycerides to yield fatty acids and glycerin (Fig. 9.2). About one-third of fatty acid production is converted to surfactants or surfactant intermediates such as soap, fatty alcohols, fatty amines, alkanolamides, and specialty surfactants such as betaines.

Fatty Methyl Esters

In the other process methanol is used to transesterify the triglycerides to yield fatty methyl esters and glycerin (Fig. 9.3). Most of the fatty methyl esters produced are converted to fatty alcohols. Also produced from methyl esters are alkanolamides, methyl ester sulfonate, various specialty surfactants, and synthetic, low calorie cooking oils (Fig. 9.4).

Natural Fatty Alcohols

Natural fatty alcohols of C12 to C18 chain lengths are produced by the hydrogenation of both fatty methyl esters and fatty acids (Fig. 9.5). These alcohols are even

TABLE 9.1
Composition of Natural Triglycerides (wt%)

Triglyceride Fat or Oil	Caprylic, C8	Capric, C10	Lauric, C12	Myristic, C14	Myristoleic, C14	Pentadecanoic, C15	Palmitic, C16	Palmitoleic, C16	Margaric, C17	Stearic, C18	Oleic, C18	Linoleic, C18	Linolenic, C18	Ricinoleic, C18	Arachidic, C20	Eicosenoic, C20	Behenic, C22
Tallow				3.2	1.0	0.4	26.4	2.6	0.9	26.9	36.7	(~1)					
Linseed							6.3			4.3	18.2	14.3	56.7				
Cottonseed			0.6				21.7			2.1	17.8	57.9					
Palm			0.9				46.6			4.1	39.3	9.1					
Soybean							10.5			3.8	23.7	55.5	6.6				
Groundnut							9.0			3.5	64.5	18.2			1.2	(0.9?)	(1.6?)
Castor							0.9				9.6	10.3		79.0			
Coconut	8.0	6.7	51.3	16.2			7.6			2.7	5.9	1.6					
Palm kernel	4.0	5.0	50.0	15.0			7.0	0.5		2.0	15.0	1.0					

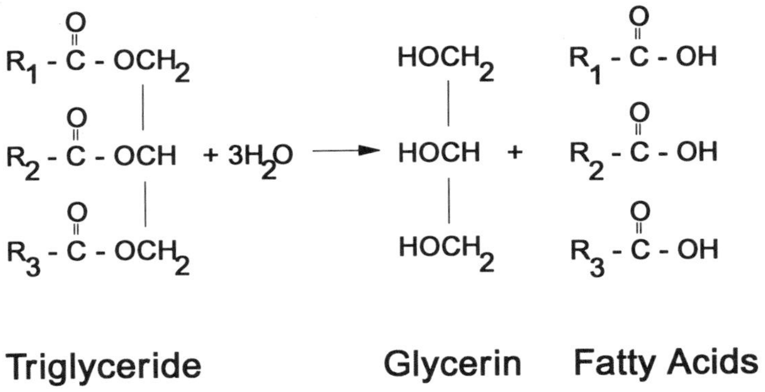

$$
\begin{array}{ccc}
R_1 - \overset{\overset{\displaystyle O}{\|}}{C} - OCH_2 & HOCH_2 & R_1 - \overset{\overset{\displaystyle O}{\|}}{C} - OH \\[2mm]
R_2 - \overset{\overset{\displaystyle O}{\|}}{C} - OCH \; + 3H_2O \longrightarrow & HOCH \; + & R_2 - \overset{\overset{\displaystyle O}{\|}}{C} - OH \\[2mm]
R_3 - \overset{\overset{\displaystyle O}{\|}}{C} - OCH_2 & HOCH_2 & R_3 - \overset{\overset{\displaystyle O}{\|}}{C} - OH
\end{array}
$$

Triglyceride   **Glycerin**   **Fatty Acids**

Fig. 9.2. Fatty acid production (fat splitting).

carbon chain length, 100% linear, primary alcohols. Fatty alcohols are very important oleochemical-based surfactant intermediates. From them are made many surfactant products including alcohol sulfates, alcohol ethoxylates, and alcohol ether sulfates (Fig. 9.6). There is even commercial production in India of alpha olefins by the dehydration of natural fatty alcohols (the Godrej-Lurgi Process).

Surfactant Intermediates Produced from Petroleum

Linear hydrocarbon chains or normal paraffins can be extracted from petroleum fractions. Kerosene and gas oil are different boiling fractions of petroleum that contain hydrocarbons of C10–C16 and higher chain lengths.

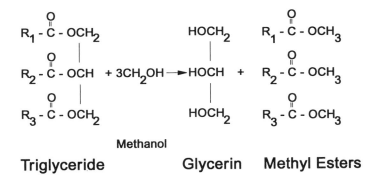

Triglyceride   **Glycerin**   **Methyl Esters**

Fig. 9.3. Methyl ester production (transesterification).

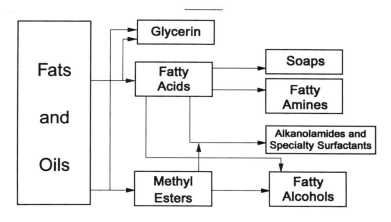

Fig. 9.4. Natural fat/oil-based surfactant intermediates.

Normal Paraffins

For surfactant production kerosene is the most important hydrocarbon source. Using molecular sieve separation processes such as MOLEX or ISOSIV, the linear or normal paraffins are separated from the branched and cyclic hydrocarbons (Fig. 9.7). Typically 20%–25% of the kerosene consists of normal paraffins in the C10–C16 chain length range. The normal paraffins are distilled into various cuts for surfactant manufacture. The branched/cyclic hydrocarbon stream or raffinate is sold as an upgraded fuel.

$$R - \overset{O}{\overset{\|}{C}} - OH \ + \ H_2 \ \longrightarrow \ ROH \ + H_2O$$

Hydrogenation of Fatty Acids

$$R - \overset{O}{\overset{\|}{C}} - OCH_3 \ + \ H_2 \ \longrightarrow \ ROH \ + CH_3OH$$

Hydrogenation of Methyl Esters

Fig. 9.5. Natural alcohol production.

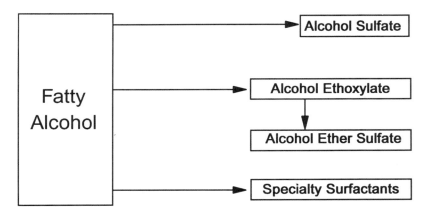

Fig. 9.6. Surfactant alcohol derivatives.

Linear Alkylbenzene

Linear alkylbenzene (LAB) is the most important surfactant intermediate made from normal paraffins and the largest-volume surfactant intermediate produced today. The major process for manufacturing LAB today is the UOP PACOL/HF process. This process involves first the catalytic dehydrogenation (the PACOL process) of n-paraffins to convert about 12% of the paraffin to internal olefin (Fig. 9.8). The internal olefins are then reacted with benzene using liquid HF as catalyst (Fig. 9.9). The liquid HF is first separated, and the remaining organic mixture of benzene, paraffin, LAB, and heavy alkylate is separated by distillation. This process produces a low 2-phenyl type LAB.

The older but less widely practiced process for producing LAB involves chlorination of the normal paraffin, followed by alkylation with benzene using

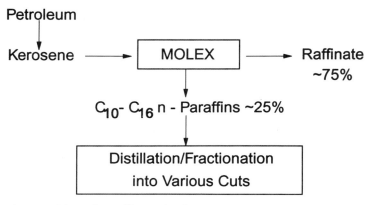

Fig. 9.7. Normal paraffin production.

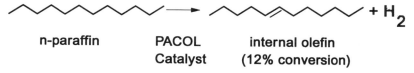

n-paraffin   **PACOL**   **internal olefin**
            Catalyst   **(12% conversion)**

Fig. 9.8. PACOL dehydrogenation of n-paraffins.

$AlCl_3$ catalyst. In both LAB processes the crude mixture of excess benzene, excess paraffin, LAB, and heavy alkylate is then separated by distillation. The newest process for making LAB is the UOP PACOL/DETAL process. In this process, the alkylation reaction takes place in a fixed bed of solid catalyst, rather than a liquid HF catalyst. The type of LAB produced is a high 20 variety.

Internal Olefins

Pure internal olefins can also be produced from linear paraffins. In the combined PACOL/OLEX process, dilute PACOL olefins are concentrated by the OLEX process to about 96% internal olefin (Fig. 9.10). These olefins can be used to alkylate benzene using either $AlCl_3$ or HF as catalyst. An older process for producing internal olefins involves chlorination of the paraffins followed by dehydrochlorination.

OXO Alcohols Based on Paraffins

Internal olefins can also be converted to OXO alcohols. In contrast to natural alcohols, OXO alcohols have both odd and even chain lengths, and they have up to 50% branching at the second carbon position.

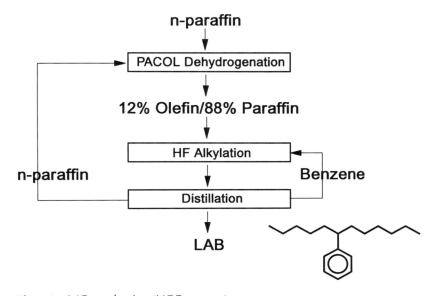

Fig. 9.9. LAB production (UOP process).

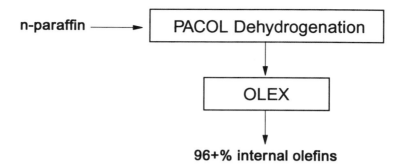

Fig. 9.10. PACOL/OLEX internal olefins.

Secondary Alkane Sulfonate

Normal paraffins of C14–C17 carbon chain lengths are used in the sulfoxidation process to make paraffin sulfonate or secondary alkane sulfonate (SAS) (Fig. 9.11).

Surfactant Intermediates Produced from Ethylene

Ziegler Ethylene Growth Processes

For surfactant applications ethylene is used as a building block to form long hydrocarbon chains. This process usually employs what is called a growth reaction to make hydrocarbon chains of C2 to C20 in length. Hydrocarbon chains are grown by adding ethylene units to an organometallic compound such as triethyl aluminum (Fig. 9.12). The ethylene units are inserted between the growing alkyl chains and the aluminum, producing trialkyl aluminum or growth product.

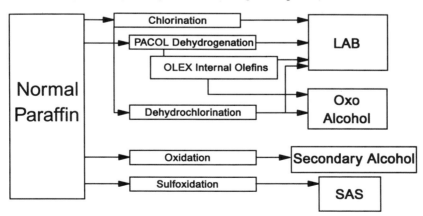

Fig. 9.11. Paraffin-based surfactant intermediates.

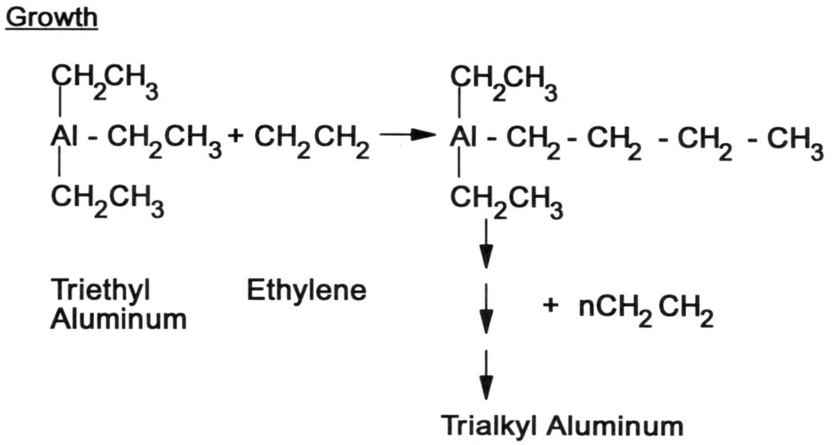

Growth

$$CH_2CH_3 \qquad\qquad\qquad CH_2CH_3$$
$$Al - CH_2CH_3 + CH_2CH_2 \longrightarrow Al - CH_2 - CH_2 - CH_2 - CH_3$$
$$CH_2CH_3 \qquad\qquad\qquad CH_2CH_3$$

Triethyl Aluminum	Ethylene

$$\downarrow$$
$$\downarrow \quad + nCH_2CH_2$$
$$\downarrow$$

Trialkyl Aluminum

Fig. 9.12. Ziegler ethylene growth process.

Ziegler Alcohols

Further processing of this growth product can yield linear primary alcohols directly or alpha olefins (Fig. 9.13). In the Ziegler alcohol process (ALFOL process) linear, even-carbon-chain fatty alcohols are produced from the growth product by controlled oxidation followed by hydrolysis (Fig. 9.14). For a given chain length, these alcohols are essentially identical to natural alcohols, having linear, even-carbon-chain-length primary structures. A stoichiometric amount of aluminum is used in this process that eventually is converted into high-purity alumina after hydrolysis.

Ziegler Alpha Olefins

In a modification of the Ziegler alcohol process, alpha olefins are produced by an olefin exchange involving ethylene with the alkyl groups on the growth product. This modification is called transalkylation. Alcohols of a C12–C14 chain length can also be produced eventually by a second transalkylation step with C12–C14 alpha olefins, followed by oxidation and hydrolysis (Fig. 9.15). But alpha olefins are the main product of this modified Ziegler process. Due to the extra transalkylation steps, there is a higher degree of branching in these alcohols (Epal process) compared with the ALFOL alcohols.

SHOP Alpha Olefins

The SHOP process employs an ethylene oligomerization reaction to make alpha olefins. In the first part linear, even-carbon-chain alpha olefins are produced (Fig. 9.16). As with other ethylene growth reactions, the olefins are produced in a broad distribution of carbon chain lengths. Some of these chain lengths are more desir-

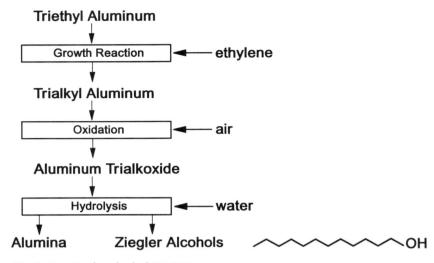

Fig. 9.13. Ziegler alcohol (ALFOL) process.

able as alpha olefin products than others, so they are separated out by distillation and sold.

SHOP Internal Olefins and Modified OXO Alcohols

In the second part of the SHOP process, alpha olefins of the less desirable chain lengths are converted to linear internal olefins in a complicated process called isomerization/disproportionation/metathesis. The internal olefins produced in this process have both odd and even chain lengths in the range of C10 to C14. This makes them very useful for the production of alcohols and for LAB. Internal olefins

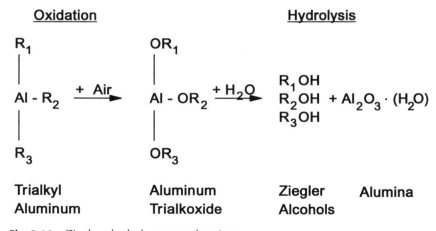

Fig. 9.14. Ziegler alcohol process chemistry.

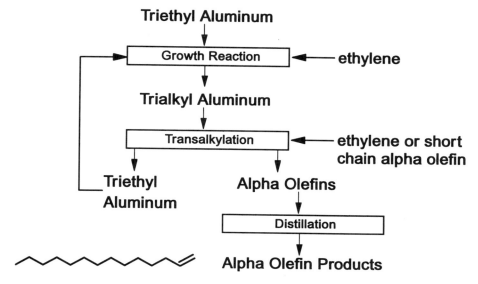

Fig. 9.15. Ziegler alpha olefin process.

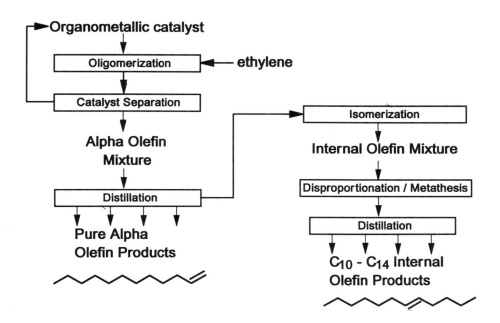

Fig. 9.16. SHOP olefin processes.

from the SHOP process are converted to alcohols in the Modified OXO process, which produces OXO alcohols having only 20%–25% branching (Fig. 9.17).

Fatty alcohols, internal olefins, and alpha olefins are all useful surfactant intermediates (Fig. 9.18). Fatty alcohols are converted to alcohol sulfate, alcohol ethoxylate, alcohol ether sulfate and fatty tertiary amines. Internal olefins are used to make modified OXO alcohols or to make LAB. Alpha olefins are used to make fatty amines, alpha olefin sulfonate, and OXO alcohols.

Ethylene Oxide

It should also be noted that although ethylene oxide is used to make the hydrophilic part of an ethoxylated surfactant molecule, which is not within the scope of this chapter, it is nevertheless produced from ethylene. This makes ethylene doubly important as a feedstock for some alcohol ethoxylates.

Surfactant Intermediates Produced from Propylene

The fourth basic raw material for producing hydrocarbon chains is propylene. Surfactant intermediates based on propylene are characterized by a high degree of methyl branching. To a certain extent its utilization for surfactant hydrocarbon chains in household products has declined over the years. This is because the high degree of branching slows the rate of biodegradation.

Propylene Oligomers—Highly Branched Olefins

Propylene is oligomerized by a Lewis acid catalyzed process to produce highly branched mono-olefins. Under the acid-catalyzed oligomerization conditions many different olefin structures are formed of all chain lengths, but the greater percentages are in multiples of three carbons, that is, C9, C12, C15, and so on (Fig. 9.19).

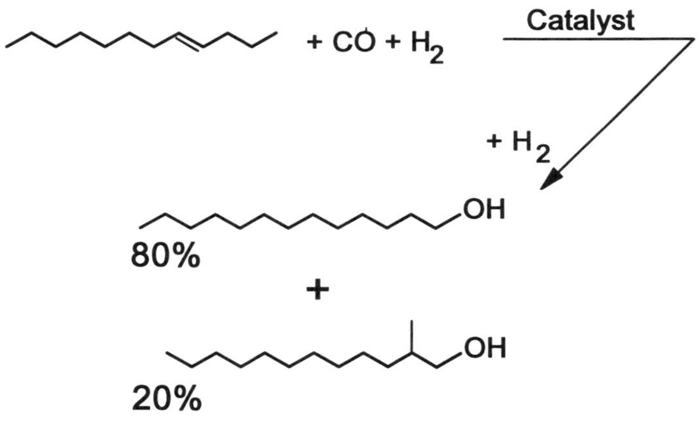

Fig. 9.17. Modified OXO alcohol process.

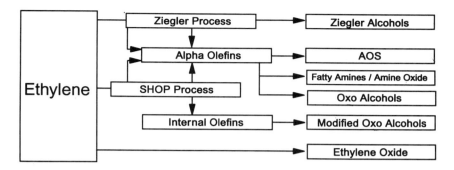

Fig. 9.18. Ethylene-based surfactant intermediates.

The olefins are then fractionated by distillation. As with other olefins, propylene oligomers can be used in various reactions.

Alkyl Phenol, Dodecyl Benzene, and Isotridecyl Alcohol

The fraction consisting primarily of propylene trimer can be used to alkylate phenol, which is then converted to alkylphenol ethoxylate. The fraction consisting primarily of propylene tetramer is used to produce branched alkyl benzene, which is converted to branched alkylbenzene sulfonate. Propylene tetramer can also be converted to an OXO alcohol, as in the case of isotridecyl alcohol (Fig. 9.20).

Propylene Oxide

Propylene is also used to make propylene oxide, which is used in the production of EO/PO block copolymer surfactants.

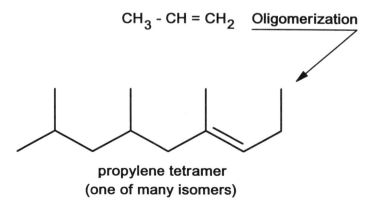

$$CH_3 - CH = CH_2 \quad \text{Oligomerization}$$

propylene tetramer
(one of many isomers)

Fig. 9.19. Propylene oligomerization.

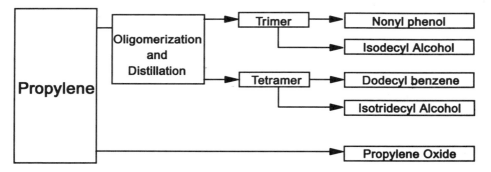

Fig. 9.20. Propylene-based surfactant intermediates.

Structural Characteristics of Detergent Alcohols

As has been noted previously, the same type of surfactant intermediates can be produced from different raw material sources. Detergent alcohols are the prime example of multiple sourcing because they can be made from all four raw material sources. However, the structural characteristics such as carbon chain distribution and the type and degree of branching will depend on the raw material source. Table 9.2 lists these differences in more detail.

Natural and Ziegler Alcohols

Natural and Ziegler alcohols are essentially equivalent in physical properties and performance characteristics, provided the carbon chain distributions are the same. Both are even-carbon-chain, linear, primary fatty alcohols. When both types of alcohols have equal carbon chain distributions and purity they will exhibit similar sulfation and viscosity characteristics of the alcohol sulfate, ethoxylate, and ether sulfate derivatives.

OXO and Modified OXO Alcohols

OXO and modified OXO alcohols are also primary alcohols. They have both odd and even chain lengths with a linear side chain at the second carbon position. The degree of branching is about 50% for OXO and 20%–25% for modified OXO. This branch-

TABLE 9.2
Structural Comparison of Detergent Alcohols

Alcohol Type	Raw Material	Carbon Chain Distribution	Percent Linear
Natural	coconut oil PKO	C10–C18 even only	100%
Ziegler (ALFOL)	ethylene	C2–C20 even only	98%
Modified OXO	ethylene	C12–C15 even and odd	80%
Regular OXO	ethylene, n-paraffin	C12–C15 even and odd	50%
Isotridecyl	propylene	C13 average	0%

ing is readily seen in the GC traces of the alcohols (Figs. 9.21 and 9.22). It should be emphasized that this type of branching poses no problem for biodegradability.

Isotridecyl Alcohol

Isotridecyl alcohol is also a primary alcohol, averaging C13 in chain length with a 100% branched structure. The branching in this case consists of several methyl side groups along the chain.

Performance Effects of Chain Distribution and Branching

The performance characteristics (foam and detergency) of an alcohol-derived surfactant depend strongly on the carbon chain length distribution and, to a lesser extent, on the branching content. Formulation properties, such as viscosity in liquid formulations, will depend on both branching content and carbon chain distribution. Table 9.3 summarizes these effects for AS and AES derivatives.

The carbon chain distributions of OXO and modified OXO alcohols are inherently different from natural or Ziegler alcohols because of the presence of odd carbon chains. Nevertheless, at the same average carbon chain length, the alcohol sulfate, ethoxylate, and ether sulfate derivatives of the modified OXO alcohols will exhibit similar detergency performance. The foam stability of OXO ether sulfates in mixed active dishwashing liquids is also generally equal to the performance of corresponding linear alcohol derivatives in those formulations. The high level of branching (up to 50%–60%) in some OXO alcohols results in their ethoxylates having a higher level of free alcohol and a broader EO distribution.

Formulation properties (that is, viscosity and solubility) of alcohol derivatives are strongly dependent on the branching content of the starting alcohol. Modified OXO and OXO alcohol derivatives exhibit lower viscosities in liquid formulations compared with the natural or Ziegler alcohol derivatives. For alcohol sulfates and

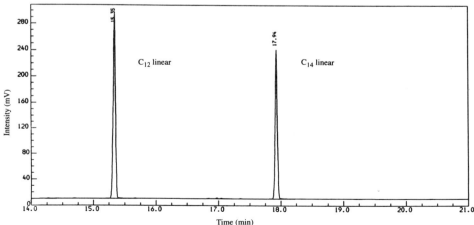

Fig. 9.21. Capillary GC trace for C12–C14 Ziegler alcohol.

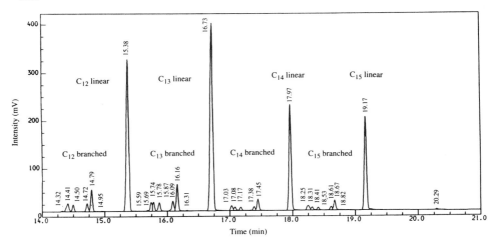

Fig. 9.22. Capillary GC trace for C12–C15 modified OXO alcohol.

ether sulfates used in shampoo formulations, the natural or Ziegler alcohols are preferred due to their higher formulation viscosity characteristics. On the other hand the lower-viscosity behavior of OXO alcohol ether sulfates makes them preferred for certain concentrated light duty liquids.

Economic Considerations

The hydrocarbon chain usually makes up the largest and most expensive part of a surfactant molecule. Thus the basic raw material costs exert a strong influence on the price of surfactants. The relative raw material costs and supply availability for oleochemical oils, petroleum, ethylene, and propylene help to determine the market prices for surfactants and surfactant intermediates.

Over the years kerosene (for n-paraffins) has generally been the lowest-cost raw material. As a result, where they can be substituted freely, surfactants based on n-paraffins have been somewhat more cost competitive.

TABLE 9.3
Effects of Surfactant Alcohol Structural Characteristics
on Alcohol Derivative Performance

Performance Effect	Carbon Chain Distribution	Branching Content
Detergency	Optimum ave. chain length gives best detergency	Little effect
Foam stability	Optimum ave. chain length gives best foam	More branching decreases foam slightly
Viscosity	Longer chain length increases viscosity	More branching lowers viscosity of AS, AES
Solubility	Longer chain length decreases solubility	More branching increases solubility

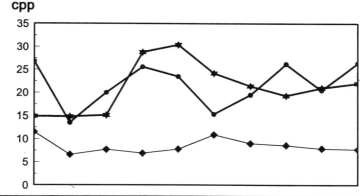

	1985	1986	1987	1988	1989	1990	1991	1992	1993	1994
Coconut Oil ●	26.8	13.5	20.0	25.6	23.5	15.3	19.5	26.2	20.5	26.4
Ethylene ★	14.9	14.8	15.1	28.8	30.4	24.2	21.4	19.3	21.0	22.0
Kerosene ◆	11.5	6.6	7.7	6.9	7.8	10.9	9.0	8.6	7.9	7.7

Fig. 9.23. Average annual prices in cents per pound for coconut oil, ethylene, and kerosene.

Natural oils have in the past been somewhat volatile in pricing, but recently their prices have stabilized due to increased production of oils, especially palm and palm kernel oils. Ethylene pricing is particularly important to the price of alcohol ethoxylates because ethylene is the basis not only for the production of alcohol but also for the production of ethylene oxide. The averaged yearly U.S. prices in cents per pound for ethylene, coconut oil, and kerosene for the last decade are summarized in Fig. 9.23. At the time of this writing (summer 1994), both ethylene and edible oils remain higher in price than kerosene. Tightened market supplies of edible oils and ethylene production problems are causing pressure to increase prices of oleochemical- and ethylene-based surfactants.

Conclusion

It is important for the detergent producer to understand the cost relationships of raw materials as well as the performance characteristics of different surfactants. During times of fluctuating raw material prices, the wise detergent producer can change formulations, within certain limitations, to take advantage of these price fluctuations or to lessen their negative impact.

It is not surprising then that today many detergent formulations are combinations of two, three, or more different surfactants where the ratios have been developed to maximize performance and minimize cost. Mixed active detergent formulations allow the manufacturer more flexibility to change these ratios, whereas a formulation based on a single surfactant is not so easily changed.

Chapter 10

Raw Material Selection for Detergent Manufacture

René Maldonado

FMC Corporation, Princeton, New Jersey, U.S.A.

Introduction

Today's detergent market has placed strict demands on the quality and specifications of raw materials for detergents. Detergents have evolved to include significant variations in composition, physical form, and dosage. For each product type there are both a well-defined manufacturing process and a list of raw material specifications that must be met. In consumer products, for example, the advent of compact or ultra detergents has placed emphasis on processes that produce concentrated and relatively dense products that deliver acceptable performance at low dosage under normal use conditions. Products like these often require raw materials that complement the manufacturing process and facilitate the production of the finished product with the desired properties.

The suppliers of raw materials have developed grades of the standard ingredients with variations intended to meet the requirements of the finished product and the manufacturing process. The following parameters are always considered for the inorganic components:

- *Assay:* Usually high assay is preferred for better quality and for extending the shelf life of the finished product.

- *Density and particle size distribution:* Product segregation and flow properties are in part determined by the density and particle size distribution of the solid ingredients, which also play a role in the absorptivity of surfactants, and other liquids.

- *Friability:* The tendency of a solid material to crumble or be easily pulverized is called friability. This property may affect the particle size distribution in the finished product and affect product segregation, flow properties, and absorptivity. Friability is an important consideration during product processing and shipment.

- *Hydration characteristics:* Water is present in most detergent products, including detergents in powder form. Often a solid ingredient will bind some of that water to form hydrates. The rate of hydration and the stability of the resulting hydrated species affect the caking and flow properties of powder detergents and the stability of liquid products.

- *Chemical stability:* Detergent ingredients must be compatible with each other and must be stable enough to withstand the manufacturing process. This is especially critical in liquid detergents and in powder detergents containing bleach, enzymes, or high alkalinity.

The selection of raw materials for detergent manufacture can be illustrated from our experience as a long-term supplier of detergent ingredients. The properties and use recommendation for the various types of phosphates, carbonates, and oxygen bleaches we supply are discussed in the sections that follow.

Phosphates

Phosphate salts are used as builders in detergents, where they provide water softening, alkalinity, and soil suspension and dispersion. They are also excellent processing aids during detergent manufacture, providing benefits such as the absorption of liquid surfactants and the binding of free water. There is a broad selection of phosphate chemicals for use in detergents and other related applications (Table 10.1). It should be kept in mind, however, that the choice of phosphate ingredients for some applications is at the present time limited by legislation in some parts of the United States and Europe.

The most commonly used phosphates for detergent applications are sodium and potassium salts of pyrophosphate and tripolyphosphate. Some of the chemical properties of these compounds are listed on Table 10.2. All these salts give solutions of alkaline pH and are effective in chelating both calcium and magnesium ions in water. The potassium salts are significantly more soluble in water than the sodium salts and are preferred for many liquid detergent applications. Sodium tripolyphosphate (STPP) is by far the most important phosphate for detergent

TABLE 10.1
Phosphate Products and Applications

Phosphate	Salt	Applications
Ortho	Sodium and potassium (mono-, di-, and tribasic)	Metal cleaning and treatment, janitorial, water treatment, hazardous metals fixation
Pyro	Tetrasodium and tetrapotasssium Sodium acid	Water treatment, laundry detergents Acid cleaners
Tripoly	Sodium (several grades) Potassium Sodium hexahydrate	Laundry and dishwash detergents I & I Cleaners
Glassy phosphates	Sodium chain lengths: n = 6 (Sodaphos®), n = 13 (Hexaphos®), n = 21 (Glass-H®)[a]	Water treatment, bath salts, clay manufacturing, ore drilling mud
Phosphoric acids	85%–117% H_3PO_4, technical and electronic grades	Metal treatment, electronics, acid cleaners, hazardous metals fixation

[a]Hexaphos, Sodaphos, and Glass-H are registered trademarks of FMC Corporation.

TABLE 10.2
Chemical Properties of Phosphate Salts

Properties	pH (1% Solution)	Solubility (g/100 g H$_2$O @ 25°C)	Sequestering Capacity[a] (g Ca/100 g Phosphate)
Pyrophosphates			
Tetrasodium pyrophosphate (TSPP)	10.3	8	15.2
Tetrapotassium pyrophosphate (TKPP)	10.5	187	12.2
Tripolyphosphates			
Sodium tripolyphosphate (STPP)	9.9	15	10.9
Potassium tripolyphosphate (KTPP)	9.6	193	9.0

[a]Conditions: calcium ion electrode, 150 ppm hardness, alkaline pH, room temperature.

applications. Some of the detergent benefits derived from STPP include hardness control, alkalinity and buffering, soil dispersion and peptization, processing aid for powder manufacturing (absorbs surfactants and binds moisture), and the control of the rheology and stability of liquid detergents.

STPP Physical Form

In selecting a grade of STPP for a particular application there are several parameters to choose from. Among these are granularity (powder or granular grades of various densities), crystalline phase (Phase I, Phase II, or hexahydrate forms), and moisture content (dried or moisturized). Understanding these parameters is important because they control the behavior of STPP in the presence of water during detergent manufacture and in the finished product.

Anhydrous STPP crystals can have one of two crystalline structures, referred to as either Phase I or Phase II crystals (1). Phase II crystals have a more stable structure than Phase I crystals. As a result, when Phase II STPP is placed in contact with water, it hydrates at a slow rate and forms relatively large crystals of the hydrated salt, STPP•6H$_2$O (Fig. 10.1). The Phase I form of STPP, on the other hand, hydrates relatively fast to provide a relatively large quantity of small hexahydrate crystals (2). The presence of a small level of STPP•6H$_2$O seed crystals (which can be achieved by premoisturizing STPP with as little as 0.5% water) will accelerate STPP hydration and make the Phase II form behave somewhat similar to Phase I material (3). Commercial STPP is always a combination of Phase I and Phase II STPP with or without moisture added. Any STPP with more than about 10% Phase I crystals is usually referred as Phase I STPP.

The importance of phase and moisture content of STPP is clearly illustrated in the case of detergents in slurry form. Table 10.3 shows the initial viscosities measured in a home laundry detergent slurry containing 15% STPP. When STPP of

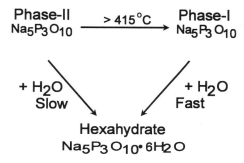

Fig. 10.1. Crystalline phases of sodium tripolyphosphate and their hydration.

very low Phase I content was used, an unstable slurry with relatively low viscosity was obtained. Here, hydration of STPP occurred relatively slow with little water bound during processing. The few hexahydrate crystals that eventually formed were relatively large in size and difficult to suspend in the slurry. On the other hand, when STPP of high Phase I content was used, a homogeneous and stable product of high viscosity was obtained. In this case, STPP hydrated quickly during processing to provide a large number of small hexahydrate crystals. Hydration of STPP into solid particles of STPP $\cdot 6H_2O$ results in the binding of a significant quantity of water (up to about 29 g of water for every 100 g of STPP). This alone would result in an increase in the viscosity of the slurry system. Viscosity is also directly proportional to the quantity of particles in the system, which in this case are mainly crystals of STPP$\cdot 6H_2O$.

If low–Phase I STPP is premoisturized (1.8% water in the example in Table 10.3), the slurry obtained is somewhat similar to that obtained using high–Phase I STPP. This effect is further demonstrated by measuring the viscosity of a laundry slurry as a function of time (Fig. 10.2). The system using moisturized STPP gives a slurry of relatively high viscosity, which does not change much on aging. On the other hand, the slurry made using dry STPP starts out with a very low viscosity. As discussed before, the effect of premoisturizing STPP is the formation of seed crystals of STPP$\cdot 6H_2O$, which will help promote hydration of the remainder of the STPP as soon as it comes in contact with water.

TABLE 10.3
Effect of STPP Phase and Moisture Content on Home Laundry Detergent Slurry

STPP Phase I, Content (%)	STPP Moisture, Content (%)	HLD Slurry,[a] Viscosity (cps)
7	0.03	2,000
72	0.02	7,300
7	1.87	6,500

[a]HLD slurry composition: 0.5% sodium carboxymethylcellulose, 3% sodium carbonate, 15% sodium tripolyphosphate, 12% linear alkylbenzenesulfonate (C_{13} LAS), 3% alcohol ethoxylate ($C_{9-11}EO_{2.5}$), 3% alcohol ethoxysulfate ($C_{12-15}EO_3S$), balance water.

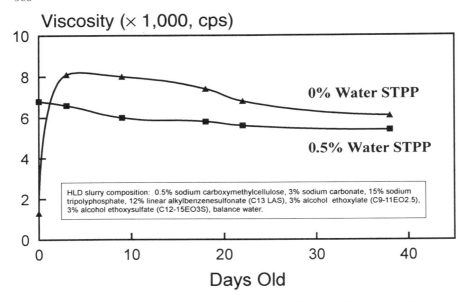

Fig. 10.2. Aging of home laundry detergent slurry made with dry or moisturized STPP of 3% Phase I content.

The effect of moisture and phase content of STPP on the *stability* of slurries is illustrated in the formulation of automatic dishwash slurries containing 25% STPP (Fig. 10.3). The slurries were formulated using STPP of various Phase I and moisture contents. They were then monitored for physical stability, which was taken as the number of days that they remained smooth and homogeneous before separating into two or more distinctive phases. By using STPP of relatively high Phase I content and adding at least 0.5% moisture, the stability of the slurry was maximized (Fig. 10.3).

STPP Selection

Table 10.4 lists the most common grades of STPP in terms of their densities. Most of these grades are available as either the Phase I or the Phase II crystalline form. In the agglomeration of home laundry detergent (HLD) powders, the density of granular STPP has an impact on the density of the finished product and its "absorptivity," which is defined as the maximum amount of nonionic surfactant that can be added and still obtain a dry, free-flowing powder. Light-density granular STPP has a density of about 0.5 g/cc and is able to absorb almost 18 g of nonionic surfactant per 100 g of phosphate. On the other hand, dense granular STPP (with a density of 1.0 g/cc) has an absorptivity of just 12 g of nonionic surfactant per 100 g of phosphate. The ability of STPP granules to hold surfactants is associated with the presence of microscopic pores on the STPP particles.

The types of STPP recommended for the different detergent applications are summarized in Table 10.5. In traditional home laundry powder manufacturing,

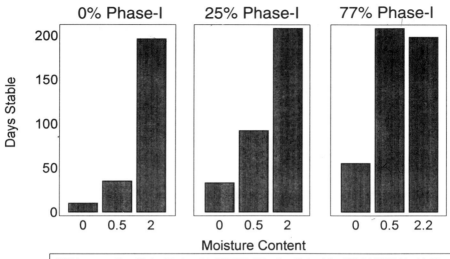

Fig. 10.3. Stability of automatic dishwashing detergent slurry made with STPP of various Phase I and moisture levels.

most of the ingredients are mixed in a crutcher to provide a slurry that is then dried in a spray tower to yield a granular product of relatively low density. This manufacturing method requires ingredients that will hydrate fast into a slurry that is smooth and has the proper viscosity for easy pumping. The ingredients must be chemically stable to the conditions found in the crutcher and in the spray tower and must yield a finished product that is acceptable in terms of density, solubility, flowability, and other predefined physical properties. In this application, moisturized STPP powder of Phase I crystalline structure is preferred. It hydrates quickly and yields a smooth slurry that, on drying, produces a detergent bead that is crisp and free flowing and does not tend to cake in storage.

TABLE 10.4
STPP Typical Grades: Typical Densities and Absorptivities

STPP Granularity	Bulk Density (g/cc)	Absorptivity (g of Surfactant/100 g STPP)*
Light density	0.51	18
Medium density	0.70	12
Regular density	0.93	8
Dense	1.01	12
Powder	0.85	< 6

*Absorption of nonionic surfactant (Triton X-100 from Union Carbide) measured with a Brabender Visco-Corder, Model VC-3, at 125 rpm.

TABLE 10.5
Phosphates for Detergent Applications

Detergent Form	Manufacturing Process	Phosphate Recommended
Regular home laundry powders	Spray drying Dry neutralization	STPP Phase I powder STPP Phase I powder and granular (minor)
Ultra home laundry powders	Spray drying plus agglomeration	STPP Phase I powder or granular
Regular automatic dishwash powders	Agglomeration	STPP Phase I powder or granular
Laundry or automatic dishwash slurries	Mixing	STPP Phase I powder
Laundry liquids	Mixing	KTPP and/or TKPP

Dry neutralization could be used for the manufacture of regular or ultra laundry detergents (4). In this method, the acid form of an anionic surfactant is sprayed directly onto an alkaline powder, where it is neutralized and absorbed together with any excess water. The alkaline powder usually consists of sodium phosphates and/or sodium carbonate. Compared to spray drying, this method reduces manufacturing costs and provides product with improved appearance. Powder STPP is the most effective grade of STPP for this application. It can be used in combination with granular STPP in applications in which nonionic surfactants are also part of the formulation.

In many processes for the manufacture of ultra or compact home laundry powders, a small portion of the product is still spray dried, but the bulk of the production may involve agglomerating liquid components (that is, nonionic surfactants or liquid silicates) with the powders (phosphates, zeolites, carbonates, and the spray-dried component). Post addition may also be part of this process. The physical properties of the solid ingredients are especially critical during the agglomeration step. Phase I STPP is again preferred for this application. The selection of STPP granularity is based on the best compromise on the need for density and absorptivity.

Automatic dishwash powders rely mainly on agglomeration technology. Many of the requirements discussed for the manufacturing of ultra home laundry powders also apply here. The absorptivity of nonionic surfactants, however, is not as important, because their level in these products is usually very small. One of the main considerations is the uptake of water from the agglomerating agent, usually liquid sodium silicate. The selection of STPP is based primarily on its capacity to hydrate quickly, to retain enough water without caking or lumping, and to form a finished product that dissolves easily. Phase I STPP, in powder or granular form and containing a small amount of added moisture, is usually selected for this application.

Detergents in slurry form can be formulated for laundry, dishwashing, and industrial and institutional applications. In these products, most of the inorganic salts are dispersed in an aqueous solution of surfactants and other ingredients. This process incorporates many of the same raw material requirements found in the crutcher step of traditional (spray-dried) home laundry powder manufacturing. Again, the ingredients must be stable, hydrate quickly, and yield a smooth slurry. In this case, however, the slurries must have long-term stability in terms of composition, rheology, and physical appearance. Phase I STPP with a small level of moisture is preferred for this application; it hydrates quickly into micron-size crystals during the manufacturing process.

Products in gel or liquid form require ingredients that remain dissolved and stable in the finished detergent. Potassium salts of phosphate are ideal for this application because of their high solubility. Potassium tripolyphosphate and potassium pyrophosphate are used throughout. Sodium salts of these ingredients can be used as long as other sources of potassium ions (such as potassium hydroxide or potassium carbonate) are included in the formulations.

Sodium Carbonate

Like phosphates, sodium carbonate is a multifunctional detergent ingredient. It has many of the detergent attributes discussed for STPP: hardness control (which in this case occurs by precipitation rather than by complexation of calcium and magnesium ions), source of alkalinity, filler, carrier, and agglomeration aid for powders. In traditional spray-dried detergents, sodium carbonate functions mainly as a filler, alkalinity source, and builder. In spray drying plus post addition manufacturing technology, sodium carbonate provides alkalinity and acts as a builder and carrier for liquid ingredients. The same attributes are important for dry-blending and agglomeration technologies, but in the latter case sodium carbonate also acts as an agglomeration aid.

Several grades of sodium carbonate are commercially available. These are defined according to the density of the granular products: fine powder, very-light-density granular, light-density granular, medium-density granular, dense granular, and extra-dense granular. The absorptivities of the various grades are associated with the densities of the products and the size of the microscopic pores on the granular particles (Table 10.6). The absorptivity numbers range from 23% nonionic surfactant for very-light-density granular to 8% nonionic surfactant for extra-dense granular sodium carbonate. For applications involving agglomeration with nonionic surfactants, a compromise must be made in terms of product density and absorptivity, as was the case with phosphates. The light grade of sodium carbonate (density = 48 lb/ft^3) works well in most applications. This material has a high nonionic surfactant absorptivity and hydrates and dissolves quickly in water.

TABLE 10.6
Properties of Commercially Available Soda Ash

Properties	Soda Ash Grade					
	Fine	Very Light	Light	Medium	Dense	Extra-Dense
Bulk density (lb/ft^3)	51	43	48	50	59	65
Size						
+100 mesh	11	78	87	83	88	90
−200 mesh	48	4	1	4	1	1
Pore volume (cc/g)	0.14	0.33	0.26	0.22	0.14	0.14
Absorptivity						
(% nonionic surfactant)	15	23	20	18	13	8
Crystal morphology	Needle-like (from Sesqui process)					Blocky

For most applications involving the formation of a slurry (i.e., the crutcher operation before spray-drying), any of the grades of soda ash listed in Table 10.6 will perform well.

Oxygen Bleaches

The three most important oxygen bleaches in solid form for detergent applications are sodium perborate monohydrate, sodium perborate tetrahydrate, and sodium carbonate peroxide. These compounds are important in the formulation of color-safe bleach detergents for laundry, and bleach formulations in which reactivity with enzymes or other ingredients needs to be minimized. The major detergent benefits of oxygen-based bleach ingredients are in assisting stain removal, whiteness maintenance, and the control of spots and film on glasses in dishwash applications.

Table 10.7 summarizes the major attributes for each kind of bleach ingredient. Of the two sodium perborates, the tetrahydrate has the lowest active oxygen content and also the highest density. The monohydrate is preferred for use in today's detergents because it has a relatively high active oxygen content, a high rate of dissolution, and does not form lumps in the finished product. Stabilized sodium carbonate peroxide is sodium carbonate containing hydrogen peroxide in place of water of hydration. The example listed here has been coated with a proprietary technology to make it stable in detergent formulations (5). It has an active oxygen content of 11.5% and a density of 60 lb/ft^3, which is higher than either of the two perborate products. It is possible to prepare sodium carbonate peroxide with active oxygen higher than 11.5%, but that may compromise its stability and flow characteristics.

The oxygen bleach compounds discussed here are suitable for most powder detergent applications. Post addition is recommended during processing to preserve stability. It is also important to eliminate or minimize the level of free water and the presence of some heavy metals to avoid decomposition.

TABLE 10.7
Properties of Detergent Bleach Ingredients

Properties	Sodium Perborate Tetrahydrate	Sodium Perborate Monohydrate	Stabilized Sodium Carbonate Peroxide
Active oxygen (%)	10.1	15	11.5
Bulk density (lb/ft^3)	52	38	60
Process application	Post addition	Post addition	Post addition

References

1. Dymon, J.J., and A.J. King, Structure Studies of the Two Forms of Na Tripolyphosphate, *Acta Cryst. 4:* 378 (1951).
2. Troost, J.J., Crystal Growth of Sodium Tripolyphosphate Hexahydrate from Aqueous Solutions, *Crystal Growth 13/14:* 449 (1972).
3. Cassidy, J., R. Pals, and H.C. Harris, U.S. Patent 3,054,656 (1962).
4. Keast, R.R., E.H. Krusius, and J.S. Thompson, Detergent Formulations by Dry Neutralization, *Soap Chem. Specialties* 45–48(1967).
5. Copenhafer, W., B.A. Guiliano, W.A. Hills, C.V. Juekle, and S. Tomko, U.S. Patent 5,194,176 (1993).

Chapter 11

Formulation of Household and Industrial Detergents

K. Lee Matheson

VISTA Chemical Co./D.A.C. Industrie Chimiche S.p.A., Milan, Italy

A broad selection of surfactants is available to the detergent manufacturer for producing household and industrial detergents. The selection of these surfactants is guided by the performance and physical properties of the surfactants themselves and by the desired characteristics of the finished products.

Types of Surfactants

Anionic

The four general types of surfactants include anionic, nonionic, cationic, and amphoteric (Fig. 11.1). Anionic surfactants make up the largest group of surfactants in volume. Their hydrophilic character comes from the presence of a large ionic head, usually a sulfate or sulfonate group. Anionic surfactants include linear alkylbenzene sulfonate (LAS), alcohol sulfate (AS), alcohol ether sulfate (AES), alpha olefin sulfonate (AOS), and paraffin or secondary alkane sulfonate (SAS). Many other anionic surfactants are produced, but they tend to be smaller-volume specialty products.

Most anionic surfactants exhibit good foam performance; however, they differ in raw material source, cost, water hardness sensitivity, and mildness, among other things. They are used in all types of detergent applications with a few exceptions where foam is undesirable. Most anionic surfactants are produced as the sodium salts, but other metal ion salts as well as ammonium and various amine salts are produced.

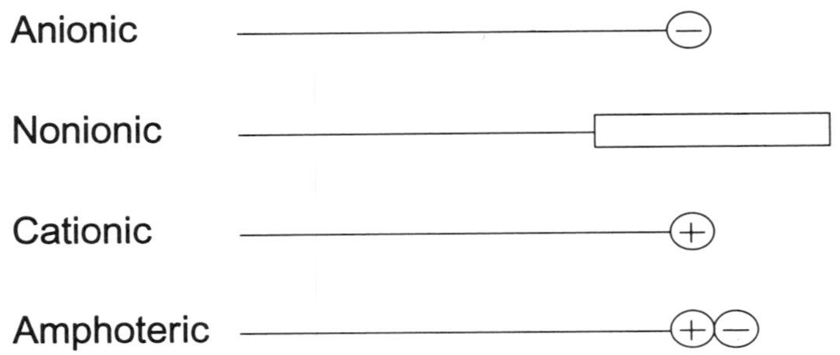

Fig. 11.1. Four types of surfactants.

Linear alkylbenzene sulfonate (LAS) is the largest-volume commodity surfactant used today (Fig. 11.2). The combination of good foam and detergency performance at low cost makes it very useful in many different types of formulations. Because LAS is one of the easiest of the anionic surfactants to sulfonate in-house, many detergent plants do their own sulfonation of linear alkylbenzene and save on the cost. In neutralized form LAS is commercially available as 50% and 60% slurries and as dried flakes. LAS is also one of the few surfactants commercially available in a stable acid form of 97% active.

Alcohol sulfate (AS) is a good surfactant for high-foaming applications in personal care products and also gives good detergency performance at the optimized carbon chain length (Fig. 11.3). Lauryl alcohol sulfate is commercially available as a 28% active aqueous solution and as dried flakes.

Alcohol ether sulfate (AES) is also a high-foaming surfactant. It is made from alcohol ethoxylate, which has an average of one to three ethylene oxide groups (Fig. 11.4). The presence of the ethoxyl groups makes this surfactant milder than alcohol sulfate and more resistant to water hardness. Alcohol ether sulfate is commercially available as a 28% active aqueous solution and as a 70% active paste. In the North American market a 60% active solution in water/ethanol is also commercially available.

Alpha olefin sulfonate (AOS) is actually a mixture of several structures produced by the SO_3 sulfonation of alpha olefins (Fig. 11.5). It is somewhat milder and more resistant to water hardness than either AS or LAS. AOS is used primarily in the United States and in Japan. AOS is generally available as a 40% aqueous solution, but dried flakes are also possible.

Paraffin sulfonate or *secondary alkane sulfonate* (SAS) is produced and consumed mainly in Europe (Fig. 11.6). SAS is a good-foaming anionic surfactant with excellent solubility properties for liquid detergent applications. Unlike most other sulfonates or sulfates, SAS is produced not in a thin film sulfonation reactor with SO_3 but by a sulfoxidation process. It is generally commercially available as a 60% active paste, but 30% active solutions and dried flakes are also possible.

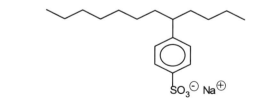

Foam	Cost	Hardness Sensitivity	Mildness
Good	Low	Moderate	Poor

Fig. 11.2. Linear alkylbenzene sulfonate (LAS).

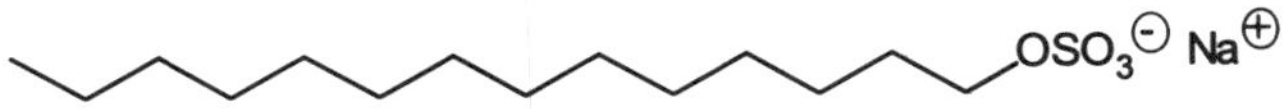

	Foam	Cost	Hardness Sensitivity	Mildness
	Good	High	Poor	Poor

Fig. 11.3. Alcohol sulfate (AS).

Nonionic

Nonionic surfactants have no charge on the molecule. Their hydrophilic character comes from the presence of a number of ether oxygens or hydroxyl groups. The largest-volume nonionics are those that are produced using ethylene oxide such as alcohol ethoxylates, alkylphenol ethoxylates, and EO/PO block copolymers. These are all available as 100% active material. They will be liquids or solids, depending on the chain structure and chain length. Other nonionic surfactants of appreciable volume include alkanolamides, alkylpolyglycosides, and alkylglucamides.

Alcohol ethoxylates (AE) are the most numerous nonionic surfactants (Fig. 11.7). With all the combinations of alcohol chain length, alcohol type, and ethylene oxide chain length there are many different types possible. Because of their good cleaning performance and insensitivity to water hardness they are widely used in laundry powders and liquids. They are not considered good foamers, but small amounts may be used in hand dishwashing products for grease cutting.

Alkylphenol ethoxylates (APE) are somewhat similar in performance to alcohol ethoxylates (Fig. 11.8). They provide excellent performance at low cost, and they are

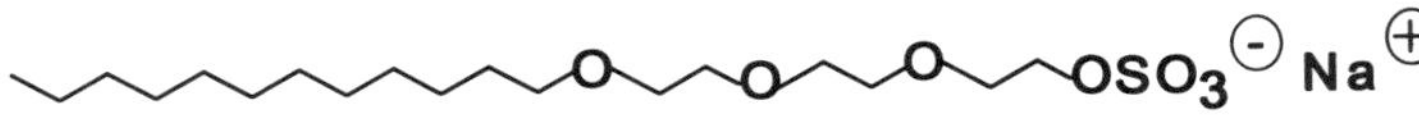

	Foam	Cost	Hardness Sensitivity	Mildness
	Good	High	Good	Good

Fig. 11.4. Alcohol ether sulfate (AES).

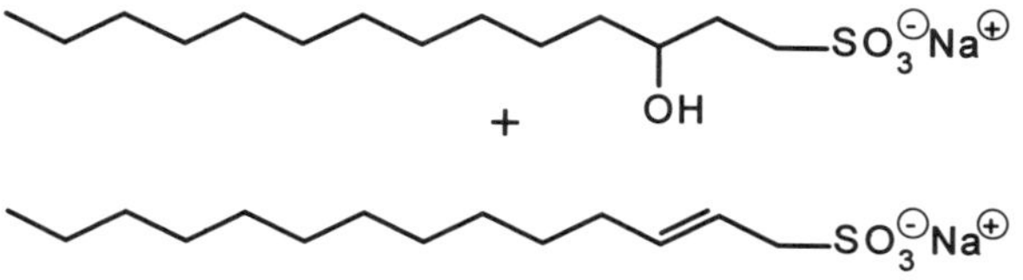

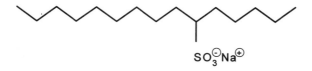

| | | Hardness | |
Foam	Cost	Sensitivity	Mildness
Good	Moderate	Good	Moderate

Fig. 11.5. Alpha olefin sulfonate (AOS).

widely used in industrial detergents and some household detergents. However, concerns about their rate of biodegradability have curtailed their use, especially in Europe.

EO/PO block copolymers are also nonionic surfactants that find use in applications in which foam is undesirable (Fig. 11.9). They provide surface tension lowering, wetting, improved rinsing action, and low foam. Many different ratios of EO and PO are produced.

Alkanolamides are nonionic surfactants in which the monoethanol or diethanol groups are generally not sufficiently hydrophilic to make these surfactants very water soluble by themselves (Fig. 11.10). However, alkanolamides are widely used as foam-boosting additives in combination with other surfactants in hand dishwashing liquids and in shampoos.

Alkylpolyglycosides and alkylglucamides belong to another group of nonionic surfactants that derive their hydrophilic character from the hydroxyl groups of sugars (Figs. 11.11 and 11.12). They are relatively more expensive than other major nonionic or anionic surfactants, but their use is growing in Europe, especially for hand dishwashing liquids.

| | | Hardness | |
Foam	Cost	Sensitivity	Mildness
Good	Moderate	Moderate	Moderate

Fig. 11.6. Secondary alkane sulfonate (SAS).

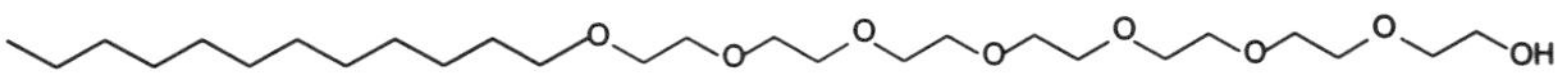

Foam	Cost	Hardness Sensitivity	Mildness
Moderate	Low	Good	Poor

Fig. 11.7. Alcohol ethoxylates (AE).

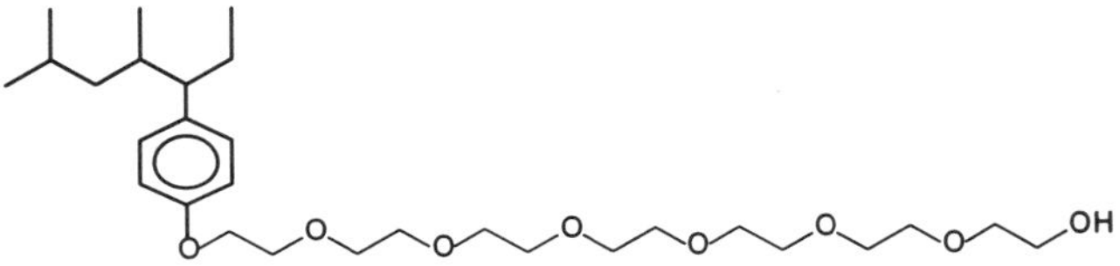

Foam	Cost	Hardness Sensitivity	Mildness
Moderate	Low	Good	Poor

Fig. 11.8. Alkyl phenol ethoxylates (APE).

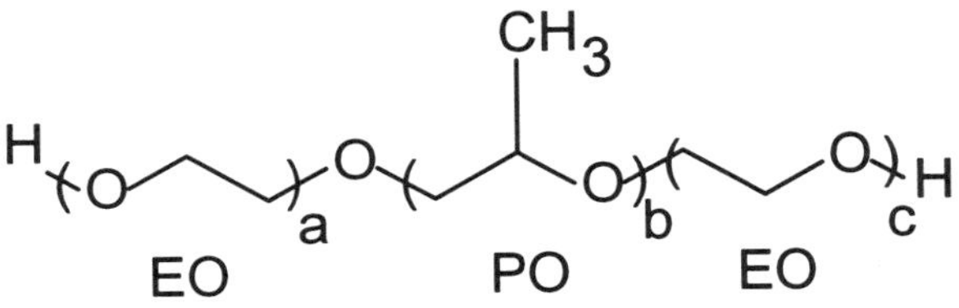

Foam	Cost	Hardness Sensitivity	Mildness
Low	Moderate	Good	Moderate

Fig. 11.9. EO/PO block polymers.

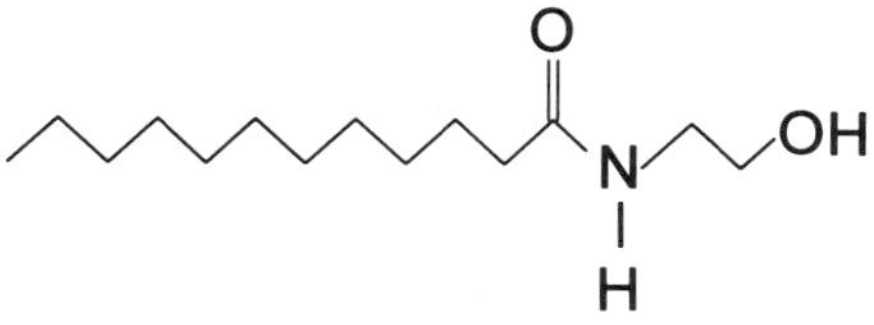

Foam	Cost	Hardness Sensitivity	Mildness
Good (Booster)	High	Good	Moderate

Fig. 11.10. Alkanolamide.

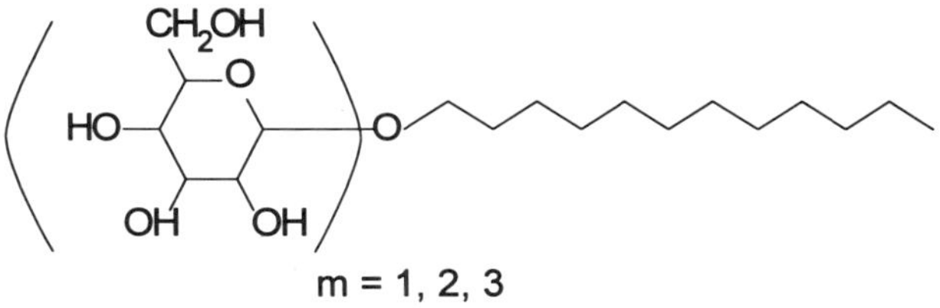

m = 1, 2, 3

Foam	Cost	Hardness Sensitivity	Mildness
Good	High	Good	Good

Fig. 11.11. Alkylpolyglycoside (APG).

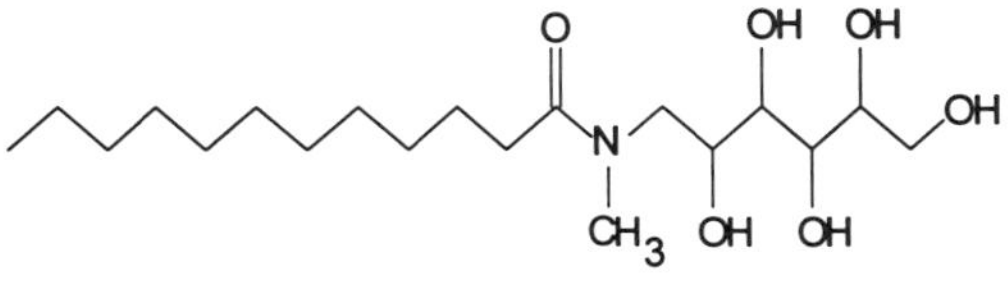

Foam	Cost	Hardness Sensitivity	Mildness
Good	High	Good	Good

Fig. 11.12. Alkylglucamide.

Cationic

Cationic surfactants have a large cationic group, usually a quaternary ammonium salt (QUAT), for their hydrophilic head group, and usually one or two long alkyl chains for their lipophilic tail(s). Generally cationic surfactants are used not for their detergent properties but for fabric softening, anti-static properties, surface conditioning, and bactericidal effects.

Ditallow dimethyl ammonium chloride (DTDMAC) has been perhaps the largest-volume cationic surfactant. The combination of a positive charge and two long alkyl tails is ideal for absorbing onto negatively charged cloth surfaces (Fig. 11.13). The positive charge neutralizes static charge and the greasy tails lubricate the fibers of the cloth to give a softening feel. This has been widely used as a rinse-added fabric softener. However, due to solubility and viscosity reasons, the liquid formulations containing this material are only 5–10% active, and there has been some concern over its environmental acceptability.

Recently new cationic surfactants have been developed to replace DTDMAC for European fabric softener formulations. These are generally called *ester QUATS*. The presence of ester groups on the alkyl chains facilitates the biodegradation of the molecule. Figure 11.13 shows one of several possible structures.

Amphoteric

Amphoteric surfactants can have under certain pH conditions both a positive and a negative charge on the molecule. Generally speaking, they are small-volume, higher-cost specialty surfactants (Fig. 11.14). However, lauryl or coco amidopropyl betaine is becoming more popular as a component in hand dishwashing liquids and shampoos. The combination of good foam performance, solubility in hard water, and skin mildness makes betaines very useful in these formulations.

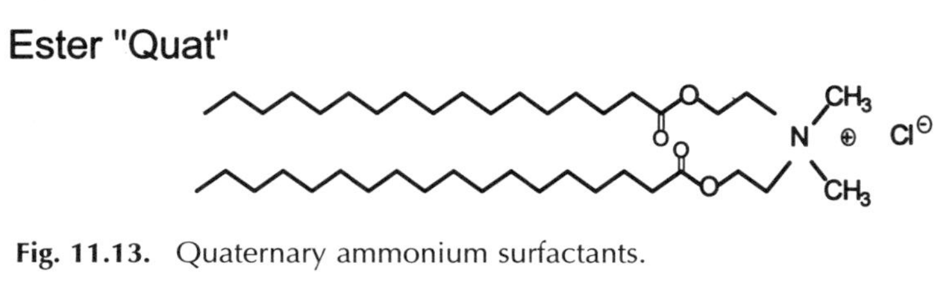

Traditional "Quat"

Distearyl Dimethyl Ammonium Chloride

Ester "Quat"

Fig. 11.13. Quaternary ammonium surfactants.

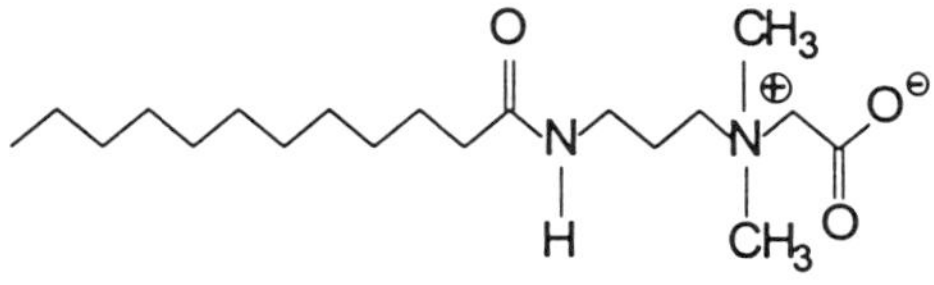

Foam	Cost	Hardness Sensitivity	Mildness
Good	High	Good	Good

Fig. 11.14. Lauryl amidopropyl betaine.

Household Detergent Product Formulations

Laundry Powders

Laundry powders are produced in two general types: conventional low-density powders and compact powders. Conventional spray-dried laundry powders typically have a density of 0.3 g/cc with an active content of 15%–25%, a builder content of 20%–40%, a sodium sulfate content of 20%–40%, and a dosage level of 1 cup. The sodium sulfate is used mainly as a processing aid, although it provides some benefit in the wash water to increase ionic strength. By comparison, the compact powders have density levels of from 0.5 to 1.0 g/cc, active levels from 15%–40%, generally very little sodium sulfate, and dosage levels as low as 0.25 cup. Table 11.1 compares the general compositions of conventional and compact types of laundry powders in the United States.

Active systems for laundry powders can be based on a single surfactant such as LAS or alcohol ethoxylate, but for most U.S., European, and Asian powders mixed active systems are used. For consumers who prefer low-foaming products, LAS/AE or mixtures with soap or some other antifoam agent are used. For consumers who prefer high-foaming products, then LAS/AES or LAS/AES/AS mixtures are preferred.

Laundry Liquids

Laundry liquids have been produced for many years, and they represent a large part (> 30% in the United States) of the total laundry detergent market (Table 11.2). They are appreciated by consumers for their convenience in use and ease of application to stained fabrics. The majority of laundry liquids are isotropic liquids in which all the surfactants and other additives are solubilized in an aqueous solution. Laundry liquids tend to be mixed active systems, and they frequently contain enzymes. Generally some kind of hydrotrope is used to preserve liquid

TABLE 11.1
U.S. Laundry Powder Formulations (Moderate Foaming, Nonphosphate)

	Low Density	Compact High Density
Surfactants	15%–25%	25%–40%
LAS/AS/AES		
LAS/AE		
AE/AES		
Zeolite	10%–20%	25%–35%
Carbonate	15%–25%	20%–30%
Silicate	2%–10%	2%–10%
Sulfate	20%–30%	0%–5%
Bleach additive	0%–5%	0%–5%
Water	5%–10%	5%–10%
Other ingredients	1%–3%	1%–3%
Antiredeposition agent		
Enzyme		
Optical brightener		
Perfume		
Density, g/cm^3	0.3%–0.4%	0.5%–0.9%

stability and to keep viscosity low. The formulations may be built with phosphate or citrate or tartrate, or they may be unbuilt. As will be explained later, newer, more concentrated liquids are appearing in the market.

Hand Dishwashing Liquids

Hand dishwashing liquids contain mixed active systems designed to produce lots of stable foam. Typical mixed active systems are LAS/AES/alkanolamide or AES/AS with a foam booster (Table 11.3). In Europe the newer, more concentrated dishwashing liquids contain high levels of AES with betaine or alkylglucamide as foam booster in place of alkanolamide.

TABLE 11.2
U.S. Laundry Liquid Formulations

Surfactants	20%–40%
LAS/AS/AES/AE	
AE/LAS	
LAS/AE	
Soap	0%–5%
Builders	0%–10%
Hydrotropes	5%–10%
Other	1%–2%
Enzymes, bleach, optical	
brightener, perfume, coloring	

TABLE 11.3
U.S. Dishwashing Liquid Formulations

Surfactants	15%–35%
AES/AS	
LAS/AES	
Foam stabilizers	2%–5%
Alkanolamide	
Amine oxide	
Betaine	
Hydrotropes	5%–10%
Other	0.5%–1%
Perfume	
Coloring	

Bars and Pastes

Laundry detergent bars and dishwashing pastes are an important part of the detergent market in Latin America. Typical bar formulations contain 10% to 30% surfactant and from 10% to 30% builder along with 20% to 50% filler. The builder may be phosphate or soda ash (Table 11.4). Dishwashing pastes contain similar active levels but usually low levels of phosphates and 30% to 40% water. The main surfactants used in bars and pastes are branched alkylbenzene sulfonate (ABS) and also linear alkylbenzene sulfonate (LAS). Both surfactants offer good detergency and foam characteristics, but they are different in bar-processing characteristics. With ABS the bars harden quickly after processing. Linear alkylbenzene sulfonate tends to produce a softer bar, but this problem can be overcome by changing the

TABLE 11.4
Laundry Bar and Paste Formulations

	Bars	Pastes
Surfactants	10%–30%	15%–30%
ABS		
LAS		
Builders	15%–35%	15%–35%
STPP		
Na_2CO_3		
Other solids	20%–50%	5%–20%
Clay		
Na_2SO_4		
$CaCO_3$		
Water	5%–20%	25%–50%
Other	1%–2%	0.5%–1%
Antiredeposition agent		
Perfume		
Coloring		
Optical brighteners		

formulation and adding bar-hardening additives. This difference can be understood by comparing the phase characteristics of ABS and LAS.

Take, for example, a typical bar mixture of 25% surfactant, 65% inorganic salts, and 10% water. With LAS there is a strong tendency for the water to remain in contact with the surfactant in a gel-like liquid crystalline structure. This retards the hydration of the inorganic salts and results in a softer or plasticized bar. On the other hand, with ABS there is less tendency to form liquid crystal structures and a greater tendency for the ABS to precipitate as solid surfactant. The water then quickly hydrates the inorganic salts and the bar hardens more quickly. As mentioned before, the addition of bar-hardening additives is required for LAS bars. These additives include clays and magnesium, calcium, and aluminum salts, which change the phase characteristics of LAS in the bar formulation.

Major Trends in Household Detergent Products

During the last few years several important trends have influenced household detergent products. These trends have affected not just the composition of these products but also the product packaging as well.

Compact Laundry Powders

The trend toward compact laundry powders began in Japan several years ago. Now throughout much of Asia, Europe, and the Western Hemisphere the compact powders have replaced much of the conventional low-density powders in the marketplace. The product changes are a combination of compositional change, increases in powder density, powder processing changes, and changes in packaging. Some of these compositional differences have been discussed previously.

Considerable variation exists for the term compact powder, depending on the market and the product. Some compacts have only increased the powder density with the same active level. Some have increased both active and density levels. Generally these changes have required different powder-processing conditions and methods. Adjustments to spray tower conditions can result in powder densities of up to 0.5–0.6 g/cc, but for higher densities newer processing methods have been developed.

Concentrated Laundry Liquids

Laundry liquids have been changing from isotropic aqueous liquids of 20%–35% active level (using hydrotropes) to more concentrated liquids having a solids level greater that 40%. Two types of concentrated liquids are being developed. The first type is called a structured liquid and has been in use now for at least a year in the United States and longer in Europe, as of 1994. These liquids contain up to 40% surfactant in the liquid as a dispersed lamellar liquid crystalline phase. They may also contain builder salts dispersed in the liquid. The technological achievement for these liquids is that they keep such high levels of surfactant and builder solids dispersed in a pourable liquid form without the whole mixture turning into a gelatinous paste. Unfortunately, in the U.S. market this product was not successful, and has since been taken off the market.

The other new type of concentrated liquid under development is a nonaqueous liquid in which little or no water is present. The surfactants, builders, and enzymes are dissolved/dispersed in an alcohol ethoxylate or polyethylene glycol solvent. The elimination of water has advantages for preserving the activity of detergent enzymes in the formulation. By the end of 1995, these products were withdrawn from the market place.

Environmentally Friendly Product Packaging

Consumers around the world have taken a greater interest in protecting the environment, and detergent manufacturers have responded by changing not only detergent compositions but also the detergent packaging as well. In the past, the large cardboard boxes of low-density detergent powder and plastic bottles of liquid detergent generated large amounts of solid waste that had to be disposed of in municipal landfills. Due to the negative publicity generated by so much solid waste with brand names on the label, detergent producers had to react.

One approach was to use recycled cardboard and plastic in their original product containers, and this is noted on the product labels. Refill packaging is another way to avoid generating so much solid waste. The first time consumers buy the product they buy the product in its normal container (bottle or box). When the consumers have used up the first container of product, they then buy a refill pack, which has the powder or liquid packaged in a flexible plastic container. The consumers then pour the contents of the refill pack into the original product container. The empty refill packs are discarded, but they generate much less solid waste than original plastic bottles or cardboard box containers.

A variation of the liquid refill pack has been developed for rinse-added fabric softener liquids. The original rinse-added liquid formulation based on DTDMAC was only about 5% active and packaged in a large 5-liter plastic bottle. Now European detergent manufacturers have developed a dilutable 15% liquid formulation. The consumer buys this product in a refill pack and dilutes one part product to two parts water into the original plastic bottle.

Selection of Environmentally Friendly Ingredients

Another trend is the replacement of certain detergent ingredients by those which are perceived to be better for the environment. During the last three years the detergent market in Mexico has converted from branched chain dodecylbenzene to linear alkylbenzene. This change is also taking place in other Central American and South American countries. In the U.S. market, the major detergent manufacturers have replaced STPP in their detergent powders with a zeolite/carbonate builder system. This change was partly motivated by environmental considerations and also by the desire to produce a denser powder. Another environmentally motivated change that occurred in Europe was the conversion of fabric softener products from the ditallow dimethyl ammonium chloride to the ester QUATS, as mentioned previously.

Increased Production of Oleochemical Alcohols

Another trend that has been developing over the last decade has been the construction of new oleochemical alcohol plants, especially in the Philippines, Indonesia, and Malaysia. This increased production of oleochemical alcohols has provided a greater variety of surfactant intermediates for detergent production in Southeast Asia, but also increased the worldwide availability of oleochemical-based alcohols.

Industrial and Institutional Detergents

The market for industrial and institutional (I&I) detergents includes commercial laundry, commercial dishwashing, janitorial supplies, food service equipment cleaners, transportation equipment cleaners, dry cleaning, metal cleaners, and carpet cleaners. There are several important differences between the I&I market and the household detergent market. In contrast to the household detergent market, which has relatively few producers, there are many more producers of I&I cleaners. Products can be highly specialized formulations to perform specific cleaning tasks. Service is very important to the I&I market. The I&I detergent producer may provide specialized training to its customers' employees for the specific cleaning operations to be performed. Sometimes the I&I detergent producer sells or leases specialized equipment that is used in the cleaning process. The detergent producer usually also maintains and repairs that equipment. Also, in the I&I market the detergent products are usually handled and used by trained personnel. As a result the operating conditions for I&I detergents (temperature, pH, and so on) can be much more severe, and the chemical compositions much more harsh than those typically encountered with household detergents.

Surfactants Used in I&I Detergents

Surfactant consumption in the I&I market amounts to about one-tenth of the total consumption for both Europe and the United States. The types of surfactants used are generally the same ones used in household cleaning products. However, in many applications the surfactants are used only as auxiliary agents while the real cleaning is done by high levels of alkalinity (or acid in some cases), high temperatures, high-pressure water, chemical bleaching agents, and so on. The surfactant may be present in the formulation for reasons other than detergency, such as foam control, bactericidal action, hydrotrope or coupling ability, rinsability, emulsification, or some other property.

In addition to the large-volume surfactants that have previously been discussed, other specialized surfactants may also be used to provide a specific function. These include, but are not limited to, the following: phosphate esters (for low foam and solvent/water coupling), tall oil fatty acid soaps (for emulsification), imidazoline derivatives (for coupling and performance enhancement), alkyl diphenylene oxide disulfonates and alcohol ether sulfonates (for bleach stability), alkylnaphthalene sulfonates (for solids dispersion, coupling, and alkaline stability), and amine oxide (for foam stabilization).

Hard Surface Cleaners

Hard surface cleaners of various types are used in janitorial cleaning: floor cleaners and strippers, wall cleaners, drain cleaners, porcelain fixture cleaners, glass cleaners, disinfectants, and the like. Generally the levels of surfactant in these products are 5% active or less, with the rest of the formulation being alkaline builders, solvents, and water. The main surfactants used are alkylphenol ethoxylate, alcohol ethoxylate, and linear alkylbenzene sulfonate.

Commercial Dishwashing Detergents

Commercial dishwashing is practiced in restaurants, hospitals, cafeterias, and so on. On the larger scale it involves the use of special machines that provide a prewash, a wash, and one or more rinsing steps. The detergent is primarily made up of alkaline inorganics, with perhaps 1%–2% surfactant, typically a PO/EO block copolymer. The function of the surfactant is to improve rinsing action in order to avoid water spots. The surfactant must not foam itself and must act as a defoamer to break up foam caused by food soils. Anionic surfactants are generally not used in machine dishwashing detergents due to their higher foaming characteristics.

However, on a smaller scale, some commercial dishwashing is done by hand, and the volume of surfactant consumed in these manual dishwashing products is relatively large compared to the machine dishwashing detergents. In these products the surfactants used are LAS, AE, AES, AOS, alkanolamides, and betaines.

Commercial Laundry Detergents

Commercial laundry cleaning is performed for hospitals, hotels, restaurants, military bases, nursing homes, prisons, and so on. The laundry may include bed linen, table linen, uniforms, towels, and furniture coverings. The washing process bears little resemblance to the household laundry process. Large-capacity machines use presoftened water, higher wash temperatures, longer batch times, and many more steps in the washing cycle. Soiled fabrics are sorted by type and washed in separate groups. The machines may have microprocessor control capability so that washing conditions, detergent composition, and detergent concentration can all be varied and selected at different steps in the cycle depending on the type of soiled fabrics.

The surfactants used in commercial laundry detergents for the U.S. market are primarily nonionics, such as APE and AE; some LAS and soap are also used. Also for fabrics such as towels, QUATS are used for fabric softening. In the past soap was used to a much greater extent, but this has changed in favor of ethoxylates as the use of permanent press fabrics has increased.

Dry Cleaning Detergents

For wool, silk, and certain other fabrics, water causes swelling of the fibers, which results in wrinkled, shrunken cloth. Dry solvents have the dual advantage that they do not swell the fibers (to cause wrinkling) and that they dissolve oil-soluble stains at low temperature. The dry cleaning process involves the use of specially designed washing machines to wash the clothes with solvent and surfactant, to separate and recover the solvent where feasible, and to dry the clothes. Small amounts of water may be added or may already be present in the formulation to help remove water-soluble stains.

Dry cleaning detergents are generally liquids containing solvent and high levels (up to 50%) of surfactants. The surfactants aid in soil removal and solubilize any water present. The surfactant system is usually a mixed active system, with linear alkylbenzene sulfonate (various alkanolamine salts) being the main surfactant used. Also in use are alkylphenol and alcohol ethoxylates, alkanolamide, and sodium dialkyl sulfosuccinate.

Food Processing Equipment Cleaners

Food process equipment cleaners are used for cleaning in dairies, meat packing plants, bakeries, canneries, beverage plants, and the like. In general, the equipment (including tanks and pipes) is cleaned in place. In most cases foam is not wanted. As with hard surface cleaners and commercial dishwashing detergents the level of surfactant in these cleaners is low, with most of the cleaning power coming from high alkalinity, mechanical action, and high temperature. The surfactants provide rinsability, foam control, and sanitizing. Some of the major surfactants used include EO/PO block copolymers, alkylphenol ethoxylates, linear alkylbenzene sulfonates, alcohol ethoxylates, phosphate esters, and QUATS.

Transport Vehicle Cleaners

Transport vehicle cleaners are used for washing trucks, buses, aircraft, trains, and ships. The surfactants used in these types of detergents are usually alkylphenol ethoxylate and linear alkylbenzene sulfonate. Alkanolamide and amphoteric surfactants are also used in combination with LAS and APE in cleaning situations in which foam is important. For example, for cleaning large vertical surfaces of trucks and aircraft, the clinging foam is important to hold the surfactant in contact with the soiled vertical surface. Where aluminum surfaces are involved, the pH of the formulation should be less than 12.

The market for car washing products is fairly developed in the United States. There are different types of washing processes and various types of cleaners developed for different parts of the car. As a result, many different types of surfactants are used in these formulations. These include LAS (the largest volume), APE, AES, AE, AOS, alkanolamides, and phosphate esters. Also quaternary ammonium surfactants (coconut range) are used in spray-on-type car wax.

Carpet Cleaners

Carpet cleaners are used to clean wall-to-wall carpeting in place. Several different cleaning methods are employed. The hot water extraction method uses primarily low-foaming surfactants such as alkylphenol and alcohol ethoxylates. The various shampoo methods use primarily lauryl alcohol sulfate, AOS, and LAS. Other foam-modifying surfactants are used. Some carpet shampoos may also contain a polymer resin that, when applied to the carpet fibers, acts like a soil repellent.

Conclusions

Household detergents have undergone more change in the last 5 to 6 years than in the last 20 to 30 years. Detergent formulations have become increasingly more complex. Detergent producers are looking to new special ingredients such as bleach additives and enzymes mixtures to obtain detergency performance improvements. Detergent products are being formulated and packaged in ways to minimize their effects on the environment. The field of I&I detergents is sufficiently large and complex that it merits consideration as a separate discussion topic. Unfortunately, space has permitted only a very brief look at the formulations, and the interested reader is well advised to consult the published literature available on I&I cleaners or to contact some of the I&I producers directly.

Chapter 12

Analysis of Detergent Formulations

George T. Battaglini

Stepan Company, Northfield, Illinois, U.S.A.

Introduction

The complete analysis of detergent formulations can be and often is a time-intensive undertaking. Even in laboratories experienced in the field, gross characterization will take 20 to 40 hours of labor, and a general analytical laboratory can be expected to spend much longer in consultation and in setting up apparatus for the multiple determinations required.

Nonetheless, performing the analysis is worthwhile for those who need to know market trends in various locations, and for those who wish to compete in the detergent field. Not to be forgotten, either, is the importance of analysis in synthesis and process development support, quality control, environmental safety, and general troubleshooting. Although it is generally not possible to prepare a performance match for a formulated detergent on the basis of analysis alone, the skilled formulator will find that the results of the analysis will give helpful clues and guidelines.

This chapter outlines strategies that may be employed for analyzing formulations including detergents for laundry, shampoos, hand and machine dishwashing, hard surface cleaners, liquid hand soaps, fabric softeners, toothpastes, and bar soaps. Some typical formulations for a laundry detergent, a shampoo, and a liquid hand soap are given in Table 12.1. A brief description of the many methodologies actually used is given.

The methods include classical wet techniques as well as some of the very powerful instrumental approaches now available. It is acknowledged that instrumental approaches can be prohibitively expensive, particularly to the smaller laboratory. It also must be stated that a great merit of instrumental methods is that they can do what classical techniques do but faster and better, and that they can do with ease many things that classical techniques can do only with great difficulty, if at all. It may, therefore, be cost-effective and necessary in the long run for those laboratories heavily engaged in detergent analysis to make a some kind of commitment to instrumentation.

A brief discussion of some of the basic principles behind a few of the instrumental approaches is given, but a rigorous description of them is beyond the scope of this work. Refer to standard textbooks or contact instrument manufacturers directly for more information.

TABLE 12.1

Typical Formulations

A. Laundry Detergents
 1. Anionics, such as LAS, tallow ether sulfate, sulfonated methyl esters, soap, and SLS (rare)
 2. Nonionics, such as nonyl phenol or fatty alcohol ethoxylates
 3. Builders, such as citrate, sodium tripolyphosphate, sodium carbonate, sodium silicate, and EDTA
 4. Phosphate substitutes, polyacrylates and zeolites
 5. Hydrotropes, usually sodium xylene sulfonate
 6. Antiredeposition aids, sodium carboxymethyl cellulose
 7. Fabric softeners
 8. Bleach, such as perborate or percarbonate and bleach activators
 9. Optical brightener, dye, or perfume
10. Glucoamines—P&G
11. APGs—alkyl polyglycosides
B. Hair Shampoos
 1. Anionics, such as alcohol ether sulfates (AES) and alcohol sulfates (AS)
 2. Amphoterics, such as alkyl quaternary betaines
 3. Foam boosters, usually alkanolamides
 4. Viscosity builders, usually inorganic salts
 5. Conditioners
 6. Propylene glycol, glycerine
 7. Opacifiers, dyes, or perfumes
 8. Biostats, such as p-hydroxybenzoic acid
 9. Emulsifiers such as Stepan TAB-2
C. Hand Dishwashing Detergents
 1. Anionics (LAS, AES, AS, AGES, SME—Mg, Na, or NH_4 salts)
 2. Alkanolamides or amine oxides to improve foaming
 3. Hydrotropes (SXS, ethanol, urea)
 4. Sulfosuccinates (rare)
 5. Betaines
 6. APGs—alkyl polyglycosides
 7. Sarcosinates (rare)

Flow diagrams are presented to give the analyst step-by-step guidelines for formulations analysis. These are given in several figures (beginning with Fig. 12.1) throughout the chapter illustrating the characterization and quantitation of the various kinds of anionic, nonionic, cationic, and amphoteric surfactants commonly found in formulated products. It should be noted that a number of yes/no options are found in the diagrams. In Fig. 12.1, for example, if the answer to the question "Solids + Water = 100%?" is "Yes," then we proceed to the question "Separations Necessary?" If the answer is "No," then a GC analysis for solvents is indicated.

Initial Examination

An initial examination is generally in order before beginning an analysis, as shown in Fig. 12.1. This includes the asking of key questions such as the following: What is the product's use? What does the label say? Are patents or ingredients listed on

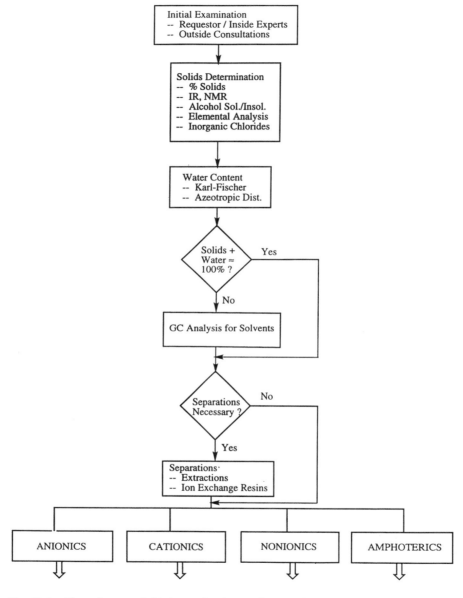

Fig. 12.1. Flow diagram: Initial examination and separations.

the label? Does in-house or outside literature reveal that there is a previous analysis? Can a typical recipe for a formulation be found in technical trade journals? What are typical formulations? Is a complete analysis needed, or is a partial analysis satisfactory? In any case, it is advisable that consultations with the requester of the analysis and with any in-house experts be carried out before starting.

It is important to stress that a representative sample must be obtained at every stage; otherwise subsequent analyses will be pointless and misleading. Obtaining a representative sample from a liquid or a paste is a relatively simple matter, but a more rigorous procedure is needed with powders or bar products. Suitable methods for obtaining representative samples are described in ASTM standards (3–5).

The analysis is generally begun by determining the percentage of solids. For liquids and pastes, this is done by heating about a gram of sample in a tared flask for 2 h at 50°C under vacuum, with care taken that the vacuum oven is vented until foaming stops. For solid products, heating about a gram of sample in a tared aluminum dish for 2 h at 105°C should suffice. This gives us not only a measure of the amount of solids but also a sample of the solids for further analysis.

We next run infrared (IR) and/or nuclear magnetic resonance (NMR) spectra on the solids and look for functional groups or chemical shifts. This was actually done in the author's laboratory on a light-duty liquid detergent.

Infrared spectroscopy deals with the vibration of atoms within the functional groups of a molecule and the radiant energy that is absorbed. The two protons and the carbon of the methylene group, for example, can have up to six fundamental stretching and deformation vibrations, as depicted in Fig. 12.2 (4). When a functional group that is vibrating at a given frequency is impacted by infrared light of the same frequency, energy is absorbed, resulting in a "peak" in the spectrum at that frequency. This is the case, for instance, when a methylene group that is "rocking" at 720 cm-1 is impacted by infrared radiation at the same frequency.

In the IR spectrum of the solids of the previously mentioned light-duty liquid detergent (Fig. 12.3), we see absorbances at about 1,208 and 1,027 wavenumbers (cm^{-1}) due to asymmetric and symmetric sulfur-oxygen bonds, at about 1,077 and 930.3 cm^{-1} due to an ether linkage, at about 1,480 cm^{-1} due to a hydrocarbon, and at 721.8 cm^{-1} due to a long methylene chain. The rather sharp band at about 1,436 cm^{-1} coupled with the broad bands from about 3,500 to 3,000 cm^{-1} are indicative of the presence of the ammonium ion. All of this tells us that the product most probably contains an ammonium alcohol ether sulfate.

Nuclear magnetic resonance (5) is possible because the nuclei of some isotopes spin on an axis. A charged spinning nucleus generates a magnetic field called the nuclear magnetic moment. The basis of NMR is dependent on the interaction of the small nuclear magnetic field with an externally applied magnetic field. When nuclei are placed in a magnetic field, the nuclei behave as small magnetic dipoles along their spin axes which orient themselves with or 180 deg against the field (Fig. 12.4). Note that slightly more dipoles (about 10 ppm) are aligned with the field than against it. The energy of the nucleus is lower when the dipole is aligned with the

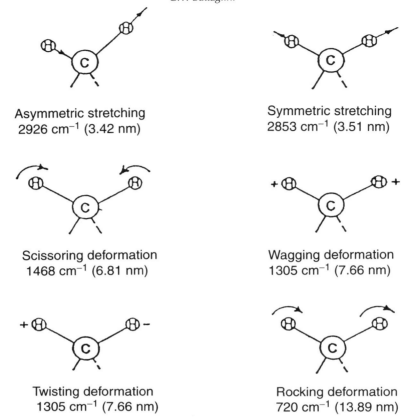

Fig. 12.2. Fundamental stretching and deformation vibrations for the methylene group. (Reproduced with permission from Vandenberg et al. [4]).

field than when it is opposed. The difference in these energies is the basis of NMR. When the correct radio frequency (rf) energy is applied to a sample in a strong magnetic field, energy is absorbed, causing lower-energy nuclei to jump to the higher energy level. The reverse also happens, and the entire process is repeated. This sequence of "excitation" followed by "relaxation" is termed *resonance* and is responsible for the peaks in the NMR spectrum.

The NMR spectrum of the solids of the aforementioned light-duty liquid detergent (Fig. 12.5) is even more definitive. In addition to the chemical shifts at about 4.2 and 4.0 ppm due to ammonium alcohol ether sulfates and ammonium alcohol sulfates, respectively, there are shifts at about 3.40 ppm due to amine oxide and about 3.25 due to a betaine. The ratio of these latter peaks to each other, incidentally, gives us an estimate of the molar ratio of amine oxide and betaine present. There are also shifts at about 3.5–3.8 ppm due to ethylene oxide (EO). The cluster of peaks between about 6.9 and 7.85 ppm is typical for the aromatic sub-

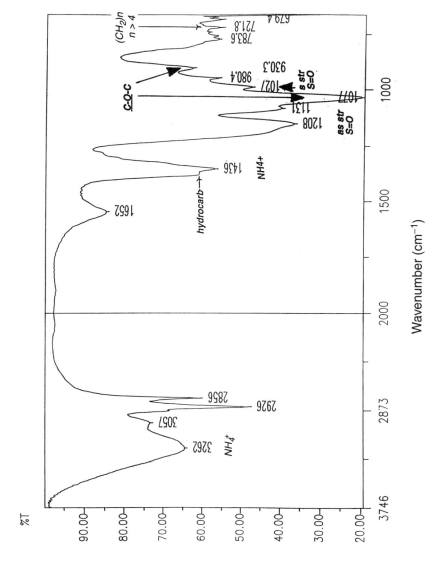

Fig. 12.3. FTIR spectrum of a commercial light-duty liquid dishwashing detergent (vacuum solids); germanium cell using heated HATR at 60°C; 100 scans at 8 cm^{-1} resolution.

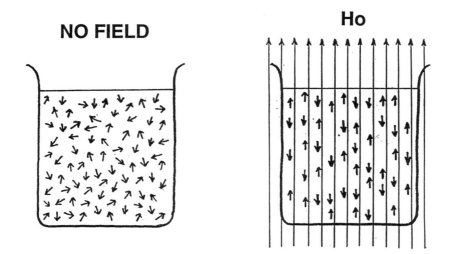

Fig. 12.4. Effect of an external magnetic field on nuclei. (Reproduced with permission from Shoolery [5]).

stitution pattern for xylene sulfonates, but does not indicate the presence of linear alkylbenzene sulfonates (LAS). The peaks at about 2.3 and 2.6 ppm are indicative of the methyl protons on the xylene of xylene sulfonate, and the ones at about 0.9 ppm are due to methyl protons. The peaks between about 1.2 and 1.8 ppm are due to methylene protons, indicating a long hydrocarbon chain. Water is indicated by the peak at about 4.9 ppm, and undeuterated methanol (an impurity in the deuterated methanol NMR solvent) by the rather large peak at about 3.35 ppm. Peak identification is made easier by spiking with known species. Additionally, chemical shifts will be induced in amine oxides and betaines by pH changes. Consultation with spectral libraries and experts is, of course, very helpful in any spectral interpretation.

Determination of percentages of alcohol solubles and insolubles (for example, by ASTM D820) (6) on the solids gives us a measure of organics and inorganics and enables us to have samples of each for possible further analysis. The analysis involves placing the sample in denatured ethanol, boiling for a short time, and filtering through a tared crucible. Most, if not all, of the organics are soluble in the alcohol, whereas all of the inorganic matter (sodium chloride excepted) is alcohol insoluble. Incidentally, if ammonium or amine salts are suspected to be present, these will have to be converted to their sodium salts by the addition of sodium hydroxide to a pH of about 10. Percent inorganic on the sodium salts may also be obtained by heating the solids to 600°C.

Elemental analysis (carbon, hydrogen, nitrogen, sulfur, oxygen, and so on) on the solids or on the as-is product may be needed. Instrumental analysis by atomic absorption, inductively coupled plasma, or X-ray diffraction may be needed, but

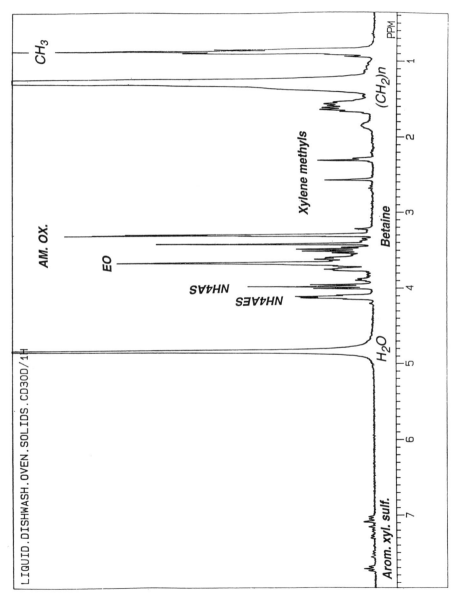

Fig. 12.5. Proton NMR spectrum of a commercial light-duty liquid dishwashing detergent (vacuum solids).

often an inexpensive determination of sulfur or nitrogen will confirm the presence of an anionic or a cationic surfactant.

Ionic chlorides, sulfates, carbonates, and so on may be determined on the solids or on the as-is product by titration, spectrographic, ion chromatographic, or X-ray fluorescence methods. My laboratory has found potentiometric titrations, with silver nitrate using a silver/silver chloride electrode for chloride, with lead perchlorate using a lead ion selective electrode (ISE) for sulfate, and with hydrochloric acid using a glass/calomel electrode for carbonates, to be effective.

There may be volatile matter (such as water and/or organic solvents) present in the product. Water can be determined by the well-known Karl Fischer titration as outlined, for example, in ASTM D1568 (7). A potentiometric method is preferred in my laboratory for high moisture levels, whereas a coulometric method has been found to be more effective at low moisture levels. Azeotropic distillation with mixed xylenes, as described in ASTM D3673 (8), has also been found to be effective at higher water contents.

If the total of solids plus moisture is less than 100 percent, organic solvents are probably present. These are usually solvent alcohols such as isopropanol, methanol, ethanol, butanol, and ethylene or propylene glycol, which can usually be determined by gas chromatography. In a typical analysis (9), the sample is diluted to a known volume with an internal standard, such as tertiary butanol, and injected into a programmed gas chromatograph equipped with a packed or a capillary column. A standard mixture of solvent alcohols and standard is also injected into the chromatograph to obtain relative response factors for the alcohols. A typical standard chromatogram, with chromatographic conditions and peaks for methanol, ethanol, isopropanol, and tertiary butanol, is given in Fig 12.6. Chromatograms depicting the determinations of ethanol and of ethanol and isopropyl alcohol are

Column: 6 ft. x 1/8 in. OD stainless
        steel packed with Porapak Q
Inj. Port Temp: 150 deg C
FID Temp: 200 deg C
Temp. 1: 150 deg C
Temp. 2: 200 deg C
Rate: 6 deg C per min.

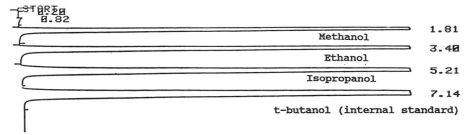

Fig. 12.6. Typical GC chromatogram of standards.

given in Fig. 12.7. Percent alcohol in the sample is calculated by comparing the data from the sample and standard.

Separations

It may be necessary to separate the organics from each other. This can be done by liquid/solid extraction using solvents other than ethanol (for example, methanol) or by liquid/liquid extraction of aqueous solutions with chloroform or ethyl acetate. It can also be done by ion exchange. Figure 12.8 shows how anionics, cationics, amphoterics, and nonionics may be separated from each other by ion exchange (10). Here we see a methanolic solution of sample being passed through three ion exchangers. The first is an anionic exchanger of the chloride ion form onto which anionics are adsorbed. The second is another anionic exchanger (this one of the hydroxide ion form) onto which amphoterics are held. The third is a cationic exchanger in the hydrogen ion form that holds cationics. Nonionics pass through completely. The first column is stripped with ethanolic ammonium hydroxide to isolate anionics as their ammonium salts, the second column is stripped with ethanolic HCl to isolate amphoterics as their hydrochlorides, and the third column is extracted with ethanolic HCl to isolate cationics as their hydrochlorides. Each of these fractions can be saved for later analysis.

Open-column silica gel chromatography, in which the sample is separated based on polarity by elution with increasingly polar solvents, can also be used (11).

Once again, IR and NMR can be used to characterize the isolated fractions. Figure 12.9, for example, shows an IR spectrum of a quaternary ammonium compound (quat) used as a disinfectant. Although the quat functionality cannot be discerned by IR, we see that this compound has an aryl group by absorbances in the 650–1,000 cm^{-1} region. Absorbances near 720 and 700 cm^{-1} are indications of the presence of a benzyl group. In fact, this compound is BTC 8358, a benzalkonium chloride.

Figure 12.10 is an NMR spectrum of the two principal components (sodium alkene monosulfonate and sodium hydroxyalkane monosulfonate) of sodium alpha olefin sulfonate (AOS). For the alkene sulfonate, the peak at about 2 ppm is due to protons on carbons adjacent to the double bond, and those at about 5.4 ppm are due to protons on carbons at the double bond. For both components, the peaks at about 2.8 ppm are due to protons on the carbon adjacent to the sulfonate group, whereas the peaks at 3.5 ppm are due to the proton on the hydroxyl group on the hydroxysulfonate (12).

Anionics, nonionics, cationics, and amphoterics may be characterized by other means.

Anionics

Anionics (Fig. 12.11), for example, are most often quantified by titration against a standard cationic such as Hyamine 1622.

The preferred method in my laboratory is a modified version of the potentiometric titration described in ASTM D4251 (13). This is preferred because it is fast

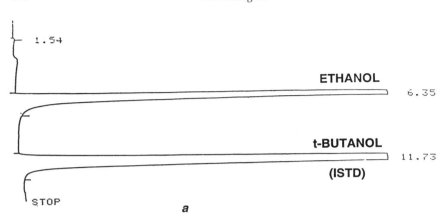

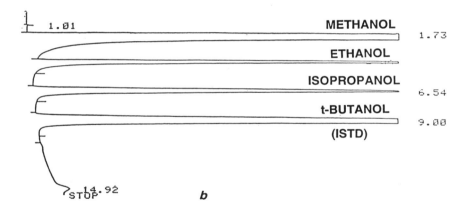

Fig. 12.7. Typical GC chromatograms: determination of (a) ethanol and (b) methanol, ethanol, and isopropanol. Note: The dates and instruments on which the sample chromatograms were run are different, accounting for the varying retention times observed.

and precise and avoids the use of hazardous solvents. In this approach, a detergent sample containing anionic active matter is titrated potentiometrically in an aqueous medium with a standard solution of a cationic titrant (Hyamine 1622) using a nitrate ion selective electrode (ISE). The titration involves the reaction between the cationic quaternary ammonium titrant and the anionic surfactant to form a complex that precipitates. The electrode responds to the increasingly lower concentration of anionic surfactant in solution as the titration proceeds.

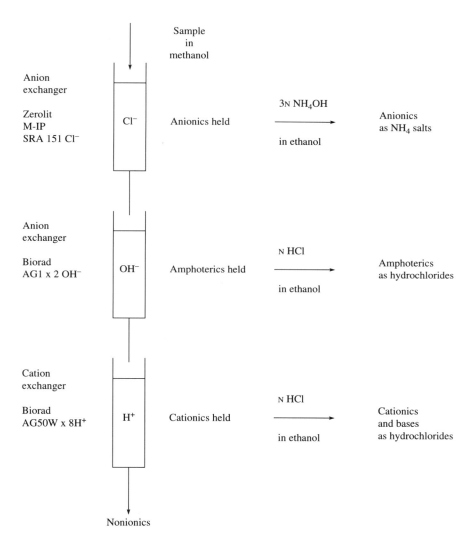

Fig. 12.8. Ion exchange scheme for separation of surfactant types. (Reproduced with permission from Gabriel [10]).

Typical titration curves are given in Figures 12.12 and 12.13 for the titration of an alpha sulfo methyl ester (Alpha-Step ML-40) at low and high pH. Alpha sulfo methyl esters generally contain a significant amount of an alpha sulfo carboxylate (14), shown in formula 2 of Fig. 12.13. When the titration is done at pH 2.5 as shown in Fig. 12.12, only the sulfonate groups are titrated because the carboxylate becomes protonated and is therefore not anionic. Note that the milliequivalents per gram (meq/g) given in Fig. 12.12 is 1.121. When the titration is done at pH 9.5 as

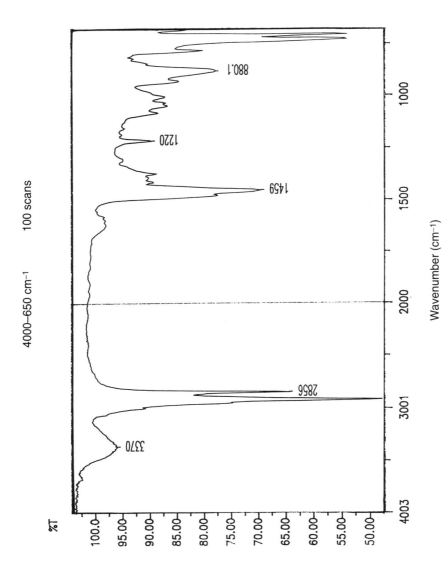

Fig. 12.9. FTIR spectrum of a quaternary ammonium antimicrobial agent.

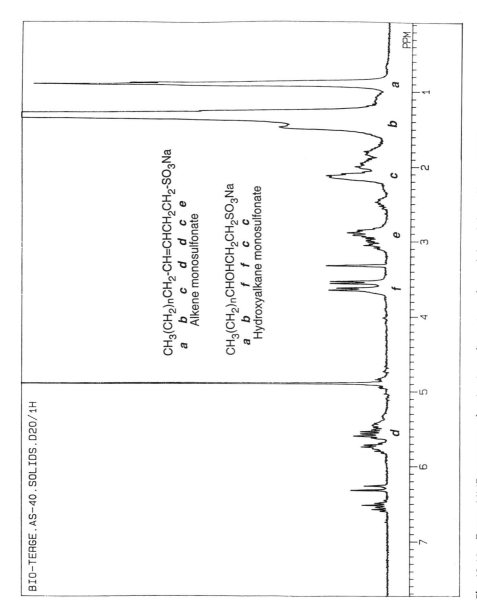

BIO-TERGE.AS-40.SOLIDS.D20/1H

$CH_3(CH_2)_nCH_2\text{-}CH\text{=}CHCH_2CH_2\text{-}SO_3Na$
 a   b   c   d   c   e
 Alkene monosulfonate

$CH_3(CH_2)_nCHOHCH_2CH_2SO_3Na$
 a   b   f   f   c   c
 Hydroxyalkane monosulfonate

Fig. 12.10. Proton NMR spectrum of active ingredients in sodium alpha olefin sulfonate (AOS).

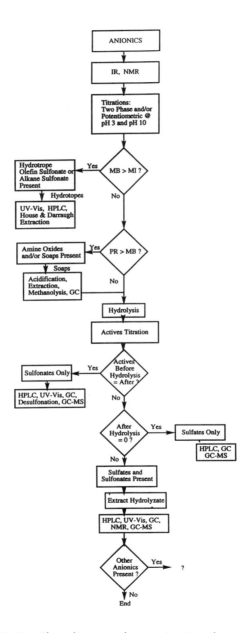

Fig. 12.11. Flow diagram: characterization of nonionics.

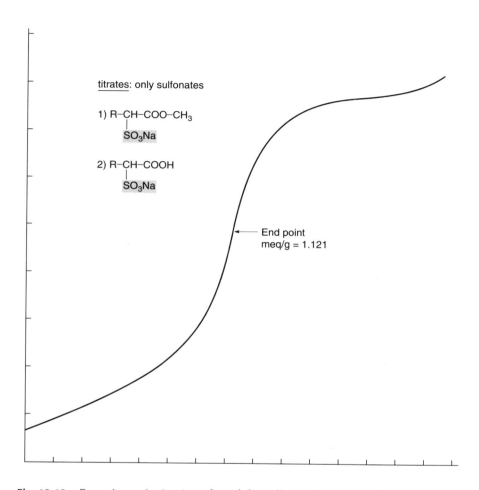

titrates: only sulfonates

1) R–CH–COO–CH₃
    |
    SO₃Na

2) R–CH–COOH
    |
    SO₃Na

← End point
meq/g = 1.121

Fig. 12.12. Potentiometric titration of an alpha sulfo methyl ester at pH 2.5.

shown in Fig. 12.13, on the other hand, the carboxylate is no longer protonated and is available for titration as an anionic surfactant. Observe that the meq/g in Fig. 12.13 is 1.321. The difference in meq/g between the two titrations is attributable exclusively to the carboxylate function. In this case, therefore, quantitation of both molecules by two simple titrations is possible. In general, if the titration at high pH gives a higher result than that at low pH, then amine oxides and/or soaps (carboxylates) are present.

Other titrations involving Hyamine 1622 are so-called two-phase (organic/aqueous) titrations involving a visual indicator system (usually acidic methylene blue (MB) (15), acidic dimidium bromide-disulphine blue mixed indicator (MI) (16-17), basic bromcresol green (BCG) (18), or basic phenol red

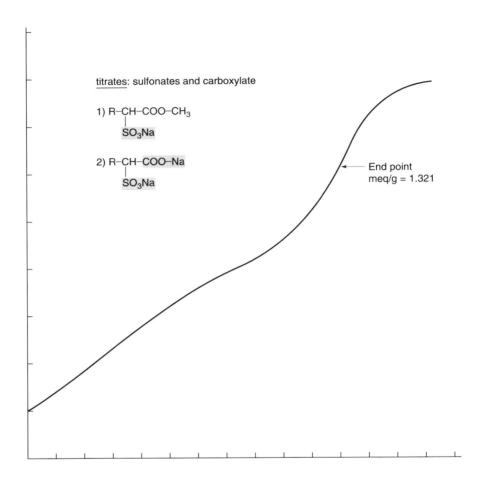

titrates: sulfonates and carboxylate

1) R–CH–COO–CH₃
        |
     SO₃Na

2) R–CH–COO–Na
        |
     SO₃Na

End point
meq/g = 1.321

Fig. 12.13. Potentiometric titration of an alpha sulfo methyl ester at pH 9.5.

(PR) (14). The organic solvent used is usually chloroform. The titrations involve the addition of Hyamine 1622 to a mixing cylinder containing chloroform, an aqueous solution of sample, and the desired indicator, and then shaking the stoppered cylinder. Hyamine 1622 is added incrementally and the cylinder is shaken until the desired transfer of colors between the two phases is observed. These titrations are less desirable because they are rather slow, are subject to operator end-point interpretation, and use a hazardous solvent. The MB and PR titrations are subject to positive interference when hydrotropes (for example, low molecular weight sulfonates) are present, but the potentiometric, MI, and BCG titrations are not.

When hydrotropes are not present and the results using MB or MI are lower than those using BCG or PR, this is an indication that amine oxides or soaps are present.

Performing these titrations before and after acid hydrolysis (for example, refluxing for 2 h in the presence of 1N sulfuric acid) gives a measure of organic sulfonate and organic sulfate content. This is because the C-S (carbon-sulfur) bond of sulfonates is relatively stable to acid hydrolysis, whereas the C-O-S (carbon-oxygen-sulfur) bond of sulfates is not.

If the titration result before hydrolysis equals that after hydrolysis, only sulfonates are present. If the result after hydrolysis is zero, all of the active is hydrolyzed, meaning that only sulfates are present. If the result before hydrolysis is greater than the result after hydrolysis and the result after hydrolysis is greater than zero, both sulfonates and sulfates are present.

Many sulfonates, including LAS, are characterizable by IR (19), HPLC (20), and NMR. Figure 12.14, for example, an IR spectrum of the solids of a sodium LAS, shows bands at 1,189 and 1,046 cm^{-1} (asymmetric and symmetric S=O stretching vibrations), 1,011 (phenyl carbon to sulfur), and 834 (para disubstituted benzene) cm^{-1} normally found in linear alkylbenzene sulfonates. In Fig. 12.15 we see how HPLC has been used not only to characterize LAS but to delineate the carbon number and phenyl alkyl distributions of the parent alkylate, all without a desulfonation step. Sulfonates can also be characterized by GC, and NMR after desulfonation and extraction of the resulting hydrocarbon with petroleum ether. Figure 12.16 (21) shows a gas chromatogram of an LAS alkylate denoting the peaks due to the performance influencing 2-phenyl isomers. The amount of 2-phenyl isomer in LAS is important in that high 2-phenyl content (e.g. 20%–30%) results in slightly lower foaming and viscosity, whereas low 2-phenyl content (e.g. 10%–20%) means that the product has slightly higher foaming and viscosity. The higher viscosity may necessitate the use of a hydrotrope to improve solubility.

The NMR spectrum in Fig. 12.17 shows aromatics at 7–8 ppm, methylenes at 1–2 ppm, and methyl protons at about 0.8 ppm. The small peaks between 2 and 3 ppm are attributable to methyne protons on carbons attached to the benzene ring. The proton NMR spectrum given in Fig. 12.18 is interesting in that it shows peaks at about 3.55 ppm due to the methylene protons adjacent to the nitrogen and at about 4.1 ppm due to the methylol protons on the triethanolamine cation (TEA+). The small peaks at 3.30 and 3.95 ppm are from protons associated with free amines. These, coupled with the peaks between 7 and 8 ppm assigned to a para disubstituted benzene, are typical of a triethanolamine alkybenzene sulfonate. Desulfonation, incidentally, is usually effected by reacting the sample with 85% phosphoric acid at high temperature. When the desulfonation is carried out at low temperatures with added phosphoric acid, however, the method may be applied to mixtures of LAS, alcohol sulfates, and alcohol ether sulfates without prior separation (22–25). This is because the "cracking" of alkyl chains seen at high temperature is now minimized while the C-S and C-O-S bonds are still cleaved. Aromatic sulfonates can be examined by UV-Vis spectroscopy (26), with a peak maximum appearing at 220–225 nm.

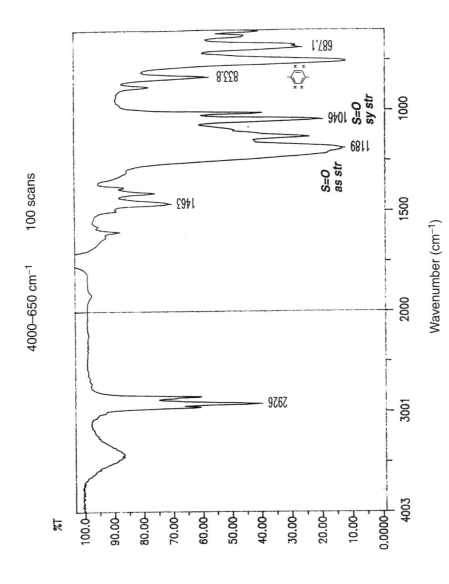

Fig. 12.14. FTIR spectrum of sodium LAS solids.

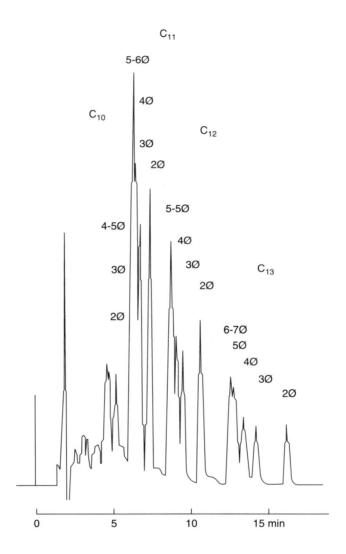

Fig. 12.15. HPLC chromatogram of a linear alkylbenzene sulfonate. Conditions: column, Hitachi Gel 3053 5 μm, 4.6 mm i.d. × 150 mm; eluent, 0.1 M sodium perchlorate in acetonitrile/water (45/55); flow rate, 1.0 mL/min; pressure, 100 kg/cm^2, column temperature, 40°C; detector, UV 225 nm, 0.16 AUFS; injection volume, 5 μL; sample concentration, 0.1%. Ø is the position of the phenyl group from the terminal methyl group on the chain. (Reproduced with permission from Nakae et al. [20]).

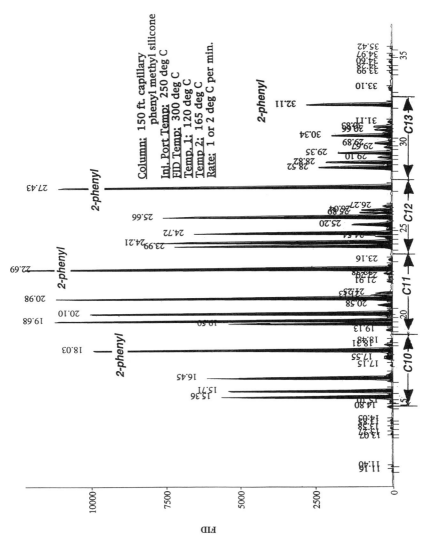

Fig. 12.16. GC chromatogram of an LAS alkylate.

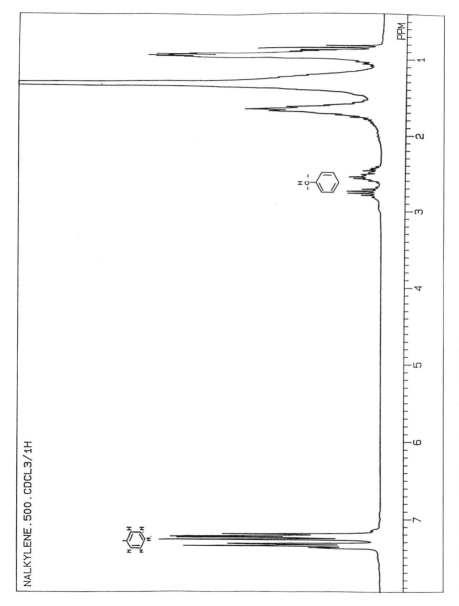

NALKYLENE.500.CDCL3/1H

Fig. 12.17. Proton NMR spectrum of an LAS alkylate.

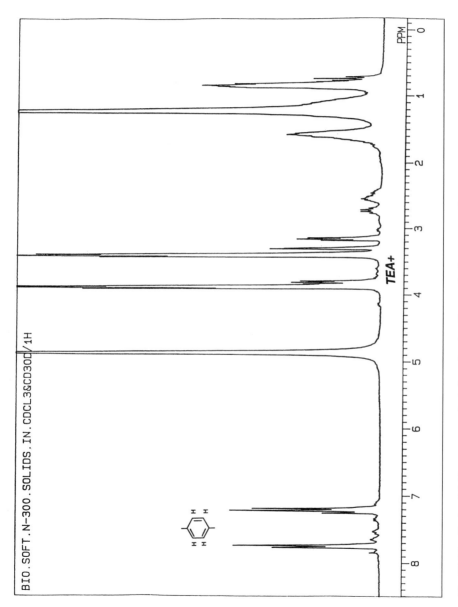

Fig. 12.18. Proton NMR spectrum of triethanolamine linear alkylbenzene sulfonate.

Alcohol sulfates can also be characterized by IR, NMR, and HPLC. In Fig. 12.19 we see an IR spectrum for a sodium lauryl sulfate giving typical asymmetric and symmetric S=O stretching absorbances at 1,216 and 1,073 wavenumbers, plus a methylene rocking vibration at about 720 cm^{-1}. Aromatic sulfates, such as alkyl phenol ether sulfates, can be characterized by UV-Vis (26), but alcohol ether sulfates generally cannot.

Sulfonates and sulfates can be analyzed in their acid forms by titrating a methanolic solution potentiometrically with cyclohexylamine (27), as depicted in Fig. 12.20. Two inflection points are shown. The first inflection represents the neutralization of strong acids, such as sulfonics and alkylsulfurics, and the first hydrogen of sulfuric acid. The second inflection represents the neutralization of the second hydrogen of sulfuric acid. The amount of sulfonic or organosulfuric acid is calculated based on the titrant volume of the first inflection minus that between the two inflections. The amount of sulfuric acid is calculated from the titrant volume between the two inflections, which is the amount of base required for the neutralization of the bisulfate anion.

A number of other anionic surfactants, including phosphate esters, fluorosurfactants, sulfosuccinates, alkane sulfonates, olefin sulfonates, alkyl glyceryl ether sulfonates, alkyl isethionates, and taurides can also be found. These can be characterized by IR and/or proton and carbon-13 NMR, HPLC, and thin-layer chromatography (TLC), which is a powerful tool in the hands of experienced analysts for the general analysis of a mixture of surfactants (28). Figure 12.21, for example, shows how secondary alkane sulfonates (SAS) and alpha olefin sulfonates (AOS) may be distinguished from each other by TLC. AOS is composed of three component types: alkene monosulfonates, alkane hydroxy sulfonates, and disulfonates. These all clearly show up as well resolved spots in the chromatogram on top for AOS. SAS, however, is composed of only two component types: alkane monosulfonates in which the sulfonate group is substituted onto secondary carbons all along the chain, and alkane disulfonates. These show up as two spots on the bottom chromatogram, with the second one being rather elongated because of the broad substitution of the sulfonate group on the secondary carbons. The chromatograms were prepared by spotting 0.1 µL of about 4.5% active solutions dissolved in 70/30 (volume/volume) isopropanol/water on to 10 by 20 cm. TLC plates coated to a thickness of 250 microns with silica gel (no binder) and 5% ammonium sulfate. The separations were done using a mobile phase composed of 50/55/6 volume/volume/volume chloroform/isopropanol/0.1N sulfuric acid.

Nonionics

Nonionic detergents (Fig. 12.22) can be characterized by IR and/or NMR spectroscopy on the isolated nonionic fraction. Although nonionic in charge and, therefore, isolatable by mixed bed ion exchange, alkanolamides are sometimes not referred to as such in surfactant literature because they function more as foam

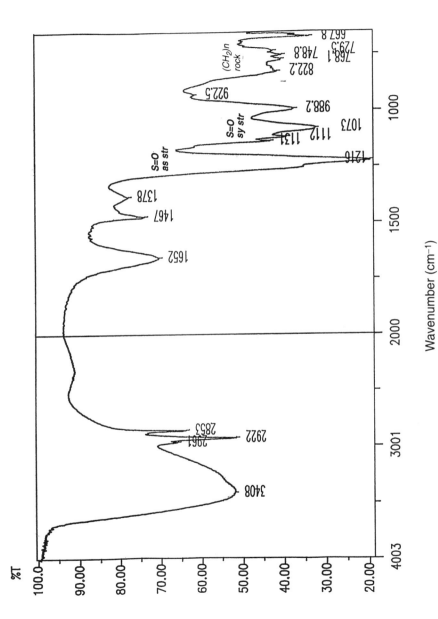

Fig. 12.19. FTIR spectrum of sodium lauryl sulfate; 100 scans at 8 cm^{-1} on 60°C heated HATR dissolved in 1:1 H$_2$O/3A alcohol; evaporation approximately 2 h.

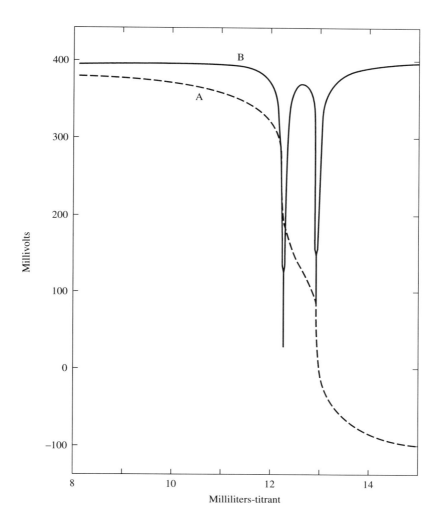

Fig. 12.20. Potentiometric titration of a detergent intermediate containing alkyl-benzene sulfonic acid and sulfuric acid. Solvent: methanol. Titrant: 0.1 N cyclo-hexylamine. A = differentiating titration curve; B = the first derivative curve of curve A. (Reproduced with permission from Yamaguchi [27]).

stabilizers and viscosity builders than as detergents. They may be characterized and/or quantified, however, by IR, HPLC, proton or carbon-13 NMR, GC after methanolysis, and nonaqueous titration. The IR spectrum seen in Fig. 12.23 for a fatty diethanolamide shows absorbances at 1,621 cm^{-1} due to amide carbonyl, 1,054 due to the primary alcohol stretching vibration, and at about 720 cm^{-1} due to the long methylene chain.

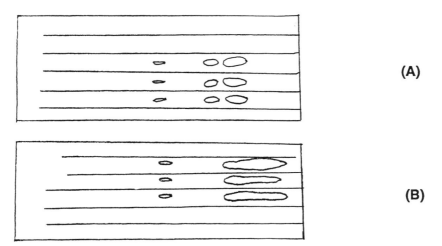

Fig. 12.21. Thin-layer chromatograms of sodium alpha olefin sulfonate (A) and secondary alkane sulfonate (B).

The NMR spectrum in Fig. 12.24 once again shows peaks between 1 and 2 ppm due to methylenes. It also has peaks at about 1.6 ppm from methylene protons on the carbon beta to the amide carbonyl. The peaks at about 2.45 ppm are due to protons on the carbon adjacent to the carbonyl, and the peaks at about 2.75 ppm are from the protons on the carbon alpha to the nitrogen of any free diethanolamine (DEA) present. The peaks at about 3.5 ppm and at about 3.7 ppm are due to the methylene protons adjacent to the nitrogen and to the methylol protons, respectively, on both the amide and free DEA. The methanolysis is accomplished by refluxing the amide in methanol in the presence of a sulfuric acid catalyst, resulting in a conversion of the amide into methyl esters, which are then extracted and chromatographed, and free amine. Figure 12.25 shows a typical chromatogram of a methyl ester derived from a coconut oil diethanolamide. As expected, we see a carbon chain distribution typical of coconut oil methyl esters.

Wimer (29) demonstrated that amides dissolved in acetic anhydride can be titrated by perchloric acid in acetic acid. This is because perchloric acid exhibits increased acidic behavior in the presence of acetic anhydride, making titration of weakly basic amides possible. Amines are also titrated in acetic anhydride, so a correction must be made for them. This is accomplished by titrating a second sample with perchloric acid, using glacial acetic acid as the sample solvent (amines are titratable in this system, whereas amides are not).

Alkoxylated fatty alcohols and alkoxylated alkyl phenols may be characterized by HPLC, IR, NMR, refractometry, and possibly high-temperature GC or GC-MS. UV-VIS cannot be used on fatty alcohol alkoxylates. Figure 12.26 gives a gas chromatogram of a C12-14-16 alcohol ether with an average of 3 moles of ethoxylation, and Fig. 12.27 shows an HPLC plot of a fatty alcohol ethoxylated with an average 9 moles of ethylene oxide (EO). The latter chromatogram was obtained

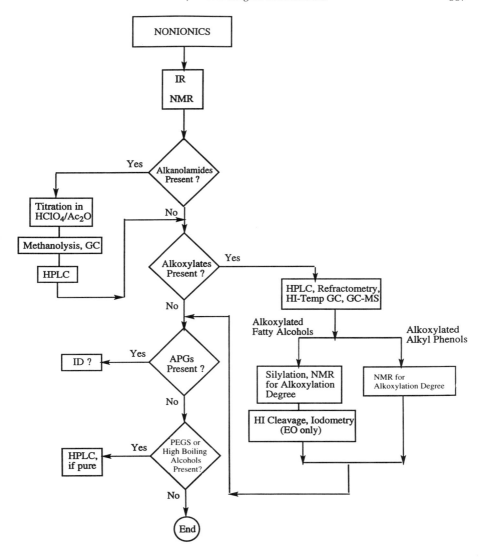

Fig. 12.22. Flow diagram: characterization of nonionics.

using a UV detector after derivatization with phenyl isocyanate. Figure 12.28 shows the HPLC analysis of an ethoxylated (average 12 moles) nonyl phenol, also using a UV detector. Derivatization was not needed in this case because the molecule has a UV absorbing aromatic ring.

Quantitation of ethylene oxide or propylene oxide (PO) content can be done by iodometry after cleavage with hydriodic acid (30). In this method, the EO or PO chains are cleaved with hydriodic acid. The unstable 1,2 diiodoalkane formed

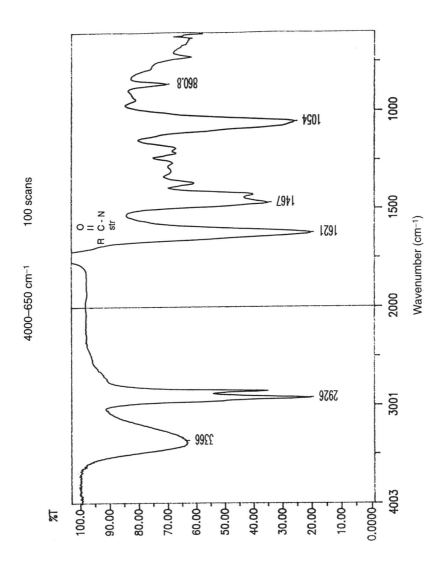

Fig. 12.23. FTIR spectrum of a lauric diethanolamide.

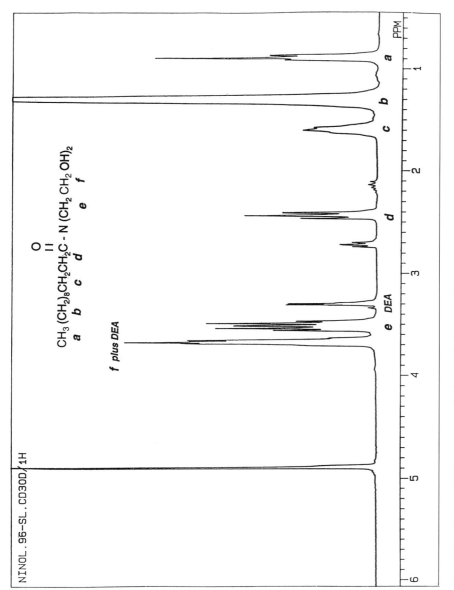

Fig. 12.24. Proton NMR spectrum of a lauric diethanolamide.

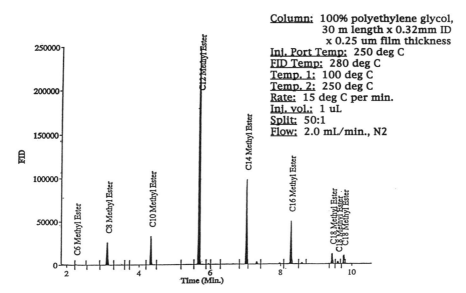

Fig. 12.25. GC chromatogram of a methyl ester derived from a cocamide DEA.

quickly decomposes to form ethylene (or propylene) and iodine. EO or PO content is determined by titrating the amount of free iodine formed with standard sodium thiosulfate. Hydrophobe carbon chain distribution is determined on the alkyl iodides formed in the cleavage by extracting them with petroleum ether after the titration, concentrating them, and analyzing them by GC. They may also be analyzed by proton NMR without cleavage. A spectrum of a 10-mole (average) nonyl phenol ethoxylate is given in Fig. 12.29. Again, the hydrocarbon protons appear between 0 and 2 ppm and aromatics between 6.7 and 7.5 ppm. The peaks at about 4.1 ppm are from methylene protons alpha to the phenyl group, those at about 3.8 ppm are due to methylene protons beta to the aromatic, whereas those around 3.6 ppm are from the protons that are part of the ethylene oxide units. It should be noted also that the peaks due to terminal methyl protons at less than 1.0 ppm are almost as great as those due to methylene protons between 1 and 2 ppm, indicating that the nonyl side chain is highly branched. Integration of the peaks, incidentally, enables us to determine the degree of ethoxylation.

Other nonionic surfactants commonly found in formulated products include alkyl polyglucosides; ethylene and polyethylene glycols; propylene and polypropylene glycols; polyethylene glycol esters; fatty-fatty esters; mono-, di-, and triglycerides; glycerine; amine ethoxylates; sorbitan ester derivatives; and amide ethoxylates. They may be characterized by silica gel open-column chromatography (11), titration, elemental analysis, hydroxyl value, GC, HPLC, TLC, supercritical fluid chromatography (SFC), refractometry, cloud point, iodine value, and various kinds of spectrophotometry (31).

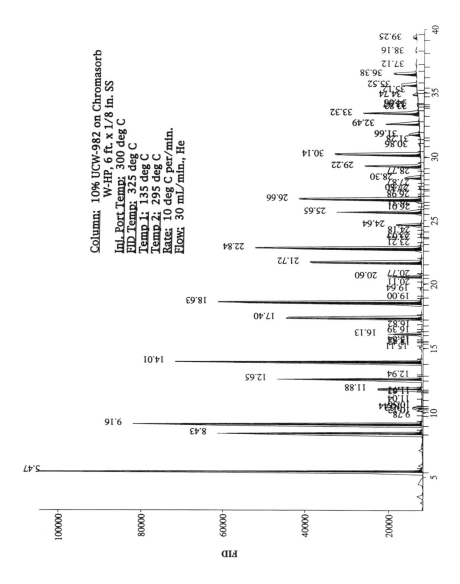

Fig. 12.26. GC chromatogram of a C12-C14-C16 ethoxylated (average 3 moles EO) fatty alcohol.

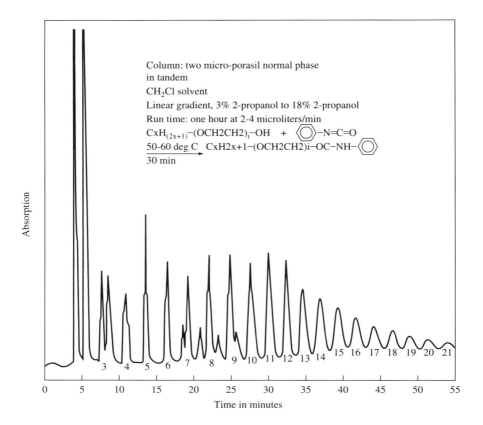

Fig. 12.27. HPLC chromatogram of an ethoxylated (average 9 moles EO) fatty alcohol.

Amine oxides are sometimes classified as nonionic or amphoteric surfactants. They behave as nonionics only at pH greater than 7. At lower pH they are protonated and behave as cationic surfactants. Further characterization information on amine oxides is given later in the chapter in the discussion on amphoterics.

Cationics

Fabric softener, disinfectant, and hair care quaternary ammonium compounds (quats) are the categories of cationic surfactants (Fig. 12.30) commonly used. They can be characterized by IR, NMR, HPLC, and sometimes UV-Vis.

In my laboratory, pyrolysis GC is used to characterize disinfectant quats, whereas saponification GC is employed for fabric softener quats. In pyrolysis GC, long-chain quats decompose in the hot injection port of the chromatograph to form tertiary amines and alkyl, benzyl, or ethylbenzyl chlorides. The homolog

Conditions

Column: Ultrasphere Cyano (Altex) - 15 cm.

Solvent A: 95% hexane; 5% IPA
Solvent B: 90% IPA; 10% H2O

Gradient: 30% B to 70% B in 20 minutes--Curve 7

Flow: 1 ml/min
Run Time: 25 min
Equilibration time: 7 min

UV Detector at 280 nm; 0.1 AUFS

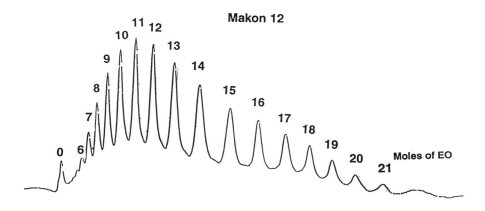

Fig. 12.28. HPLC chromatogram of a nonyl phenol ethoxylate (average 12 moles EO).

distribution of the QUAT is determined (Fig. 12.31) from the carbon chain distribution of the resulting tertiary amines, whereas the ratios of alkyl, benzyl, and ethylbenzyl chlorides are determined directly from the gas chromatogram. Amine oxides, by the way, also decompose in the injection port to form alpha olefins and dialkylhydroxyl amines, and the distributions of the resulting alpha olefins and unreacted parent tertiary amines are determined. In saponification GC, the sample is saponified by refluxing with methanolic sodium hydroxide, acidified with hydrochloric acid, and then converted to the methyl esters by refluxing with BF3/methanol. The methyl esters are then extracted and analyzed by GC.

Quantitation of quats can also be done by potentiometric (32) or two-phase (33) titration with standard sodium lauryl sulfate (SLS). In the potentiometric method, an aqueous solution of QUAT is adjusted to pH 10.5 to avoid positive interference from amine hydrochlorides and titrated potentiometrically with the SLS using the

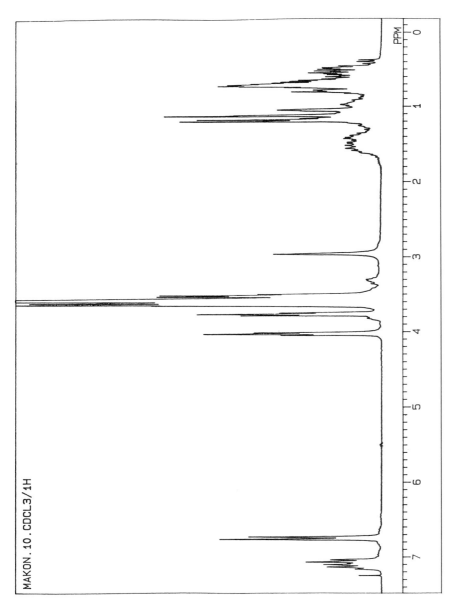

Fig. 12.29. Proton NMR spectrum of a nonyl phenol ethoxylate (average 10 moles EO).

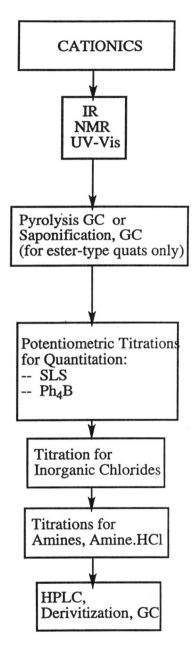

Fig. 12.30. Flow diagram: characterization of cationics.

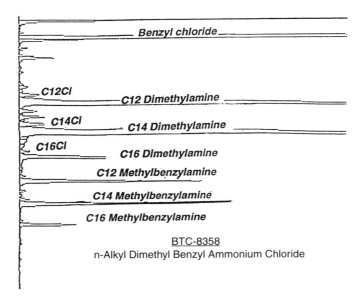

Fig. 12.31. Pyrolytic breakdown and GC chromatogram of breakdown products for an antimicrobial quaternary ammonium compound.

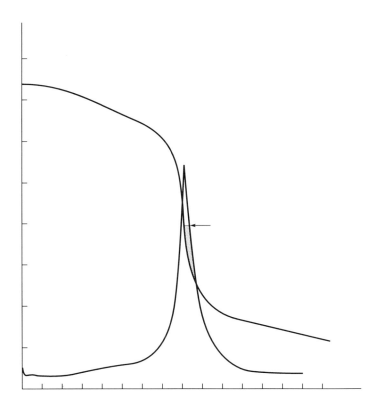

Fig. 12.32. Potentiometric titration of an antimicrobial quaternary ammonium compound with 0.02 N SLS.

nitrate ion selective electrode. The titration has to do with the formation of a water-insoluble complex between the quat and SLS. The end point is read at the inflection point of the S-shaped curve caused by the electrode's response to the increasing concentration of SLS, as shown in Fig. 12.32. In the two-phase titration, SLS is added incrementally to a mixing cylinder containing chloroform as the organic phase and aqueous bromphenol blue buffered to a pH of about 11 as the indicator, stoppered, and shaken. The end point is reached when the first purple color appears in the aqueous top layer. Once again, the potentiometric approach is preferred because it is easier, faster, safer, and more precise.

Quats can be assayed by titration with sodium tetraphenyl boron (34), which reacts quantitatively with quats to form a water-insoluble complex. Sucrose is added to the sample to reduce coagulation of the precipitate, and 2,7-dichlorofluorescein is used as the indicator. The titration proceeds through a "pink" or "peach" to a sudden "creamy yellow" end point as the last of the quat reacts. A potentiometric version of this titration has been found to be successful (35). Quats having an inorganic halide as the counter-ion can be quantified by potentiometric titration with silver nitrate, but corrections must be made for amine hydrohalides and inorganic halides.

Figure 12.33 shows the NMR spectrum of Accosoft 501, a fabric softener quat. Most notable are the peaks at about 1.8 ppm due to the CH_2CH_2N beta to the vinyl group, at about 2.2 ppm due to the protons on the carbon next to the carbonyl, and those due to ethylene oxide protons at about 3.6 ppm.

Cationic surfactants can be further characterized by HPLC, derivatization-GC, and UV-Vis (for aromatic and imidazolinium quats). In Fig. 12.34, the peak maximum at about 235 nm is due to the imidazoline ring shown in the structure. Measurement of the net absorbance at the peak maximum permits quantitation of the quat.

Amphoterics

Amphoterics (Fig. 12.35) may be characterized by IR and/or NMR. Although there are a number of betaines classified as amphoteric surfactants, only the carboxybetaines are widely used. They may be characterized by IR, NMR, HPLC, and pyrolysis GC. Figure 12.36 is an IR spectrum of a cocamidopropyl betaine. Most notable here are the absorbances near 1,633, 1,548, and 722 cm^{-1} due to amide carbonyl, carboxylate, and a long methylene chain, respectively. In the NMR spectrum shown in Fig. 12.37, the peaks at about 1.6 and 2.2 ppm are due to the methylene protons on the carbons beta and alpha, respectively, to the carbonyl. Those near 2.0 ppm are caused by the protons on the carbon beta to the amido nitrogen and to the quaternary nitrogen, and those near 2.25 ppm are due to the protons alpha to the amido carbonyl. The peaks at 2.9 and 4.0 ppm are due to impurities. Those near 3.25 ppm are from the methyl protons attached to the quaternary nitrogen, the ones near 3.65 are due to the methylene protons on the carbons alpha to the quaternary nitrogen and to the amido nitrogen, and the peaks at about 3.8 ppm are from the protons on the methylene between the quaternary nitrogen and the carboxylate carbonyl.

Betaines may also be quantitated by HPLC or by potentiometric or two-phase titration with SLS at low pH, at which they behave like cationic surfactants. In blends with anionic surfactants, however, they will negatively interfere with low pH

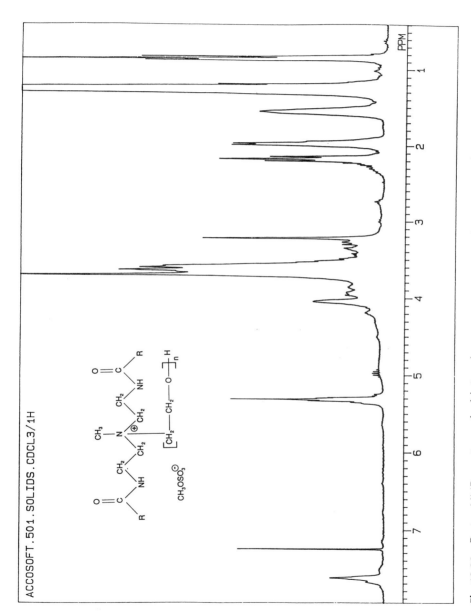

Fig. 12.33. Proton NMR spectrum of a fabric softener quaternary ammonium compound.

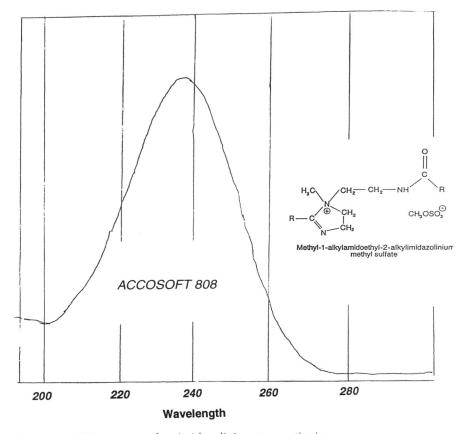

Fig. 12.34. UV spectrum of an imidazolinium-type cationic.

titrations, yielding a low anionic result. Potentiometric titration using the nitrate ISE at low and high pH, in fact, gave us a measure of the total of amine oxide plus betaine in the previously mentioned light-duty liquid detergent analysis.

Amine oxides can be isolated by extraction from alcohol-water with chloroform at high pH. If true nonionics are present, however, these must first be extracted at a low pH before the amine oxide is extracted at high pH. Amine oxides may be characterized by NMR, GC, and possibly IR. In the NMR spectrum given in Fig. 12.38, the peaks near 1.8 ppm are due to the methylene protons beta to the nitrogen and the singlet at about 3.1 ppm comes from the methyl protons on the carbons attached to the nitrogen. The peaks around 3.25 ppm are due to the methylene protons on the carbon alpha to the nitrogen.

They may also be quantified by potentiometric or two-phase titration with SLS at low pH or by potentiometric titration with alcoholic hydrochloric acid before and after addition of methyl iodide, as described by Metcalfe (36).

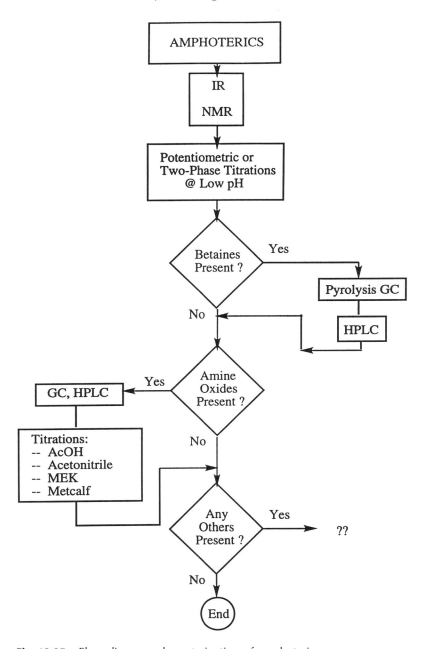

Fig. 12.35. Flow diagram: characterization of amphoterics.

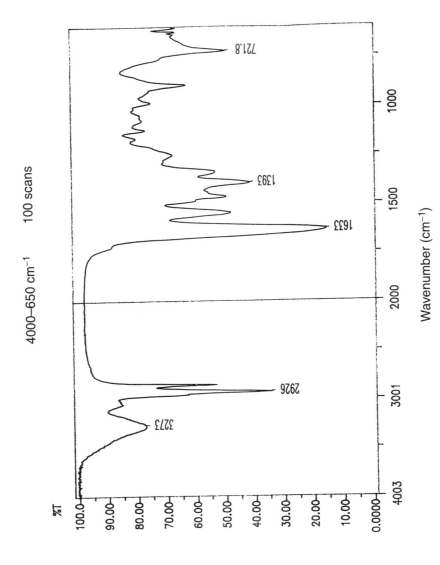

Fig. 12.36. FTIR spectrum of a cocamidopropyl betaine.

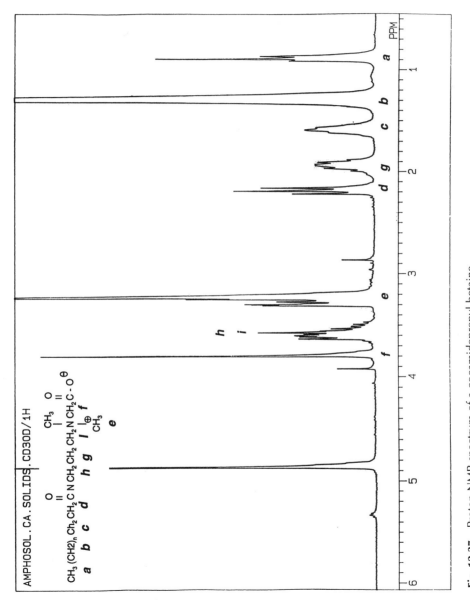

Fig. 12.37. Proton NMR spectrum of a cocamidopropyl betaine.

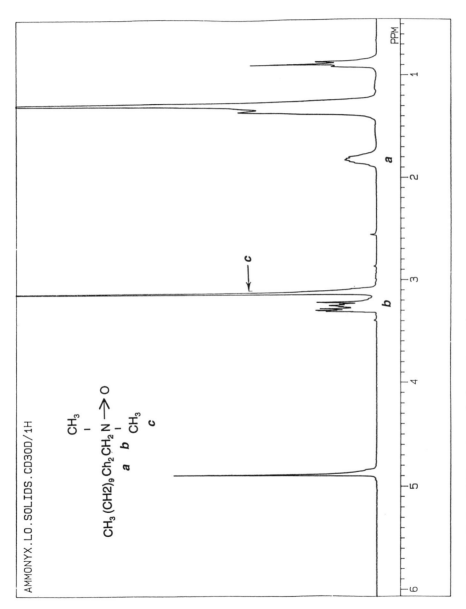

Fig. 12.38. Proton NMR spectrum of a lauramine oxide.

Newer Approaches

New approaches in the field of formulated detergent analysis appear regularly. This includes reports on the uses of HPLC, nuclear magnetic resonance (NMR), fast atom bombardment tandem mass spectroscopy (FAB/MS/MS), capillary electrophoresis (CE), near infrared (NIR), and Fourier transform IR (FTIR).

HPLC is being used in the analysis of surfactant mixtures (Wiley, J., Liquid Chromatographic Analysis of Surfactant Mixtures, Unilever Research, presented at 1993 AOCS National Meeting, Anaheim, California). This includes the separations of mixtures of betaines and lauryl ether sulfates, amine oxides and betaines, fatty acid soaps and fatty isethionates, and ether sulfates and LAS. For surfactant mixtures with carbon chain lengths from C8–18, a reverse phase column (for example, C2 or C6) with a relatively low retention has been found to be helpful to elute all components isocratically (that is, using a single mobile phase).

Most surfactants are not volatile enough to be analyzed by the most common configuration of mass spectroscopy, gas chromatography–mass spectroscopy (GC-MS), without some kind of derivitization or degradation to convert them to volatile compounds. With specialized instrumentation, however, it is possible to dispense with both gas chromatography and derivatization in the analysis of anionic surfactants by the use of FAB/MS/MS (37). The technique has been applied to alkyl sulfates and sulfonates, LAS, hydrotropes, alkyl ether sulfates, olefin sulfonates, fatty acid soaps, N-acylated amino acids, and sulfosuccinates.

The term *capillary electrophoresis* describes a process based on the migration of ions and/or charged particles in narrow bore capillaries under the influence of an electric field. The capillaries are filled with liquids without supporting solid or gel media, usually in free electrolyte solutions (38). Cationic and anionic detergents behave differently from each other in that, although both are subject to electroosmotic flow that pushes them toward the negative end of the capillary, cations will move much more quickly (have greater electrophoretic mobility) because of their positive charge. The lower molecular weight cationics will elute first. The mobility of the anionics is retarded, however, because of their greater attraction to the positive side of the capillary, and the higher molecular weight anionics will elute first. The separations of alcohol sulfates and alcohol ether sulfates (39), of alkybenzene sulfonates (40), and of alkane sulfonates and alcohol sulfates (41) have been reported.

Another example of the use of FTIR in formulated detergent analysis is given in Fig. 12.39 (42). The top spectrum is that of 100 mg of a powdered laundry detergent that was dissolved in 2 mL of water, deposited on a preheated (60°C) attenuated total reflectance (ATR) crystal, evaporated, and scanned 50 times in 1 min by FTIR. Here we see only the major inorganic component, sodium carbonate. A similar amount of the same product was filtered from a few milliliters of ethanol. The filtrate was deposited to the ATR, evaporated, and scanned in the same way. The lower spectrum reveals the presence of alcohol sulfate at about 1,230 and 1,040 cm^{-1} and ethoxylate at about 1,100 cm^{-1}.

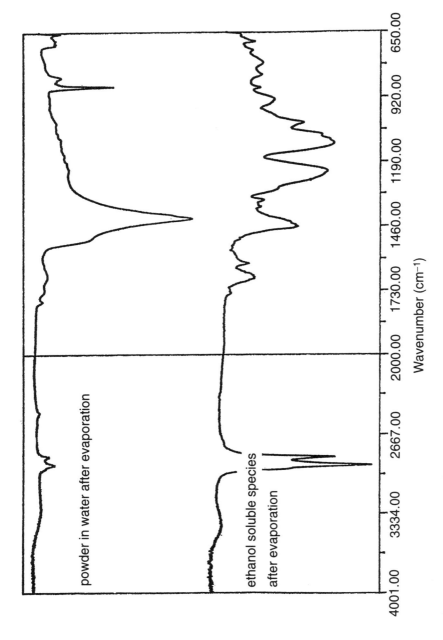

Fig. 12.39. FTIR spectrum of (top) a powdered laundry detergent and (bottom) the ethanol soluble portion thereof. (Reproduced with permission from DeSalvo [42])

Among the advantages that FTIR has over the older dispersive IR operating in the conventional mid-infrared range of 4,000 to 400 wavenumbers is speed. The complete spectrum can be measured on an FTIR instrument in the time it takes to measure just a few wavenumbers on a dispersive instrument. Multicomponent quantitative analysis by FTIR requires a considerable amount of calibration work, with the number of calibration standards required increasing greatly with the number of components present. Fuller (43) discusses a method for the quantitation of surfactants in a laundry detergent that requires 21 reference samples.

Summary

A number of approaches to formulated detergent analysis currently being used have been discussed or mentioned in this chapter. These include so-called "wet" methods such as extraction; titration by visual, optical, and potentiometric techniques; and spectroscopic approaches such as infrared (IR), ultraviolet/visible (UV/VIS), and fast atom bombardment mass spectrometry (FAB/MS/MS). They also encompass separation methods such as ion exchange, thin-layer (TLC), high-pressure liquid (HPLC) and gas liquid chromatography (GLC or GC).

As Table 12.1 indicates, modern detergent formulations tend to be rather complex. This means that highly sophisticated methods may be required to render a meaningful analysis. It is believed that the most dramatic development lies in the area of multicomponent analysis without the need for prior separation. This has been observed most notably using HPLC, FTIR, and TLC. For these reasons, the field of surfactant analysis is dynamic and growing rapidly, and constant surveillance of the literature is mandatory if one is to stay current.

Acknowledgments

I wish to acknowledge with gratitude the contributions of the laboratories of the Stepan Company in this work. Thanks are due especially to the following Stepan personnel: Stacey Greenhill, Jean Goze, Jeff Strnad, Dennis DeSalvo, Ellen Novak, and David McGregor. Their help in manuscript preparation and review, spectral and chromatographic interpretation, and preparation of the figures and flow diagrams has been invaluable.

References
1. American Society for Testing and Materials, ASTM D 1568, Standard Test Methods for Sampling and Chemical Analysis of Alkylbenzene Sulfonates, in *1994 Annual Book of ASTM Standards, 15.04,* American Society for Testing and Materials, Philadelphia, 1994, pp. 145–150.
2. American Society for Testing and Materials, ASTM D 501, Standard Test Methods of Sampling and Chemical Analysis of Alkaline Detergents, in *1994 Annual Book of ASTM Standards, 15.04,* American Society for Testing and Materials, Philadelphia, 1994, pp. 46–70.
3. American Society for Testing and Materials, ASTM D 460, Standard Test Methods for Sampling and Chemical Analysis of Soaps and Soap Products, in *1994 Annual*

Book of ASTM Standards, 15.04, American Society for Testing and Materials, Philadelphia, 1994, pp. 9–27.

4. Vandenberg, J.T., et al., *An Infrared Spectroscopy Atlas for the Coatings Industry,* Federation of Societies for Coatings Technology, Philadelphia, 1980, chap. 1.

5. Shoolery, J., *A Basic Guide to NMR,* Varion Associates, Palo Alto, California, 1972, pp. 6–8.

6. American Society for Testing and Materials, ASTM D 820, Standard Test Methods for Chemical Analysis of Soaps Containing Synthetic Detergents, in *1994 Annual Book of ASTM Standards, 15.04,* American Society for Testing and Materials, Philadelphia, 1994, pp. 93–101.

7. American Society for Testing and Materials, ASTM D 1568, Standard Test Methods for Sampling and Chemical Analysis of Alkylbenzene Sulfonates, in *1994 Annual Book of ASTM Standards, 15.04,* American Society for Testing and Materials, Philadelphia, 1994, pp. 147–148.

8. American Society for Testing and Materials, ASTM D 3673, Standard Test Methods for Chemical Analysis of Alpha Olefin Sulfonates, in *1994 Annual Book of ASTM Standards, 15.04,* American Society for Testing and Materials, Philadelphia, 1994, pp. 378–379.

9. Wefer, E., *Solvent Alcohol Determination in Surfactants,* Stepan Company Analytical Method 410-0, 1991.

10. Gabriel, D.M., Specialized Techniques for the Analysis of Cosmetics and Toiletries, *J. Soc. Cosmet. Chem.* 1974, *25:* 33–48.

11. Rosen, M.J., A General Method for the Chromatographic Separation of Nonionic Surface Active Agents and Related Materials, *Anal. Chem. 35:* 2074 (1963).

12. Cross, J., ed., *Anionic Surfactants-Chemical Analysis, 8:* 157,172, Marcel Decker, New York, 1977, chap. 4.

13. American Society for Testing and Materials, ASTM D 4251, Standard Test Method for Active Matter in Anionic Surfactants by Potentiometric Titration, in *1994 Annual Book of ASTM Standards 15.04,* American Society for Testing and Materials, Philadelphia, 1994, pp. 447–449.

14. Battaglini, G.T., J.L. Larsen-Zobus, and T.G. Baker, Analytical Methods for Alpha Sulfo Methyl Tallowate, *J. Amer. Oil Chem. Soc. 63:* 1073–1077 (1986).

15. Epton, S.R., New Methods for the Rapid Titrimetric Analysis of Sodium Alkyl Sulfates and Related Compounds, *Trans. Faraday Soc. 44:* 226–230 (1948).

16. Reid, V.W., G.F. Longman, and E. Heinerth, Determination of Anionic Active Detergents by Two Phase Titration, *Tenside 9:* 292 (1967).

17. Reid, V.W., G.F. Longman, and E. Heinerth, Determination of Anionic Active Detergents by Two Phase Titration (II), *Tenside 3–4:* 90 (1968).

18. Lew, H.Y., Analysis of Detergent Mixtures Containing Amine Oxides, *J. Amer. Oil Chem. Soc. 41:* 297 (1964).

19. American Society for Testing and Materials, ASTM D 2357, Standard for Qualitiative Classification of Surfactants by Infrared Absorption, in *1994 ASTM Annual Book of ASTM Standards 15.04,* American Society for Testing and Materials, Philadelphia, 1994, pp. 264–266.

20. Nakae, A., K. Tsuji, and M. Yamanaka, Determination of Alkyl Chain Distribution of Alkylbenzene Sulfonates by Liquid Chromatography, *Anal. Chem. 53:* 1818–1821 (1981).

21. American Society for Testing and Materials, ASTM D 4337, Standard Test Methods for Analysis of Linear Detergent Alkylates, in *1994 Annual Book of ASTM Standards 15.04,* American Society for Testing and Materials, Philadelphia, 1994, pp. 468–474.

22. Lew, H.Y., "Acid" Pyrolysis—Capillary Chromatographic Analysis of Anionic and Nonionic Surfactants, *J. Amer. Oil Chem. Soc. 44:* 359–366 (1967).

23. Lew, H.Y., Some New Developmements in Surfactant Analysis, *J. Amer. Oil Chem. Soc. 49:* 665–670 (1972).

24. Denig, R., Surfactant Characterization by Pyrolysis Gas Chromatography. I. Nonionic and Anionic Surfactants (in German), *Tenside 10:* 59–63 (1973).

25. Denig, R., Characterization of Surfactants by Pyrolysis Gas Chromatography (in German), *Fette, Seifen, Anstrichm. 76:* 412–416 (1974).

26. Llenado, R.A., *Surfactant Science Series; Detergency 20:* 102–104 (1987).

27. Yamaguchi, S., Nonaqueous Titrimetric Analysis of Detergent Intermediates, *J. Amer. Oil Chem. Soc. 55:* 359–362 (1978).

28. Schmitt, T.M., *Surfactant Science Series: Analysis of Surfactants, 40,* Marcel Decker, New York, 1992, pp. 118–120.

29. Wimer, D.C., Potentiometric Determination of Amides in Acetic Anhydride, *Anal. Chem. 30:* 77 (1958).

30. American Society for Testing and Materials, ASTM D 2959, Standard Test Method for Ethylene Oxide Content of Polyoxyethylated Nonionic Surfactants, in *1994 Annual Book of ASTM Standards 15.04,* American Society for Testing and Materials, Philadelphia, 1994, pp. 313–315.

31. Cross, J., ed., *Nonionic Surfactants: Chemical Analysis,* Marcel Decker, New York, 1987, parts B and C.

32. American Society for Testing and Materials, ASTM D 5070, Standard Test Method for Synthetic Quaternary Ammonium Salts in Fabric Softeners by Potentiometric Titrations, in *1994 Annual Book of ASTM Standards 15.04,* American Society for Testing and Materials, Philadelphia, 1994, pp. 542–545.

33. Epton, S.R., A Rapid Method of Analysis for Certain Surface-Active Agents, *Nature 160:* 756 (1947).

34. Metcalfe, L.D., R.J. Martin, and A.A. Schmitz, Titration of Long-Chain Quaternary Ammonium Compounds Using Tetraphenyl Boron, *J. Amer. Oil Chem. Soc. 43:* 355–357 (1966).

35. Wang, C.N., L.D. Metcalfe, J.J. Donkerbroek, and A.H.M. Cosijn, Potentiometric Titration of Long Chain Quaternary Ammonium Compounds Using Sodium Tetraphenyl Borate, *J. Amer. Oil Chem. Soc. 66:* 1831–1833 (1989).

36. Metcalfe, L.D., Potentiometric Titration of Long Chain Amine Oxides Using Alkyl Halide to Remove Tertiary Amine Interference, *Anal. Chem. 34:* 1849 (1962).

37. Lyon, P.A., and W.L. Stebbings, Analysis of Anionic Surfactants by Mass Spectroscopy/Mass Spectroscopy with Fast Atom Bombardment, *Anal. Chem. 56:* 8–13 (1983).

38. Foret, F., and P. Bocek, Capillary Electrophoresis, in *Advances in Elecrophoresis,* edited by A. Chrambach, M.J. Dunn, and B.J. Radola, VCH, 1989, vol. 3, p. 273.

39. Goebel, L.K., and H.M. McNair, Separation of Ethoxylated Alcohol Sulfates by Capillary Electrophoresis Using Indirect UV Detection, *J. Microcol. Sep. 5:* 47–50 (1993).

40. Desbene, P.L., C. Rony, B. Desmazieres, and J.C. Jacquier, Analysis of Alkylaromatic Sulphonates by High Performance Capillary Elecrophoresis, *Journal of Chromatography 608:* 375–383 (1992).

41. Chen, S., and D.J. Pietrzyk, Separation of Sulfonate and Sulfate Surfactants by Capillary Electrophoresis: Effect of Buffer Cation, *Anal. Chem. 65:* 2770–2775 (1993).

42. DeSalvo, D.P., Surfactant Analysis Using a Thermal Horizontal ATR Accessory, *American Laboratory,* 32C–32D (1994, March).

43. Fuller, M.P., *Real Time On-line Quantification of Surfactants,* Nicolet Application Note, AN-8715.

Chapter 13

Performance Evaluation of Household Detergents

Arno Cahn and George C. Feighner

ARGEO, Inc., Pearl River, New York, U.S.A.

Introduction

This chapter discusses the performance evaluation of household detergents. The major product categories are laundry detergents, light-duty liquids, hard-surface cleaners, autodish detergents, and fabric softeners. Fabric softeners are of course not strictly detergents, but they are a large enough product category to deserve inclusion in this discussion.

Household products, like other consumer products, are used under a myriad of different conditions that are too complex to permit a complete description or, in fact, a complete understanding. When users assess the performance of such products—if indeed they assess it at all—they generally do so based largely on subjective criteria, including aesthetic properties such as fragrance and color, as well as ease of use and, perhaps, an occasional outstanding (or dismal) result that happens to have been observed and noted.

In the laboratory, the goal is to evaluate objectively. With a few exceptions, the goal of laboratory evaluation is to provide an objective—ideally numerical—performance assessment, with the intentional exclusion of all subjective factors.

Performance evaluation is conducted on different types of products and in different contexts:

- On finished products to establish performance compared with other products, perhaps to substantiate advertising claims
- On experimental formulations in the course of their development
- On both types to determine the effects of different external factors (such as storage conditions)

The ideal performance test is:

- Reproducible within a range of precision
- Verifiable in different laboratories
- Accurate—predicts the outcome of future evaluation results
- Convenient
- Rapid

The first characteristic is self-evident. Without reproducibility, the test is not reliable and the information it provides is of little value. If the results are reliable, they are also verifiable; that is, they are transportable from one laboratory to another.

By *verifiable* we mean that the test method produces identical results in different laboratories. To ensure verifiability, it is necessary to fix as many experimental conditions as possible to produce a numerical result that can be substantiated anywhere—with the critical proviso that the identical conditions apply.

As we will see later, we go to great lengths to ensure that all conditions are identical. The procedures developed by the American Society for Testing and Materials (ASTM) and the Chemical Specialties Manufacturers Association (CSMA) illustrate the point. As far as humanly possible, all aspects of the evaluation are specified. Even so, in recognition of the complexity of the real situation, these procedures are often referred to as guidelines rather than methods.Testing under conditions that are fixed and not necessarily realistic is not unique to household products. In the well-known EPA mileage test in the United States, the efficiency of an automobile in terms of miles per gallon is measured under a set of fixed conditions. A single result may or may not predict actual efficiency for the specific conditions under which a car is driven in real life.

The third characteristic of the ideal performance test, accuracy, also calls for some elaboration. It is almost synonymous with predictiveness. Performance evaluation is most often conducted in a given context and with a given purpose. At each stage, in this context, the objective in conducting a performance evaluation in the first place is to predict the outcome of the next stage.

The next stage depends on the specific context in which the work is done. For someone developing consumer products, it is part of the general development scheme. Such a scheme might include the following stages:

- Laboratory development and evaluation of prototype formulations
- Pilot-scale validation
- Small-scale consumer tests
- Larger-scale consumer tests
- Experimental plant run validation
- Test market
- Full-scale marketing

This is an illustrative example only. Some stages may be reversed in certain cases and some schemes comprise additional stages.

Evaluation is required at the end of each stage, with the implicit objective of anticipating the outcome of subsequent stages. Of course, the progression is not always linear and disciplined. Implicitly, however, we assume that a favorable outcome—a good evaluation result—will carry through to a later stage down the line, though perhaps with a diminished numerical value.

For example, when we conduct a small-scale, paired-comparison, in-home consumer test of a laboratory-prepared prototype *vs.* a reference product, we expect—on the strength of the accuracy of our laboratory evaluation—that panelists will prefer the product that proved superior in the laboratory performance evaluation. If this expectation is borne out by the results, we acquire some measure of confidence that at least some of the preference will carry through to the marketplace.

For a producer of raw materials that are intended for use in formulated product, the development scheme is different in its details and might include the following stages:

- Laboratory development
- Semi-works manufacture
- Market development
- Experimental plant run validation
- Full-scale production

Again, this is an illustrative example only. The first exposure of a new material occurs in the market development stage when experimental samples are submitted to potential customers. Obviously, the performance assessment by the seller must be accurate in that it must predict the outcome of the evaluation by the potential buyer. It is a critical step in the progression to ultimate commercialization. The performance assessment must, in fact, be not only accurate but also verifiable.

The requirement for accuracy makes for a tall order, if not an impossible one. Consider, for the case of a laundry detergent, the variety of use conditions—soil, fabric load, detergent concentration, washing regimen, and the assessment of the final result by the consumer—and it becomes clear that it is highly unlikely that all conditions can be anticipated by the results of a laboratory test. A common approach to maximizing accuracy/predictiveness is to include a reference standard in the evaluation test.

Because we have gone to such lengths in rigidly specifying every aspect of the performance test, we have removed ourselves from the real world and have in a sense traded accuracy, as defined earlier, for a measure of verifiability. Further, when we include a reference standard in our experimentation, we have traded an absolute result for a relative value.

The last two parameters of the ideal laboratory performance test—convenience and speed—are not quite as essential as the first three but are nonetheless of some importance. An inconvenient and cumbersome protocol cannot truly qualify as a screening test. Speed is important because experimental results are most useful while the experiment is still "hot." If it takes three months to complete an evaluation, chances are that the reason for doing it in the first place will long since have been forgotten.

In sum, in the evaluation of household products—indeed, in any numerical performance evaluation that deals with systems that in real life are highly complicated—our best method is to aim at reproducibility/verifiability first and trade absolute predictiveness for a relative value.

The foregoing discussion has been somewhat abstract. Yet it is essential to have some understanding of the nature and meaning of screening tests to assess the results of specific tests.

Evaluation of Laundry Detergents

Laundry detergents make up by far the largest category of household products. Their evaluation illustrates perhaps most clearly the complicated nature of the test situation. We shall consider laboratory screening tests, a somewhat more lifelike laboratory evaluation, and a practical evaluation method.

The factors that determine detergency performance are as follows:

- Water hardness
- Washing temperature
- Type, amount, and distribution of soil
- Type of substrate (fabric)
- Type and concentration of detergent
- Cloth/wash liquor ratio
- Extent of mechanical agitation
- Washing time

Laboratory Measurement of Detergency (Soil Removal)

The history of attempts to measure soil removal in the laboratory is long, spanning well over four decades. It is summarized and discussed admirably by Kissa (1).

The general problems here are considerable, starting with the nature of test soil. The earliest experiments with a synthetic soil concentrated heavily on carbon black as the model soil, often applied to test fabric out of a chlorinated hydrocarbon solvent. This is an unrealistic soil and an equally unrealistic method of application. Methods of applying a soil by solid transfer from soiled polyurethane cubes in a stirred chamber (accelerator) in a manner more representative of real life have been developed and are claimed to lead to good results in reasonable time (2).

Soil removal is generally determined by measuring the increase in reflectance of soiled cloth after washing. This measurement also suffers from shortcomings, notably the nonlinear relationship between reflectance values and the amount of soil present on the fabric. The reason lies in the difference in both reflectance and light scattering exhibited by soils of different particle size. One approach to taking these factors into account is the well-known Kubelka-Munk equation:

$$\frac{k}{s} = \frac{(1-r)^2}{2r}$$

where k is the absorption coefficient, s is the scattering coefficient, and r is the percentage of reflectance in a situation in which monochromatic light is used. But even this equation assumes a uniform particle size distribution of soil.

There are alternative methods of assessing soil removal. With fatty soil, it is possible to extract the fabric and determine residual soil gravimetrically. This is obviously quite cumbersome.

With the advent of tracer technology, another approach to estimating residual soil on cloth became available. Most of the early experiments in the 1940s and 1950s were carried out with what turned out to be impure materials. Only the combination of radiochemical purity and analytical purity, generally established by gas chromatography, proved to be capable of providing meaningful results. This combination was introduced and used with considerable success by Gordon and Shebs of the Shell Development Company and is summarized by Shebs (3). For oily soils, β-emitters, ^{3}H, ^{14}C, ^{35}S, ^{45}Ca, ^{36}Cl, and ^{32}P were considered. As γ-emitters, ^{22}Na, ^{51}Cr, ^{59}Fe, ^{57}Co, ^{125}I, ^{131}I, and ^{152}Eu are useful. In practice, tritium, ^{14}C, and ^{35}S are the preferred radionuclides for organic soil. Labeled inorganic soil is prepared by irradiating clay.

The methodology that has been developed also includes highly automated arrangements for sample handling and radiation counting. It requires, of course, a considerable investment in apparatus and the appropriate licenses from the nuclear regulatory authorities.

It should be noted, however, that from a real-life viewpoint, reflectance differences track the appearance of the garment and hence are closer to the assessment that is made by the ultimate consumer.

This has been the briefest of surveys of this challenging area. In terms of current practice, the general procedure, specified in the ASTM D-3050 Guideline, is to wash artificially soiled test swatches in a miniature washing machine and read the increase in reflectance on an appropriate instrument (4).

Artificially soiled test swatches can be prepared in-house or purchased from a number of commercial suppliers. At minimum, two types of soil are generally used: an oily type (a combination of dust and sebum is most common) and a particulate type, generally a clay soil, ground into the fibers.

For U.S. conditions, three types of fabrics are generally sufficient: cotton, cotton/polyester, and polyester. In some parts of the world, nylon is an important part of apparel construction, and there it may also serve as a test fabric. The three fabrics are uniformly soiled with the test soils.

The most commonly used miniature washing machine in the United States is the Terg-O-Tometer (5), which consists of a series of stainless steel beakers in which a reciprocating agitator simulates the action of a top-loading agitator washing machine of the type most common in the United States (Fig. 13.1). The beakers—the "pots"—are thermostatted in a water bath.

An alternative design, the Launderometer, is frequently used in Europe. This is a closed beaker with ballast balls that is subjected to agitation. It simulates the

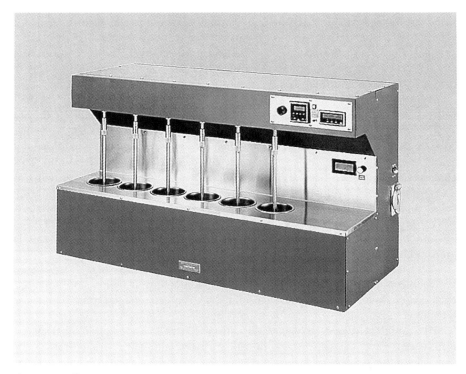

Fig. 13.1. The Terg-O-Tometer.

action of a tumbler machine of the type most common on the European continent. A representative test protocol (6) for evaluating detergency is shown in Table 13.1.

The percentage reflectance (L scale) on each swatch is read before laundering and again after the laundered swatches have been allowed to dry in air. Results are reported as the change in L value of the six soiled and the three clean cloths.

TABLE 13.1
Conditions for a Terg-O-Tometer Evaluation of Detergency

Temperature	105°F
Water hardness	150 ppm, Ca:Mg = 2:1
Product concentration	Recommended use level
	(generally in the range of 1.5 to 2.5 g/L)
Terg-O-Tometer speed	100 cpm
Dissolve time	2 min
Washing time	10 min
Rinsing time	5 min
Soil cloth load	Two swatches each of Dust/Sebum soiled cotton (C),
	cotton/polyester durable press (CPE), and polyester (PE)
	Two swatches each of ground-in-clay soiled C, CPE, and PE
	One swatch each of unsoiled C, CPE, and PE

Treatment of the results depends on the relative importance attached to dust/sebum and ground-in-clay soils. In the absence of any weighting, the increase in reflectance of the three fabrics and two soils and the loss of reflectance of the unsoiled cloths are summed algebraically to provide the Detergency Total, which can serve as a convenient aid in comparing results for different products.

Commercial reflectometers in the United States are marketed by Hunter Associates Laboratory, Inc. (Reston, VA) and Gardner Laboratory, Inc. (Bethesda, MD). These instruments permit a reading of whiteness/grayness on the L scale, of the red hue on the red/green a scale and the yellowness of fabric on the yellow/blue b scale. A filter can be inserted to eliminate the effect of fluorescent whitening agents present in the test detergent.

Overall, this procedure is straightforward and relatively simple, and it yields reasonably reproducible results. It is useful when the task is to measure the contribution to detergency of a new chemical or to investigate the effect of an experimental variable, such as the concentration of a particular material.

Detergency evaluation in the Terg-O-Tometer is, however, a screening test with rigorous control of the experimental variables. It is therefore far removed from real-life situations in which most variables are uncontrolled, subjective, and random. A comparison of the test parameters in this screening test with those in real life is given in Table 13.2.

The table emphasizes again that a Terg-O-Tometer screening test differs in almost every one of its conditions from what might be expected in an in-home use test by consumers.

The Terg-O-Tometer test and an evaluation in the field represent the extremes. Some approaches toward a middle ground are possible. One such procedure is the bundle test.

Detergency by Bundle Testing

The general procedure is to provide panelists—generally a family—with two identical sets of articles (bundles) and, after use, washing the bundles in a washing

TABLE 13.2
Comparison of Terg-O-Tometer Conditions with Field Conditions

Condition	Terg-O-Tometer	Field (Real Life)
Soil	Fixed, synthetic	Random, natural
Soiling	Uniform	Random
Fabric types	Three	Multiple/random
Distribution of fabrics	Uniform	Random
Wash load	Controlled	Random
Washing temperature	Controlled	Random
Agitation	Controlled	Random
Cloth/wash liquor	About 1:65	1:15 to 1:20
Detergent concentration	Controlled	Random
Assessment	Instrumental/objective	Subjective/random

machine in the laboratory with two different test detergents followed by evaluation of the efficacy of each test detergent by comparing the cleanliness of matched articles from the two bundles. The washed bundles are then returned to the panelists. The procedure is repeated for a number of cycles—at least 10—after which the set of articles is given a final evaluation by a larger group of evaluators (7).

In a bundle test the nature and type of soiling are natural and random and the assessment is visual. Visual assessment is generally conducted by a panel of trained evaluators and often under controlled lighting conditions. The ASTM Guideline specifies that visual comparisons be paired, that is, that the operator choose one type of article from the two test bundles and may prefer one over the other. Preference is determined under two kinds of light—"northern daylight," either natural or artificial, and incandescent light. Although this method of assessment is more natural than an instrumental approach, it still is not quite as random and subjective as that in in-home use tests. There it might well be conducted at the location of the washing machine, which often is in a relatively dark place.

Visual assessment, even by trained operators, requires careful controls and observation of panelist techniques. Panelists must be trained to be able to explain their preferences, to ignore extraneous factors, to make forced choices, and not to peek at the coded identifications. In the test, it is important to weed out pairs of garments with unequal staining. The overall result is usually reported as a score, which is an attempt to quantify an inherently qualitative situation. Use of statistics to calculate the 95% confidence intervals for differences, for example, can help place the score in an unbiased perspective. Also, instrumental reflectance readings of typical garments can help relate evaluation by trained operators to what untrained consumers might prefer.

Because unmatched stains are excluded from bundle tests, and matched staining almost never occurs, the bundle test becomes in most cases a measure of redeposition. A test to speed up natural redeposition has been developed (8). Here naturally soiled paired articles are laundered together with clean swatches of several fabric types and are evaluated for soil redeposition. The number of cycles that can be run in each working day is determined only by the required wash-dry cycle time and the supply of soiled laundry. Because up to 20 cycles may be desirable before assessing the pair of test products as equal or different, the savings in elapsed time can be considerable.

To solve the problem of unmatched stains, the split garment protocol is used. In this test, the whole garment or stained areas in each of the two garments are divided equally. The bundles of half-garments are used for paired laundering with two test detergents.

In general, the bundle test is best limited to a comparison of two products, unlike the Terg-O-Tometer procedure, which can handle—at least in theory—any number of products in a single test. It is correct to point out that the result of a bundle test is strongly influenced by the redeposition performance of the test detergents.

The large detergent manufacturers conduct pseudo-bundle testing on a routine basis. They collect soiled laundry items from a large group of panelists, wash and evaluate the collected items in their laboratories, and thus gain a near-realistic impression of the performance of different test products. A somewhat simpler, and hence considerably more affordable, approach is the T-shirt panel.

Detergency by T-Shirt Panel

This test is similar to a bundle test except that T-shirts are the sole items in this bundle. Generally, a group of employees is supplied with clean T-shirts, which are worn for a specified period of time, washed by the testing laboratory, and returned clean to the panelists for another wear cycle. Assessment may be visual by a panel of trained evaluators, or it can be done instrumentally, taking care to select the right areas for the instrumental headings. Reflectometers cover only a very small area, the careful selection of which influences the final test results. For a one-time comparison of the detergency of two different products, the T-shirts can be split after wearing and each half can then be washed in one of the two test products.

Limitations of Detergency Testing

Several conclusions emerge from the foregoing discussion. For one, it is clear that the closer one comes to simulating a real-life, random situation, the greater becomes the need to apply statistics to both test design and evaluation of results.

When laboratory testing is fine-tuned in the interest of precision, we can sometimes detect relatively small differences between different test products. With the appropriate mathematical techniques, these differences may even be statistically significant. As we approach the real-life situation, these differences may no longer be observable. If product A is superior to product B in laboratory screening, it is acceptable that both products may prove to be nearly equal in subsequent, less-controlled tests. On the other hand, reversals (that is, product B proves to be superior to product A) cannot be accepted easily. In this case, it is appropriate to identify the factors causing the reversal and to reexamine the screening test procedure.

Then, too, it is evident that the predictive content of the different test methodologies is still relatively limited, especially with respect to the ultimate consumer reaction to a test detergent. Reliable predictions are limited, at best, to the stage in which consumers test different products in-house. Even at this stage, the effect of aesthetics such as fragrance and color must be factored out.

In real life, of course, aesthetic factors have been known to outweigh truly technical aspects. Many a technically inferior detergent has found consumer acceptance principally on the basis of an attractive fragrance.

In real life, too, there are the effects of advertising, price, position on the shelf, and other imponderables, not the least of which has to do with changes in lifestyle. In the United States and other countries with a large female workforce, the number of women who see the whitest wash on the block as their life's fulfillment is diminishing rapidly. For them, the process of washing is satisfaction enough at the expense of concern for the end result. The gap from a Terg-O-Tometer result to market share is large.

Redeposition in Laundering

A portion of the soil that is removed from soiled articles in a washing process can redeposit back onto washed fabric or, what is worse, on relatively clean white fabric. Redeposition, as this process is referred to, is a complex phenomenon that involves a series of kinetic interactions going in opposite directions. A good mathematical description is therefore also quite complicated.

Empirically, it is possible to obtain a fix on the magnitude of redeposition by either of two approaches: a deposition test, in which soil is added directly to a wash load containing clean (white) items only, or a redeposition test, in which some clean, white fabric is included with the wash load and a measurement is taken of the loss of reflectance brought about by soil redeposited after it has been deterged off the soiled parts of the wash load. A representative protocol (9) for the deposition test is shown in Table 13.3.

Evaluation of redeposition can also be conducted in a Terg-O-Tometer detergency test by including some clean swatches of the different test fabrics in the wash load. When these clean swatches are carried through a number of wash cycles, it is possible to measure 5-cycle, 10-cycle, or any other multicycle cumulative redeposition.

As noted earlier, redeposition also emerges as an important factor with increasing number of cycles in bundle testing and in the T-shirt panel.

Two special cases of redeposition deserve mention. The first is that of dye transfer. In this case, it is not deterged soil that is redeposited but dye that has been leached from a colored article and then dyes either uncolored articles or else modifies the shade or hue of other colored articles in the wash load.

The procedure for testing for dye transfer (8) is similar to that of redeposition. Either a soluble dye is added to the wash solution or some articles with poor or no

TABLE 13.3
Conditions for Terg-O-Tometer Evaluation of Soil Deposition

Temperature	105°F
Water hardness	150 ppm, Ca:Mg = 2:1
Product concentration	Recommended use level
	(generally in the range of 1.5 to 2.5 g/L)
Terg-O-Tometer speed	100 cpm
Dissolve/disperse time	2 min
Washing time	10 min
Rinsing time	5 min
Cloth load	Two swatches each of clean, prewashed C, CPE, and PE
Soil load	3.09 g of Dust/Sebum emulsion, 1.81 of clay slurry
(dispersed in the detergent solution)	
Number of wash-dry cycles	Five

dye fastness are included in the wash load, and then the loss of reflectance of clean articles included in the wash load is measured.

A second form of deposition that affects laundering results is the encrustation of fabrics with limestone deposits from the reaction of sodium carbonate in the detergent formulation with soluble minerals in water. This deposition occurs along with, and interacts with, the redeposition of soil. To minimize encrustation, detergent formulations contain additives that modify the crystallization behavior of $CaCO_3$. Table 13.4 shows the conditions of evaluating fabric encrustation.

Swatches of prewashed cotton cloth are laundered consecutively for 10 cycles in the Terg-O-Tometer with hard water solutions of the test detergents. The swatches are oven dried and calcium carbonate deposition is measured after 0, 1, 5, and 10 cycles by determining the amount of calcium in an EDTA titration of HCl extracts of the cotton swatches.

The efficacy of additives designed to prevent or reduce encrustation can be tested on a formulation shown in Table 13.5.

Evaluation of Stain Removal

In addition to generalized soiling, worn fabrics often exhibit localized soiling, that is, staining. The evaluation of stain removal is similar to the laboratory evaluation of detergency in that it uses artificially stained test swatches and measures the increase in reflectance of the swatches after washing. Generally, the change in reflectance measured on the black-white L scale will tell the whole story, but sometimes it is necessary to measure differences in the a and b color scales to understand color changes and especially changes in stain hue. The Terg-O-Tometer may be used to wash the stained articles/cloths. For a more realistic set of parameters—cloth/liquor ratio, agitation, and others—the swatches are washed in a regular household washing machine (11).

To obtain a reasonable overall picture of stain removal efficacy, it is necessary to test a range of stains, at least 12 or so, and desirably an even larger number. The selection of particular stains depends to some extent on the assumed strengths of the test products. Certain stains are known to respond well to certain ingredients. Thus

TABLE 13.4

Screening Test for Fabric Encrustation

Temperature	100°F
Water hardness	300 ppm, 2/1 Ca/Mg
Detergent concentration	2 g/L
Terg-O-Tometer speed	100 cpm
Detergent dissolve time	2 min
Wash time	10 min
Rinse time	5 min
Wash load	12 3" x 4" clean cotton #400 swatches, fresh coded swatches being added to replace those removed for analysis after one and five cycles
Drying conditions	170°F in circulating oven for 15 min between wash cycles

TABLE 13.5
Test Formulation for Anti-encrustation Agents

LAS	12% (100% active)
Nonionic	6%
Na silicate, RU as is	10%
Na carbonate	25%
Zeolite A	25%
CMC	1%–2%
Anti-encrustation additive (e.g., polymer)	1%–4%

grass stains, blood, and other proteinaceous stains are removed more completely by products containing a proteolytic enzyme. Nonionic surfactants are effective on oily hydrocarbon stains. Bleaches, such as hydrogen peroxide (or sodium perborate), are generally effective on certain, bleachable, stains but can also set some others.

Pretreatment—rubbing the stained fabric with full-strength detergent prior to washing—generally makes a positive contribution to stain removal. Here the effects vary depending on the particular stain/fabric combination.

The preparation of uniformly stained swatches and control of their storage are critical to success in evaluating relative stain removal performance. Various techniques are used. Because by definition the term stain refers to a soiled spot rather than overall soiling, the usual approach is to stain a spot of a size that corresponds approximately to the field of vision of the reflectometer. However, it is difficult to produce a large number of uniform stain spots. Good experimental results can be obtained with other staining techniques, notably those in which the whole swatch is soiled. After drying, it is important to control oxygen-initiated aging processes. Storage under nitrogen and limiting storage time in general have been found to be effective in minimizing changes in the aged stains.

It is possible to wash a whole series of stained swatches in a single washing machine run. Clean ballast load is added to make up a medium-size wash load to take up the color bodies that are removed and to prevent redeposition from confounding the test results. In a variation of this protocol, the swatches are attached (sewn or stapled) at one corner or one edge to a larger sheet of ballast load. This is intended to expose stained swatches to the normal agitation in the washing machine and prevent them from becoming enmeshed in the ballast load.

Numerical values of stain removal are usually obtained by measuring reflectance of the stained swatch before and after laundering in much the same manner as it is done in detergency testing. Generally speaking, the change is positive. Certain stain/detergent combinations, notably tea and alkaline detergents, lead to a negative change; that is, the stain is set and appears darker after laundering than before. Blood stains and peroxide, especially with bleach and activator, show a smaller increase in reflectance than similar combinations without bleach. Tea stains are a special case.

For purposes of comparing the relative stain removal efficacy of a series of products, these measurements are perfectly adequate. However, for predicting consumer response, the procedure has some shortcomings. The principal drawback is that even after a positive change in reflectance, a noticeable residual stain may be left on the fabric. In terms of accuracy, that is, in terms of predicting what the ultimate consumer sees, this can be misleading. In other words, a marked increase in reflectance after washing could be accompanied by a residual stain that in real life would make the garment unwearable. Another common stain problem is exemplified by a colorless, yet obvious, salad oil stain on a dark blue or green cotton-polyester tablecloth.

As a solution to this dilemma, the concept of the Stain Removal Index (SRI) has been put forward (12). This quantity concentrates on the contrast between a residual stain and an adjacent clean area, the latter represented by including clean, unstained cloth into the test load. The numerical expression of the SRI is as follows:

$$SRI = 100 - [(L_c - L_w)^2 + (a_c - a_w)^2 + (b_c - b_w)^2]^{1/2}$$

where L is the reflectance on the black/white scale, a is the reflectance on the red/green scale, b is the reflectance on the yellow/blue scale, c is the unstained fabric, washed under the test conditions, and w is the stained fabric, washed under test conditions.

Another realistic procedure involves the AATCC (American Association of Textile Chemists and Colorists) Stain Release Replica (13), shown in Fig. 13.2.

The chart shows a series of gray stains of graduated intensity, from 0 (lightest, no residual stain) to 5 (darkest), and serves as a reference for grading the intensity of residual stain color after washing. With a panel of three to six evaluators, it is possible in practice to assign a value precise to 0.1 unit. This kind of precision is needed in the salad-oil/dark-tablecloth stain/fabric combination.

The foregoing discussion summarizes the evaluation of what can be termed fabric detergency. The discussion provided some specific test protocols and also dealt with the limitations of screening test and the gap between an individual screening results and real life assessment.

Experience has shown that a meaningful overview of the relative performance of a series of test products can be obtained by subjecting them to different screening tests under a wide range of conditions, such as water hardness, temperature, and concentration. Combining the screening results of detergency, redeposition, and stain removal provides a series of overall performance numbers. These numbers—obtained by a simple summation, average, or a weighted average—hide much of the fine structure and the sensitivity of individual results. At the same time, the array of performance numbers is closer to an assessment under real-life conditions.

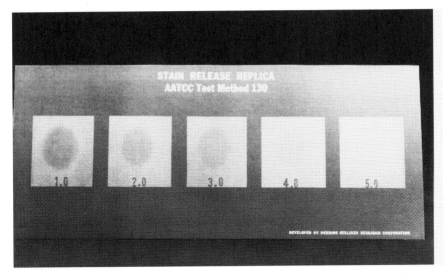

Fig. 13.2. The AATCC Stain Release Replica. Reproduced by permission of the American Association of Textile Chemists and Colorists, P.O Box 12215, Research Triangle Park, NC 27709, as used in their Method 130, Soil Release: Oily Stain Release Method.

Evaluation of Fabric Softeners

Fabric softeners confer a soft hand to washed fabrics and reduce the static charge that is built up when a wash load is dried in a tumble dryer. Softness is needed particularly for cotton fabrics, and prevention of charge build-up is needed at low humidity especially for synthetics, such as nylon, polyester, and acrylics.

Fabric softeners are marketed in three forms: rinse-cycle-added liquid products, softener-impregnated substrate sheets that are added to the dryer, and combination products providing both detergency and softening, the so-called softergents.

Fabric softeners function by depositing active material, typically a cationic surfactant, on cloth. Because these cationics contain one or more hydrophobic chains, they can, under certain circumstances of repeat usage, reduce the water-absorbing capacity of cloth, making it waterproof. The three important performance attributes to be tested are therefore softening, static charge reduction, and re-wet characteristics (water absorption).

Softening

To date a truly reliable objective procedure that would measure softening instrumentally has not been found. For this reason, tactile evaluation by a panel of evaluators is the protocol of choice (14, 15). Because cotton is the fabric most in need of softening, the test substrate generally is cotton terry cloth toweling. A load of toweling is washed in a washing machine, dried (generally in a dryer but also, if desired, on a line), and subjected to tactile evaluation by a panel. An effort is made to expose the different evaluators to areas of the cloth that have not previously been touched by other evaluators.

The degree of softness (or harshness) is rated on a scale. One common scale ranges from 1 to 5 (5 being the softest). The ratings are established for each series of test products. After evaluating the series, panelists first select the harshest and softest of the cloths. Thus the ranking of different products here is limited to an individual test. To establish a relationship from one test to another, a reference product is generally included in each test series.

Softness scales other than 1 to 5 have also been used by different organizations, such as 0 to 10 (10 is the softest) and −5 to +5 (negative numbers are harsh, positive numbers are soft).

It is evident that the selection of the scale influences the interpretation of the results. In the absence of a negative dimension, an unsoftened cotton towel would receive a rating of 0, and all products providing even the slightest reduction in hardness would qualify for a positive assessment in the sense that they would be given something like a "slight" or "marginal" softening—but softening nevertheless.

Attempts have been made to introduce internal standards, much in the manner in which the AATCC gray scale provides some reference points. However, in general these efforts have not improved significantly on the results that are obtained by a panel of trained evaluators.

Along a different approach, the "fluffiness" of a collection of softened towels can be taken as a measure of softening. Here the difference in height of a stack of softened towels and that of a stack of unsoftened towels is taken to demonstrate the efficacy of softening. Such comparisons have appeared in TV commercials but have not made it into the literature as a reliable test method.

Static Control

Synthetic fabrics—polyester, nylon, and acrylics—can develop a significant electrical charge in the course of drying in a household dryer. Fabric softener products function to reduce the charge buildup. The assessment of static control consists of drying a bundle of items constructed of synthetic fibers, a high-static bundle, and measuring the charge, either instrumentally or subjectively (15).

One protocol for the instrumental assessment of static control involves the use of a Simco Electrostatic Locator (16), shown in Fig. 13.3. Readings are taken on this instrument as it is brought within a specified distance from a freshly dried article that is hanging flat over a wooden rack.

The other involves the use of a Faraday cage (Fig. 13.4), which consists of two cages, one inside the other, that are insulated from each other. The cage is connected to a voltmeter, which reads out the voltage as each article is removed singly from the cage. The total of the individual readings represents the charge on the whole load.

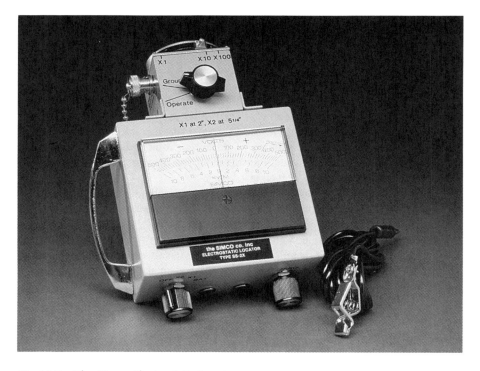

Fig. 13.3. The Simco Electrostatic Locator.

In the subjective evaluation, note is taken of the cling, crackle, and static as each article of the load is removed from the dryer. Again, a subjective rating scale of 1 to 5 is employed, but—in contrast to the fabric softener scale—a rating of 1 designates no cling, crackle, or static, and at the other hand of the scale a 5 designates heavy cling, crackle, and static.

Measurements of static control—and indeed, measurement of fabric softening efficacy—are strongly affected by the relative humidity and temperature at the point of evaluation. Ideally, therefore, these two parameters should be kept constant. At the very least, they should be recorded.

Re-wet/Water Absorbency

Excessive adsorption of cationic fabric softeners can reduce the water absorbency of fabric. In the extreme, the fabric becomes waterproof. It is of interest, therefore, to test the water absorbency of the softened cloth. This is a relatively simple wicking test. Two strips, 2 in. × 5 in., are cut from softened terry toweling, marked 1 cm from the lower edge, and immersed to that mark in a solution of a water-soluble dye that is not substantive on cotton. After 6 min, the strips are removed and the

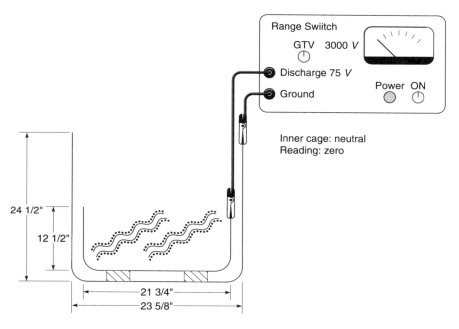

Fig. 13.4. A Faraday cage.

migration of the red-dyed water up the strip is recorded. The more water-absorbent the fabric, the greater will be the migration of water (14, 15).

Evaluation of Light-Duty Liquids (LDLs)

Light-duty liquids are a subcategory of hard surface cleaners and are used in washing dishes by hand. Much of the cleaning work is therefore done by mechanical means—the sponge, a dishcloth, and the like. The soil is generally taken to be fatty in nature, even though proteinaceous and starchy components are also present. Critical for assessment of product performance by the consumer are the generation and persistence of a generous level of foam covering the dish pan. As a result, all procedures for the performance evaluation of LDLs measure the persistence of foam in the presence of an increasing load of fat.

Foaming

In evaluating the foaming performance of light-duty liquids, several parameters are of importance:

- The nature of the soil
- The uniformity of soil on plates

- The water temperature—generally 105°–120°F in U.S. usage.
- Product concentration—generally 1.5 g/L in the United States.
- The uniform generation of initial foam levels.
- The assessment of the foam end-point.
- Most critically, the extent of physical/mechanical input.

Foam end-points are usually defined in terms of a fraction of the surface covered by foam. One-half is a number that can be measured visually with a reasonable degree of precision. Because operators vary in the strength they apply to the dishwashing task, it is customary in a multidish test to rotate operators after each has washed a set number of plates.

There are three realistic procedures for evaluating light-duty liquids. The most lifelike of these tests is the cafeteria test, in which a set of naturally soiled dishes is washed by hand in a dish pan to a no-foam end-point. Next in realism is the plate-washing test in which regular-size dishes soiled with a fatty test soil are washed by hand in the laboratory.

The third of these tests is the so-called mini-plate, which simulates manual dishwashing operation. In this procedures, small watch glasses coated with fat are "washed" with a camel hair brush (17).

Outside the arena of reality, many options exist for generating foam by various means of agitation and introducing increments of fatty soil. The methods of foam generation in different procedures are as follows:

- A high-speed Waring blender
- A piston with a perforated head plate, reciprocating in a cylinder (18)
- Continuous, gentle inversion of stoppered glass cylinders
- The action of the agitator in a Terg-O-Tometer

Options for the incremental addition of fatty soil include a pellet of solidified fat, a volume of liquefied fat, and swatches of fabric coated with a greasy soil. Greasy soils include commercial vegetable shortening, shortening "fortified" with free (tallow) fatty acid, and others.

Each of these approaches possesses both advantages and disadvantages. Each can yield consistent results in an assessment of a series of test products. In other words, consistent, but relative, results are obtainable by any of the above protocols provided, again, that all experimental details are observed religiously and—for an indication of realism—a commercial reference standard is included.

Emulsification

In addition to maintaining a stable foam in the presence of increasing quantities of fatty soil, light-duty liquids are capable, to various degrees, of emulsifying greases and oils. On a macro scale, emulsification converts fats and oils into smaller aggregates. In reality, this performance attribute comes closer to "cleaning" soiled dishes.

Emulsification can be tested in a number of ways. Two useful protocols involve evaluation of the stability of the emulsion and the process of removing fat from a substrate.

The general approaches are simple. To determine the capability of a material to form a stable—generally oil-in-water—emulsion, one shakes a two-phase system consisting of water and an oily layer marked with an oil-soluble dye until a single phase is formed. When the emulsion splits, two layers are formed and the dye is concentrated in the upper oil layer. The height of the upper layer is measured as a function of time (and possibly storage temperature).

More pertinent to the dishwashing process is a test protocol that measures the removal of fat or oil from a substrate as the result of the emulsifying action of the test product. In this protocol, microscope slides are coated with fat containing an oil-soluble dye and are immersed into a solution of the test product. Removal of the fatty coating to a predetermined end-point is recorded as a function of time (and possibly other test parameters such as temperature and product concentration). Generally, the end-point is taken when the immersed portion of the slide is clean. The purpose of the dye is to facilitate identification of the clean end-point. When the removal of fat is determined gravimetrically, no dye is needed.

The protocol can be run in one of two ways. In the first, the slides are at rest so that fat removal is strictly a function of the chemical contribution of the test product. In the second option, the slides are agitated gently, for example, by repeated dipping into the test solution.

Evaluation of Hard Surface Cleaning

Different segments make up the hard surface cleaner category (excluding products for hand and machine dishwashing) depending on the task these products are intended to perform. The principal categories include all-purpose cleaners, bathroom (tub-tile) cleaners, glass cleaners, and toilet bowl cleaners.

In their respective end-uses, these products encounter different soils and different substrates. All-purpose cleaners can be expected to encounter both greasy and particulate soils on a variety of substrates. Bathroom cleaners in general aim at removing soap scum from both tiles and metal surfaces. Glass cleaners deal with an essentially greasy soil on a very specific substrate. Toilet bowl cleaners are formulated for removing both organic soil as well as preventing the buildup of a metal ion-induced ring that forms at the water level.

The general approach to the evaluation of hard surface cleaners is to simulate the application of product and mechanical energy input of actual use. The most universally used instrument for this purpose is the Gardner Straight Line Washability Tester (19), in which a weighted applicator wetted with a solution of a test product is passed in a reciprocating motion over a substrate coated with a soil appropriate to the product category substrate (Fig. 13.5).

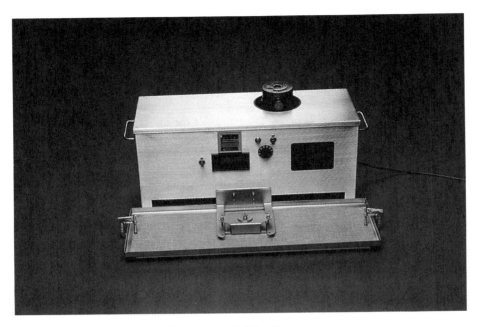

Fig. 13.5. The Gardner Straight Line Washability Tester.

Apart from the factors that are to be tested—the composition and concentration of the test products—the results obtained in this test depend critically on the nature of the soil, soil uniformity, and the substrate.

Realistic soils have been found useful, such as dust/sebum as representative of finger prints, crayon for soil on walls, and dirty motor oil for soiled metal surfaces. A standard ASTM protocol has been developed specifically for evaluating cleaning performance of ceramic tile cleaners (20).

"Bathroom" test soil is complicated, consisting primarily of calcium stearate in a mixture with synthetic sebum, Mg and Fe stearate, carbon black, and dried potting soil that is mixed again with sodium stearate. While hot, this soil is applied to hot ceramic tiles and the tiles are then baked for 1 h at 70–80°C. After the tiles are cooled, they are "cleaned" with a sponge moistened with test product. The number of scrubbing cycles is established independently to coincide with 75% soil removal by commercial products. The tiles are rinsed, air dried, and read on a photovolt reflectometer. Alternatively, the cleaned tiles can be rated to the nearest 0.5 rating on a scale of 1 (very little removal) to 5 (soil removal is virtually complete) by a panel of least eight independent judges.

The actual test protocol is replete with specifics (source of soil components, soil applicator, baking time and temperature, sponges, and so on) and illustrates once again the extent to which these factors must be specified to ensure verifiability by different laboratories.

For the purpose of developing relative performance, there is some leeway in the selection of test parameters. Thus a simpler simulated bathtub soil consisting of calcium stearate and finely divided charcoal has been disclosed. The soil is sprayed onto the substrate tiles, which are subsequently baked at 180°C for 20 min (21).

The role of penetration and subsequent emulsification in the removal of solid soil by hard surface cleaners has been studied in a protocol in which a combination of lanolin, cetyl alcohol, and stearine grease served as the test soil deposited on metal coupons. These were then immersed into the test solutions without mechanical agitation and the cleaning efficacy was determined by weight loss (22).

Evaluation of Automatic Dishwashing Detergents (ADWs)

Automatic dishwashing detergents also represent a special case of hard surface cleaners. In contrast to other near-universal cleaning operations, automatic dishwashing is characterized by the following parameters:

- Higher washing temperature
- Greater mechanical energy input
- More aggressive chemicals in the product
- Relative absence of particulate soil
- Ease of evaluating performance by the ultimate user

In many ways, testing for ADW performance is the most realistic of the test protocols discussed so far; the apparatus is identical to the machine used by consumers. The evaluation criterion—the appearance of glasses—parallels very closely the assessment that a consumer will make in home use. This assessment is the only realistic part of the method. The soil is a standard mixture that is distributed uniformly over a set of clean plates. This ASTM test protocol is designated as a "screening test that will not necessarily predict performance under all end-use conditions" (23).

The food soil in this test consists of a mixture of nonfat dried milk and margarine with optional addition of wheat-based cooked cereal. The food soil is manually distributed over six dinner plates. The products are evaluated by rating (a minimum of eight) glass tumblers for spotting and filming. The tumblers are viewed in a light box with three internal black surfaces and fluorescent light passing upward through the tumblers. The rating scale is 1 to 5, with 1 being best and 5 worst. For spotting, a 1 signifies no spots, and a 4 indicates that about one-half of the surface is covered. For filming, a 1 signifies no filming, a 4 represents moderate filming, and 5 is heavy filming. For meaningful comparisons, multiple cycles (from 5 to 15) are required, using newly soiled plates in each cycle. A photographic scale can serve as a helpful reference.

Water hardness, water temperature, the distribution of dinnerware and tumblers within the machine, agitation, and the performance characteristics of individual machines are among the factors that need to be taken into consideration in running this test protocol.

In this test protocol, spotting and filming of glassware is taken as an indication of dishwashing performance. This is a reasonable assumption. It leaves some doubt about the cleaning performance on actual soils. Many attempts have been made to develop a reproducible quantitative method involving naturally soiled dishes, but none has made it into the active literature, although in-house methods of this type may be practiced in several laboratories.

The difficulty, of course, is to develop a standard realistic soil—a "Sunday dinner," as it has been referred to in some places—and also to develop a reproducible means of reading the results, ideally by an instrumental method. We know that tough soils, such as dried-on egg and certain starch residues, are difficult to remove. The problem is how to weight these specific performance factors within the framework of an overall assessment.

An interesting recent publication (24) describes an approach toward an instrumental method for assessing starch residues using the anthrone color reaction and a starch soil on glass plates. Both the glass plates and the application of the starch soil borrow from the procedures of thin-layer chromatography.

Pattern Removal

Because of the presence of aggressive—alkaline—ingredients in ADWs, there is a possibility that some of the decorations on fine china will be attacked over the course of repeated dishwashing cycles.

The test method for assessing tableware pattern removal (25) consists of repeated immersion of some standard patterned tableware in solutions of the test product for 2 h at 205–211°F (96–99.5°C), rubbing the pattern vigorously with muslin and assessing visually both the residual pattern and the muslin cloth on a scale of 0 (no pigment removal) to 5 (essentially complete pigment removal).

Summary

This chapter has discussed some theoretical aspects of the nature of screening tests and their relation to evaluation under realistic field conditions. It has also reviewed some specific methods for the evaluation of the product categories that make up the household product area. Most of the protocols are characterized by precise and detailed specifications of the test parameters in order to ensure verifiability. The individual test methods can be relied on to provide a series of internally consistent results that are not, however, necessarily related to the real world. To relate one test series with another, the inclusion of a reference standard is generally a good practice. When a commercial product is chosen as reference standard, it provides a reality check on a series of internally consistent, but not necessarily realistic, results.

References

1. Kissa, E., in *Detergency—Theory and Technology*, Surfactant Science Series, edited by W.G. Cutler and E. Kissa, Marcel Dekker, New York, 1987, vol. 20, pp. 2–81.

2. *Ibid.,* p. 36.
3. Shebs W.T., in *Detergency—Theory and Technology,* Surfactant Science Series, edited by W.G. Cutler and E. Kissa, Marcel Dekker, New York, 1987, vol. 20, pp. 126–187.
4. D 3050, Standard Guide for Measuring Soil Removal from Artificially Soiled Fabrics, in *Annual Book of ASTM Standards,* American Society for Testing and Materials (ASTM), Philadelphia, PA, 1993, pp. 326–327.
5. Manufactured by United States Testing Company, 1415 Park Avenue, Hoboken, NJ.
6. ARGEO, Inc., Detergency Screen[SM] protocol.
7. D 2960, Standard Test Method of Controlled Laundering Test Using Naturally Soiled Fabrics and Household Appliances, in *Annual Book of ASTM Standards*, American Society for Testing and Materials (ASTM), Philadelphia, PA, 1993, pp. 312–315.
8. Feighner, G.C., *J. Am. Oil Chem. Soc. 61(10):* 1645 (1984).
9. D 4008, Standard Test Method for Measuring Anti-Soil Deposition Properties of Laundry Detergents, in *Annual Book of ASTM Standards,* American Society for Testing and Materials (ASTM), Philadelphia, PA, 1993, pp. 425–427.
10. D 5548, Standard Guide for Evaluating Color Transfer or Color Loss of Dyed Fabrics in Home Laundry, in *Annual Book of ASTM Standards,* American Society for Testing and Materials (ASTM), Philadelphia, PA, 1994, pp. 589–590.
11. ARGEO, Inc. See also D 4265, *Annual Book of ASTM Standards,* 1993, pp. 453–458.
12. Neiditch, O.W., K.L. Mills, and G. Gladstone, *J. Am. Oil Chem. Soc. 52(12):* 426 (1980).
13. Available from American Association of Textile Chemists and Colorists, PO Box 12215, Research Triangle Park, NC 17709.
14. D 5237, Standard Guide for Evaluating Fabric Softeners, in *Annual Book of ASTM Standards,* American Society for Testing and Materials (ASTM), Philadelphia, PA, 1993, pp. 545–549.
15. D-13 Fabric Softeners, in *Detergents Division Test Methods Compendium,* 2nd edn., Chemical Specialties Manufacturers Association, Washington, DC, pp. 35–44.
16. Simco Co., 2257 North Penn Road, Hatfield, PA 19440.
17. Anstett, R.M., and E.J. Schuck, presented at the 57th Annual Spring Meeting of the American Oil Chemists' Society, Los Angeles, CA, 1966.
18. DNA (German Standards Organization), Method DIN 53902.
19. Paul N. Gardner Co., Pompano Beach, FL 33060.
20. D 5343, Standard Guide for Evaluating Cleaning Performance of Ceramic Tile Cleaners, in *Annual Book of ASTM Standards,* American Society for Testing and Materials (ASTM), Philadelphia, PA, 1993, pp. 550–554.
21. Procter & Gamble, European Patent 0,227,195.
22. Cox, M.F., *J. Am. Oil Chem. Soc. 63(4):* 559–565 (1986).
23. D 3556, Standard Test Method for Deposition of Glassware During Mechanical Dishwashing, in *Annual Book of ASTM Standards,* American Society for Testing and Materials (ASTM), Philadelphia, PA, 1993, pp. 358–359.
24. Rietz, M., K. Rubow, and M. Teusch, *SOFW Journal 119(6):* 340–348 (1993).
25. D 3565, Standard Test Method for Tableware Pattern Removal by Mechanical Dishwasher Detergents, in *Annual Book of ASTM Standards,* American Society for Testing and Materials (ASTM), Philadelphia, PA, 1993, pp. 362–363.

Chapter 14

Sulfonation and Sulfation Processes

Norman C. Foster

The Chemithon Corporation, Seattle, Washington, U.S.A.

Introduction

Sulfonation and sulfation are major industrial chemical processes used to make a diverse range of products, including dyes and color intensifiers, pigments, medicinals, pesticides and organic intermediates. Additionally, almost 500,000 metric tons per year of lignin sulfonates are produced as a by-product from paper pulping. Petroleum sulfonates are widely used as detergent additives in lubricating oils. However, the majority of the 1.6 million metric tons of sulfonates and sulfates produced annually in the United States (1) are used as surfactants in laundry and in consumer products applications. This chapter focuses only on commercial techniques for production of detergent range sulfonates and sulfates.

Basic Chemistry

Although sulfonates and sulfates are similar in structure, there are important differences. Figure 14.1 shows the reaction to produce a sulfonate. Sulfur trioxide (SO_3) reacts with an organic molecule—in this case an alkyl benzene—to form a sulfur-carbon bond. One of the characteristics of this process is that the resultant alkyl benzene sulfonic acid is a stable molecule.

Sulfation, on the other hand, involves forming a carbon-oxygen-sulfur bond as shown in Fig. 14.2. The resultant alcohol sulfuric acid is not hydrolytically stable. Unless neutralized, it decomposes to form sulfuric acid and the original alcohol.

Because they are stable, sulfonic acids can be isolated, stored and shipped as an article of commerce. Sulfates, due to their instability, are available only as neutral compounds. This stability difference in the products of reaction with SO_3 also has a pro-

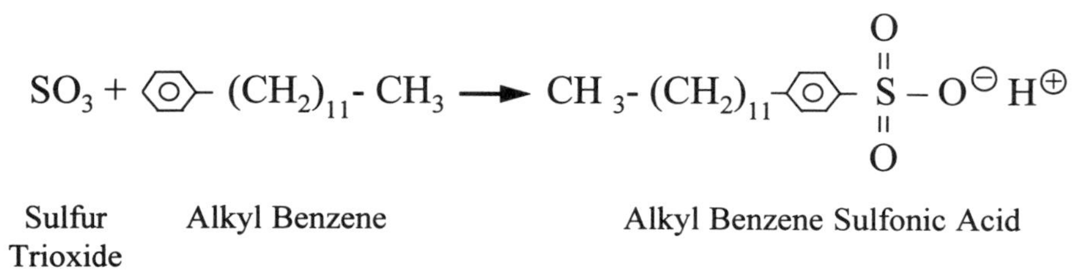

$$SO_3 + \langle\!\bigcirc\!\rangle\text{-}(CH_2)_{11}\text{-}CH_3 \longrightarrow CH_3\text{-}(CH_2)_{11}\text{-}\langle\!\bigcirc\!\rangle\text{-}\overset{\displaystyle O}{\underset{\displaystyle O}{\overset{\|}{\underset{\|}{S}}}}\text{-}O^{\ominus}H^{\oplus}$$

Sulfur   Alkyl Benzene   Alkyl Benzene Sulfonic Acid
Trioxide

Fig. 14.1.   Sulfonation.

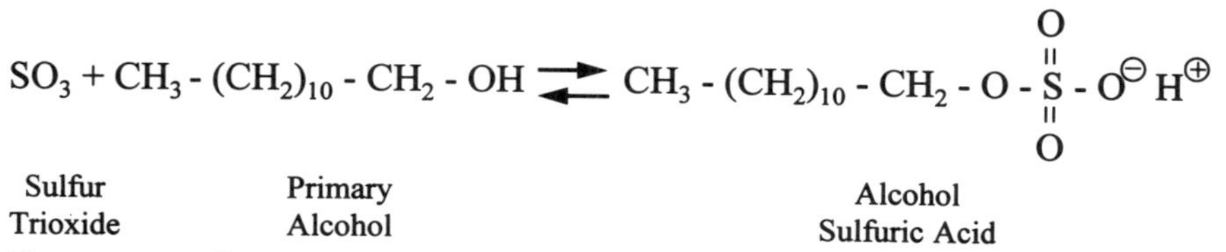

$$SO_3 + CH_3 - (CH_2)_{10} - CH_2 - OH \rightleftharpoons CH_3 - (CH_2)_{10} - CH_2 - O - \overset{\overset{O}{\|}}{\underset{\underset{O}{\|}}{S}} - O^{\ominus} H^{\oplus}$$

Sulfur	Primary	Alcohol
Trioxide	Alcohol	Sulfuric Acid

Fig. 14.2. Sulfation.

found impact on the choice of process used to produce sulfonates or sulfates. Some processes such as oleum sulfonation cannot be used to make alcohol sulfates containing a low level of inorganic sulfate. However, others such as sulfamic acid sulfation cannot be used to make sulfonic acids.

SO_3 is an aggressive electrophilic reagent that rapidly reacts with any organic compound containing an electron donor group. Sulfonation is a difficult reaction to perform on an industrial scale because the reaction is rapid and highly exothermic, releasing approximately 380 kJ/kg SO_3 (800 BTUs per pound of SO_3) reacted (2). Most organic compounds form a black char on contact with pure SO_3 due to the rapid reaction and heat evolution. Additionally, as shown in Table 14.1, the reactants increase in viscosity between 15 and 300 times as they are converted from the organic feedstock to the sulfonic acid (1). This large increase in viscosity makes heat removal difficult. The high viscosity of the formed products reduces the heat transfer coefficient from the reaction mass. Effective cooling of the reaction mass is essential because high temperatures promote side reactions that produce undesirable by-products. Also, precise control of the molar ratio of SO_3 to organic material is essential because any excess SO_3, due to its reactive nature, contributes to side reactions and by-product formation. Therefore, commercial-scale sulfonation reactions require special equipment and instrumentation that allow tight control of the mole ratio of SO_3 to organic material and rapid removal of the heat of reaction.

Historically, the problem of SO_3 reactivity has been solved by diluting and/or complexing the SO_3 to moderate the rate of reaction. Commercially, the diluting or complexing agents (Fig. 14.3) include ammonia (sulfamic acid), hydrochloric acid (chlorosulfuric acid), water or sulfuric acid (sulfuric acid or oleum) and dry air (air/SO_3 film sulfonation). Control of the ratio of SO_3 to organic raw material is vital in achieving the desired product quality with the use of any of the agents. Additionally, these processes require heat removal to maintain product quality. As we examine each of these industrial processes, we will see how they have been engineered to achieve these requirements.

TABLE 14.1 Viscosity Increase on Sulfation

Feedstock	Feed viscosity (cp)	Acid viscosity at 40–50°C
Linear alkyl benzene	5	400
Branched alkyl benzene	15	1000
Ethoxylated alcohol	20	500
Tallow alcohol	10	150
Alpha alefins	3	1000

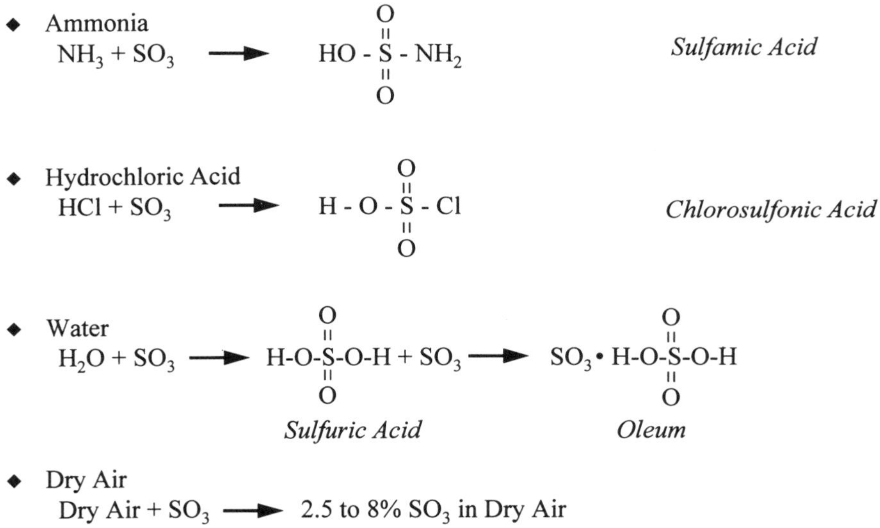

◆ Ammonia

$NH_3 + SO_3 \longrightarrow$ HO - $\underset{\overset{\|}{O}}{\overset{\overset{O}{\|}}{S}}$ - NH_2 *Sulfamic Acid*

◆ Hydrochloric Acid

$HCl + SO_3 \longrightarrow$ H - O - $\underset{\overset{\|}{O}}{\overset{\overset{O}{\|}}{S}}$ - Cl   *Chlorosulfonic Acid*

◆ Water

$H_2O + SO_3 \longrightarrow$ H-O-$\underset{\overset{\|}{O}}{\overset{\overset{O}{\|}}{S}}$-O-H $+ SO_3 \longrightarrow SO_3 \cdot$ H-O-$\underset{\overset{\|}{O}}{\overset{\overset{O}{\|}}{S}}$-O-H

Sulfuric Acid   *Oleum*

◆ Dry Air

Dry Air $+ SO_3 \longrightarrow$ 2.5 to 8% SO_3 in Dry Air

Fig. 14.3.   Agents to reduce SO_3 reactivity.

Commercial Sulfonation Processes

Sulfamic acid (NH_2SO_3H) is used to sulfate alcohols and ethoxylated alcohols to form an ammonium neutralized salt. A typical reaction is shown in Fig. 14.4. The reaction goes directly to the ammonium salt of the alcohol sulfuric acid. Sulfamic acid is an expensive reagent, costing approximately U.S. $0.51 per pound of reactive SO_3. Sulfamic acid sulfation is a mild and specific sulfating reagent suitable for making ammonium neutralized alcohol ethoxylates. Another major advantage of sulfamic acid is that it selectively sulfates alcohol groups and will not sulfonate aromatic rings. Therefore, its major use is sulfation of alkyl phenol ethoxylates. This specificity prevents formation of mixed sulfate-sulfonate compounds. Sulfamic acid is easily handled and reacts stoichiometrically with the alcohol or ethoxy alcohol. It readily adapts to making small quantities of material in low cost batch equipment.

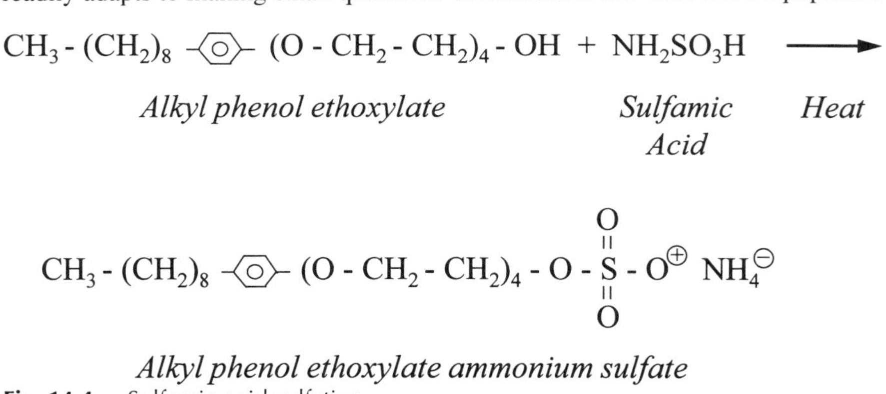

$CH_3 - (CH_2)_8$ ─⟨O⟩─ $(O - CH_2 - CH_2)_4 - OH + NH_2SO_3H \longrightarrow$

Alkyl phenol ethoxylate   *Sulfamic   Heat*
                                              Acid

$CH_3 - (CH_2)_8$ ─⟨O⟩─ $(O - CH_2 - CH_2)_4 - O - \underset{\overset{\|}{O}}{\overset{\overset{O}{\|}}{S}} - O^{\oplus} \ NH_4^{\ominus}$

Alkyl phenol ethoxylate ammonium sulfate

Fig. 14.4.   Sulfamic acid sulfation.

Chlorosulfuric acid ($ClSO_3H$) is also widely used to produce alcohol sulfates, alcohol ether sulfates, dyes and dye intermediates. Figure 14.5 shows a typical reaction. Note that as the reaction moves to completion, hydrochloric acid (HCl) is released. This acid must be scrubbed or otherwise recovered. Chlorosulfuric acid is an expensive source of SO_3 although it is approximately one-half the cost of sulfamic acid. The cost per pound of reactive SO_3 is U.S. $0.255. It is a rapid, stoichiometric reactant. However, it is still more expensive than other sources of SO_3. It is also corrosive, is a hazardous chemical to handle and liberates HCl as a by-product during the reaction. The HCl can be recovered by scrubbing the off-gas stream with water, or neutralized by scrubbing the off-gas with a dilute basic scrubbing solution. In either case, additional equipment and complexity are added to the process.

Sulfuric acid (H_2SO_4) and oleum ($SO_3 \cdot H_2SO_4$) are widely used as sulfonating agents. Oleum is used to sulfonate alkyl benzene and sulfate fatty alcohols for heavy duty detergents. The reaction is shown in Fig. 14.6. It is an equilibrium process because water is formed in the reaction and the resultant water dilutes the oleum and/or sulfuric acid. The sulfonation reaction stops when the sulfuric acid concentration drops to approximately 90%. This "spent" acid may be separated from alkyl benzene sulfonic acid to produce a product, which, on neutralization, contains a relatively low level (6–10%) of sodium sulfate. When fatty alcohols are sulfated, the spent acid cannot be separated. It must be neutralized with the alcohol sulfuric acid to make a product containing a high level of sodium sulfate. Oleum is relatively inexpensive—about U.S. $0.153 per pound of reactive SO_3. Oleum sulfonation can be operated as either a batch or continuous process. This process has the dual advantage of low SO_3 cost and low capital equipment cost. However, it has the disadvantage of being an equilibrium process which leaves large quantities of unreacted sulfuric acid. This waste acid must be separated from the reaction mixture and subsequently disposed.

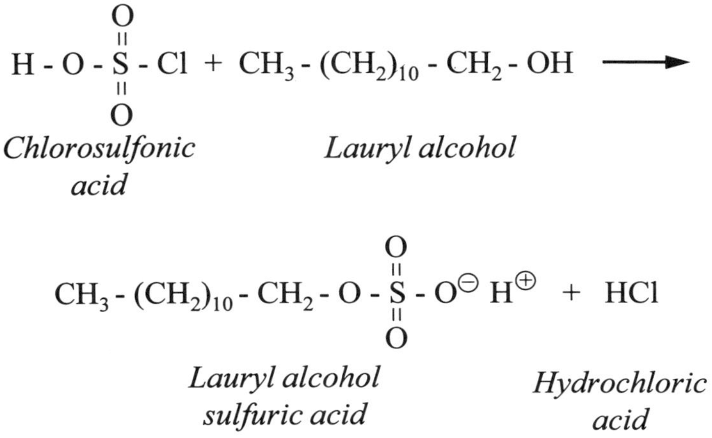

$$H-O-\overset{\overset{O}{\|}}{\underset{\underset{O}{\|}}{S}}-Cl \;+\; CH_3-(CH_2)_{10}-CH_2-OH \;\longrightarrow$$

Chlorosulfonic   *Lauryl alcohol*
acid

$$CH_3-(CH_2)_{10}-CH_2-O-\overset{\overset{O}{\|}}{\underset{\underset{O}{\|}}{S}}-O^{\ominus}\,H^{\oplus} \;+\; HCl$$

Lauryl alcohol   *Hydrochloric*
sulfuric acid   *acid*

Fig. 14.5. Chlorosulfonic acid sulfation.

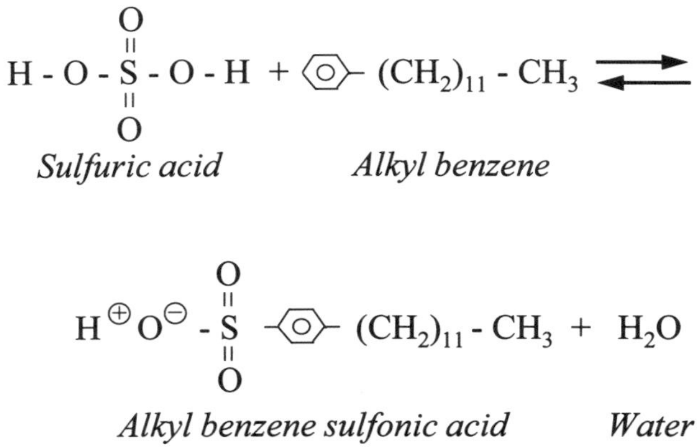

$$H - O - \underset{\underset{O}{\|}}{\overset{\overset{O}{\|}}{S}} - O - H \; + \; \langle\!\bigcirc\!\rangle\!- (CH_2)_{11} - CH_3 \; \rightleftharpoons$$

Sulfuric acid   Alkyl benzene

$$H^{\oplus} O^{\ominus} - \underset{\underset{O}{\|}}{\overset{\overset{O}{\|}}{S}} \; \langle\!\bigcirc\!\rangle\!- (CH_2)_{11} - CH_3 \; + \; H_2O$$

Alkyl benzene sulfonic acid   Water

Fig. 14.6.   Sulfuric acid/oleum.

In North America, the expense of disposing the "spent" sulfuric acid has become so high that the process economics are now questionable.

Sulfonation with sulfuric acid is a special case of oleum sulfonation. Because the sulfonation reaction stops when the acid concentration in the reaction mixture drops to less than approximately 90%, sulfonation of detergent feedstocks with sulfuric acid is not normally practiced. Today, sulfuric acid sulfonation is used principally for production of hydrotropes by azeotropic reaction with benzene, toluene or xylene. In this special process, the water formed during the reaction is removed by azeotropic distillation of the water and unreacted feedstock. The water is then separated from the immiscible organic feedstock which is returned to the reaction vessel. Because water is removed, the reaction may continue to completion.

The air/SO_3 sulfonation process is a direct process in which SO_3 gas is diluted with very dry air and reacted directly with the organic feedstock. The source of the SO_3 gas may be either liquid SO_3 or SO_3 produced by burning sulfur. As shown in Fig. 14.7, the reaction of gaseous SO_3 with organic material is rapid and stoichiometric. It is complicated by the possibility of side reactions, and therefore tight process control is essential. The cost for liquid SO_3 is US $0.09 per pound of reactive SO_3; whereas SO_3 from sulfur burning is US $0.02 per pound of reactive SO_3. The air/SO_3 sulfonation process has the lowest SO_3 cost of any sulfonation process and is extremely versatile, producing very high quality products. However, it is a continuous process best suited to large production volumes. In addition, it requires expensive precision equipment and highly trained operating personnel.

As previously mentioned, a commercially successful sulfonation process requires reaction of SO_3 with the organic feedstock under tightly controlled conditions. Figure 14.8 illustrates the level of control demanded by the air/SO_3 sulfonation process (3). This illustration shows the production of 1,4-dioxane, an undesirable by-product formed during the sulfation of ethoxylated alcohols. The 1,4-dioxane formed is a func-

$$SO_3 + CH_3-(CH_2)_{10}-CH_2-(O-CH_2-CH_2)_3-OH \longrightarrow$$

Sulfur trioxide *Ethoxylated lauryl alcohol*

$$CH_3-(CH_2)_{10}-CH_2-(O-CH_2-CH_2)_3-O-\overset{O}{\underset{O}{\overset{\|}{\underset{\|}{S}}}}-O^{\ominus}\ H^{\oplus}$$

Ethoxylated lauryl alcohol sulfuric acid

Fig. 14.7. Air/SO_3.

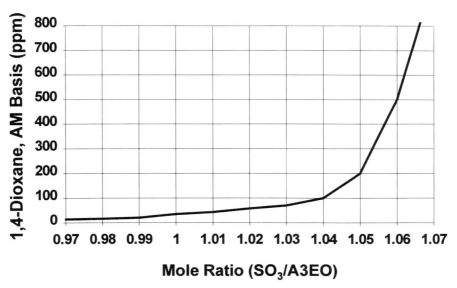

Five minutes acid recycle residence time at 30°C and 2.5% SO3 inlet gas concentration
1,4-Dioxane, AM Basis (ppm)
800
700
600
500
400
300
200
100
0
0.97
0.98
0.99
1
1.01
1.02
1.03
1.04
1.05
1.06
1.07
Mole Ratio (SO3/A3EO)

Fig. 14.8. 1,4-Dioxane *vs.* mole ratio.

tion of mole ratio (kg moles of SO_3 per unit time fed to the reactor divided by kg moles of feedstock per unit time fed to the reactor). As the mole ratio of SO_3 to organic feedstock increases, the level of dioxane in the product remains relatively low at 20–30 ppm. A critical point of oversulfation occurs at a mole ratio of approximately 1.03. Once oversulfation occurs and the mole ratio exceeds 1.04, dioxane production increases rapidly to values measured in hundreds of parts per million. Similar adverse responses are observed with product color or levels of unsulfonated or unsulfated (free oil) materials in the product. Clearly, the sulfonation process must be controlled to within 1% of the desired mole ratio to achieve excellent product quality. Other important process variables are reaction temperature, SO_3 gas concentration, time to neutralization, neutralization pH and neutralization temperature. These variables also influence product quality although the effect is not as dramatic as the effect of mole ratio.

The profound effect of mole ratio on product quality implies that an air/SO_3 sulfonation reactor must be designed to ensure that the mole ratio is equally and constantly maintained at all points in the reactor. I have coined the phrase "micro scale" mole ratio control to describe this condition. Micro scale control means that the reactor has been designed and calibrated so that the same mole ratio is held constant at every point at a cross section through the linear flow axis of the reactor. All molecules of feed see exactly the same quantity of SO_3. This is different than the macro mole ratio control which is the overall mole ratio of organic feedstock and SO_3 fed to the reactor. Macro control is determined by the plant's control system. It is imperative that the equipment be capable of both macro and micro mole ratio control because sulfonation is such a critical reaction with respect to mole ratio control. A 1% variation in the mole ratio can spell the difference between a world-class product and off-specification material.

Choosing a Sulfonation Process

The choice of sulfonation process depends on many factors. One of the most important is the nature of the desired products and their required quality. Some processes are very versatile while others produce only a few types of products. Each process produces slightly different products. For example, the sulfamic acid process produces only ammonium sulfates from alcohols or ethoxylated alcohols. Another example is the presence of a minimum of 8% sulfate in sodium alkyl benzene sulfonates made with oleum. Some processes such as the air/SO_3 process are capable of sulfating or sulfonating a wide variety of feedstocks and producing excellent quality products from all of them.

A second factor to consider in the choice of sulfonation process is the required production capacity. The sulfamic acid process is a batch process suitable for production of small quantities of material. The air/SO_3 process is a large-scale continuous process best suited to 24 hours per day, seven days per week manufacture of tons of product per hour. The chlorosulfuric acid and oleum processes can be run as either batch or continuous processes.

Reagent cost may have a major impact on choosing a process. The air/SO_3 process has the lowest cost per pound of SO_3 reacted, whereas the sulfamic acid process has the highest. For large-scale commodity production, the air/SO_3 process clearly has an advantage. However, for small-scale production of a high value specialty product this advantage may be outweighed by other considerations such as initial equipment cost and the necessity for continuous operation.

The process equipment cost is an important factor to be considered in the choice of a sulfonation process. You must look at the installed cost of the system, tankage and required safety systems. The equipment cost is almost exactly the inverse of the reagent cost. Here, the air/SO_3 process is highest in cost while the simple batch sulfamic process is lowest. Other processes are intermediate.

The final factor to consider in the choice of sulfonation processes is the cost of waste disposal. The chlorosulfuric acid and oleum sulfonation processes produce large by-product streams of either hydrochloric acid or sulfuric acid. These by-products must be recovered and sold, or disposed of as a waste. Waste disposal can have a significant impact on the profitability of these processes because the necessary equipment can be costly and the disposal costs can be high.

Table 14.2 shows the trends in sulfonation plants in the United States (1). The air/SO_3 process has rapidly overtaken the oleum process as the predominant choice. This is the result of several trends. The first is the waste disposal cost for the spent sulfuric acid from the oleum process. The second is the desire of many processors to avoid storing a hazardous material such as oleum. The third is the move toward compact detergent products which reduces or eliminates the sodium sulfate content of detergent products. The oleum process adds a large quantity of sulfate to the products and, for many applications, this is not acceptable. Finally, the air/SO_3 process is capable of making a broad range of very high quality products.

Commercial Scale Sulfonation Equipment

Sulfamic Acid Sulfation Equipment

Figure 14.9 illustrates the equipment used for sulfamic acid sulfation. This batch process is run in a stainless steel or glass lined, air tight, stirred tank reactor. The reactor has heating and cooling coils and provision for weighing in the organic reactant and the sulfamic acid. Before the reaction starts, air is purged from the reactor with dry nitrogen and the reaction is run under a blanket of nitrogen. The organic is weighed into the reactor and a 5% molar excess of sulfamic acid is then added. The reactor is purged and blanketed with dry nitrogen to remove oxygen. The reactants are heated to 110–160°C and held at this temperature for approximately 90 min. The

TABLE 14.2 Estimated U.S. Sulfonation Plants

	1980		1985		1990	
Plant Type	No. of Plants	Capacity*	No. of Plants	Capacity	No. of Plants	Capacity*
SO_3	44	630.0	62	855	41	1,016
Oleum	51	720.0	50	605	39	548
Chlorosulfonic acid	5	35.0	6	50	6	59
Sulfoxidation	1	0.45	2	1.35	1	0.45
Totals	101	1,385.45	120	1,511.35	87	1,623.45

*Million kg/y. *Source:* Knaggs (7).

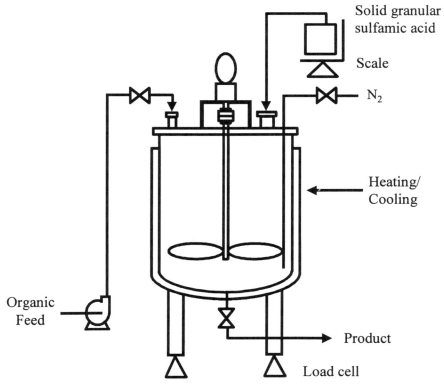

Fig. 14.9. Sulfamic acid sulfation.

products are then cooled to 70°C and water or alcohol are added to dilute the product. As previously mentioned, an ammonium salt is the direct reaction product, thus no neutralization step is required.

Chlorosulfonic Acid Sulfation Equipment

Chlorosulfuric acid can be used to sulfonate in either a batch or continuous process. For the batch process, illustrated in Fig. 14.10, the equipment is a glass lined, stirred, sealed reactor with heating and cooling jackets. The reactor must be fitted with a glass lined absorber to remove the HCl gas evolved in the reaction. A slight vacuum is usually pulled on the reaction vessel to enhance HCl removal. The liberated HCl gas is absorbed into water to make a dilute HCl solution. In operation, the alcohol or ethoxy alcohol feedstock is charged to the reactor and chlorosulfuric acid is gradually added. A good refrigeration system is required for heat removal because the reaction is exothermic. The reaction mass must be kept at approximately 25°C to avoid side reactions and color body formation, and to minimize foaming. The rate of addition of chlorosulfuric acid is adjusted to ensure that this temperature is not exceeded. Immediate neutralization is required once the reaction is complete.

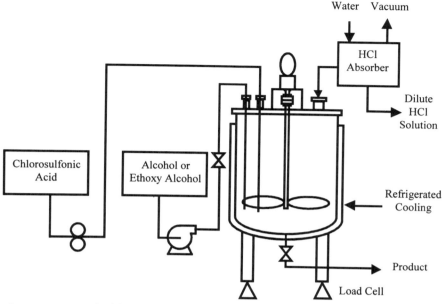

Fig. 14.10. Batch chlorosulfonic acid sulfation.

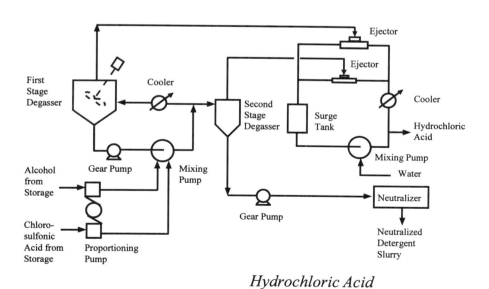

Sulfation

Hydrochloric Acid Absorption System

Fig. 14.11. Continuous alcohol sulfation with chlorosulfonic acid.

Chlorosulfation can also be continuous. Figure 14.11 shows a typical flow sheet for a continuous chlorosulfuric acid sulfation process. In this application the alcohol and chlorosulfuric acid are added into a mixing zone, combined and sent to a degasser. A slight vacuum is pulled on the degasser to assist separation of HCl from the reaction products. The disengaged sulfonic acid is sent through a heat exchanger to remove the heat of reaction and recycled back to the mixer to cool the process. A portion of the reaction mixture is sent to a second degasser where the HCl separation is completed. The HCl is continuously absorbed into water and the acid mixture is continuously neutralized. Several companies including Henkel use continuous chlorosulfation technology for making detergent actives. The process is economically viable if a source for the HCl is available and if the product, which contains some residual chloride ion, is acceptable.

Oleum and Sulfuric Acid Sulfonation Equipment

Oleum and sulfuric acid can be used to sulfonate aromatics and alcohols in either batch or continuous equipment. For detergent alkylates, the batch equipment is very similar to other processes. As shown in Fig. 14.12, the required equipment is a stirred, sealed, glass lined or stainless steel kettle with a provision for heating and cooling. The detergent alkylate is first added to the reaction vessel, then the oleum is slowly added over a period of several hours. The reaction is highly exothermic and the oleum addition rate is determined by the ability to remove the heat of reaction. The temperature should be maintained below 35°C for optimum product quality. Frequently, the heat of reaction is removed by pumping the reaction mixture through an external heat exchanger. Because it is an equilibrium reaction, except for the special case of azeotropic sulfonation of hydrotropes with sulfuric acid, a large surplus of sulfuric acid forms. When the sulfonation reaction is complete, the sulfuric acid may be separated from the sulfonated detergent alkylate by adding water. The water addition (typically about 10% by weight of the reaction mixture) causes a phase separation to occur between the sulfonic acid and the diluted sulfuric acid. The separation usually takes place in a separate, glass lined vessel and occurs over a period of

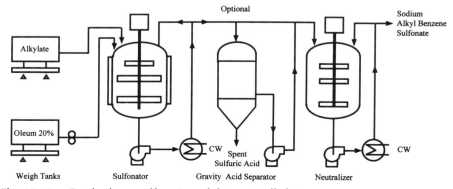

Fig. 14.12. Batch oleum sulfonation of detergent alkylate.

about 10 h. Materials of construction are crucial because the dilution process makes sulfuric acid which is in a very corrosive temperature and concentration range. After separation, the sulfonic acid may be neutralized with aqueous sodium hydroxide, usually in a separate neutralization vessel. Including neutralization, total batch time is 15–20 h. The product contains about 15% sodium sulfate after neutralization if the acid is separated, and about 60% sodium sulfate if not. Without separation, the product's application is limited to low active, traditional detergent powders in which the large content of sodium sulfate is used as a filler.

In the special case of azeotropic sulfonation of toluene, cumene or xylene with 98% sulfuric acid to form hydrotropes, a reflux condenser is added at the top of the reactor. The condenser separates the unreacted feed from the water produced in the reaction. The water is removed from the condenser, and the feed is refluxed back to the reactor. Because the water is removed, the reaction proceeds to completion and a large excess of sulfuric acid is not required. Typical equipment for hydrotrope sulfonation is shown in Fig. 14.13.

The invention of the process for continuous oleum sulfonation was the foundation of The Chemithon Corporation in 1954 (4,5). Figure 14.14 shows a flow sheet for continuous oleum sulfonation of detergent alkylate. In this process, alkyl benzene is mixed and reacted with oleum in a recycle loop where the reaction mixture is cooled by recycling it through a heat exchanger. Typical reaction temperatures in the recycle loop are 38–54°C. The mixed acid products are digested in a plug flow reactor and then dilution water (approximately 13% by weight) is added in a second mixing loop. A second heat exchanger removes the heat of dilution. The diluted sulfuric acid and alkylbenzene sulfonic acid are separated in a continuous settler. The sulfonic acid is then continuously neutralized with aqueous sodium hydroxide solu-

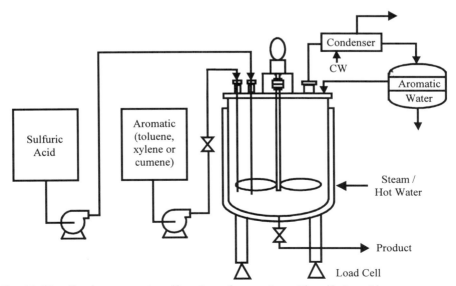

Fig. 14.13. Batch azeotropic sulfonation of aromatics with sulfuric acid.

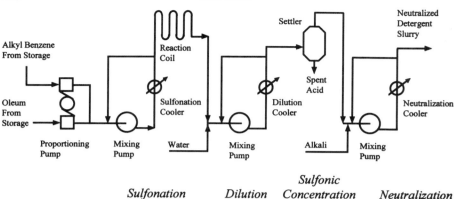

Sulfonation   Dilution   Concentration   Neutralization

Fig. 14.14.   Continuous high active alkylate sulfonation with oleum.

tion in a third, cooled mixing loop. Sulfate levels as low as 10% in the final product can be achieved in this equipment. Total processing time is less than 1 h. The equipment can also be used to sulfate detergent range alcohols for use in laundry powders if a high level of sulfate in the product can be tolerated. The sulfates are present because the separation process cannot be used when sulfating alcohols. If water is added to the alcohol sulfuric acid/sulfuric acid mixture, the alcohol sulfuric acid immediately hydrolyzes. Additionally, mixed products containing both alkyl benzene sulfonates and alcohol sulfates can be manufactured as shown in Fig. 14.15. In this case alkylate is sulfonated first, followed by sulfation of the alcohol. Some of the excess sulfuric acid from the sulfonation stage is used to react with the alcohol, and the combined mixture is immediately neutralized. Because no separation stage is used, the resultant product is high in sulfate.

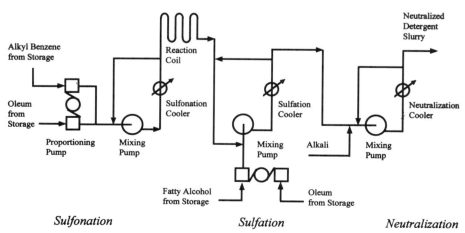

Sulfonation   *Sulfation*   *Neutralization*

Fig. 14.15.   Continuous tandem type sulfonation-sulfation with oleum.

Air/SO₃ Sulfonation Equipment

Four possible sources of SO_3 gas used for an air/SO_3 sulfonation system are as follows:

1. Sulfuric acid plant converter gas.
2. SO_3 from boiling concentrated oleum.
3. Liquid SO_3.
4. Sulfur burning in equipment specifically designed to produce SO_3 gas for sulfonation.

Converter gas from a sulfuric acid plant contains 10–12% SO_3 and appears to be a potential SO_3 source for sulfonation. There are several problems with using a sulfuric acid plant as an SO_3 source for sulfonation. Nevertheless, such an arrangement has been commercially installed and at first glance appears to be an attractive, low cost method of supplying SO_3 gas to a sulfonation plant. Physical location is a limiting factor because the sulfonation plant must be installed as closely as possible to the sulfuric acid plant converter. In addition, the sulfonation plant can run only when the sulfuric acid plant is running.

There are three other more subtle difficulties when using a sulfuric acid plant as an SO_3 source for sulfonation. First, the SO_3 gas at approximately 18% concentration must be diluted to the normal range for sulfonation (typically 4–7%). An auxiliary air supply must be installed, which adds expense and complexity. Second, because sulfuric acid absorption towers are used for air drying, the air/SO_3 from a sulfuric acid plant has a higher dew point (typically –35°C) than that required in a sulfonation plant (typically –60 to –80°C). The high dew point causes product quality problems in the sulfonation process and accelerates corrosion of the process equipment. Third, the pressure of the air/SO_3 from the sulfuric acid plant is usually not sufficient to overcome the pressure drop of the sulfonation system. Compressing the air/SO_3 from the converter is not trivial because it requires a high alloy compressor to withstand the corrosive environment created by the wet air/SO_3 stream. This problem can be overcome, but the solution is not inexpensive. Considering all of the problems inherent in utilizing the converter gas stream from a sulfuric acid plant, the conclusion is that it is technically feasible. However, this choice adds significant operational difficulties and does not result in a major cost savings over installing a complete sulfur burning sulfonation plant.

Another possible source of SO_3 for sulfonation is produced by boiling oleum to produce gaseous SO_3 which is then blended with dry air. It is practically limited to locations where fresh oleum can be received, and depleted oleum returned by pipeline. Compared with sulfur burning, this process somewhat reduces the equipment requirement. However, it still requires an air supply system, an oleum boiler and an SO_3 metering system. Unlike a sulfur burning plant which generates its own heat for air dryer regeneration, this air supply system requires an external source of heat which adds extra utility expenses. Also, significant safety hazards are associated with handling concentrated oleum. Such an installation may be economical for a few site locations, and at least one is commercially operating in North America.

Some of the first air/SO$_3$ sulfonation plants installed were based on the use of liquid SO$_3$. These plants require an air supply system identical to the system described below for a sulfur burning plant except that they also require an external heat source for air dryer regeneration. In addition, a liquid SO$_3$ plant requires an SO$_3$ storage system. This storage system is usually a large 20,000–80,000 kg storage tank located in a heated room and maintained at a temperature of about 40–43°C. Heating the SO$_3$ storage room can be a significant cost in colder climates. In case of SO$_3$ leaks, the room must be sealed and should have provision for scrubbing any SO$_3$ that escapes into its atmosphere. In the sulfonation process, the liquid SO$_3$ is metered from the storage tank into a steam heated vaporizer where it is evaporated and mixed into the dried air stream from the air supply system. From this point on, the process is identical to a sulfur burning air/SO$_3$ sulfonation plant, described below. A liquid SO$_3$ storage and metering system is shown in Fig. 14.16. Because of the rigorous storage requirements imposed by the hazardous nature of liquid SO$_3$, the installed cost for a liquid SO$_3$ sulfonation facility is close to that for a sulfur burning installation.

There are significant safety advantages to a sulfur burning system. With sulfur burning air/SO$_3$ sulfonation processes, the only SO$_3$ on site is the small quantity of dilute gaseous material in the process piping between the converter (SO$_2$ to SO$_3$) and the sulfonation reactor. Even in the world's largest sulfonation plant (20,000 kg/h active production), this amounts to only about 100 kg of dilute SO$_3$ gas. The sulfur burning process is much safer than transporting, storing and handling tank truck (18,000 kg) or rail car (72,000 kg) quantities of oleum or liquid SO$_3$. Sulfonation equipment based on liquid SO$_3$ has become increasingly undesirable for the following reasons:

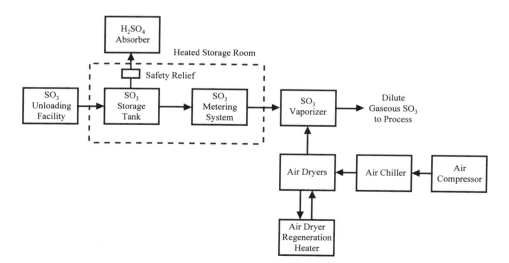

Fig. 14.16.   Dilute SO$_3$ gas from liquid SO$_3$.

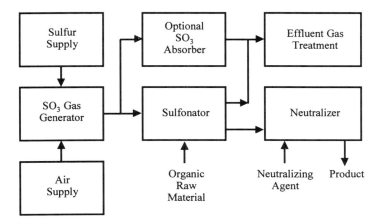

Fig. 14.17. Continuous air/SO$_3$ sulfonation.

1. Safety concerns.
2. Liquid SO$_3$ is unavailable in many parts of the world.
3. Sulfur is readily available worldwide.
4. Sulfur is relatively inexpensive.

The remainder of this section, therefore, is confined to the description of a state-of-the-art sulfur burning, air/SO$_3$ sulfonation plant.

The basic plant package for a sulfur burning, air/SO$_3$ sulfonation installation includes a sulfur supply system, air supply system, SO$_3$ gas plant system, SO$_3$ absorber system, sulfonator, neutralizer, effluent gas clean-up system, control system and motor control center. This system is shown as a block diagram in Fig. 14.17. Additional equipment can be added as required to manufacture specialty products, such as a hydrolyzer for alpha olefin sulfonate production, re-esterfication and bleaching equipment for methyl ester sulfonate production, dioxane strippers for production of ethoxy alcohol sulfates containing ultralow dioxane levels (less than 10 ppm), and dryers for producing concentrated, pure detergent active. Capacities for commercial sulfur burning air/SO$_3$ sulfonation units range from 250 to 20,000 kg/h of 100% detergent active. Typically, an air/SO$_3$ sulfonation plant is designed to sulfonate with approximately 4–7% (volume) SO$_3$. However, if ethoxylated alcohol is to be used as a major feedstock, the SO$_3$ gas concentration may be as low as 2.75% SO$_3$ in air to the sulfation systems.

As Fig. 14.18 illustrates, the process air is first compressed to a pressure of approximately 1 kg/cm^2 (15 psig) using either a rotary compressor or a high efficiency centrifugal compressor. In larger sulfonation plants—capacities greater than 4000 kg/h—the process air capacity should be adjustable without venting between approximately 60 and 100% of full capacity (while maintaining efficiency). This ability conserves electric power if the plant is operated at partial capacity.

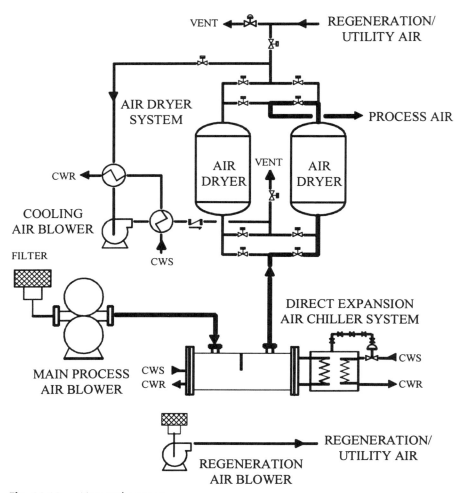

Fig. 14.18. Air supply system.

A direct expansion air chiller vessel cools the compressed air to a temperature of 7°C (45°F). The chilled air is then dried in dual, automatic pulseless desiccant-type air dryers to a dew point of –80°C (–112°F). The lack of a pulse in the air stream is vital to product quality. If the air pulses, there will be a momentary interruption in mole ratio which results in a loss of product quality. The result is product with high free oil, dark color and possibly other undesirable by-products. The dual air dryers are equipped with 11 individual bubble tight control valves. The control valves ensure absolutely no interruption of process air when changing dryers and a smooth pressure transition when the dryers switch to the regeneration cycle. A cooling air blower supplies cool air to the regenerated (off-line) air dryer to reduce the bed temperature to an acceptable level prior to bringing the regenerated air dryer back on-line. This is important—if the dryer bed is too hot, the process air dew point will be

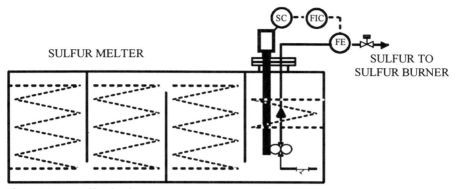

Fig. 14.19. Sulfur feed system.

higher than desirable due to the increased equilibrium concentration of water in the hot air exiting the dryer.

A regeneration/utility air blower supplies cooling air to the double pipe gas coolers used in the gas plant. Hot air (200°C) leaving the low maintenance double pipe coolers regenerates the off-line air dryer. During the air dryer cooling cycle in smaller plants, the regeneration/utility air vents to the atmosphere or is sent to other processes where the heat in the air can be used. In larger plants (greater than 4000 kg/h) the heat may be recovered as steam. In this case, the gas plant is modified so that it produces steam instead of hot air and a portion of this steam heats the regeneration air.

In the sulfur supply system shown in Fig. 14.19, bright Frasch or by-product sulfur flows by gravity to one of two gear pumps. Typically dual strainers, pumps, and flow meters are provided to minimize downtime during maintenance. The sulfur passes through a highly accurate mass flow meter prior to delivery to the sulfur burner. Because the mole ratio is the most important process variable, control of the sulfur flow to the sulfur burning air/SO_3 gas plant is one of the two most important process control functions in the plant. The entire sulfur supply system is steam jacketed because the sulfur has a freezing point of 112.8°C. Sulfur is normally handled at its viscosity minimum, which is 136°C. A steam and condensate system supplies the necessary tracing circuits for the sulfur supply piping.

In the SO_3 gas generator, illustrated in Fig. 14.20, the metered sulfur is delivered to the refractory-lined atomizing sulfur burner where combustion with the dry process air generates sulfur dioxide (SO_2). The atomizing burner ensures instant, constant sulfur burning. The sulfur dioxide gas leaving the burner is cooled to 420°C and delivered to a three-stage vanadium pentoxide catalytic converter, where the gas is filtered and converted to sulfur trioxide (SO_3). The conversion efficiency of the converter is between 98.0 and 99.5%. Prior to entering the inlet mist eliminator, the sulfur trioxide gas leaving the converter is cooled to nearly ambient temperature in an SO_3 double pipe cooler followed by an SO_3 water cooled heat exchanger. The inlet mist eliminator removes traces of sulfuric acid or oleum from the cooled SO_3 gas stream. Heat recovered from the SO_2 cooler converter interstage coolers and SO_3 cooler is used to regenerate the air dryers, eliminating the need for an external heat

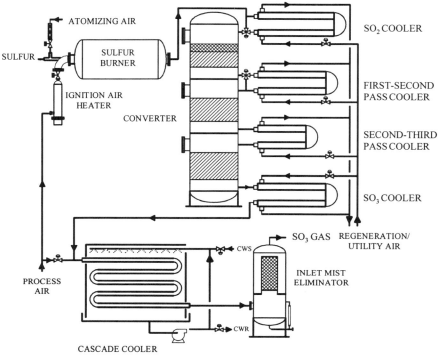

Fig. 14.20. SO₃ gas generator.

source. Heat recovered from the SO$_2$ and first to second pass interstage cooler can also be used to generate steam in the optional heat recovery system. In state-of-the-art sulfonation plants, computer-aided hot gas piping design eliminates troublesome hot gas expansion joints in the gas plant. The enhanced piping design improves plant reliability significantly and eliminates a possible source of inadvertent SO$_2$ or SO$_3$ discharge into the environment.

An in-line electric ignition air heater warms the combustion air stream for sulfur burner preheating and ignition. Features such as high-temperature gas diverting valves and converter heat hold can reduce the time required to achieve stable conversion of SO$_2$ to SO$_3$ so that sulfonation can begin. These features significantly reduce the time required for gas plant start-up. Production of SO$_2$, which must be scrubbed from the exhaust gas prior to discharge into the atmosphere, is minimized, therefore reducing plant operating costs.

The SO$_3$ absorber unit, shown in Fig. 14.21, is capable of treating the total output from the SO$_3$ generation system to form 98% sulfuric acid. It is a convenience in plant start-up, shutdown and product changeover. It is also recommended in situations in which frequent power failures can interrupt production. With this equipment, the plant can be restarted without forming off-specification product. If this unit is not used, the sulfonation reactor itself must be used to scrub SO$_3$ from the process gas during gas plant start-ups. This is an undesirable procedure, however,

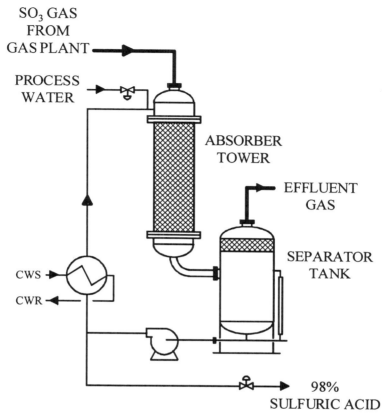

Fig. 14.21. SO$_3$ absorber system.

because the gas plant has not stabilized during the start-up process and the mole ratio of SO$_3$ to feedstock is unknown. The practical solution to this problem is to purposely undersulfonate alkylbenzene during the start-up. The off-specification material made in this start-up procedure can be stored and later blended back into the product. The SO$_3$ absorber system is almost a necessity if only alcohol feed-stocks are run in the plant, because alcohol sulfuric acids cannot be stored and reused.

SO$_3$ mixed in air enters the absorber column where it contacts 98% sulfuric acid. The SO$_3$ is absorbed into the acid, which separates from the remaining air in the scrubber body. A mesh pad mist eliminator removes entrained acid mist from the air as it exits the vessel. Water is added to control the concentration of the sulfuric acid as it flows through the acid circulation system. An instrument system with dual conductivity sensors holds the acid concentration at 98% by controlling water addition. A heat exchanger in the absorber circulation loop removes heat of dilution.

The sulfonator is the heart of a sulfonation plant. Sulfonic acid forms in the sulfonator when an SO$_3$-in-air mixture is injected into the reactor simultaneously with the desired organic feed under carefully controlled conditions of mole ratio, SO_3 gas concentration and temperature. Potential organic feedstocks include alkylates, alcohols, ethoxylated alcohols, methyl esters and alpha olefins.

Several types of commercial sulfonators are available. In broad classifications they are film reactors (6–9), including the Chemithon Annular Falling Film Reactor (10–12), the Ballestra (13), IIT and Siprec multitube film reactors; the Chemithon dispersed phase or jet reactors (14–16); and stirred tank or cascade type reactors (17). Film reactors are the most common in detergent processing for consumer products, especially for production of cosmetic quality materials from oleo chemical feedstocks. In a film reactor, the organic feedstock is extruded onto the wall of the reactor (reaction surface) as a continuous film. Organic feed rate to the reactor vessel is measured by a highly accurate mass flow meter and controlled by a variable speed driven gear pump. The proper organic feed rate is based on the preset sulfur-to-organic mole ratio. The SO_3, diluted with very dry air, flows over the film of organic material. The SO_3 diffuses into the organic film and reacts to form a sulfonic acid. In almost all commercial reactors, both the organic and SO_3 flow concurrently from the top of the reactor to the bottom. The heat of reaction is removed by cooling water which flows through cooling jackets underneath the reaction surface of the reactor.

Figure 14.22 shows a Chemithon Annular Film Reactor. This patented reactor (10–12) is unique in many ways. It employs interchangeable, factory calibrated organic metering flanges. These flanges ensure that the flow of organic feed to any two points on the reaction surface is equal, within ±1%. When combined with the geometric design of the reactor that similarly controls the flow of SO_3 to the reac-

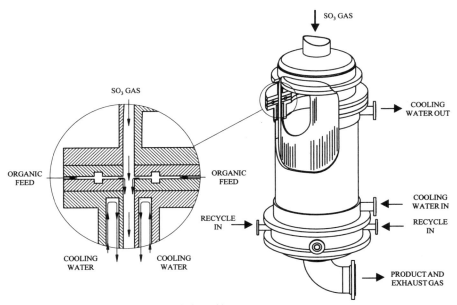

Fig. 14.22. Chemithon annular falling film reactor.

tion surfaces, these features ensure that mole ratio control is maintained on a micro scale throughout the sulfonator. Cooling jackets on the reactor remove most of the heat of reaction. Additionally, the patented recycle system allows the reactor to overcome the increase in viscosity of the sulfonic acid as its level of sulfonation increases. The product exiting the reactor is instantly quench cooled by removing the acid, pumping it through a heat exchanger to cool, then returning it to the bottom of the reactor. This cooling process reduces the time that the sulfonic acid is held at an elevated temperature and results in better product quality. A side benefit is that the reactor is more compact—less than 2 m in height—and therefore considerably less expensive to install.

Figure 14.23 shows this reactor installed in a typical sulfonator. Upon exiting the reactor, the spent gas is separated from the sulfonic acid recycle stream in the liquid separator and cyclone vessels. Sulfonic acid product discharges from the recycle stream at a controlled rate, maintaining continuity of the quantity of material in the reactor system. The acid product from the reactor can then be fed directly to the digestion and hydration system (or optional degasser system) where reaction with absorbed SO_3 is completed. Hydration water is injected and mixed with the sulfonic acid, leaving the digesters to remove anhydrides.

Figure 14.24 shows a typical multitube sulfonation reactor (9,13,18). The multitube reactor is the other commonly employed film sulfonator design. In this reactor the organic feed is distributed among a number of parallel reaction tubes, 25 mm in diameter, 7 m long. The tubes are arranged like the tube bundle in a vertical shell and tube heat exchanger. The organic and SO_3 gas flow concurrently down the reactor tube, react and exit the bottom of the reactor into a separator vessel. The heat of

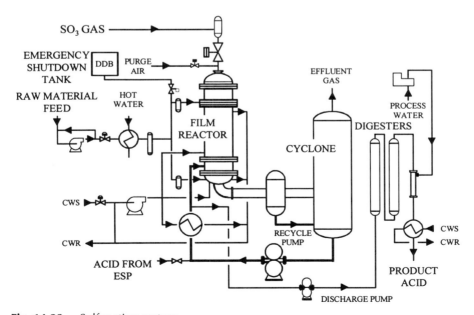

Fig. 14.23. Sulfonation system.

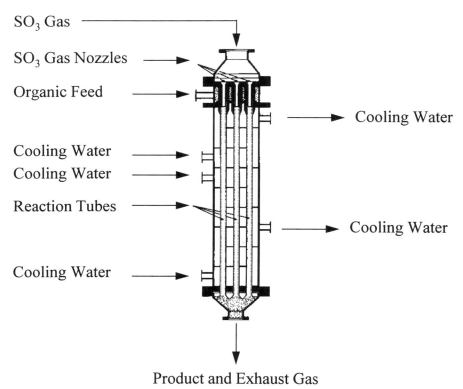

SO$_3$ Gas

SO$_3$ Gas Nozzles

Organic Feed

Cooling Water

Cooling Water

Cooling Water

Reaction Tubes

Cooling Water

Cooling Water

Product and Exhaust Gas

Fig. 14.24.   Ballestra multitube film sulfonator.

reaction is removed by cooling water which flows through the reactor jackets. The approximate residence time of the acid from the top of the reactor through the separator and cyclone and to the neutralizer is 2–3 min.

Stepan Company (7) patented a multitube reactor in 1965. They currently operate a large number of these reactors worldwide. Similar reactor designs are manufactured by other firms (19,20).

A continuous neutralizer system is illustrated in Fig. 14.25. The neutralizer combines sulfonic acid or organo-sulfuric acid with a neutralizing agent, additives, and diluent (water), in a dominant bath neutralization. The result is a solution of neutral active matter (slurry or paste) of the desired composition and pH. Caustic soda, usually 50 wt% NaOH, is the most common neutralizing agent. However, caustic potash, aqueous ammonia, triethanolamine and other agents are compatible with the neutralizer. A recycle loop circulates neutral slurry through a heat exchanger to remove heat of neutralization, mixing and pumping. Individual metering and/or gear pumps feed sulfonic acid, neutralizing agent, dilution water, buffer solution and any additives into the loop to mix through a high shear mixer. A positive displacement pump circulates paste through the process heat exchanger, and a large portion of this stream recycles back to the mixer. A pressure control valve allows product to leave the recycle system as feed enters. A pH sensor is placed in the recycle line near the

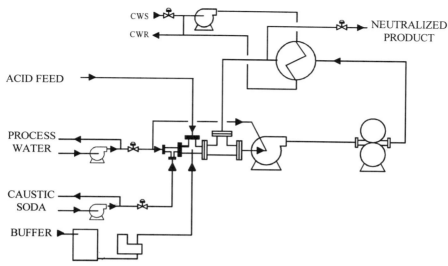

Fig. 14.25. Neutralizer system.

inlet to the mixer. The in-line pH sensor measures an accurate pH on dilute slurries and produces a single-valued output signal that is proportional to pH on concentrated pastes. The signal from the sensor feeds to the pH control system in either case and controls the flow of the neutralizing agent into the process. Use of a buffer promotes stability in the pH control loop. Cooling water is also recirculated and automatically adjusted to maintain a desired temperature at the process heat exchanger inlet. Therefore, higher melting point products can be processed without plugging the heat exchanger during start-up.

Effluent process gases leaving the sulfonation system or SO_3 absorber are virtually free of residual SO_3, but contain any unconverted SO_2 gas and entrained particulate anionic materials (acidic mists of sulfonic and sulfuric acids). This gas stream is not

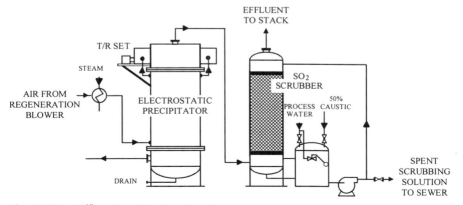

Fig. 14.26. Effluent gas treatment system.

suitable for direct discharge to the atmosphere. The recommended clean-up system, shown in Fig. 14.26, employs an electrostatic precipitator designed to collect particulate mists and a packed tower scrubber to absorb SO_2 gas in a dilute caustic. Final effluent gases cleaned to contain less than 5 ppm_v SO_2 and less than 10 mg/m^3 combined sulfate mist and neutralized organic mist are suitable for discharge to the atmosphere.

Sulfonation Plant Control

A typical sulfonation plant control system consists of two components that are integrated into a single system: main instrument control panel (ICP) and a motor control center (MCC). When properly designed and installed, this system not only allows operation of the sulfonation plant but also improves the operators' understanding of the plant's operation through graphic interfaces. Better control results in improved product consistency and quality.

The ICP integrates the loop control, logic control, data acquisition and operator interface into a single system. The functions include: analog control or PID control loops; discrete control, e.g., to automatically sequence valves in the air dryers, and when diverting gas from the sulfonator to the SO_3 absorber; interlocks to reduce operator errors that could result in damage to the equipment or environmental releases; alarm monitoring; and recipes that automatically enter variables used to calculate set points for organic feed and the neutralizer raw materials.

The hardware can be as simple as individual electronic controllers or as complicated as integrated, plant-wide distributed control systems. Naturally, more complex and expensive control systems allow more control functions and permit data logging. They also reduce the dependence on operator knowledge to produce top quality products.

As a cost-effective option which allows maximum control at minimum cost, many sulfonation plants employ a PLC-based control system in which the operator interaction is accomplished by at least two, redundant PC-based workstations. These workstations connect directly to the controllers and supply the operator with information necessary to operate the plant. Normally, a graphic display package is included in the control system software. It shows graphic representations resembling a Process & Instrumentation Diagram, which displays the main operating variables and indicates which motors are operating. Typical displays include: sulfur metering system; air dryer system; SO_3 gas generator system; sulfonator system; neutralizer system; and effluent gas treatment system. The system is also capable of tracking trends in selected variables, logging alarms and collecting data for historical logs.

The motor control center (MCC) includes a main power disconnect, a lighting transformer, motor starters, variable frequency speed controllers, and disconnects as required for the equipment.

Future Trends

The detergent market is rapidly changing as a result of three interrelated drivers. The first is the "green" movement in which consumers are demanding environmentally friendly products and packaging. The second is the move toward compact detergents which eliminate large bulk fillers from detergent formulations. The third is a desire on the part of consumers for "natural" products which are perceived as purer and less harmful to the environment. All of these forces are moving detergent producers to modify their processes and their choice of feedstocks.

The rapid increase in the market share of compact detergents indicates that large quantities of sulfates are no longer acceptable in detergent actives. This trend, in conjunction with the increased cost of spent sulfuric acid disposal, has helped eliminate oleum sulfonation as a viable technology for detergent manufacture. This change has been accentuated by a move toward the use of natural feedstocks such as oleo chemical-based alcohols and methyl esters. Because the feedstocks are made from renewable resources, they are perceived by consumers to be somehow cleaner and purer than petrochemical-based raw materials. The net result is that oleum sulfonation and sulfation technology is increasingly being replaced by air/SO_3 technology.

The purity issue is being hastened by consumers' interest in products containing low levels of impurities or by-products. Colorless products carry the perception of purity and are therefore currently favored by consumers. Again the movement is toward very light colored materials which are typically air/SO_3 sulfated alcohols and alcohol ether sulfates. Manufacturers of other feedstocks such as linear alkylbenzene have responded to this challenge by introducing new grades of feedstocks. These new feedstocks are specially prepared so that color formation on sulfonation via air/SO_3 is minimized. Some of these feedstocks will produce sulfonates which rival alcohols for lightness of color.

Another by-product which has received wide spread attention in Europe is 1,4-dioxane. This by-product is formed during sulfation of ethoxylated alcohols. The mechanism for its formation has been studied extensively and most sulfonation equipment suppliers can guarantee 1,4-dioxane levels of less than 30 ppm on a 100% active basis in ether sulfates without stripping. The most obvious strategy in dealing with 1,4 dioxane is to avoid forming it. However, if you have an existing plant or if undetectable levels are required, equipment is available to strip the neutral product to remove any traces of 1,4-dioxane. This technology allows existing equipment to be run at full capacity while permitting manufacturers to produce essentially 1,4-dioxane-free product.

The attention focused on compact detergents has created an interest in very high active sulfonates and sulfates. If active levels are sufficiently high, these products can be directly agglomerated to form compact laundry detergents without spray drying. This ability is a major advantage because it allows manufacture of a finished laundry formulation in low cost equipment—without the energy and environmental penalties of spray drying. Chemithon has developed high active neutralization technology capable of producing 70–80% active neutralized sulfonates or sulfates from any of the normally used detergent feedstocks.

These high active pastes can then be dried to very low moisture levels (≤5%) in either a wiped film evaporator or in the Chemithon Turbo Tube™ dryer. This equipment removes moisture from the detergent paste to produce a detergent noodle, needle, pellet or powder which can be agglomerated into the finished detergent formulation or shipped to the formulator for further processing.

Additional Equipment

The Chemithon Turbo Tube™ dryer (patent pending), shown in Fig. 14.27, is a heated, vacuum flash device in which the volatile components of a surfactant slurry are removed in a nondusting manner. The process is carried out under carefully controlled conditions of temperature and pressure to ensure that product quality of sensitive surfactants is not compromised during the removal process. Because of the low production costs of this new technology, the surfactant manufacturer can supply a low cost, inexpensive dried detergent active to the formulator. Transportation costs are significantly reduced as well, because water and other possible diluents are removed prior to shipping. The formulator can use this source of active to customize the final formulation for his particular market using inexpensive agglomeration equipment.

Production of sodium alpha sulfo methyl ester (SASME) is the final advance in sulfonation technology discussed in this chapter. Because of the vast quantities of methyl ester produced worldwide [estimated to exceed 540,000 tons in 1995 (21)], there is widespread interest in using this raw material for detergent production. The Chemithon Corporation has developed a new bleaching process for production of low disalt, light colored SASME. Chemithon has applied for patents to cover this new process. The process, shown in Fig. 14.28, combines hydrogen peroxide, methanol and digested methyl ester sulfonic acid. The hydrogen peroxide bleaches the methyl ester, while the methanol is essential to prevent hydrolysis of the methyl ester to a disalt and methanol. Reducing disalt is important because disalt decreases solubility and detergency of SASME in hard water and shortens the shelf life of detergent formulations. Even if no methanol is added to the process, methanol forms from degradation of the methyl ester under the acidic bleach conditions. The use of

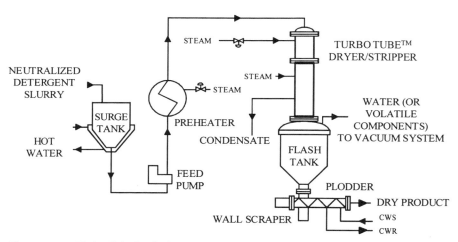

Fig. 14.27.   Turbo Tube™ drying system.

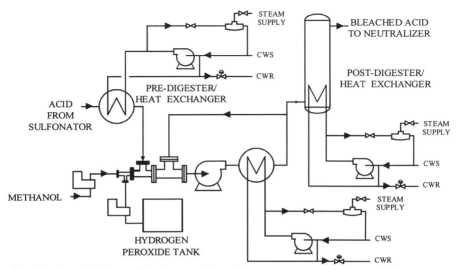

Fig. 14.28. ASME esterification and bleaching system.

the methanol prevents this and makes it possible to bleach the methyl ester to a color of less than 50 Klett in a single step, without using sodium hypochlorite. It simultaneously keeps the disalt levels below 5% on an active basis.

The SASME product leaving the neutralizer contains approximately 65–70% solids and 10–20% methanol. Methanol and water are removed from this material in the Turbo Tube™ Dryer, separated and recycled back to the process. The dried SASME is pumped from the dryer and cooled. The "needle" making apparatus forms the material into small uniform particles ("needles") containing less than 0.5% methanol and less than 5.0% water. This system is also applicable to processing other detergent actives such as alcohol sulfates or linear alkylbenzene sulfonates into dry detergent needles.

The past 50 years have seen the rise of synthetic, anionic detergents as a major item of commerce. The processes used to manufacture these detergents continue to improve as the consumer market drives producers to offer cleaning products with lower cost and improved performance. This trend will continue as consumers broaden their definition of performance to include environmental and purity concerns.

References

1. Knaggs, E.A., private communication (1994).
2. *Sulfan,* Allied Chemical Corp (1959).
3. The Chemithon Corporation, Application of Pilot Studies to Minimizing 1,4-Dioxane Production During Sulfonation of Ethoxylated Alcohols, Seattle, WA, 1988.
4. Brooks, U.S. Patent 3,024,258 (1962).
5. Brooks, U.S. Patent 3,058,920 (1962).
6. Falk, R. Taplin, U.S. Patent 2,923,728 (1960).
7. Knaggs, E.A. and Nussbaum, M., U.S. Patent 3,169,142 (1965).

8. Van der May, U.S. Patent 3,501,276 (1970).

9. Lanteri, A., U.S. Patent 3,931,273 (1976).

10. Brooks, U.S. Patent 3, 257,175 (1966).

11. Brooks, U.S. Patent 3,427,342 (1969).

12. Brooks, U.S. Patent 3,350,428 (1967).

13. Moretti, F., Canadian Patent Publication 1,144,561 (1983).

14. Brooks, U.S. Patent 4,113,438 (1978).

15. Brooks, U.S. Patent 4,185,030 (1980).

16. Brooks, U.S. Patent 4,311,552 (1982).

17. Herman de Groot, W., *Sulphonation Technology in the Detergent Industry,* Kluwer Academic Publishers, Dordrecht, The Netherlands, 1991, p. 5.

18. Ibid, p. 5.

19. IIT, Busto Arsizio, Italy.

20. Siprec, Milano, Italy.

21. Kaufman and Ruebush, R.J., *Proceedings World Conference on Oleochemicals Into the 21st Century,* Applewhite, Thomas H., ed., American Oil Chemists' Society, Champaign, IL, 1990, p. 18.

Chapter 15

Drying and Agglomeration Processes for Traditional and Concentrated Detergent Powders

Icilio Adami and Franco Moretti

Ballestra S.p.A., Milan, Italy

Detergent Powders

Classification and Physical Properties

Since the introduction of synthetic detergents, the powder type has been adopted as the most advantageous physical form to comply with manufacturing and end-user requirements. The preference for the powder form is greatly responsible for the continuously growing success of powdered detergents worldwide. Consumers expect an economically priced good detergent to have the following:

- Good overall performance (suitable foaming, fabric protection, stain removal, and other properties)
- High solubility and dispersability
- Appealing shape and fragrance
- Absence of negative side effects on skin and fabrics

These are ensured when the detergent powder has these characteristics:

- Correct balance of surfactants, builders, and other additives
- Components of good quality
- Free-flowing and homogeneous granulometry
- Correct perfuming (quality and quantity)

The various manufacturing techniques developed by the industry have been based on precise market trends, and their profitability for the producers has been well proven.

Independently from the production method, detergent powders can be classified according to type of duty, formulation structure, and physical characteristics.

Type of Duty

Examples of classification based on the product application and type of item to be washed are illustrated in Figs. 15.1 and 15.2, respectively.

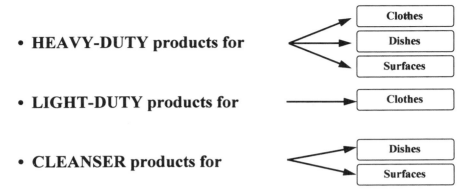

• **HEAVY-DUTY products for**
  - Clothes
  - Dishes
  - Surfaces

• **LIGHT-DUTY products for**
  - Clothes

• **CLEANSER products for**
  - Dishes
  - Surfaces

Fig. 15.1.   Detergent powders classification according to product application.

Formulation

The formulation is a topic that can be discussed by considering factors such as required properties (determined by the specific duty), type and range of components, and balance of components. Independently from the required level of performances, all detergent powders are formulated with these main components:

• Surfactants
• Builders
• Bleaches
• Fillers
• Specific additives

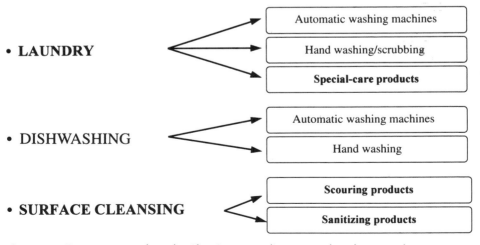

• **LAUNDRY**
  - Automatic washing machines
  - Hand washing/scrubbing
  - **Special-care products**

• **DISHWASHING**
  - Automatic washing machines
  - Hand washing

• **SURFACE CLEANSING**
  - **Scouring products**
  - **Sanitizing products**

Fig. 15.2.   Detergent powders classification according to goods to be treated.

TABLE 15.1
Formulation of European Detergent Powders

Component (%)	Laundry Powders		Dishwashing Powders	Scouring Cleansers
	Automatic Washing	Hand Washing		
Anionic surfactants	8–12	15–25	0–2	1–5
Nonionic surfactants	5–11	0–5	2–5	0–2
Builders plus cobuilders	30–45	25–40	55–65	10–15
Bleaches plus activators	15–25	0–5	—	0–5
Fillers	5–10	15–30	10–30	70–85
Additives	3–8	0–3	0–5	2–10

Examples of widely used European detergent powder formulations are reported in Table 15.1.

For the same type of application, the formulation can be quite different to fulfill the washing standard of a particular market area. The influence of local washing conditions on the formulation and the main factors characterizing the washing habits of European, American, and Japanese markets are listed in Tables 15.2, 15.3, and 15.4 together with the basic formulation for regular and compact washing powders.

These examples show how delicate and important is the component's balance to ensure the desired level of performances and characteristics for all types of detergent powders.

TABLE 15.2
Washing Conditions on Different Continents

Washing Conditions	USA/Canada	Japan	Western Europe
Washing machine type	Agitator	Impeller	Drum
Heating coil	No	No	Yes
Fabric load (kg)	2–3	1–1.5	3–4
Total water consumption (L), regular heavy cycle	140	150	120
Washing time (min)	10–15	5–15	60–70 (90°C), 20–30 (30°C)
Washing, rinsing, and spinning time (min)	20–35	15–35	100–120 (90°C), 40–50 (30°C)
Washing temperature (°C)	Hot: 50 Warm: 27–43 Cold: 10–27	10–40	90 60 40 30
Water hardness (ppm $CaCO_3$)	100	50	250
Recommended detergent dosage (g/L)	1–3	1–3	8–10
(g/kg fabric)	35–50 (no bleaching)	30–40 (no bleaching)	60–80

TABLE 15.3
Formulation of Traditional Detergent Powders

Component (%)	Traditional Powders			
	Europe	United States	Far East	Japan
Surfactants				
Anionic	8–12	10–20	15–25	20–22
Nonionic	3–5	0–5	0–5	0–3
Builders				
plus cobuilders	35–40	25–45	35–50	30–35
Bleaches				
plus activators	15–25	0–5	0–10	0–10
Fillers	15–25	20–30	5–30	30–35
Additives	2–7	0–3	0–3	0–5

The performance evaluation of detergent powders is controlled by measuring the following in-use properties:

- Washing efficiency
- Foaming power
- Wetting power
- Antiredeposition action
- Biodegradability
- Toxicity

Because these properties depend on local washing habits and rules, the standard quality of a product can be defined more conveniently by evaluation of the chemical and physical characteristics of the same.

Physical Characteristics

The physical characteristics of a detergent powder are factors determining performance and marketing appeal of the product. As an example, the physical characteristics of European laundry powders are summarized in Table 15.5.

TABLE 15.4
Formulation of Concentrated Detergent Powders

Component (%)	Concentrated Powders			
	Europe	United States	Far East	Japan
Surfactants				
Anionic	5–12	15–25	15–30	30–35
Nonionic	6–10	5–10	8–12	3–5
Builders				
plus cobuilders	25–40	40–60	20–30	40–45
Bleaches				
plus activators	18–25	0–5	5–15	5–10
Fillers	5–12	0–5	0–10	0–3
Additives	3–10	1–3	3–8	3–6

TABLE 15.5
Formulated Laundry Powders: Physical Characteristics and Use Indications

Physical Characteristics	Laundry Detergent Type		
		Automatic Washing Machine	
	Hand Washing	Regular	Compact
Bulk density (g/L)	280–450	350–550	700–900
Granulometry (microns)			
Ø > 1,000	3–4	5–8	3–6
500 < Ø < 1,000	45–48	38–42	45–47
250 < Ø < 500	35–40	48–52	38–43
Ø < 125	4–6	1–6	2–5
Dry residue (%)	75–85	80–85	75–80
pH (1-g/L solution)	8.5–9.5	8.5–10.5	9.5–10.5
Suspended solids (%)		25–29	20–27
Chemical oxygen demand (COD)		40–55	35–50
Dissolved organic carbon (DOC)		11–15	11–18
Biodegradability OCDE-ST		44–52	50–53
% total		> 90	>90
Washing temperature range (°C)	15–(30)–40	40–60–(90)	40–60
Recommended amount (mL/kg load)	150–250	260–450	120–130
Load size (kg)	0.5–1.0	1.0–4.5–5.0	2.0–2.2–2.5

There is a wide range of powdered products available worldwide, but they all fall into two main categories: traditional and concentrated. Traditional powders with standard quality (physical properties, chemical composition, and level of performances) are well established and widely accepted (for the specific duty) in the considered market area. The newer concentrated (compact) powders have higher bulk density and total active ingredient content and perform efficiently.

Industrial laundry powders, scouring powders, sanitizers, and other products tailored for specific uses have characteristics that don't fall into the framework of the traditional or concentrated classification.

Bulk density of a detergent powder, the most important physical property for consumers, is determined by particle porosity, bed porosity, particle sphericity, and particle size distribution.

All these properties depend on how the various components have been linked or built up by the production process. To obtain change in product bulk density, it is crucial to adopt a production path capable to control (and possibly to modify) the mechanism of particle formation.

The bulk density, as a result of chemical, physical, and mechanical properties, indicates how the process characteristics affect the obtainable product shape and properties.

Manufacturing Processes

Spray Drying

Spray-dried detergent powders represent the major part of the world detergent market and have certain advantages over the competing products made by different drying processes, but also show some limitations both in manufacturing and use.

The various sections constituting the spray-drying process are outlined in Fig. 15.3.

Manufacturing Process Description

The liquid components (received in drums or in bulk and then stored in tanks) are proportioned and mixed with the solid components (received in bags or in bulk in special containers and stored in silos) to form a homogeneous slurry.

Such slurry having different viscosity and concentration according to the specific formula is pumped at high pressure (up to 100 bar) and sprayed through special nozzles into a cylindrical tower (Fig. 15.4), where a stream of hot air is conveyed. In most cases the air stream flows countercurrently to the product in order to ensure high thermal efficiency and controlled drying.

The option of co-current drying is substantially limited to the different "drying profile," which results in more regular and resistant hollow beads originated by initial expansion and superficial drying when the slurry droplets meet high-temperature air in the top of the tower.

In this case, while continuing the downstream flow, the product drying proceeds in contact with lower air temperature. Cocurrent drying has lower heat efficiency and it is mainly used for drying high active thermosensitive products of low bulk density (200 g/L, approximately).

The dried product in form of hollow beads is collected at the bottom of the spray-drying tower and subsequently cooled and crystallized into the air-lift conveying system by means of a cold airstream. After the air-lift, the base powder is sieved, perfumed, eventually mixed with other thermosensitive components (such as perborate and enzymes) or minor additives (perfume or speckles), stored in silos or buggies, and finally conveyed to the packaging machines.

Raw Materials Handling

Various options of raw material handling/storage and distribution are available in the market and are chosen according to well-defined requirements depending on product formulation. Figure 15.5 illustrates a typical scheme of bulk solid raw materials handling for plants larger than 3 to 5 tons/h in capacity.

The chemicals used in detergent manufacturing are generally of very limited danger or environmental impact. Nevertheless, some, such as caustic, inorganic acids, concentrated fine chemicals, and enzymes, do require special safe handling and processing procedures as specified by the raw materials suppliers.

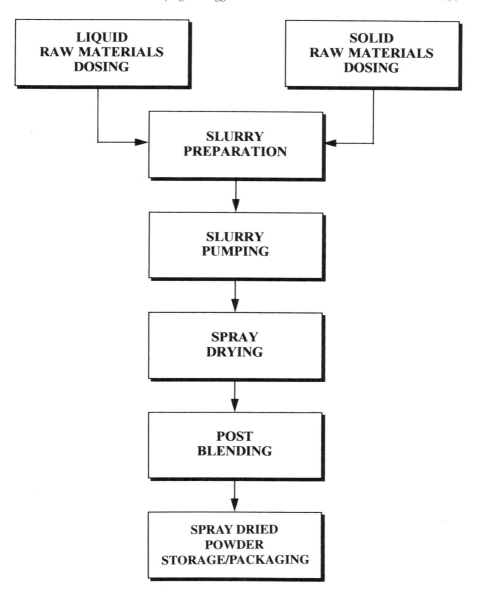

Fig. 15.3. Spray drying plant block diagram.

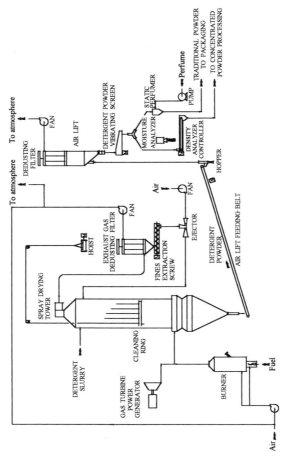

Fig. 15.4. Spray-drying plant.

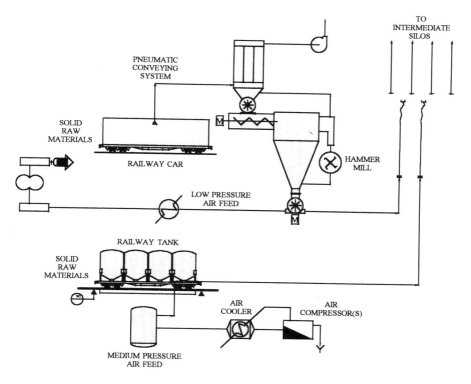

Fig. 15.5. Spray drying: solid raw materials handling.

Slurry Preparation

Continuous Systems

The Ballestra Dosex illustrated in Fig. 15.6 is a typical continuous slurry preparation system. Each single component is weighed in cycles ranging from 30 s to a few minutes, variable according to the required production rate.

The present technology based on load-cell scales and real net weight measure ensures better than 0.1% weighing accuracy for each single component.

For certain specific applications, the modern in-line mass flow measurement is also applied for additional process control with the advantage of in-line combined detection of density, concentration, and viscosity of some critical component.

The quantity of each component required by the product recipe is set and controlled by a computer program. The number of the proportioning units depends on the number of components to be dosed; in some cases small components, before their proportioning in these groups, are premixed in predetermined ratios and dosed as a single premix.

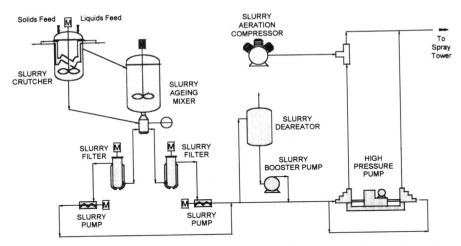

Fig. 15.6. Spray drying: Ballestra Dosex continuous slurry preparation and pumping system.

Continuous slurry preparation systems are used for medium (6–10 ton/h) or large (higher than 10 ton/h) production capacities and batch systems for small-size plants (maximum 4 ton/h).

Slurry Making

A continuous screw conveyor collects and premixes the preweighed solids before conveying them to the slurry crutcher.

The slurry crutcher also receives the liquids in a regular and uniform flow from a damper that collects the various feeds.

When the product formula includes sulph(on)ated anionics and soap, fatty acids and sulphonic acid are neutralized with the required alkali in a mixer before being fed to the slurry crutcher. In some cases, when no side reactions are expected from other components, the acids can be fed and neutralized directly into the slurry crutcher, which in this case must be constructed in corrosion-resistant 304-stainless steel. The slurry crutcher is a high-speed mixer designed for fine dispersion and perfect homogenizing of the mixture. Proper crutching operation also avoids agglomeration and formation of solid lumps, which could clog the lines feeding the spray nozzles.

From the crutcher, the slurry is transferred by overflow to an aging vessel where the mixture is further homogenized during a controlled residence time, which is set according the required hydration degree of inorganic salts such as soda ash, sodium sulfate, and sodium tripolyphosphate present in the formula.

Filtering and Pumping

The concentrated, viscous slurry is fed through magnetic filters followed by special self-cleaning filters designed to remove any possible solid particles that could damage or clog the pumps and the spray nozzles.

Positive-displacement booster pumps with single-acting plunges feed at the appropriate positive suction head, special high-pressure pumps. These pumps are designed to handle viscous and abrasive slurries at a pressure up to 100 bar and to keep it constant at the required preset value.

Batch System

The batch slurry preparation is recommended only for plants of low/medium capacity and may have a wide range of configuration (from manual to fully automatic). Automatic systems can also be supplied where the components are weighed automatically into the crutcher through intermediate hoppers.

After dosing and mixing in the special crutcher the slurry is transferred into an intermediate aging and storage vessel, from which it is continuously filtered and pumped to the spray tower.

The slurry preparation system is dimensioned to perform a complete batch, including the neutralization of sulfonic acid and/or fatty acids in a time of 20 to 30 min.

Spray Drying Nozzles Circuit

The slurry is pumped to a circuit installed in the upper part of the spray tower on which the spraying nozzles are connected with shutoff and service valves designed for local or remote control operation. The size and number of nozzles depend on the plant capacity and required product granulometry.

The nozzles are constructed with high-hardness material, as they must stand abrasive slurries with long operation times. The spray is generally of the hollow-cone type. Nozzle sizes vary in diameter from 3 to 5 mm with 50- to 70-deg spray angles.

Spray Tower

The special design of the hot air distribution chamber allows operation with high differential temperatures hot air inlet temperature up to 400°–500°C and exhaust air outlet temperature down to 85°–90°C with consequent optimum thermal efficiency. Spray towers are fitted with inspection doors, explosion-proof doors, and inspection holes for lighting and inspection.

Special cleaning devices, air brooms and blade rings, can also be used during the running of the plant to avoid product buildup on the tower walls.

Multicyclones and high-efficiency sleeve filters, placed on the air outlet line or on top of the tower, recover the fines, which are continuously collected and recycled into the tower.

All the operating conditions are automatically adjusted by setting the proper parameters of fuel consumption, air flow, slurry concentration, and all pressures/temperatures at the required optimum value.

Hot Air Generation

The characteristics of the powder beads are determined by direction, velocity, and temperature of the hot air stream.

The hot air is generated in furnaces designed to obtain a smokeless combustion. The most commonly available fuels, ranging from natural gas to heavy fuel oils, can be utilized without causing inconveniences or affecting the whiteness of the finished products.

The hot air is conveyed to the lower or upper part of the spray tower according to the selected circuit. Two fans, determining the pressure conditions of the air circuit, are installed for regulation of the airflow passing through the tower.

Cocurrent Drying Circuit

The hot air is conveyed to the top of the tower, where a distributor provides a vertical flow from top to the bottom of the tower.

In this manner, the particles to be sprayed by the nozzle and the drying air proceed in the same direction (cocurrently) to the bottom cone of the tower from where the product is discharged.

In the lower part of the tower, the powder separates from the air, which is sucked at the peripherical part of the chamber by the final fan toward the dedusting system.

The cocurrent operation is only adopted to obtain 200 g/L low bulk density products sensitive to the high temperatures (high-active products).

Countercurrent Drying Circuit

This system is used for the detergent powders with a bulk density in the range of 200–450 g/L.

Hot air is conveyed with a special distribution ring located at the bottom and from there up to the top of the tower coming into contact countercurrently with the slurry spray falling from the top.

Finished Product Handling

The detergent powder is discharged from the tower at a temperature of $60°–70°C$ and is conveyed on a belt to a continuous crystallization unit, the air lift. In the air lift the detergent is conveyed upward by a flow of ambient air that cools it down by completing drying and initiating surface crystallization.

This way the product is also lifted up to a height as to allow the next operations by a gentle gravity-discharge avoiding as much as possible any breaking of the hollow beads.

The transport air is sucked through a sleeve filter before being discharged to the atmosphere. The separated fines are reblown into the spray tower.

The detergent powder finally collected in the air-lift bottom cone is discharged into a sieve to remove any coarse agglomerated/wet material (usually 1%–2% of the total product) before perfuming, eventual post-addition and packaging. The coarse material recovered from the sieve is reprocessed via a separate dry or wet mixing/milling directly into the slurry-making section.

Exhaust-Air Treatment

A high-efficiency filter (in some cases installed downstream from a battery of cyclones) equipped with high-temperature resistance sleeves recovers the fines entrained by the exhaust air; fine particles are continuously collected and recycled in the spray tower, thus ensuring dust-free operation and avoiding manual handling.

The exhaust air is sent to the atmosphere with a final dust content lower than 5 mg/m^3.

Energy Recovery

To cope with ever-increasing energy costs, spray-drying plants are designed to increase their overall energy efficiency. There are four basic approaches to achieve this result.

The first involves the designing of the hot air inlet to the spray dryer to improve the heat efficiency inside the drier. Maximum temperatures have to be carefully considered to avoid product deterioration.

The second method involves the partial recovery of the heat that is normally exhausted to the atmosphere. The system works in the following manner: the air at the outlet of the tower at about 100°C flows through a separation system where the entrained powder particles are separated out to a high degree. The airstream is then divided into two separate streams. The first stream, about 50% of the total, is recycled into the tower through a special system in which it is mixed with fresh air coming from the atmosphere. The second stream is discharged to the atmosphere. The heat recovery is obtained by preheating the air going to the hot-air generation.

The third method increases the partial recovery of the heat that is normally exhausted to the atmosphere. The air at the outlet of the tower is exhausted with a fan, which sends it to the primary separation system. The air is then split into two separate streams. The first stream, as described before, is recycled to the tower. The second stream is sent to a special double-purpose exchanger/scrubber, which simultaneously recovers part of the heat and scrubs out the powder traces not trapped in the primary, wet scrubber powder-separation unit. The recovery of heat is obtained either by preheating the air going to the hot-air generator or by cycling water from the heat exchanger to the slurry preparation unit. The water sent to heat the exchanger accomplishes two purposes. It improves heat-exchange efficiency and traps any residual detergent powder particles that may still be present in the air to be exhausted to the atmosphere. The scrubbing liquid, in form of concentrated solution, is recycled in the slurry-making unit. The advantages of this system are as follows:

- Total heat recovery in the first airstream being directly recycled in the tower. The amount of the recycle flow is directly proportional to the air temperature set at tower inlet and to the outlet temperature from the hot air generator.
- Partial heat recovery in the second airstream not recycled in the tower.
- Scrubber action air pollution prevention.

- Recovery of powder not retained by primary separator for direct recycling to the tower.
- Heat recovery in the exchanger/scrubber.

The fourth approach is an interesting option for the reduction of energy costs in a large detergent spray-drying plant. The system is based on the use of a gas turbine for the cogeneration of electric power with complete heat recovery in the spray tower of the gas turbine exhaust gases. The option is of special interest for factories in which the overall electric load is high due to the presence of other process units, packaging lines, utilities, and so on. In this case the gas turbine will generate all the electric power necessary for the complete factory plus an eventual excess to be sold to the external network.

The process described refers to the basic state of the Art for spray-dried detergent powders, and it is still applied worldwide for the major part of powder products of low- and medium-bulk density.

As evidenced by the equipment description and operating parameters, the limits of this processing route are intrinsically connected to the energy requirements, both in terms of electric power and evaporation heat.

In recent years detergent manufacturers have been focusing on output optimization and energy consumption reduction in spray-drying operations by doing the following:

- Increasing slurry concentration (up to 75%, compared to the old levels of 60%–65%)
- Operating with higher air temperature (up to 500°–550°C, compared to the old 300°C) and installing heat recovery or cogeneration systems
- Applying, wherever feasible, post-addition or post-mixing of dry components

Each of these actions can increase output capacity by up to 100% and save energy up to 30% with respect to the traditional process.

Important limitations of the spray-drying process are due to its environmental impact. The use of special high-efficiency dedusting systems, scrubbers, and filters is required, but some problems, such as the removal of organic fumes from the tower exhaust, remain substantially unsolved. For this reason detergent manufacturers are reconsidering the option of operating spray dryers in a closed circuit (Fig. 15.7).

This application is rather new in detergent manufacturing and can cut by more than 50% the energy input per ton still running the same base plant.

Spray drying of detergent slurry is the most widely used process for the production of low to medium bulk density (250–450 g/L) powders, total active content up to 40% and moisture content in the range of 2% to 10%. These characteristics can be achieved by controlling the main operating parameters (spray pressure, slurry composition, and drying-air temperature) and require a total energy input in the range of 300–330 kW/ton, with a specific production yield of 0.5–1.0 ton/m^2 of tower section.

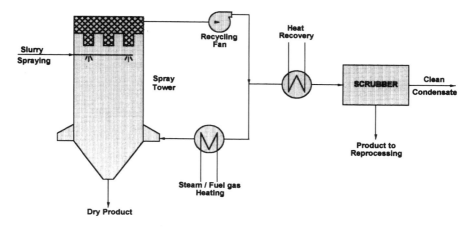

Fig. 15.7. Spray drying: superheated steam drying.

Fluid-Bed Drying

The principle of fluid-bed drying (Fig. 15.8) is based on the application of the fluidization technique to detergent components. These later are fluidized in a bed of solid particles where drying air is flowing upward. This airflow, passing through the interstices of the bed, produces a frictional resistance comparable to the weight of the bed; under these conditions, even a small increase of the airflow rate is enough to lift the solid particles and to keep them floating on a film of mixed solid and gas phase.

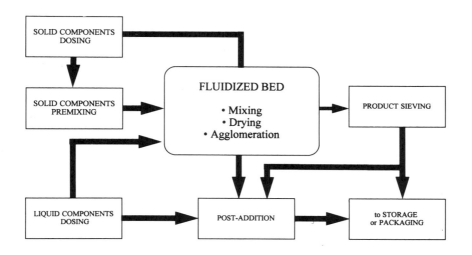

Fig. 15.8. Fluid bed drying: detergent powder production, process block diagram.

The heating of the airstream, in the fluid-bed drying of detergents, can be done before the entrance of the air in the fluid bed or directly inside it by means of heat exchangers directly in contact with the fluid.

The water to be evaporated in fluid-bed drying of detergents comes from two possible sources: water already contained in the solid to be fluidized and water sprayed on to the fluid bed (together with other components). In the first case, the incorporation of water has occurred on a previous step of mixing and agglomeration (this is separately described). The second possibility, combining drying with agglomeration in the same process step, is the most common.

When liquids are sprayed into the fluidized powder, agglomeration is started and at the same time the product undergoes a classifying effect. The agglomerated particles, rich in liquids, among which is water to be partially removed by evaporation, move to the bottom part of the fluidized bed, and along this migration they are dried, because of the contact with upward-flowing hot air, without interruption of the fluidization. The unagglomerated particles, due to their low bulk density, are carried to the upper part of the fluidized bed, where the contact with the finely sprayed liquid is enhanced, and therefore agglomeration is ensured.

Fluidization depends on many factors that in turn depend on characteristics and demand of the product to be processed. To ensure thermal efficiency and no side effects, the operating conditions in fluid-bed drying should be accurately selected due to the strong dependence of the heat transfer coefficient on the gas velocity.

For the whole fluid-bed detergent drying, the various operations and equipment involved can be outlined as follows.

Ingredients Dosing

Each solid component is individually dosed with dosing belts or loss-in-weight devices. The dosed streams are then premixed and conveyed to the fluid-bed inlet duct by premixing screws or bucket elevators to ensure constant and uniform feed of the powdered ingredients. All the dosing and transport equipment have to be connected to the dedusting net.

The liquid components are dosed with piston or positive displacement gear or lobe-type pumps and sprayed onto the fluid bed with double-fluid-type nozzles, which ensure uniform distribution of the liquid droplets to facilitate their contact with the fluidized solids. (Fig. 15.9)

Drying and Agglomeration

As already indicated, drying and agglomeration take place in the fluid-bed dryer/cooler fed with a stream of powderized components and sprayed streams of liquids. The fluidizing air provides for mixing, fluidization, agglomeration, and drying. Although many types of fluid-bed dryers are commercially available, for detergent production the horizontal and vibrating type are the most widely used (Fig. 15.10). The profile of a fluid-bed dryer shows the importance of the air

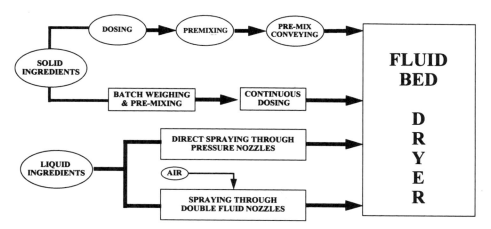

Fig. 15.9. Fluid-bed dryer: solid and liquid dosing.

distribution grid in optimizing heat transfer rate, product particle size, and product entrainment in exhaust air.

There are many types of air distribution plates, but for detergent processes the perforated plates, with variable hole diameter and superficial density, are the most widely used.

The design of the air distributor plate also includes a variable overflow device to regulate the fluidized-bed depth depending on the process demand; usually the

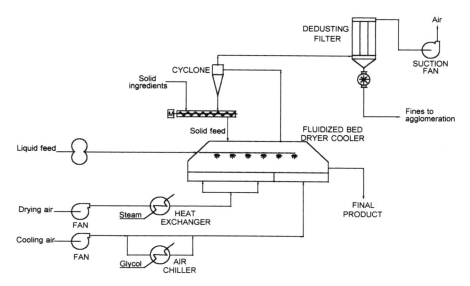

Fig. 15.10. Fluid-bed drying plant.

detergent products are treated in a fluid-bed dryer for not more than 5–10 min and the bed thickness is kept in the range of 5–20 cm.

Exhaust Gas Treatment

A fluid-bed dryer must separate the amount of powder entrained by the exhaust air leaving the contact zone. A disengagement chamber is designed to take into account the correlation between itself and the superficial velocity and the particle entrainment rate. The disengagement chamber provides for the first recovery of the entrained particles, mainly fines, overdried, or not-granulated product, while further exhaust air treatment is performed with cyclones and final sleeve filter. While the cyclones recover the product in the form of fines, the filter ensures the degree of cleaning necessary to allow exhaust air discharge to the atmosphere.

The recovered product from the cyclones is directly collected and conveyed to the fluid bed together with the feed of solid component, while the fines from the filter are usually collected separately before being fed again to the fluid bed.

Fluid-bed drying finds application mainly for powders of medium density, bulk density, and low to medium active ingredient content.

Agglomeration Equipment

Agglomeration can be described as the physical buildup of powdered and liquid components into a granular and chemically homogeneous product.

The Ballestra Non-Tower processing steps for the production of detergent powder based on agglomeration is outlined in Fig. 15.11. Among these various processing steps, agglomeration represents the most important and critical operation, as it is related to the physical structure and, at the same time, to the chemical composition of the product.

The design of the ideal agglomerator for a given non-tower detergent powder production process should be based on the following:

- Short/medium residence time of the product inside the agglomerator
- Optimized and controllable energy input
- Optimized and modifiable liquid distribution system
- Provision to minimize/avoid the product buildup on the inner wall
- Mechanical reliability
- Operation flexibility
- Reduced cleaning and maintenance requirements

The basic design of multipurpose and commercially available mixers is not sufficient to ensure a full and satisfactory application in the specific field of detergent powder production. Consequently it is worth considering the comparative description of those mixers for which the application in the detergent industry is widely proven.

These mixers differ not only in mechanical design but also in the way they apply energy to the agglomeration process.

Both horizontal and vertical mixers can furnish the necessary mechanical energy to form the granule and to carry it out agglomeration, provided they have the appropriate distribution systems for the liquid and powder components. The handling of the granules in the constitution phase (agglomeration and neutralization of acid components) is extremely critical inside the agglomerator. An excess of energy can produce negative results such as overgranulation or excessive temperature rise with loss of desired physical characteristics such as flowability and bulk density. Some of the more commonly used types of agglomerators are illustrated in Figs. 15.12, 15.13, 15.14, and 15.15.

Considering, for example, the Kettemix reactor (Fig. 15.12), one can view schematically the important phases of the agglomeration process that takes place in the machine. In phase A the powder components are added in a continuous, controlled fashion to the agglomerator, which renders the powder highly turbulent through the use of a propeller. In phase B the liquid components are added and placed in contact, without any decrease in stirring, with the previously energized powder. In these phases the following phenomena take place:

- Initial formation of the homogeneous product granule by mixing of the solid and liquid components
- Absorption of liquid components on the solid particles with subsequent penetration in the particle cavities themselves
- Dry neutralization of the acids present (mainly sulfonic and carboxylic acids) with sodium carbonate to form the desired amount of active anionic components

In phase C, thanks to the energy input provided by the combined action of the agitator blades and the internal walls of the machine, agglomeration is completed and the product is discharged and made available for the successive phases of conditioning.

Concerning the energy input, 50% of the total energy is expanded during powder addition (phase A), 35% when liquids are added and react with the powders (phase B), and the remaining 15% in the final agglomeration stage (phase C).

The liquid and solid components determine for their agglomeration a specific energy requirement, which the various machines furnish in qualitatively different ways.

In parallel to these considerations on the working principle of the various machines, it is important to emphasize how each step is integrated in the overall process of producing powders by the non-tower process. The sequence of operations provides for the following:

- Dosing of individual solid components and their premixing
- Dosing of individual liquid components
- Granulation
- Product conditioning (partial drying and final cooling)
- Product sieving and coarse milling

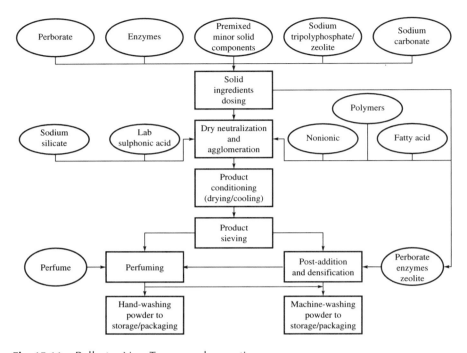

Fig. 15.11. Ballestra Non-Tower agglomeration process.

- Post-addition of thermosensitive components and coating
- Product storage/packaging

Continuous dosing of liquid and solid components requires equipment that can ensure accurate and constant flows. Figures 15.16 and 15.17 illustrate two solids and liquids dosing methods. The residence time in continuous agglomerators can range from a few seconds to minutes.

For the process technology based on fluid-bed drying, the conditioning of a granulated product is generally carried out in a horizontal vibrating or static fluid bed dryer. The first stage is used to heat the product with hot air up to 100°–160°C, and then the second stage cools the partially dried product using cool air from 5° to 20°C.

This double treatment ensures the partial elimination of water introduced in the granulation step as dilution water from such ingredients as silicates, polymers, and binders to other solid components.

This partial evaporation together with the successive cooling ensures the product's free-flowing characteristics and the homogeneity and mechanical strength of single granules. Figure 15.18 summarizes the operating conditions for fluid-bed product conditioning in a non-tower agglomeration system.

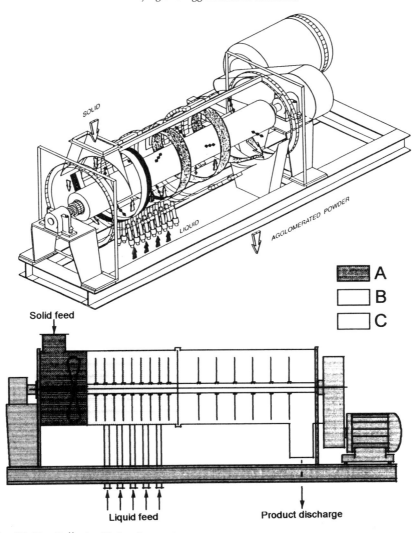

Fig. 15.12. Ballestra Kettemix reactor.

The processes of agglomeration and successive conditioning also form granules with undesirable dimensions (coarse and fines) that are selectively recovered and reintegrated in the process, as illustrated in Figs. 15.19 and 15.20.

The coarse granules formed due to overagglomeration are first sieved and then chopped up. Then they are mixed with the same solid components in the formulation, and they are added back into the agglomerator. The fines entrained in the flow of exhaust air flowing out of the fluid-bed dryer/conditioner are separated

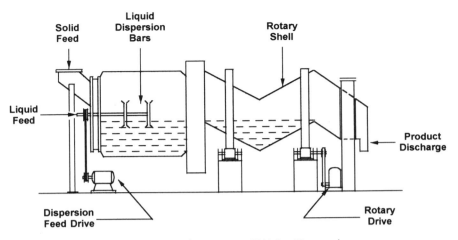

Fig. 15.13.   Zig-zag continuous agglomerator—PK Niro/Denmark.

by sleeve filters and then added back continuously to the stream of solids at the entrance of the agglomerator.

The completion of granulation, in terms of balance of the various components and of the physical properties, is obtained in a post-addition/agglomeration step. Sodium silicate, nonionics, antifoaming agents, and other liquids are added to the formula together with thermosensitive components such as perborate, percarbonate, TAED, enzymes, and coating agents (zeolites). This operation, performed in low-velocity mixers with limited energy input, completes the formula and improves the finished product physical properties with respect to narrow granule size distribution, greater free-flowing characteristics, and increase of product bulk density.

Figures 15.21 and 15.22 illustrate the variations of important characteristics, such as granulometry, bulk density, and flowability in the different phases of the agglomeration process for European compact products.

The amount of anionic surfactant introduced into the formula by neutralization during the agglomeration is limited somewhat by the thermodynamics (heat evolved by neutralization) and by the heat transformed from the supplied mechanical energy of the equipment. In any case, it is possible to produce formulations with active content (mainly anionic) higher than here indicated by the introduction (total or partial) of the same surfactants not generated by dry neutralization but by steps such as spray-drying, mixing and kneading, and active matter drying and powderizing. The use of spray drying in agglomeration is covered later separately; the description for the other two processes follows.

Mixing and Kneading

The overall processing steps for kneading/mixing subsequent milling/sieving and surface modification are shown in Fig. 15.23 and it is only used in concentrated

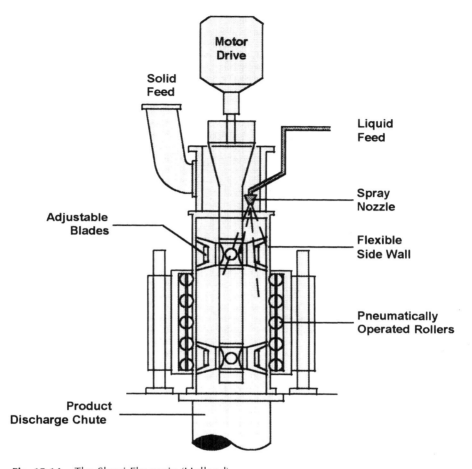

Fig. 15.14. The Shugi Flexomix (Holland).

detergent granules recently introduced on the European market. The preliminary deformable mixture of the various washing ingredients is worked into a homogeneous mass and extruded at pressures in the range from 20 to 200 bar through perforated molds with holes with a width equal to the granule size. The compacted, extruded product is cut to the desired granule size by rotating knives. The still deformable granules will be further treated in, for example, a Marumerizer (trade name, Japan), to improve the roundness of the extruded product. The anionic surfactants in the form of a concentrated paste are mixed with other main solid ingredients in a kneader and subsequently pulverized. This physical treatment results in substantial modifications to the surface and the structure of final product granules. The presence of high levels of surfactant acts as a plasticizer for the mechanical

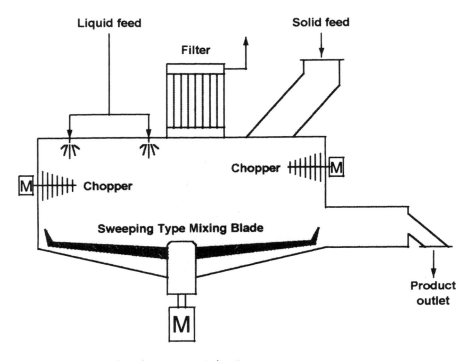

Fig. 15.15. Vertical agglomerator—Fukae/Japan.

kneading of the mass. This effect helps to compact and to shatter the mass in a way that the addition of a large quantity of water or other liquids is not necessary to act as a binder for the solid components. This non-tower technology does not contain agglomeration as a main step, and it is used only in specific and limited cases.

Ballestra Active Matter Drying Plant

The possibility of introducing anionic surfactants in the form of dry powder into the agglomeration process is important when the formulation requires active levels higher than can be obtained by other processes. An efficient and versatile system for the production of dry, free-flowing anionic surfactants (especially oleofinsulfonates and fatty alcohol sulfates) is based on the thin film drying of concentrated pastes followed by flaking and milling. The Ballestra active matter drying plant with a wiped film evaporator is shown in Fig. 15.24.

The paste-like surfactant, coming from a storage tank or directly from the neutralization unit, is preheated to a temperature of 80°–85°C and fed to the top of the wiped film evaporator. This film evaporator is equipped with a multistep heating jacket and connected with a condenser; a slight vacuum provided by a fan, mainly to convey the vapor to the condenser; at the bottom of the vertically installed

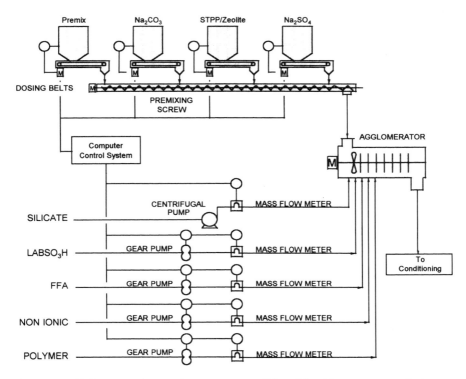

Fig. 15.16. Ballestra agglomeration process—solid and liquid components dosing.

wiped film evaporator the dried product is cooled and mechanically transformed into flakes or noodles before reaching the storage hopper.

Even when the use of the two described systems to increase the active matter content is not foreseen, the agglomeration process is considered the most suited method to produce powders of high bulk density (700–900 g/L) and high active content (20%–35%).

Spray Drying and Agglomeration

Spray drying combined with agglomeration is a very convenient and suitable process route for the production of detergent powders having high content of active ingredients and high bulk density. The coupling of the two processes ensures the listed advantages in terms of product characteristics and economy of operation:

- *Production characteristics benefits:* increased flowability, product solubility, dispersion in water, bulk density, average beads diameter, and product effectiveness

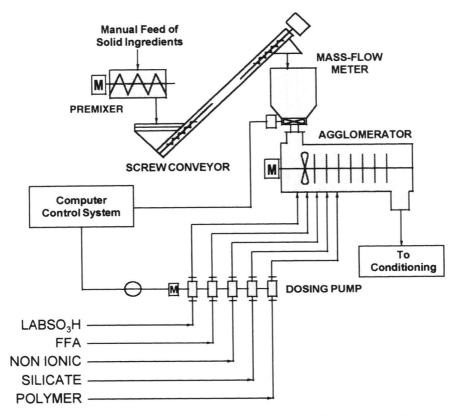

Fig. 15.17. Ballestra agglomeration process—solid and liquid components dosing.

- *Plant operation benefits:* increased production capacity and operating flexibility, integration of the features of the two basic processing systems, and reduction of energy consumption

In terms of process duty the spray drying is used to generate almost all the amount of the anionic surfactant to be incorporated in the formulation, and the subsequent step of agglomeration is required for nonionic surfactant addition and for physical modification of the final product (increase of bulk density and flowability).

The dosing system for the ingredients in the combined process is basically the same as adopted for spray drying and for agglomeration. The base powder coming from spray towers represents an important amount of the global product, and its handling and dosing are not minor factors affecting the plant operation and the final product characteristics. As the organic content of the base-powder (mainly anionic surfactants) is beyond 30%, handling is made easier by low product temperature. Therefore the processing conditions in the spray-drying section should be selected

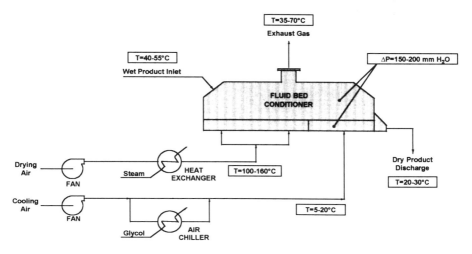

Fig. 15.18. Agglomeration fluid-bed conditioning: operating parameters.

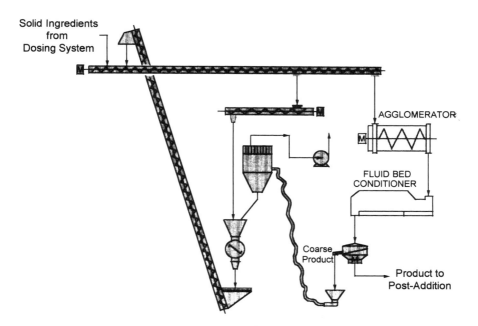

Fig. 15.19. Agglomeration process: coarse product milling and recovery.

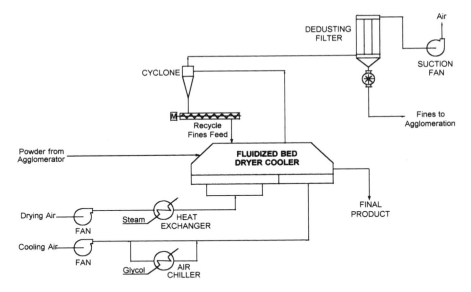

Fig. 15.20. Fines recovery from fluid bed.

for the more critical conditions necessary to dry and handle powder with high anionic surfactant content.

In practice, the combined process is operated periodically to comply with the product mix and to the formulation changeover as required by the market; therefore the intermediate storage of base powder rich in anionic surfactant is very frequent

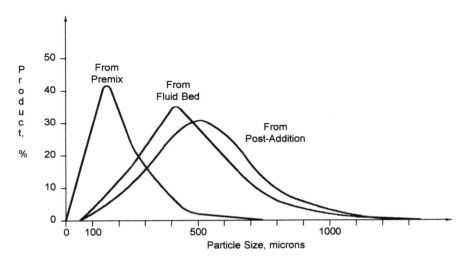

Fig. 15.21. Agglomeration: particle size distribution.

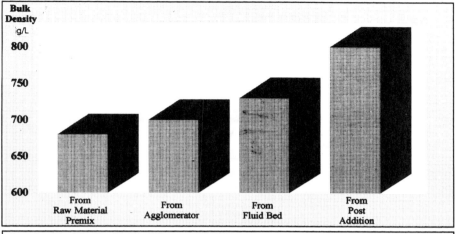

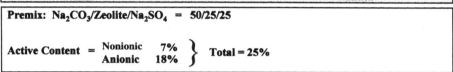

Fig. 15.22. Agglomeration: increase of bulk density.

and accepted by the producers. In this case, the base powder from the air lift is conveyed to the intermediate storage silos, from where it is transported to the feeding silos of the dosing device (either dosing belt or loss-in-weight" system) to the agglomeration step. This system allows the operation of the combined process alternatively to the traditional spray-drying plant; the operation flexibility will be according to the size of the storage silos, which are designed to store at least 8 to 10 h of production capacity of spray-dried base powder. The intermediate storage is a must when batch agglomeration is foreseen as the second step of the combined process.

The direct use of base powder coming from the spray tower involves the step of additional product conditioning. In this case, the combined use of relatively hot (40°–60°C) base powder and horizontal high-shear mixer requires a product cooling/conditioning before feeding the product to the final step of agglomeration. The combined spray-drying and agglomeration process is outlined in Fig. 15.25.

Spray Drying

Base powder rich in anionic surfactant content is produced by adopting slightly modified conditions:

- Detergent slurry concentration: 65%–70% dry matter
- Anionic active content in slurry: 20%–25%
- Hot air inlet temperature: 250°–350°C

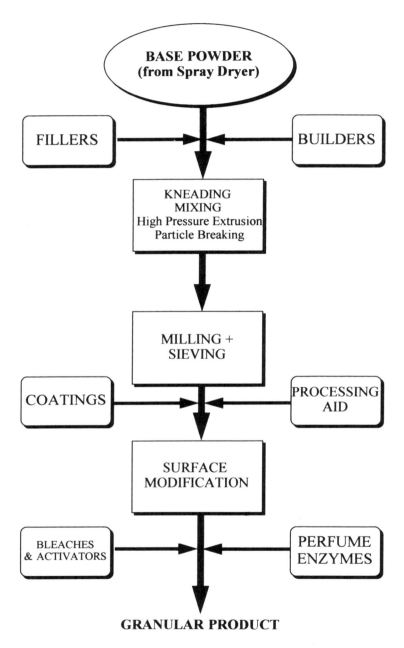

Fig. 15.23. Block diagram.

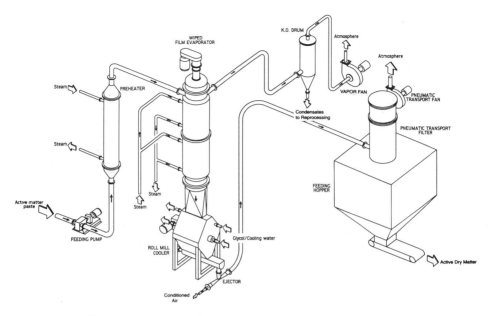

Fig. 15.24. Ballestra active matter drying plant.

- Spraying pressure: 20–40 bar
- Average air velocity through the tower: 0.3–0.6 m/s
- Exhaust air temperature: 90°–95°C

Agglomeration (First Step)

The base powder is fed together with other solid components to a high-shear mixer with characteristics already described for the non-tower agglomeration process, where liquid ingredients (such as nonionic solution, polymer solution, or sodium silicate solution) are dosed. The solid and liquid components are agglomerated, and the formulation is almost completed for the surfactants portion.

Granulation and Coating

The agglomerated product from the first step is further processed in a low-shear mixer where minor quantities of solids, mainly zeolite, are added as coating agent.

The combination of the mechanical work supplied by the mixer and the addition of solid, and sometimes liquid, components results in a granulation effect of the product. In other words, the particles are densified and made more regular, while the further coating with solids ensures free-flowness due to the noncontact of sticky particles.

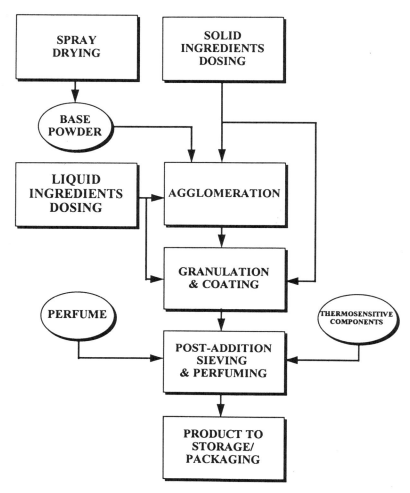

Fig. 15.25. Spray drying and agglomeration.

The final step of addition of thermosensitive components such as perborate, bleach activators and enzymes, perfuming, and sieving is usually performed in rotary blenders of the same type used in traditional spray-drying plants.

Whereas the choice of the first agglomerator is relatively wide (as described already for the non-tower process), the step of coating-densifying and final post-addition must be performed with mixers having particular characteristics and offering several advantages, such as the following:

- Increase of the ratio of post-added components versus the base spray-dried powder (with consequent increase of overall production capacity)

- Increase of the total achievable level of active matters (range and types) in the finished products
- Possibility to control final bulk density by changing ratio of product from tower and post-added component, mixer speed, and sequence of addition of the components

The key benefits of the combined spray-drying and agglomeration process are as follows:

- The possibility of increasing the nonionics percentage in the formula
- The modification of physical characteristics of the beads (mainly density and average dimensions)
- Product concept (process flexibility)
- Product effectiveness (high active concentrated product)

Table 15.6 presents the combination of a product consisting of a spray-dried base powder containing the water-rich anionic surfactants and some inorganic ingredients. This base powder is subsequently further processed in the agglomeration process (three steps: agglomeration, densification, and final blending) where nonionics and further ingredients such as zeolite, perborate, and perfume are added, resulting in a final product with a powder bulk density of approximately 850 g/L.

The average increase in bulk density through the spray drying and agglomeration process is as follows:

- From the spray tower: approximately 350 g/L
- From the agglomerator: approximately 600 g/L
- From the coating and densification step: approximately 800 g/L
- From the final blender: approximately 850 g/L

The increases in output are, respectively, 50%, 75%, 80%, and 100%.

The possibility of producing a wide range of detergent powders richer in active ingredients (potentially more efficient than traditional products) at reduced energy demand might be considered a determinant contribution in defining a process as "industrially advantageous."

The combined process fulfills the scope and offers at the same time the possibility of increasing the degree of utilization (and in turn of profitability) of existing and already depreciated spray towers with limited investment cost.

TABLE 15.6
Spray Drying and Agglomeration

Drying System	Power Input (% of Total)	Nonionic Quantity Range (%)	Powder Bulk Density (g/L)
Spray drying	60–70	0–4	250–500
Agglomeration, Densification, and Final Blending	30–40	5–15	600–800

Spray-drying combined with agglomeration produces high bulk density detergents (650–800 g/L) of high active content (25%–40%) and moisture content of 5%–10%.

Dry Mixing

The ancestors of the commercial detergent powders were produced by simple mixing of base ingredients, which were empirically formulated by dosing and proportioning components whose effects in the household cleaning and laundry were traditionally well known.

The manufacturing of detergents by dry mixing of solid ingredients is sometimes completed with the addition of limited amounts of organic/inorganic acids involving some "substantially dry" neutralization or "saponification" reaction and by adding minor quantities of liquids (such as silicates, nonionics, or perfumes) in order to improve product quality in terms of both composition and consumer appeal. Toward this aim, the basic equipment required is to be chosen among the various solid/solid and solid/liquid mixers described in the literature and available to the industry.

A number of mixers are listed here, along with their basic operating principles, key mechanical features, and applications.

Rotating Drum Mixers

This family of mixers includes all type of mixers based on the principle of partial filling of solid components into a vessel of different geometrical shape and rotating the entire body to achieve redistribution of the content by displacement and gravity fall with no or limited internal friction. They include the following:

- *Twin-shell V blenders (drum axis horizontal).* An example of this type is the PK (Patterson Kelley–Niro) mixer.
- *Large diameter/length ratio mixers with blades fixed to the drum.* These are the concrete mixer types with generally limited mixing range, segregation, poor cleanability, and long cycle times.
- *Small diameter/length ratio mixers.* This type of mixer is a special form of concrete mixers.

Volume size may vary from 0.1 to 10 m^3 and they may include internal baffles, screw feeders, bottom discharge, or filling discharge. They have the following features and advantages:

- Strong construction
- Simple operation
- Relatively low cost
- Good mixing
- Limited product buildup
- No or limited internal friction

Limits of the application and disadvantages are as follows:

- Requirement of special dedusting hoppers and/or seals during charge/discharge
- Difficult scale-up design for specific purpose from pilot size equipment
- Risk of lump formation in case of addition of liquids

The mentioned mixers are available for the powder detergent manufacturing in sizes up to 2–3 m³ of geometrical volume. The working volume is about 50% of the total or about 1 ton per batch assuming average product bulk density of 700–900 g/L.

A batch cycle, including (manual) charge of the components and discharge to product hopper, is usually completed within 20 to 30 min. This corresponds to a capacity of 2 to 4 tons/h. Energy consumption is in the range of 10–20 kW per ton of product.

Fixed-Body Mixers with Internal Rotating Shaft, the So-Called Agitator-Type Mixers

- *Cylinder with agitator arms at base axis vertical, as with the Fukae mixer* (Fukae is the name of the Japanese company manufacturing this type of mixer). In this high-speed mixer the (spray-dried) powder is initially broken down to a fine state of division; the surface-improving agent and binder are then added and the pulverized material granulated to form a final product of high bulk density. This version of the Fukae mixer process is essentially a batch process.
- *Cone with rotating screw and imposed orbital motion, axis vertical, as with the well-known Nauta-Hosokawa mixer* (Nauta-Hosokawa, the Netherlands/Japan).
- *Single-trough ribbon mixer, axis horizontal,* such as *Gardner.* The name Gardner was originally a brand name but became a general name for a ribbon mixer, and is manufactured by many mixer companies.
- *Double trough, double screw, axis horizontal.*
- *Cylindrical with rotating plows, axis horizontal.* An example is the Lödige KM, also referred to as the Lödige Ploughshare Mixer.

The unit essentially consists of a horizontal hollow static cylinder having a rotating shaft in the middle. On this shaft various plow-shaped blades are mounted. When these plow-shaped blades are rotated, horizontally upward propulsion and centrifugal force are imparted to the granules, which rotate on the surface of the inner wall of the cylindrical body of the mixer. The granules roll toward the center, and agitation is reported by centrifugal force. The shaft can be rotated at a speed of 140–160 rpm. Optionally, one or more high-speed cutters can be used to prevent excessive agglomeration, or these cutters can be used for the rapid dispersion of minor and liquid ingredients. Turning at high rpm (3,600) and mounted perpendicular on the main agitator, these so-called choppers perform a valuable function toward enhanced mixing and control of particle size and particle size distribution. The Lödige KM is totally enclosed and can be easily connected to a central dedusting system. Relatively high levels of dry-neutralized anionic active (alkylbenzene sulfonic acid with soda in powder form) can be incorporated to levels higher than 10% in the final product. There is total product discharge at the end of the batch cycle. (See Fig. 15.26.)

Limitations of Horizontal Shaft Mixers

- Relatively higher peripherical friction with high power requirements (up to 20–30 kW/ton)
- Limited heat removal efficiency generated by the power/friction and/or by dry neutralization reactions
- Relatively high ratio of equipment cost versus capacity (also depending on number/type of added options)

This family of mixers is constructed in dimensions ranging from size 20–50 Kg/batch pilot to 2–3 tons/batch production size units.

When choosing this type of mixer for its advantage in product quality and higher formulation flexibility, automate the batch feeding, as outlined in Fig. 15.27. An industrial system, based on high-efficiency mixers, can operate for a higher number of batches (up to 10–12 batches per hour, each batch with maximum capacity of 300–400 kg) for a total plant capacity of 4–5 tons/h.

In conclusion, dry mixing is the cheapest and simplest method to produce various types of products for industrial applications and is based on ingredients that do not require sophisticated processing. The application is limited by the reduced range of formulations and by the need of special predried raw materials (mainly containing anionic surfactants) to achieve the required product characteristics without further conditioning.

Detergent powders produced by the dry mixing process are characterized by low 1% to 5% surfactant content, bulk density in the range of 400–700 g/L, and moisture content lower than 5%.

Choosing the Right Manufacturing System

The selection of a detergent powder manufacturing system involves technical and economical considerations. The selection must take into consideration these basic factors:

- Type of powder to be produced
- Quality and grade availability of the key components
- Required hourly production capacity
- Product changeover frequency
- Local environmental regulations

The environmental regulations must be carefully considered as they might exclude a process that does not comply with them, although already assessed as "ideal" with respect to the other parameters.

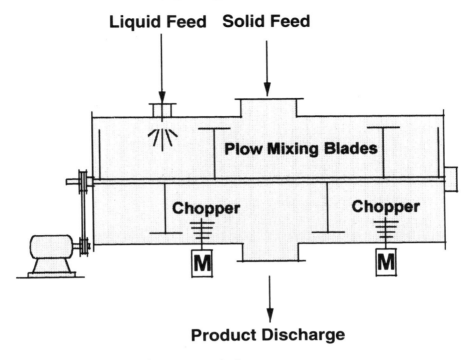

Fig. 15.26. Dry mixing: plow mixer with choppers.

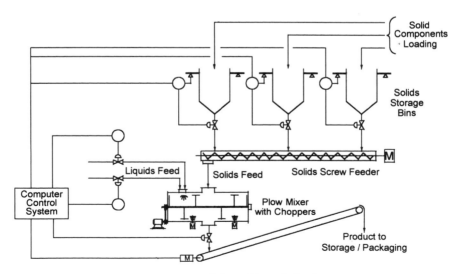

Fig. 15.27. Dry mixing: automatic solids and liquids feeding system.

Upon definition of these factors and evaluation of their impact on the project feasibility, it is possible to base the further step of process selection on parameters linked to this order of factors:

Technical Factors

- Range of active ingredients to be used
- Content of active ingredients to be incorporated
- Range and types of builders
- Range and types of specific additives
- Range and types of fillers
- Product bulk density (range and limits)
- Product granulometry (range and limits)
- Formulation flexibility

In Table 15.7 a comparative evaluation is presented at the various processes described in this text and the attainable product characteristics. A few examples of how this table should be read are given: In the spray-drying processes, for example, the range of active detergents that can easily be incorporated can be indicated with the statement medium, because nonionics may present serious problems in terms of high slurry viscosity, separation tendency before spray drying, and a blue haze in the stack of the spray tower. The level of active ingredients, assuming exclusively anionics such as alkylbenzene sulfonate, is not a problem (rating high). The product bulk density in spray drying is limited on the upper range to a maximum of about 600 g/L; the lower part of the range can be manipulated by atomizing pressure, nozzle type, orifice diameter, and notably with forced slurry (de)aeration in the pipeline system between the slurry-holding vessel and nozzle lances (rating medium). Product granulometry can be manipulated by varying the nozzle orifice diameter and spray-drying pressure (rating high). Finally, many types of formulation can be spray dried, and therefore the formulation flexibility can be rated high. The reader can now follow the various ratings for processes other than spray drying. In addition, for the agglomeration medium ratings are presented for the content of active ingredients and for the range of product granulometry. The incorporation of anionic pastes (alkylbenzene sulfonate active paste containing 40% water) may present some problems in high anionic formulations and the granulometry rating (medium) because its range and limits are strongly dependent on solid and liquid ingredients (quantity and ratio S/L and type of S and L). Coarse granules levels are expected to be higher when the ratio S/L decreases and vice versa; fines will increase as formulations contain small amounts of liquids. In spray drying and agglomeration the bulk density is rated as medium because a relevant part of the product consists of spray-dried powder having a low/medium bulk density.

TABLE 15.7
Detergent Manufacturing Systems: Obtainable Product Characteristics Comparison

Evaluation Parameter	Spray Drying			FB Drying			Agglomeration			Spray Drying and Agglomeration			Dry Mixing		
	High	Medium	Low	High	Medium	Low	High	Medium	Low	High	Medium	Low	High	Medium	Low
Range of active ingredients		1			1		2			2				1	
Active ingredients contents	2					3			1	2					3
Product bulk density (range and limits)		1			1		2				1			1	
Product granulometry (range and limits)	2				1				1		1			1	
Formulation flexibility	2					3	2			2					3

TABLE 15.8
Detergent Manufacturing Systems: Cost Comparison

	Drying System														
	Spray Drying			FB Drying			Agglomeration			Spray Drying and Agglomeration			Dry Mixing		
Cost Relevant To:	High	Medium	Low	High	Medium	Low	High	Medium	Low	High	Medium	Low	High	Medium	Low
Investment	1				2			2		1					3
Energy consumption	1				2			2			2				3
Building	1				2			2		1				2	
Off-sites		2				3		2			2				3
Ingredients			3	1					3			3	1		
Operation	1				2			2		1					3
Environmental impact	1				2			2		1					3

Economic Factors

The economic aspects of the various manufacturing routes are presented in Table 15.8. It may not come as a surprise that spray drying proper or processes containing a spray-drying component rate high on investment, energy consumption, building cost, operational costs, and environmental impact (blue haze from spray-tower stack). An important virtue is the low ingredient costs as water containing raw materials can be handled and processed.

In conclusion, for traditional fabric washing powders with bulk densities in the range of 350–600 g/L, spray drying remains the recommended process route. For companies producing mainly light-duty fabric washing products, fluid-bed granulation/drying may be recommended. For concentrated heavy-duty fabric washing powders, agglomeration/granulation and the Combex (SPD + AS) processes are recommended. Existing spray-tower operations can be complemented with an agglomeration/granulation plant to produce modern concentrated powders for the heavy-duty fabric washing market. The cheapest recommended route for industrial detergent powders, characterized by a large number of small-volume products, is the dry-mixing process.

Acknowledgment

The authors wish to express their thanks to Luis Spitz for his many comments and suggestions for improving the text and the illustrations of this contribution. Without the extensive help and perseverance of Mr. Spitz it would have been difficult to finish this job.

References

1. Ballestra S.p.A., brochures concerning manufacture of detergent powders via the spray-drying route (1990–1995).
2. Fluid-bed processing brochure, Niro Atomizer.
3. Fluid-bed technology, brochure, Ventilex, Heerde, Holland.
4. Moretti G.F., I. Adami, and F. Nava, High Purity Concentrated Anionic Surfactants from Improved Sulphonation and Vacuum Neutralization Technology, *Proceedings of the Third World Conference on Detergents,* Montreux, 1993.
5. Zig-zag continuous mixing, PK-Niro brochure, 1991; Patterson Kelley—USA, Niro Denmark joint venture.
6. De wilde, T., *The Cost-Effective and Proven Shugi Technology for High Density Detergent Manufacture,* Hosokawa Shugi, Lelystad, The Netherlands, 1992.
7. Range of production plow mixers, Gebrüder Lödige Maschinen Bau GmbH, Paderborn, FRG, 1990.
8. Herman de Groot, W., I. Adami, and G.F. Moretti, *The Manufacture of Modern Detergent Powders,* Herman de Groot Academic Publisher, Holland, 1995.

Chapter 16

Packaging of Soaps and Detergents

Luis Spitz[a] and John W. Heath[b]

[a]L. Spitz, Inc., Skokie, Illinois, U.S.A.

[b]ACMA/GD, Berkshire, England

Packaging of Soaps and Detergents

In the past, the main function of packaging of most household products was to serve as a convenient method in which to carry the products home and store them until used.

In the late 19th century, soap flakes and soap powders were introduced by all major soap manufacturers; they were packaged in vertical rectangular cartons. In 1906, Pneumatic Scale Company sold the first automatic carton-filling machine to fill cartons with soap powder manufactured by the Detroit Soap Co. Shortly after, in 1908, they completed their first foreign order to Lever Bros. in England, thus introducing automatic packaging of soap powder to Europe.

The first automatic bar soap wrapping machine was introduced in 1914 by the Package Machinery Company in the United States. The model N-1 unit wrapped Colgate's Octagon laundry soap at a speed of 75 bars/min.

Today's discriminating and demanding customers of soaps and detergents do not make the decision to purchase a particular brand of product based upon price alone. Product performance, safety, dispensing convenience, and attractive packaging all play an important role in the buyers preference for a particular household product. The awareness and concern of the environmental impact of both the product and the package are also an increasingly important factors for all customers today.

During the 1990s, there have been more new and improved products, packages and packaging systems, in all household categories, than ever before. There is intense research and development activity to reduce the amount of packaging and to increase the use of recycled materials, thus reducing the amount of waste and saving scarce resources.

As we approach the end of the 20th century, we can safely predict that there will be faster, more frequent changes in our industry. Many of the packaging machinery companies will meet this challenge with many creative innovations.

This chapter, with the aid of many illustrations, offers an overview of the most important and widely used packages and packaging machinery for bar soaps and powdered detergents.

Soap Packaging

Soap Shapes

Soap bars can be stamped into many different shapes but there are only two main categories—*banded* and *bandless*. Each category can be grouped into the following four basic shapes: *rectangular, oval, round* and *irregular.*

Classification of Soap Packages

Mass market and specialty (novelty) soaps are packaged in various distinct *single* and *multipack* styles. The most popular and widely used single packs are either *wrapped* or *cartoned* bars but there are several other types including overwrapped carton, wrapped and cartoned, fin sealed and cartoned, and stretch wrapped and labeled.

Packaging of individually wrapped and cartoned single packs into *banded* and *bundled multipacks* has become the most commonly offered package style worldwide. The most widely used market *multipacks* are the following: *banded (U-banded) wrapped and cartoned* bars; *bundle wrapped and cartoned* bars; *fin sealed*; *overwrapped carton*; *shrink wrapped*; and *wrapped and/or cartoned in a tray with a lid.*

There are only three *specialty type single packs*: *stretch wrap; pleat wrap*; and *envelope wrap.* Of the many *specialty multipacks*, the following four types are the most commonly used: *tray with lid; shrink wrap; tray with sleeve*, and *clam shell in a carton.*

Wrapping Machines

Stamped soap bars are discharged on single- or dual-lane flat belt conveyors, grouped together but not necessarily in pitch with the next group. The number of bars per group depends on the specific soap press model and bar size.

Most soap wrappers have intermittent motion pocketed timing in-feed conveyors, whereas all horizontal cartoners have continuous motion pocketed, bucket conveyors. A soap transfer (in-feed) system is therefore required to interface the soap press with the wrapping or cartoning machine.

In 1992, ACMA introduced their model 771 high speed wrapper with a single lane TH noncontact, 16 rotary suction cups soap transfer system (Fig. 16.1). This unit is the first new wrapper to appear in a decade.

The TH transfer (Fig. 16.2) is a dual-purpose system which can be used with a wrapping machine or a cartoner. It is designed to handle up to 500 bars/min (bpm) wrapping speed, thus breaking the 300 bpm speed barrier which was the maximum available until then.

This year there is a new entry in the bar soap packaging field. Binacchi & Co., is introducing three new soap wrappers. These include two models with their own non-contact oscillating pick and place type transfer systems (Figs. 16.3 and 16.4), and a high speed, 550 bars/min unit in direct combination with a soap press (Fig. 16.5). Table 16.1 lists all current ACMA, Binacchi and Carle & Montanari bar soap wrappers.

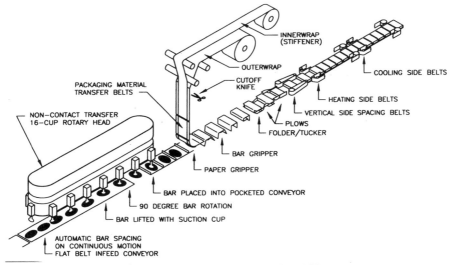

Fig. 16.1. ACMA 771 bar soap wrapper. *Source:* ACMA/GD.

Contact Type Transfer (Infeeds) for Wrapping Machines

Rotating turret transfers with vacuum cups have been in use successfully for decades. The turret(s) driven by the wrapping machine receives the stamped soap bars arriving lengthwise on a single- or dual-lane flat belt conveyor. A photocell detects the presence of the bars and signals the turret's suction cup to pick up the soaps. The turret indexes 90° and lowers the soap into the wrapping machine's pocketed in-feed timing belt.

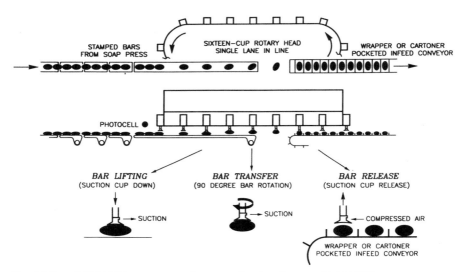

Fig. 16.2. ACMA TH noncontact bar soap feeder. *Source:* ACMA/GD.

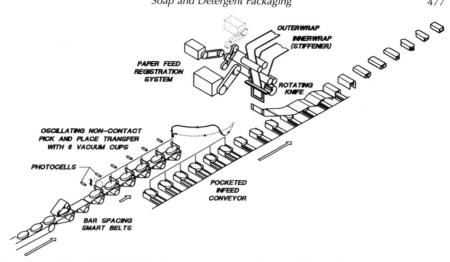

Fig. 16.3. Binacchi BSW-330 bar soap wrapper. *Source:* Binacchi & Co.

These contact type feeder transfers allow a few bars to touch each other slightly before they are lifted up, but handle most bar shapes gently without marring them. The various single and dual lane transfers available are the ACMA in-line transfer—one turret with four suction cups, the ACMA right angle transfer—one turret with four suction cups, the ACMA in-line transfer—two turrets with four suction cups each, and the ACMA right angle transfer—two turrets with four suction cups each.

Noncontact Transfers (Infeeds) for Wrapping Machines

Noncontact transfers do not allow the soaps to touch each other and assure delicate, damage-free handling of all shapes and types of bar soap products. These include the

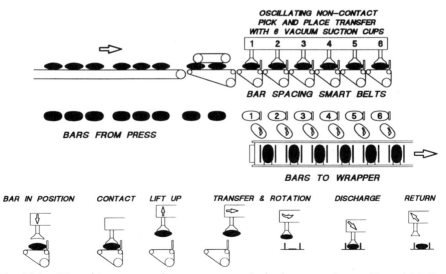

Fig. 16.4. Binacchi noncontact bar soap wrapper in-feed systems. *Source:* Binacchi & Co.

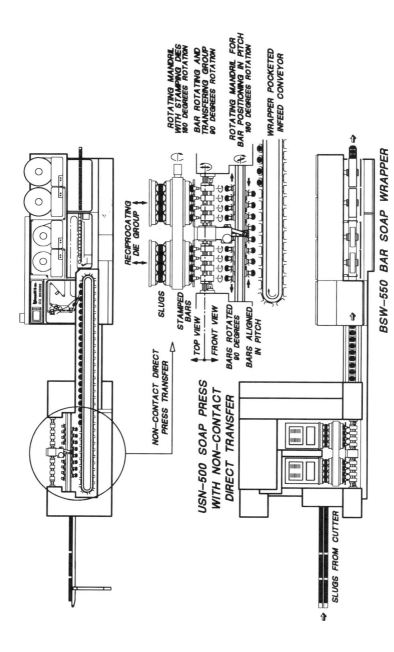

Fig. 16.5. Binacchi press and wrapper combine. *Source:* Binacchi & Co.

TABLE 16.1 Bar Soap Wrappers

Make	Model	Speed	In-Feed	Bar Soap Transfer
ACMA	711	200	Single lane in-line or right angle	Turret with 4 cups
	731	300	Single lane in-line	Flat belt
	771	500	Single lane in-line	Turret with 16 cups
	791/S	240	Single lane in-line	Turret with 4 cups
	791/D	300	Dual lane in-line	2 Turrets with 4 cups each
	791/D	300	Dual lane right angle	2 Reciprocating cups
Binacchi	BSW-220	220	Single lane in-line	Pick and place with 4 cups
	BSW-330	330	Single lane in-line	Pick and place with 6 cups or press transfer
	BSW-550	550	Single lane in-line	Press transfer
Carle &	CM-P/22	240	Single lane left angle	Turret with 5 fingers
Montanari	CM-V/33	300	Single lane left angle	Rotary head with 8 fingers

Abbreviations: Speed: wrapped bars per minute; Cups: vacuum suction cups; Fingers: mechanical grippers; Turret: contact transfer; Rotary Head: noncontact transfer; Pick and Place; oscillating noncontact pick and place transfer; Press Transfer: noncontact direct transfer from press into wrapper.

ACMA in-line transfer—one turret with 16 suction cups, (TH) (Fig. 16.2); the Binacchi in-line—oscillating pick and place twister group with four or six suction cups, (Fig. 16.4), the Binacchi direct press transfer—rotating mandrill with eight suction cups (Fig. 16.6), and the Carle & Montanari right angle transfer—one turret with five mechanical fingers.

Cartoners

Until recently, the maximum bar soap cartoning speed was limited to 300 cartons/min (cpm). Several cartoners could mechanically exceed this velocity, but they

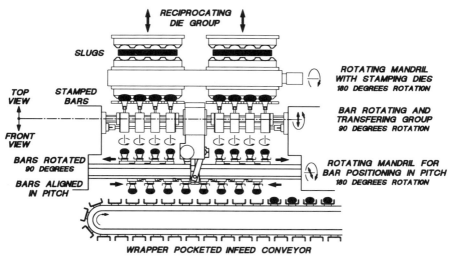

Fig. 16.6. Binacchi noncontact direct press transfer for bar soap wrapper or cartoner. *Source:* Binacchi & Co.

were limited by the speed of the available soap transfer systems. Two major developments were instrumental in breaking the 300 cpm soap cartoning speed barrier. The first was the introduction in 1989 of a novel direct transfer system offered by Binacchi as an integral part of their model USN-500 soap press, which allowed direct interface with a high speed cartoner. Several Jones OSC cartoners have been coupled to USN-500 soap presses. An extended bucket conveyor is required for proper interface as illustrated in (Fig. 16.7). The actual operating speeds reported range from 350 to 450 stamped and cartoned bars/min.

The second development was the previously described ACMA noncontact model TH transfer introduced in 1992. A few Bartelt cartoners have been fitted with

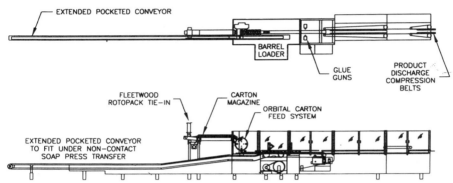

Fig. 16.7. Jones OSC bar soap cartoner. *Source:* R.A. Jones & Co., Inc.

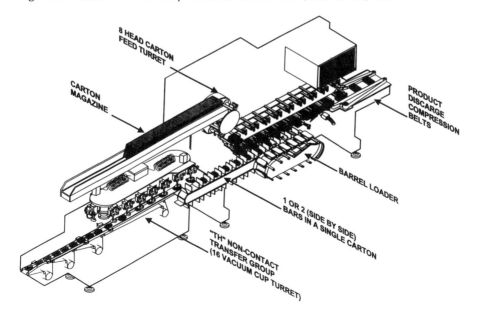

Fig. 16.8. ACMA 770/TH bar soap cartoner with a single lane in-feed. *Source:* ACMA/GD.

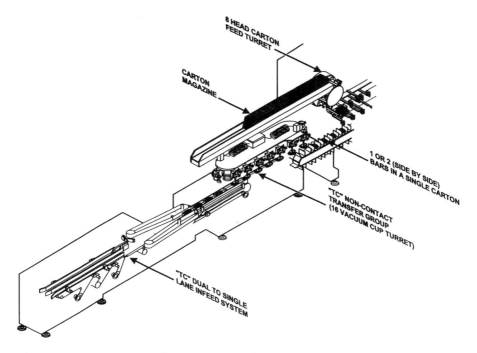

Fig. 16.9. ACMA 771/TH bar soap cartoner with a dual lane in-feed TC.
Source: ACMA/GD.

the TH transfer. ACMA's new model 770/TH (Fig. 16.8) high speed cartoner with the TH noncontact 16-cup transfer system can package 500 single soap bars per minute or alternately 250 cartons with 2 bars inside each carton positioned flat side by side. The TC system (Fig. 16.9) is a special, dual to single lane in-feed which precedes the TH group of the 771 cartoner. The TC system is applicable when the stamped soap bars leave the soap press on two discharge conveyors.

High speed cartoners above 300 cpm speeds should be equipped with large capacity carton magazines. Cartoner manufacturers offer only limited capacity extended carton magazines, while other firms specialize in larger capacity feeders which can be fitted to any cartoner. Fleetwood Systems, Inc. (United States) applied their large automatic magazine storage and feeding unit to a high speed soap cartoner in 1985. Since then, the Model 12C Rotopak carton storage and transfer unit (Fig. 16.10) has been coupled to various cartoners. The Rotopak group consists of a turret with twelve, 24 in high carton carrying pockets. Each pocket can accommodate 360 cartons of $^1/_{16}$ in thickness, or a total of 4320 cartons. The operator can feed each pocket with a complete prepackaged stack of cartons.

Table 16.2 lists all current ACMA, Bosch, CAM, IWKA, Jones and Klockner Bartelt bar soap cartoners.

Contact Type Transfers for Bar Soap Cartoners

The single and dual lane transfers available are the ACMA—one turret with four suc-

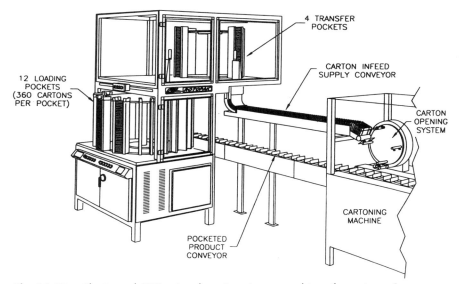

Fig. 16.10. Fleetwood 12C rotopak carton storage and transfer system. *Source:* Fleetwood Systems, Inc.

TABLE 16.2 Bar Soap Cartoners

Make	Model	Speed	In-Feed	Bar Soap Transfer
ACMA	330/S	240	Single	Turret with 4 cups
	330/D	300	Dual	2 Turrets with 3 cups each
	330/TH	350	Single	Rotary with 16 cups (TH)
	770/TH	500	Single	Rotary with 16 cups (TH)
	770/TC	500	Dual	Rotary with 12 cups (TC)
Bosch	CUK-3040	300	Single	Rotary with 24 cups (PUG-S)
Cam	HMM	200	Single	Rotary with 16 fingers (V-228)
IWKA	CPS-R	400	Single	Press transfer
Jones	Legacy CSC-4	300	Dual	2 Turrets with 5 cups each
	OSC-5	300	Dual	2 Turrets with 4 cups each
	OSC-4	350	Dual	2 Turrets with 4 cups each
	OSC4/5	450	Single	Press transfer
Klöckner-	Formula 500	500	Single	Rotary with 14 cups
Bartelt	Formula 125	200	Single	Rotary with 8 cups

Abbreviations: Speed: cartoned bars per minute; In-feed: all are in-line feed; Press Transfer: noncontact direct transfer from press into cartoner. Other abbreviations as in Table 16.1.

tion cups; the ACMA—two turret with three suction cups each; and Jones—two turrets with four or five suction cups each.

Noncontact Single Lane Transfers for Bar Soap Cartoners
These include the ACMA in-line transfer—one Turret with 16 Suction Cups, (TH); the Bosch in-line transfer—one turret with 24 suction cups (PUG-S); the CAM in-line—one turret with 16 mechanical fingers, (V-228). CAM's V-228 transfer is the

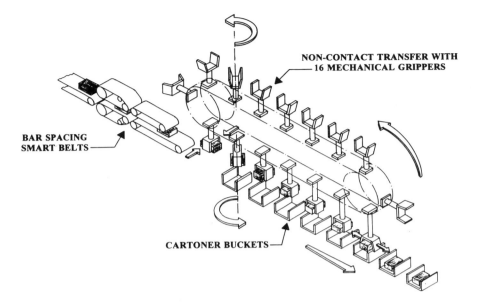

NON-CONTACT TRANSFER WITH 16 MECHANICAL GRIPPERS

BAR SPACING SMART BELTS

CARTONER BUCKETS

Fig. 16.11. Cam V-228 noncontact bar soap transfer. *Source:* Technicam S.R.L.

only transfer system which uses mechanical grippers to pick-up, rotate 90° and deposit the soaps into the buckets of the cartoner chain (Fig. 16.11); and the Klockner-Bartelt in-line transfer—one turret with 8 or 14 suction cups.

Figure 16.12 shows the most widely used vacuum suction-type transfers (not all those listed above) with bar soap cartoners.

Noncontact Soap Press Transfer Systems for Cartoners and Wrappers

The *Binacchi transfer* as an integral part of the soap press was introduced in 1989 and for the first time permitted coupling a cartoner directly with a press without the need for a separate extra cartoner bar transfer (in-feed) system. This invention allowed the 300 cpm speed limit, which was the maximum available with existing bar transfers, to be exceeded. The system designed for the Binacchi USN-500 soap press utilizes eight suction cups which hold and rotate the stamped soaps 90° and place them on a second set of eight suction cups located on a rotating mandril. The mandril with the bars rotates 180° and simultaneously moves the bars to fit the pitch of the extended pocketed conveyor. An electronic system interlocks and controls the two machines.

The new *Mazzoni LB DTU transfer* designed for the STUR model presses and Tema's SDS system for the STEMAR presses are functionally similar. They also use vacuum cups to hold and rotate the bars but they do not utilize a second transfer mandril. For more details and illustrations, please refer to the chapter on bar soap finishing lines and equipment.

The *Binacchi press transfer system* for a wrapper or carton described above, is part of the soap press and is similar to the one used to couple directly a Binacchi press with a bar soap cartoner (Fig. 16.6).

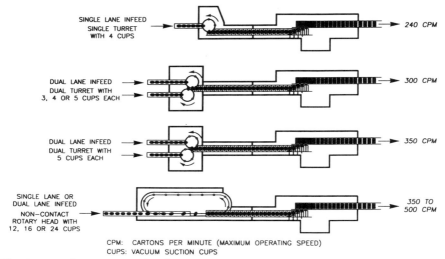

SINGLE LANE INFEED
SINGLE TURRET
WITH 4 CUPS   240 CPM

DUAL LANE INFEED
DUAL TURRET WITH
3, 4 OR 5 CUPS EACH   300 CPM

DUAL LANE INFEED
DUAL TURRET WITH
5 CUPS EACH   350 CPM

SINGLE LANE OR
DUAL LANE INFEED   350 TO
NON-CONTACT   500 CPM
ROTARY HEAD WITH
12, 16 OR 24 CUPS

CPM: CARTONS PER MINUTE (MAXIMUM OPERATING SPEED)
CUPS: VACUUM SUCTION CUPS

Fig. 16.12.   Bar soap transfers for cartoners.

Horizontal Wrappers

The use of horizontal wrappers for the single and multipack bar soap is increasing, especially for fin sealing the traditional "brick" shape, the economical laundry soaps, and the stretch film wrapping of single and multipacked ball shaped laundry soaps. A number of cosmetic beauty bars are fin sealed and then cartoned. In Europe, Imperial Leather, a major soap brand from Cussons, is fin sealed, the end seals are trimmed and the two fins folded down; then the soap is either cartoned and multipacked or only multipacked. Tear tape application is also available. Figure 16.13 illustrates a Doboy manually fed, horizontal wrapper for fin sealed style with crimped end seals. Doboy (United States) also supplies the model SBF automatic 3-belt servo-feeder with the horizontal wrappers for single bars. Figure 16.14 shows the horizontal wrapper in combination with a shrink tunnel for stretch film style wrapping. Automatic in-feed in this case is not offered.

Specialty Soap Packaging Machinery

Stretch wrapping and pleat wrapping are the two most recognizable and widely used "specialty soap" packaging styles. Traditionally, specialty soap packaging was limited mainly to cosmetics and novelty soaps.

Stretch Wrappers

Stretch wrapping is used mainly for cosmetics and novelty soap but is not limited to them. A few years ago, Lever, Proctor and Colgate introduced the stretch-wrap style for their mass marketed kidney or egg shaped soaps. This style became popular in South America first, and then moved into Asia. Many soap producers now offer this style. Stretch wrapping speeds have not reached beyond 50 wraps/min, limiting their wider use as a mass marketing packaging alternative.

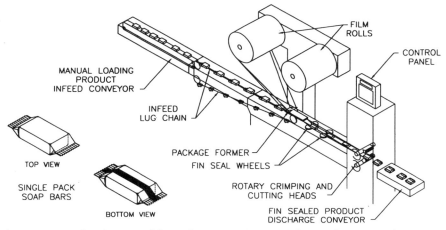

Fig. 16.13. Doboy horizontal fin seal wrapper. *Source:* Doboy Packaging Machinery, Inc.

Pleat Wrappers

While stretch wrappers can handle any shape from the very simple to the very intricate, pleat wrappers can wrap only round or oval side banded shaped soaps. The automatic stretch and pleat wrappers are supplied with automatic labelers. Alpma (Germany) and Guerze (Italy) also offer automatic bander/labeler units designed to place a full band/label all around a stretch wrapped or pleat wrapped product at a speed of 30–50 labels/min. Table 16.3 summarizes all stretch wrap and pleat wrappers offered today.

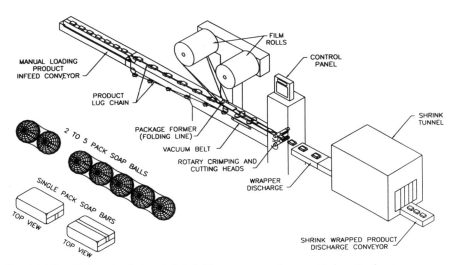

Fig. 16.14. Doboy horizontal shrink film wrapper. *Source:* Doboy Packaging Machinery, Inc.

TABLE 16.3   Bar Soap Stretch and Pleat Wrappers

Wrapper Types	Make	Model	Speed
Stretch wrappers for banded, bandless, standard, and specialty shapes	Alpma	V-64/ASCH III	25
		SDV-50	50
	Burnley	SPW-850	15
	Douglas	FCW-10	6
	Guerze	CE-15	15
		CE-25	25
		CE-50	50
Pleat wrappers for banded, round and oval shapes	Alpma	V-64/ASCH II	50
		PLV-100	100
	Burnley	PWS-750	12
	Douglas	P-8	8
	Guerze	PL-50	50
		PL-75	75
		PL-150	150

Abbreviation: Speed: maximum number of wrapped bars per minute.

Multipackers

The large supermarkets and discount warehouse club stores sell multipacked products which are placed on the shelves in their own display cases. In recent years, customer preference for the lower cost special promotion multipacked products has grown considerably. In the United States, practically all soaps are now sold in multipacks. The individually wrapped or cartoned soaps are banded or bundled in groups of two, four, six, eight or twelve bars. They are offered as "bonus packs" with one or more free bars. Multipacks are becoming popular everywhere but up to now they are smaller than in the USA.

Banders
When *banding* (U-banding), a band is adhered around three sides of a collated group of already wrapped or cartoned soap bars. Scandia's (United States) model 908 unit permits a variety of U-banding configurations as illustrated in Fig. 16.15.

Bundlers
The term *bundling* refers to the six-sided overwrapping of a group of previously wrapped or cartoned soap bars to form a multipack. BFB (Italy), Marden Edwards (England), Pester (Germany), Sollas (Holland) and others offer bundling machines which can overwrap in many configurations of up to twelve bars. The most widely used soap bundlers are made by BFB. They offer several models but the preferred one for the soap industry is the 3703BP due to its range in handling many single and double row bundling possibilities (Fig. 16.16). The Scandia Model 700 packages multiple bars in a single row (Fig. 16.17).

End Packaging

Case packers, case sealers, palletizers and stretch banders are required to complete the soap packaging line. Single and multipacked soaps are packed into the most common type pre-made corrugated RSC (Regular Slotted Container) cases. Recently

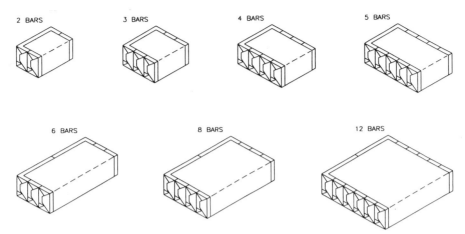

Fig. 16.15. Multipack configurations with Scandia 908 bander.

the soap industry has been converting to display style cases which are placed onto the supermarket shelves "as is." The RSC display case has a removable front panel which, once removed, exposes the product. The HSC (half-slotted box with cover) type case is used with its front panel already cut out. The tray type HSC permits maximum product display, but it has to be stretch film overwrapped for shipment.

Semi-Automatic Case Packers
These require an operator to manually form the case, place it onto the loading funnel and then allow the machine to take over. After the bars are loaded, the case is lowered 90° onto a take-away conveyor and is moved to the case sealer.

Automatic Case Packers
These perform the operation without an operator. The case packing sequence is shown in Fig. 16.18. To pack two or three deep in a case, lane dividers are required.

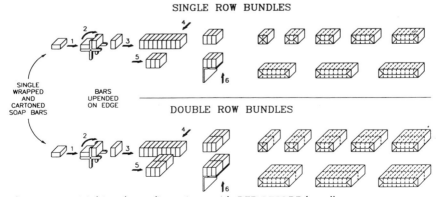

Fig. 16.16. Multipack configurations with BFB 3703BP bundler.

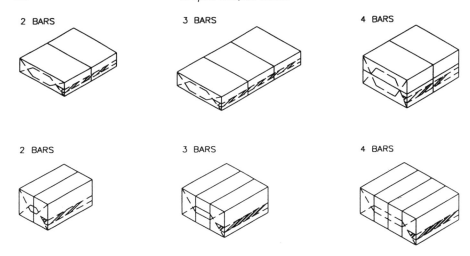

Fig. 16.17. Multipack configurations with Scandia 700 series wrapper.

Robotic Case Packers
A few of these have been installed in the United States, Europe and the Far East. Schubert (Germany) supplies most of these systems.

Wrap-Around Casers
Wrap-around case packing of soap is used only by a few firms utilizing an RSC display type case. Pak-Master (part of the SASIB/Paxall Group) pioneered the wrap-around RSC display case in 1983 for a food company, but soap applications are more recent (Fig. 16.19).

Robotic Gift Soap Packaging Line
Schubert has supplied a most interesting line for packaging various wrapped soaps into a tray with a lid. The system consists of a model SKA-D tray and lid erector, an

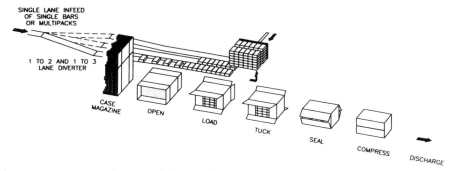

Fig. 16.18. Case packing single bars and multipacks.

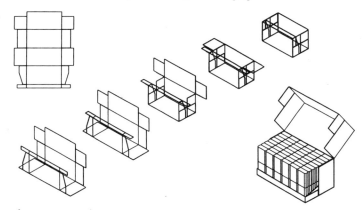

Fig. 16.19. Pak-Master wrap-around display ready-case for soap products. *Source:* The Paxall Group, Inc.

SCN-F2 pick and place robot which places the wrapped soap into trays and another SCN-F2 robot for lidding (Fig. 16.20).

Multi-Purpose Bar Soap Packaging Line

Figure 16.21 illustrates a multipurpose line and the machinery needed for wrapping and cartoning single bars, bundling or banding into multipacks and end packaging.

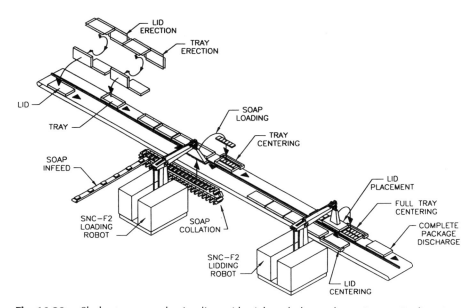

Fig. 16.20. Shubert soap packaging line with pick and place robots. *Source:* Rodico, Inc.

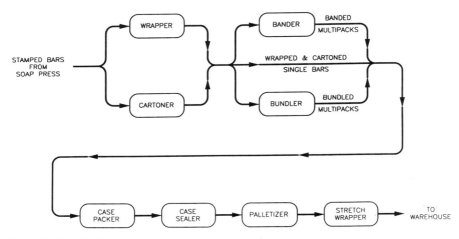

Fig. 16.21.   Multi-purpose bar soap packaging line.

Packaging Materials and Conversion Factors

Packaging materials and a proper package must provide protection against product tampering, physical damage, moisture losses, contamination, discoloration and mold growth. Technical specifications for soap wrappers are summarized in Table 16.4. Cast, co-extruded linear low-density polyethylene (LLDPE) films 17, 23, 30, and 35 microns (μm) in thickness perform very well on all stretch wrapping machines listed in Table 16.3. Biaxially orientated polypropylene (BOPP), 1.25–1.5 mil in thickness, are the most widely used films for fin seal style packaging with horizontal wrappers. To a lesser degree, mono orientated polypropylene (OPP) is also used. Low density polypropylene (LDPE) film with thickness of 1.25 mil is recommended for heat-shrinkable film wrapping.

A few useful terms and conversion factors are listed for general reference:

1. Basis Weight: In the United States, *basis weight* is the weight expressed in pounds of 500 sheets of 24 × 36 inches size paper ($24'' \times 36'' \times 500 = 3{,}000 \text{ ft}^2 = 279 \text{ m}^2$).

2. Grammage: Basis weight expressed in g/m^2 (grams per square meter).

3. Composite Structure Weight: Composite weights include, ink, overprint lacquer and heat seal coating.

4. Caliper/Point: Caliper or point is the thickness of paper or board expressed in thousands of an inch, i.e., 1 calliper or 1 point = 0.001 inch. Example: A 15-point board is 0.015 inch thick.

5. Gauge: Film thickness is expressed in terms of mils; 1 mil = 0.001 inch. For thin film, the term gauge is also used; 100 gauge = 1 mil. Example: A 90-gauge film is 0.00090 inch in thickness.

6. Micron: 1 micron (μm) = 0.001 mm; 1 millimeter = 100 microns.

7. Conversion Factors: Some useful factors are presented in Table 16.5.

TABLE 16.4 Soap Wrapper Materials

	Heat sealable					Nonheat sealable	
Wrapper type	Full size bars			Bundle tape for Multipacks	Hotel soaps		Pleat wrapped soaps
Application Available	Rolls	Rolls	Rolls	Rolls	Rolls/sheets	Sheets	Rolls/sheets
Basis Weight	Paper	Paper	Foil laminate	Paper	Paper	Paper	Tissue
lbs	45	35	38	60	55	60	15
g/m²	73	57	62	98	90	90	24
Composite Basis Weight							
lbs	58	46	51	74	68	56	16
g/m²	94	78	83	120	112	19	26
Sealing Temperature	220–470°F at 0.75 second dwell and 15 psi pressure					None	None
Coefficient of Friction (lacquer to lacquer)							
Static	0.22	0.25	0.29	0.22	0.22	0.22	0.41
Kinetic	0.21	0.22	0.20	0.21	0.21	0.21	0.35
Water Vapor Transmission Rate grams/100 in²/24 h	1.30	0.90	0.01	1.55	1.75	2.85	—
Mold inhibitors in ppm							
(BUSAN) TCTMB	1500	1500	1200	1700	1700	1700	—
ZIRAM	—	—	—	—	—	—	6000
Techpak code	FPS-905	FPS-907	FPS-909	FPS-910	FPS-900	FPS-901	FPS-912

Notes: Standard ream in the United States is 500 sheets cut to a size of 24 × 36 inches. Ream standard based upon 500 sheets on 25 × 38 inches size is used in United States for cut labels and sheets. Composite weights include: ink, overprint lacquer and hot melt heat seal coating for heat seal labels and only ink and overprint lacquer for non-heat seal labels. Sealing temperature is a range of temperatures at which materials can be heat-sealed. Coefficient of friction is a measure of slip. The higher the number, the poorer the slip. Water transmission rate (WVTR) is a measure of water vapor diffusion through the wrapper. Mold inhibitor chemical names: BUSAN (TCMTB) is 2-(thiocyanomethylthio)benzothiazole. ZIRAM is zinc dimethyldithiocarbamate + zinc 2-mercaptobenzothiazole. *Source:* Techpak, Franklin, Ohio, USA.

TABLE 16.5   Conversion Factors

From	To	Multiply by
lbs/ream	g/m^2	1.627
mils	microns	25.4
in^2/lb	m^2/Kg	0.0014
gauge	microns	0.254
reams	m^2	278.7
micron	gauge	3.937

Detergent Packaging

Introduction

The most significant recent changes in detergent packaging arise from the move away from low bulk density powders (the ratio of weight to unit volume expressed in grams/liter) to concentrated powders of higher bulk densities and, thus requiring the packaging of much heavier powders. As well as being economically advantageous, this represents an effort to pacify the Green lobby by becoming more environmentally friendly, reducing the packaging costs and materials used, while still producing similar quantities of powder with the same washing load capabilities. In the 1980s, conventional powders had a bulk density of around 300 g/L. Today, conventional powders have a bulk density ranging from 450 to 620 g/L, and concentrated (compact) powders range from 830 to over 900 g/L.

Conventional Cartons

Conventional cartons for packaging soap flakes, powders and later light powdered detergents endured for most of this century. Until the late 1960s there were many different carton sizes, weights and ratios of height to width to thickness. Each detergent manufacturer selected the weight, volume, length, width and thickness ratio which would present the best image for their product in the market place. The introduction of heavier powders meant that the packaging of powdered detergent would undergo a revolution.

Size Classification
With the introduction of the E (European) carton sizes, came a loose form of standardization; it became easier for machinery manufacturers to grade and classify their machines. These new cartons removed the thin, large faced cartons which were difficult to fill at high speeds, and in their place substituted deeper cartons with a well-proportioned open top giving rise to higher filling speeds. The E number represents that number maximum internal carton volume, e.g., E3 = 3 × 750 mL = 2250 mL. Filling speeds of up to 320 cartons/min are possible with E3 cartons, 150 cartons/min with E10 cartons and 80 cartons/min with E15 cartons. Table 16.6 illustrates that the standardization was almost purely volumetric with little regard to dimensional similarity. For some time the bulk density increased but the cartons remained approximately the same size. This meant that the individual cartons

TABLE 16.6   European E Carton Sizes for Standard Detergents

Carton size	Volume (mL)	Length (mm)	Width (mm)	Height (mm)
E1	750	116.5	42	153
E2	1500	143	53	198
E3	2250	160.5	61	230
E5	3750	191	72	273
E10	7500	250	102	330
E15	11250	274	120	375
E20	15000	283	144	398
E25	18750	283	170	425
E30	22500	295	192	430

The maximum filling volume is obtained by multiplying the E number by 750 mL. No attempt was made for dimensional similarity between the E sizes.

became heavier and heavier until the E20 size ceased to be produced in the United Kingdom because it was too heavy and too expensive for the purchaser to struggle home with. Today, with conventional powders, the E15 is the largest carton, containing 11250 mL, and weighing over 6 kg. The others in use are the E3 and E10.

About five years ago, to accommodate the ever increasing powder density, the newer VC carton range was introduced. These cartons are not as tall as their predecessors but have a similar footprint to aid in filling and to make size changing easier. They generally have an integral liner to give carton stability; VC5 sizes and over have handles for ease of carrying. The increase in bulk density has made it possible to reduce line speeds while still achieving similar tonnages. Table 16.7 shows that the standardization attempted was on dimensional similarities, especially with respect to the carton face (length) and as far as possible to the front to back dimension (width). It also gives a very rough volume proportion ranging from 810 to 890 mL for each VC number.

Carton Sealing Systems

For many years, cold glue was the traditional method for sealing conventional cartons. PVA or dextrine-based glues were applied by the roller method, shown as Item 1 in Figure 16.22. This method produces a thin film of glue evenly distributed across

TABLE 16.7   European VC Carton Sizes for Concentrated Detergents

Carton size	Volume (mL)	Length (mm)	Width (mm)	Height (mm)
VC 1.5	1233	138	74.5	120
VC 2	1620	146	74.5	149
VC 3	2682	191.5	94	149
VC 4	3475	191.5	114.5	158.5
VC 5	4320	191.5	114.5	197
VC 6	5056	191.5	143.5	184
VC 7	5880	191.5	143.5	214
VC 10	824	191.5	143.5	300

Indicative only, as individual manufacturers vary. For VC cartons, a degree of dimensional similarity between sizes is attempted with regard to cross section dimensions. To a lesser degree a volume proportion is achieved.

TABLE 16.8   U.S. Load Number Carton Sizes for Concentrated Detergents

Carton size	Volume (in³)	Length (in)	Width (in)	Height (in)
10	101	6-3/8	3	5-5/16
14	173	7-3/8	3-3/4	6-1/4
33	325	8	5	8-1/8
42	380	8-1/2	5-1/4	8-1/2
85	709	10-3/4	6	11
120	1070	11-3/4	7-3/4	11-3/4

Actual carton size may vary by product density.

the carton flaps. Until recently, this was the only sure method of achieving a sift-proof carton at medium to high speeds. A decrease in speed, coupled with advancements in electronics and control engineering such as quick operating solenoid valves, has resulted in pressure glue application systems of greater accuracy and efficiency than ever before. Cold glue applicators are now available which provide an even glue spread with good accuracy for medium speeds giving a well-sealed sift-proof carton. Similar to the roller method, cold beads of glue are spread into a thin film across the carton flaps as they are pressed together. The main drawback is that both of these cold glue systems require a long sealing belt, taking up valuable floor space. The latest development in this field is a relatively new Hot Melt system, which applies hot melt glue with great accuracy at medium to high speeds (Item 2 in Fig. 16.22). The glue is applied only in those areas necessary to produce a sift-proof carton, and not across the carton flaps completely as in the case of cold glue. This saves glue and reduces the higher cost of using hot-melt adhesive. It also obviates the requirement of all but the shortest of sealing belts. Machine cleanliness and the fact that the machine can be switched off and left without the elaborate problem of cleaning out the glue system are two further advantages.

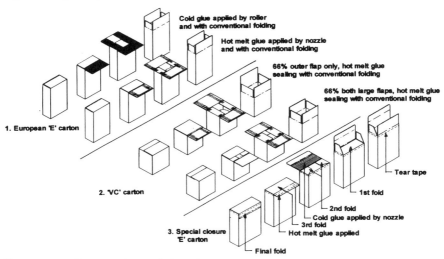

Fig. 16.22.   Carton styles and glue closures.

All these represent the conventional type carton closures. There are of course many other variations such as Item 3 (Fig. 16.22) which shows a reverse-fold carton with a mixture of cold glue applied by nozzle or by the roller method with the addition of a single bead of hot-melt glue.

New Types of Packaging

Containers and Bags

In the interests of the economy and the environment, the two most recent developments in powdered detergent packaging have been the introduction into the market of the reusable container and the economy refill bag. These reusable containers, shown in Fig. 16.23, are made of metal or plastic and have replaceable lids, some with a pouring facility built into the lid. The economy refill bag, shown in Fig. 16.24, is intended as a refill for the plastic or metal container. The theory is that there is less packaging material to be disposed of or recycled. The bag will certainly take up much less space in a landfill site than its equivalent sized carton; bags are easier to crush, and require much less material and energy than cartons in their construction. To make this system viable, each of the reusable containers should be made to last for a number of years.

Powder Dispensing and Measuring Tools

The measuring cup (Item 2, Fig. 16.25) was introduced in the 1980s. It was placed into the larger cartons of conventional powders and gave a greater degree of control to the end user. It also allowed a more accurate amount of powder to be used at each wash. Originally the measuring cups were simple plastic cups similar to slightly tapered tumblers in shape to assist stacking. Graduations were printed or molded onto the side to assist measurement. With the introduction of heavier powders, these aids have become even more important and more sophisticated in shape and complexity. Furthermore, without proper measuring, higher density powders could be easily wasted by using too much at each wash, which is bad for the environment and more costly to the user. The

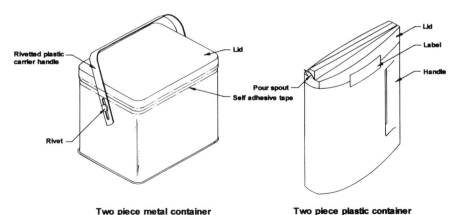

Two piece metal container **Two piece plastic container**

Fig. 16.23. Reusable containers.

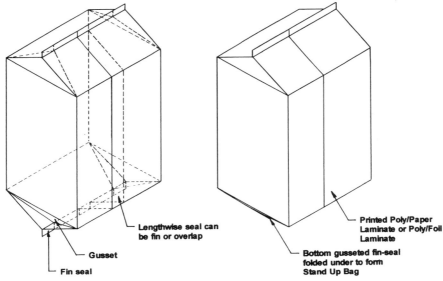

Fig. 16.24. Typical economy bags.

simple cup has evolved into a more complex plastic scoop with a handle as shown in Fig. 16.25, Item 3. The smaller amount of high density powder used for a single wash has led to the introduction of an open-ended, truncated hollow plastic sphere (Item 1, Fig. 16.25). Putting the correct amount of powder into the sphere and then placing it on

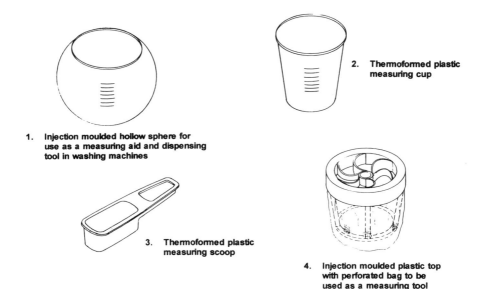

Fig. 16.25. Powder dispensing and measuring tools.

the top of the wash pile ensures that the powder will be evenly distributed throughout the wash load. The first simple hollow spheres were made from soft pliable molded plastic, in a single piece. They are now evolving into much more complicated shapes. One example is a three-piece hard plastic hollow sphere with a open fluted grid end to act as a digging tool and scoop. The second example, (Item 4, Fig. 16.25) is a hard plastic cylindrical grid with a flexible bag attached to one end. All of these measuring aids required the design and development of special devices to feed the aid into the carton or container. The reduction in cartoning speed, as discussed earlier, has made this much easier.

Carton Carrying Handles

There are a number of methods for creating a carrying handle for a carton, but they generally fall into two forms—Those that are part of the carton and those that are attached to the carton in some way, i.e., integral, or riveted onto the side of the carton. The integral handle, (Item 1, Fig. 16.26) is either placed onto, or is part of the carton as it is being manufactured. It can either be a thin plastic strip affixed to the top of the carton on the inner large flap and projects through the outer large flap when the carton is fully closed, or in some cases it can be part of the carton board itself. This type of handle generally presents no problems for the filling machine. The riveted plastic handle, (Item 2, Fig. 16.26), can either be fixed to the side of the carton before its arrival at the filling machine or it can be fixed during the erection of the carton and before filling on the actual carton filling machine. If applied before arrival at the filler, then higher filling speeds can be achieved, but the carton may be rather bulky in the magazine and may therefore

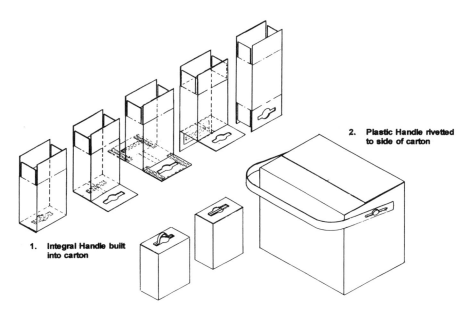

2. **Plastic Handle rivetted to side of carton**

1. **Integral Handle built into carton**

Fig. 16.26.   Two types of carton handles.

sometimes cause difficulties. If applied on the machine, it causes no actual machinabil-ity problems but, speeds are generally restricted to 100 cartons/min.

Powder Feeding Systems

Volumetric

The usual way of measuring the amount of used powder to fill the container, the car-ton or the bag is the volumetric method. This is especially the case for high speed machines. ACMA employs the system shown in Fig. 16.27 for the E5 or VC4 and smaller cartons. It uses a number of telescopic flasks with a sweep-off scraper and a lower self-aligning gate flap valve. The flap valve is held closed by a spring up against the underside of the lower telescopic element. When a carton is to be filled, the cylin-der lowers the diverter arm which forces the roller arm up the cam track thus opening the flap valve. The ACMA system for cartons larger than E5 or VC4 is shown in Fig. 16.28. The only real difference is that, because of the heavier weights of powder in the flasks, the flap closing spring has been replaced by a locking arm system. This posi-tively holds the flap valve up against the underside of the lower telescopic element. When a carton arrives to be filled, the cylinder effectively unlocks the flap and allows it to follow the cam track. The weight of powder to be deposited into the container or carton is directly proportional to the density. The accuracy of the Volumetric system depends upon the dimensional tolerances of the machine, ±1% by volume. Thus, if the density changes, so does the weight of the powder.

To achieve an efficient method of filling, some type of tracking system for the density changes must be employed and the height of telescopic flasks adjusted to suit. To avoid "light weights" if this is not done, the flasks should be set to give the correct weight when the density is at its lowest value possible. This will result in a huge "give away" as the density increases to its maximum.

The least sophisticated way of monitoring weight is the manual method. The operator periodically removes a carton, checks its weight and then either increases or decreases the length of the telescopic flasks accordingly.

This is of course very primitive and quite inaccurate because judgements of over- or underweight are based on only one carton, or at best a few random samples. It means a big "give away" to ensure that the minimum target weight is achieved.

Two automatic systems exist. The first monitors the stream of powder entering the filling head, checking the density on a regular basis by comparing the weight of a known volume with that of the target and adjusting the length of the telescopic flasks accordingly. This system is only as reliable as the sample of powder taken from the stream is representative of the total volume of powder entering the filling head. It is a known fact that the finer granules of powder migrate to the outside of the filling tube leaving the heavier granules in the middle, i.e., it is never a homoge-neous stream of powder.

The second automatic system employs the use of an in-line, high speed check weigher which weighs each individual carton. It can be done only after the carton has been closed at the top, i.e., after it has left the filling machine proper. The infor-mation gained by weighing each carton is used to adjust the length of the telescopic

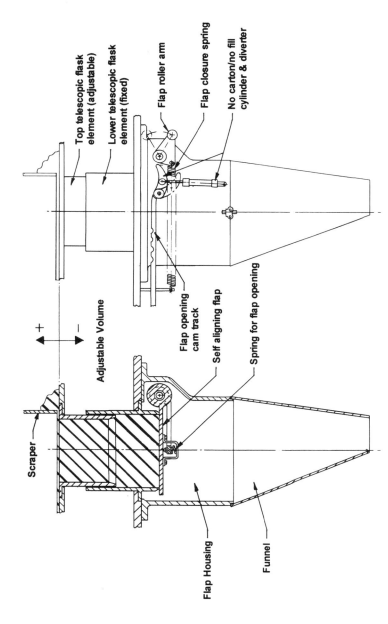

Fig. 16.27. Typical filling system for smaller cartons.

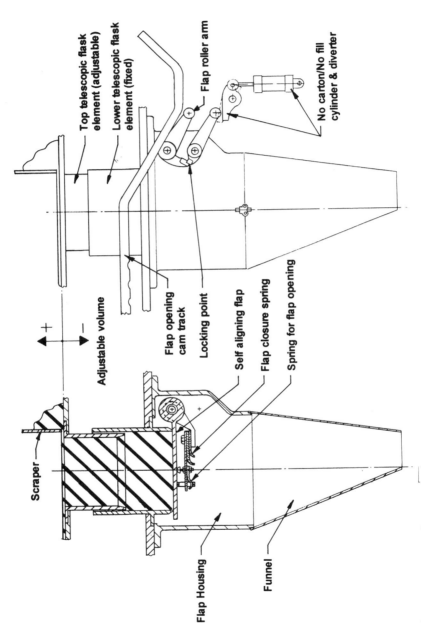

Fig. 16.28.   Typical system for larger cartons.

flask. This system is the most accurate method of checking for fluctuations, because it checks each individual carton filled with a homogeneous mixture of powder. The one drawback is the lag time because the weighing is made some distance from the filling point.

Both of these automatic systems employ the use of a computer to analyze the actual weights. Modern statistical analysis and standard deviation programs especially formulated for this purpose are used in the computer. This enables a correct adjustment to be made at the right time and gives production and quality control management access to valuable statistical information. At the same time it provides hard copy information to comply with the regulations concerning weights and measures.

Net Weigh

The normal system of net weigh employs a number of weigh scales located above and around a crossed funnel system. The simple system shown in Fig. 16.29 uses four weigh scales with two collecting funnels and a system of crossed funnels which allows two entry points to the filling area of the machine. At the correct time, the presence of the carton in a pocket signals for a weigh scale to drop its load down the collecting funnel, through the crossed funnels and into the carton. The main advantage of this system is its high degree of weight accuracy, governed by the accuracy of the weigh scales used. The main disadvantage is that it is generally much slower than its volumetric equivalent. A single weigh scale can reasonably and efficiently operate at 15–20 weighings/min. Normally, a maximum of 8 weigh scales can be placed around a filling machine, because of physical space and the ultimate length

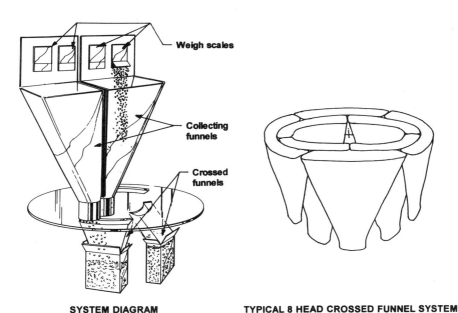

SYSTEM DIAGRAM   **TYPICAL 8 HEAD CROSSED FUNNEL SYSTEM**

Fig. 16.29.   Net weighing filling system.

of the collecting funnel. As a rule of thumb, therefore, the speed of a cartoning machine fitted with weigh scales and crossed funnels is 120–160 cartons/min. Although rather slow, this speed offers less "give away." As bulk densities increase and speed decreases, volumetric "give away" would be proportionally less because "give away" of weight would be the same.

While sometimes used with carton filling machines, this system is more often used with bag fillers.

Computer Combinational Weigher

Systems of this type employ a number of weigh scales which are continually weighing powder whose exact weight is held in computer memory. When asked to fill a carton or a bag, the computer chooses a combination of weigh scales which are the closest to the target weight when added together. Those selected scales drop their load through the funnels and into the carton or bag to be filled. Two such typical systems, the rotary and in-line systems, are shown in Fig. 16.30. The rotary system uses 15 weigh scales. The combination decided by the computer drops through the collecting funnel into the filling machine either through the crossed funnels on the carton filler or into the tube of the bag maker. The usual maximum speed for this system is 100 weighings/min, which increases to 140 weighings/min with an 18 weigh scale system. An accuracy of less than 1 g with a maximum fill of 1000 g is claimed.

The in-line system employs a slightly different method of operation. In the standard version, there are 6 weigh scales but the scales do not have to wait until a carton or bag requires filling before they drop. Instead, the weighing is deposited into one of two holding chambers, and the computer memorizes the exact weight and the location of the holding chamber. Consequently, up to 12 weighings are waiting to be used. The computer decides the combination and drops the combined load through the collecting funnels into the filling machine either through the crossed funnels on the carton filler or into the tube of the bag maker. The usual maximum for this system is 90 weighings/min, which increases to 160 weighings/min with 9 weigh scales and 18 holding chambers. Some of these systems are already in use, but, with the advent of higher bulk densities and the relative inaccuracies of the volumetric system, their applications will grow.

Carton Fillers

There are a number of carton filling machines available on the market from Europe and the United States, manufactured by firms which include the following: ACMA/GD, Ciba, G&W, Mingozzi, Pneumatic Scale, SASIB/Paxall and Senzanni. The ACMA and Paxall units will be illustrated but all the others follow similar principles.

The flow diagram in Fig. 16.31 shows the standard sections which form part of almost all carton filling machines. The steps are as follows: carton erection from a horizontal magazine; small-bottom inner flap tucking; bottom flap spreading; coding of outer bottom large flap; bottom large flap gluing; bottom flap closing; checking for carton presence, filling by volume, or weight etc.; checking for correct filling, if incorrect ejecting the carton; small top inner flap tucking; top large flap spreading; top flap gluing; top flap closing; and compression belt sealing.

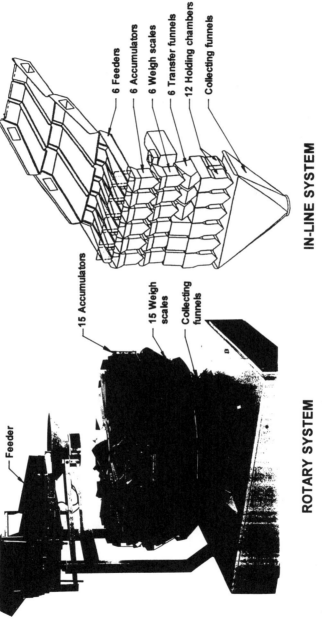

6 Feeders
6 Accumulators
6 Weigh scales
6 Transfer funnels
12 Holding chambers
Collecting funnels

IN-LINE SYSTEM

15 Accumulators

15 Weigh scales

Collecting funnels

Feeder

ROTARY SYSTEM

Fig. 16.30. Two Examples of computer combinational weighers.

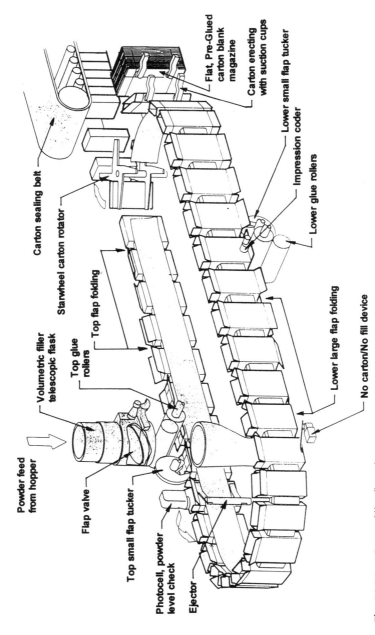

Powder feed
from hopper

Volumetric filler
telescopic flask

Flap valve

Top glue
rollers

Top flap folding

Top small flap tucker

Photocell, powder
level check

Ejector

Carton sealing belt

Starwheel carton rotator

Flat, Pre-Glued
carton blank
magazine

Carton erecting
with suction cups

Lower small flap tucker

Impression coder

Lower glue rollers

Lower large flap folding

No carton/No fill device

Fig. 16.31.   Carton filler flow diagram.

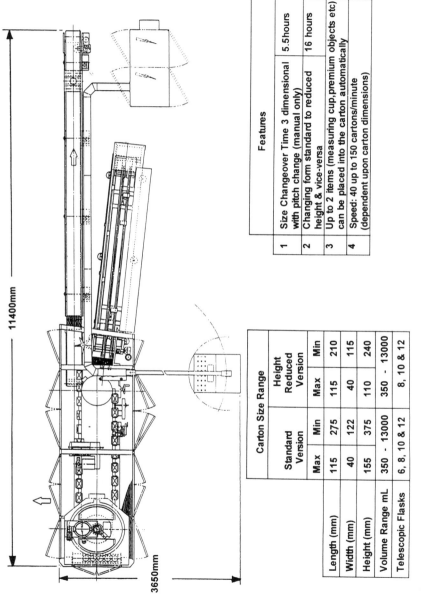

Features		
1	Size Changeover Time 3 dimensional with pitch change (manual only)	5.5 hours
2	Changing form standard to reduced height & vice-versa	16 hours
3	Up to 2 items (measuring cup, premium objects etc) can be placed into the carton automatically	
4	Speed: 40 up to 150 cartons/minute (dependent upon carton dimensions)	

	Carton Size Range			
	Standard Version		Height Reduced Version	
	Max	Min	Max	Min
Length (mm)	115	275	115	210
Width (mm)	40	122	40	115
Height (mm)	155	375	110	240
Volume Range mL	350 - 13000		350 - 13000	
Telescopic Flasks	6, 8, 10 & 12		8, 10 & 12	

11400mm

3650mm

Fig. 16.32. ACMA 956 carton filler. *Source: ACMA/GD.*

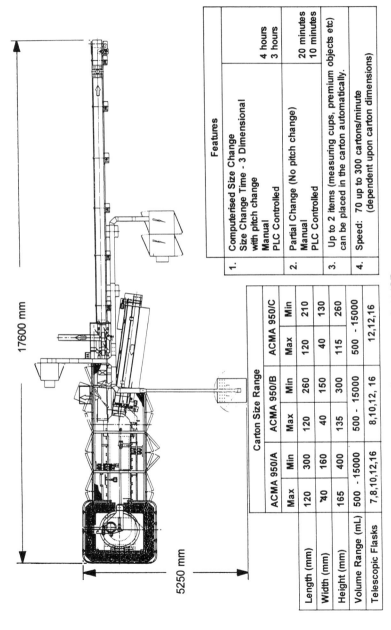

	ACMA 950/A		ACMA 950/B		ACMA 950/C	
	Max	Min	Max	Min	Max	Min
Length (mm)	120	300	120	260	120	210
Width (mm)	40	160	40	150	40	130
Height (mm)	165	400	135	300	115	260
Volume Range (mL)	500 - 15000		500 - 15000		500 - 15000	
Telescopic Flasks	7,8,10,12,16		8,10,12, 16		12,12,16	

(Carton Size Range)

	Features	
1.	Computerised Size Change	
	Size Change Time - 3 Dimensional	
	with pitch change	
	Manual	4 hours
	PLC Controlled	3 hours
2.	Partial Change (No pitch change)	
	Manual	20 minutes
	PLC Controlled	10 minutes
3.	Up to 2 items (measuring cups, premium objects etc)	
	can be placed in the carton automatically.	
4.	Speed: 70 up to 300 cartons/minute	
	(dependent upon carton dimensions)	

Fig. 16.33. ACMA 950 high speed carton filler. *Source:* ACMA/GD.

The New ACMA 950 and 956 units shown in Figs. 16.32 and 16.33 offer a wide range of possibilities. These include:

1. Large carton size range and volume range, to accommodate the new smaller sizes especially with respect to the reduction in height of the VC range of cartons.

2. Very speedy computer controlled three-dimensional size change, complete with pitch change, i.e., the biggest change in less than 3 h.

The Paxall Series 300, shown in Fig. 16.34, also offers a similar wide range and many optional extras.

Economy Bag Fillers

Throughout the detergent industry in Europe, bags are largely replacing cartons, especially for concentrated powders. They are also used extensively for conventional powders in Mexico and Latin America. They are made on machinery that has been developed or at least acquired as almost throw-away technology. Because the industry is in such turmoil, the bags could be superseded tomorrow as new marketing concepts and competition dictates. Vertical bag form-fill-seal (F/F/S) machines are offered by machinery manufacturers such as Bosch, Goglio, Gorig, Hassia, Hayssen, ICA, Triangle, and UVA. In general, the bag is formed around a mandrel after being drawn from a flat reel of film. It is sealed along the length creating a tube and then filled from the top by either weigh scales or a volumetric filler.

A gusset is then formed in each side of the tube above the detergent, and sealing bars create the top seal of the filled bag and the bottom seal of the next bag. A knife cuts off the bag from the tube at this point.

The bag materials currently in use are printed poly-paper laminates and other supported and unsupported printed films. An example of a complete economy bag line is shown on Fig. 16.35. It comprises two bag making machines, 4 weigh scale fillers and a bag transporting, top sealing and top trimming machine.

End Packaging

Until about the mid 1980s, end packaging was done by automatic or semi-automatic case packing machines. Corrugated or solid board outers were used, prefolded and glued. These were either hand or mechanically erected and presented. Later wrap-around style case systems were used with blanks as shown in Fig. 16.36. Both of these methods afforded an overall protection of the individual cartons. The outers were glue sealed with hot melt or cold glue. The disposal of large quantities of corrugated board was an increasing problem at the supermarket. Because of the need to reduce costs and for environmental reasons, end packaging was revamped. The fact that corrugated and solid board is made from wood pulp was the environmental issue that had to be addressed. As a consequence, there was a move away from board to plastic film in the form of shrink film and stretch film. Initially, in some cases a small tray was used to keep the packs together; however,

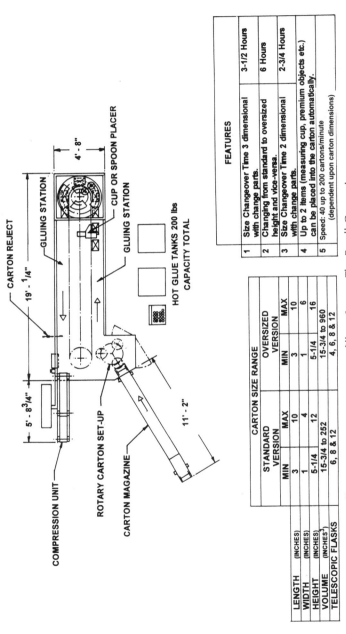

Fig. 16.34. SASIB/Paxall spectrum series 300 carton filler. *Source:* The Paxall Group, Inc.

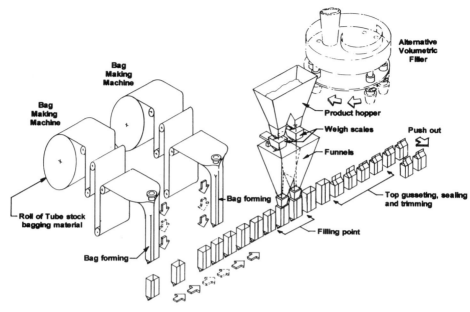

Fig. 16.35. An economy bag forming, filling and sealing line.

that was rather short-lived due to cost. When removed from the pack, the shrink or stretch film takes up much less space than corrugated cases and is disposed of easily. There are two main generic types of shrink wrapping machines, the high speed

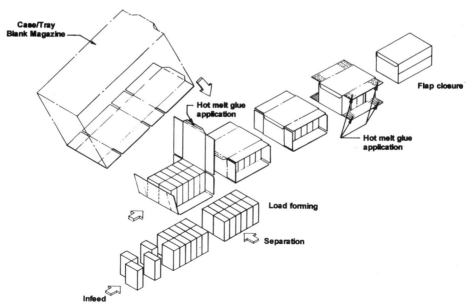

Fig. 16.36. Wrap-around case packer.

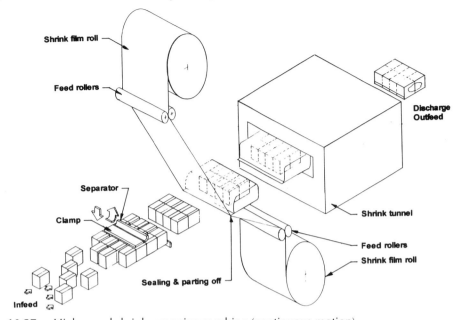

Fig. 16.37.   High speed shrink wrapping machine (continuous motion).

continuous motion machines and the medium to low speed intermittent motion machines. There are of course numerous examples of each type.

A typical high speed machine is shown in Fig. 16.37. The cartons arrive at the machine suitably divided into the required number of rows. The machine separates

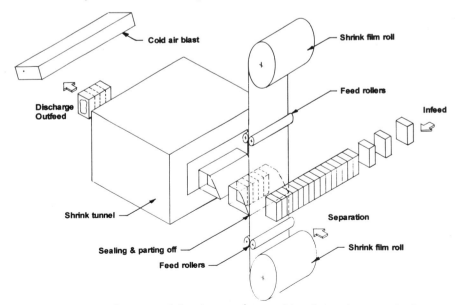

Fig. 16.38.   Low-medium speed shrink wrapping machine (intermittent motion).

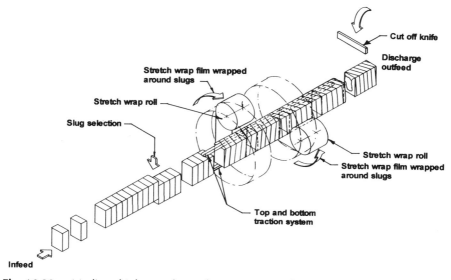

Fig. 16.39. Medium-high speed stretch wrapping machine (continuous motion).

the rows of cartons into slugs of a predetermined collation. It then subsequently wraps the slugs in shrink film and passes them through a heated tunnel, all with continuous motion. The medium speed shrink wrap machine shown in Fig. 16.38 has an intermittent motion pusher which separates the slug from a single row at right angles to the flow, then wraps the film around the slug and pushes the slug into the heated tunnel. All of the machine motions are intermittent except for the tunnel which is continuous. In the case of stretch wrapping, Fig. 16.39 shows one method of slug selection, double stretch wrapping and parting off, all with continuous motion.

Currently, the end packaging of most conventional powders is in the form of shrink or stretch wrapped packs. Some of the concentrated powders are once again being end-packaged in corrugated board. A double tray packer for the end packaging of economy bags of concentrated powder is illustrated in Fig. 16.40. There seems to be a some movement in the environmental lobby that they might again prefer to revert to corrugated board. This is possibly because corrugated board can either be recycled or it is biodegradable, whereas plastics are not.

Electronics

The mid-1980s saw a revolution in the control of packaging machinery with the arrival of the Programmable Logic Controller (PLC). Up until that time, all packaging machinery had been controlled by relay logic. Some complicated end packaging machines required a multitude of relays with the consequent reliability and maintenance problems.

The first PLCs were a little slow in their scan time and required expert programmers to overcome their inadequacies. Today's PLCs are extremely fast, reliable and relatively user friendly. They can provide all of the machine and line logical function and control, as well as fault finding with suggestions, problem identification, active machine mimics, operator message control, voice alarm systems and many other advantages. The use of

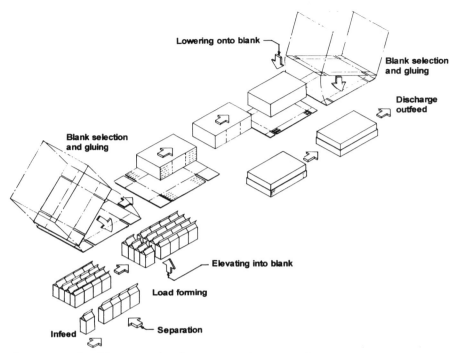

Fig. 16.40.   Double tray packer for economy bags or cartons.

mechanical timing cams operating microswitches, once a very necessary part for the timing of the then modern packaging machinery, has now been superseded by electronic timing cams as a result of the introduction of the PLC. This new system is capable of switching on and off many times during each cycle at speeds of over 500 cycles/min.

Electronic Timing Cams are so sensitive that they can detect a change of only 1°, and making an angular change takes only a few seconds. The easy interfacing of the PLC with other peripherals such as PCs, main frame computers, printers and modems makes it possible for easy access to valuable production data and information such as evaluation of downtime. These can be logged for future reference and ease of production control.

Acknowledgments

Very little has been written about soap and detergent packaging. To present a comprehensive overview, we contacted all of the major equipment suppliers in this field and requested technical data and isometric drawings of their machinery. Most of the drawings received were then modified to obtain a uniform pattern with the others and those we ourselves produced. We wish to extend our appreciation to all of those firms who assisted us and whose name is identified on each specific drawing.

References

1. Spitz, L. (1990) Soap Packaging Machinery and Materials in Soap Technology for the 1990s, edited by L. Spitz, American Oil Chemists' Society, Champaign, pp. 292–320.
2. Babber, M. (1986) The Wiley Encyclopedia of Packaging Technology, J. Wiley & Sons, New York.

Chapter 17

Process Control and Computerization

Barbara Tosco-Filtri[a] and Giuseppe Filtri[b]

[a]Logosystem USA, Inc., Indianapolis, Indiana, U.S.A.

[b]Logosystem S.p.A., Moncalieri (Torino), Italy

The need for process control stems naturally from the application of energy sources other than direct human muscular work. Once some natural, sometimes powerful phenomenon is unleashed, a device is needed to control the rate at which it occurs. The concept of power gain is implicit here, because the little effort required by the control action is, in some cases, amplified to gigantic proportions. Furthermore, the task of manually maintaining a process within desired limits can be annoying, intricate, and costly, so the idea of letting some artificial device provide the control logic comes naturally.

Controlling devices and techniques have been known since ancient times, some of them surprisingly ingenious, but it was not until the discovery of the steam engine Watt Regulator back in the 18th century that they were fully applied on an industrial scale. Even after that, the principles of feedback control, on which modern control devices are based, were not fully understood until after 1930, when control theory was established as a science.

Since then, the progress in the field has been impressive both in terms of theoretical discoveries and employment of new technologies, accounting for a vast variety of applications in practically every industrial domain. But the application of sophisticated control techniques and instrumentation devices to the soap and detergent production has been comparatively slow. Even today, most of the classical Kettle saponification plants still in operation are run manually, and until the end of the 1980s it was not uncommon, especially in comparatively low-wage countries, to see new detergent or continuous saponification plants being designed and built without any computer control.

The control systems technologies applied to chemical processes have changed with time, sometimes stepwise, reflecting the progress in the field, alongside a steady increase of the relative importance of the control systems themselves with respect to the whole projects.

Historically, the technologies applied to chemical processes can be characterized as follows:

- *Local and dedicated instrumentation.* Some measuring instruments were directly installed at floor level and allowed manual process control to be performed by locally placed attendants.

- *Large central control panels.* Sensor readings were transmitted to centralized large arrays of analog and digital displays, alarms and equipment states were flagged on graphic synoptic panels, and buttons and knobs were available in the control area to remotely adjust the process variables and the regulators settings. Control and safety logic was implemented on purposely cabled electromechanical relays. Obviously, although this technology allowed a fairly efficient remote control, it was rather inflexible, and its implementation and maintenance required a vast amount of engineering and construction work.
- *Digital control systems (DCSs).* These consisted of predominantly solid-state and computer-based systems designed to manage large quantities of analog signals. Together with DCSs, process data began to appear on screen monitors rather than on synoptic panels. This technology proved to be effective but expensive and rigid in use.
- *Programmable logic controllers (PLCs).* PLCs are industrial computers specialized to solve logic binary (Boolean) equations of the types often found in process control logic and to input and output large quantities of digital and analog signals. Using PLCs, the same control and safety logic equations once implemented by cabling networks of relays are coded in form of computer programs. Often, but not always, depending on the PLC brands, programs can be prepared and maintained simply by drawing relay circuits on the screen (as if the old technique had to be used), thus allowing plant electricians to use symbols familiar to them. The PLC translates these drawings into computer code. In spite of its strengths, PLCs are in general poorly interfaced to the human process operators, so some sort of man-machine interfaces (MMIs) have to be added and integrated. On the other hand, they are efficient, reliable, and cost-effective.
- *Networked PLCs and computer-based operator interfaces.* This architecture is the technology of today, and it will be described later in this chapter. Unlike its predecessors, systems based on this technology provide not only process control but also data and information management, thus opening new perspectives toward ever more organized production plants.

The most recent developments encompass improvements in the PLC programming tools; fast, reliable, and inexpensive computers; reliable and low-cost networks; and availability of an ever-increasing number of software packages.

These advances in technology have caused and continue to cause a dramatic shift in the attitude and, consequently, in the achievements of the providers of computer control systems. Until the mid-1980s, up to 60% to 70% of the time and effort required to design and implement a medium-size system was devoted to develop in-house general-purpose hardware platforms and basic software building blocks, leaving insufficient resources for the development of the specific application's needs. As a result, computer control systems were often perceived by production operatives as somewhat mysterious and exotic devices emanating from a distant and unrelated culture.

Today this scenario is changing rapidly. Although the applied computer technologies still include peculiar areas such as signal processing algorithms, database structures, and graphic processors, intimate knowledge of the internals of such technologies is no longer required to implement the application functions, and the center of gravity of required skills is gradually shifting from computer technology to plant engineering and process mastering.

This new technological and cultural scenario allows us to outline the conceptual stepwise approach to set up and develop the computer control system for any industrial process, including those pertaining to the fats, oils, soap, and detergent field:

1. Understand the process.
2. Generate a reasonable requirements list.
3. Generate system specifications.
4. Evaluate and select off-the-shelf hardware platforms and software packages.
5. Integrate the off-the-shelf components.
6. Perform commission and startup.

As is evident, at least in the first three steps, close cooperation between production and control systems people is required; if this is not the case, sufficient process knowledge must be available in the systems house.

Even during the integration phase, the production people can be fruitfully involved, because most of the commercially available software packages can be adapted to the application's needs by simply answering some configuration questions and filling in the blanks in some preformatted tables. In general, these operations do not require an information systems background but rather allow the direct integration of the process supplier and end user.

Taking advantage of these technologies, complete automation systems for typical soap and detergent processes can be implemented, delivering flexibility and performance that is superior to that of DCSs at about the cost of traditional synoptic panels.

As recommended by leading institutes for industrial standardization (ISO, IEEE, and so on), the control system architecture should be organized in hierarchical, functionally homogeneous layers. The lower level consists of the hardware devices and basic logic functions providing control of each individual machine and production unit (for example, all the provisions that would allow the remote, but manual, opening and closing of valves or starting and stopping motors). The second layer includes the models and algorithms that instruct the above-mentioned devices on the way to fulfill the whole production task (for example, the logic to open and close valves or start or stop motors in the sequence required by the process). On top of this structure are the networks and the data structures, which allow information pertinent to the whole factory to flow and reach the decision centers in a timely fashion (for example, production scheduling, material tracking, and inventory control).

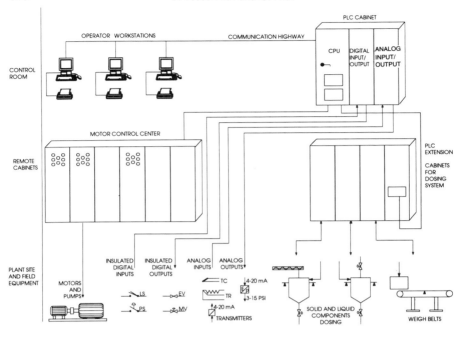

Fig. 17.1. Control system hardware architecture.

Figure 17.1 shows a typical architecture of a control system based on today's technology. The PLC hardware and its interfaces provide real-time control logic and sequencing, safety interlocks, flagging of alarms, actuation of motors and valves, and PID control of any kind of closed loops. The computer-based part of the system is mainly dedicated to recipes programming, process data archiving and reporting, and, above all, providing a friendly and accurate human interface. Since the advent of low-cost graphic stations, the operator has been allowed to issue commands by the now-popular point-and-click technique. Increasingly favored is the use of touch-screens, which allow the operator to interact with the process by touching the symbols of the various process devices as they appear on the screen. For instance, touching the symbol of a valve can make the actual valve toggle between open and close positions.

Overview of Soap and Detergent Technologies

Figures 17.2 and 17.3 summarize main plant and processing steps for the manufacturing of various soap and detergent products. It appears at first glance that the wide variety of chemical processing plants and machinery would require comparable computerization diversity. This is not the case. It is fairly important here to avoid the old misconception (justified in the past by the limited technology

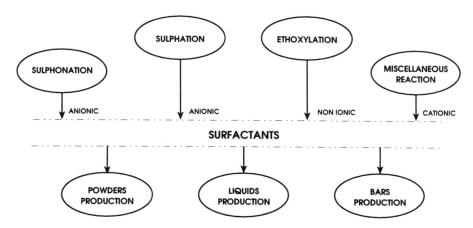

Fig. 17.2. Synthetic detergents processes.

available) of conceiving each automation anew. This significantly raises the cost and reduces the flexibility and the reliability of the system. Instead, the multiplicity and complexity should be considered from different perspectives, and the existing similarities among the various processes should be outlined to allow the application of the same automation technologies and concepts to all of them.

To this purpose, the processes considered here can be distinguished between continuous and discontinuous (batch). On the other hand, some processes are centered on chemical reactions (production of soaps and surfactants), and others are based on thermal and mechanical transformations (production of powders, liquids, and bars). Processes pertaining to the same category will share almost entirely the same control system architecture and functionality. The batch processes usually take advantage of computerization only for the orderly and friendly management of formulas, while the processes themselves are run manually using traditional control devices and instrumentation. In contrast, the continuous processes can today be designed to run almost entirely under computer control.

General Topics of Control and Computerization

Following is a brief discussion of some elements of automation control basics that outlines the building blocks of a generic soap and detergent application: remote command capability, closed loop control, and sequencing.

Remote Command Capability

This was previously done by means of traditional remote control panels, from which the operator could act on every element of the plant from a centralized position. The use of a PLC plus a graphic operator interface drastically increases the number of available command options, because it provides the operator with

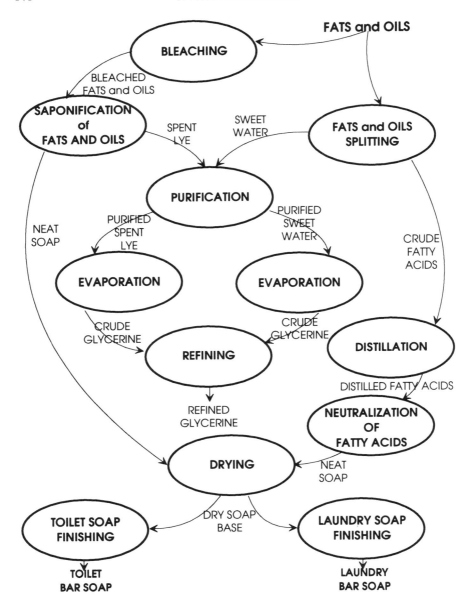

Fig. 17.3. Soap processes.

correlated information about the whole system, such as actual and historical physical data (temperatures, pressures, and so on), therefore improving the effectiveness at which the plant can be run even in manual mode. In practice, the operator sits in front of a computer monitor showing all, or at least the most important parts, of the plant and the values of the related process variables (see, for example, Fig. 17.4). When the operator determines that a change must be carried out to run the process properly (for example, stopping the flow of a certain ingredient), he or she selects the appropriate device using the keyboard's arrow keys, mouse, or other equivalent device and operates on the graphic symbol as if operating on the device itself. The control system is structured in a way to follow these remote commands closely, while the operator observes on the screen the effects of his or her action (for example, the valve pointed to turned from green to white, reflecting its change from open to close, and the flow meter reports a zero flow for the ingredient the operator wanted to stop).

Closed Loop Control

Electrical or pneumatic closed loop regulators have been used for a long time to maintain within limits pressures, temperatures, levels, and so on. The improvement brought in by PLC technology, in which the control algorithms are implemented via software, consisted at first of easier coding of the control parameters, entered in numeric form. More recently, interfacing the PLC with a PC-based console has made possible an even greater tuning capability, allowing trained operators to adjust on-line the characteristics of the loops. Figure 17.5 illustrates a typical screen used for fine-tuning the PID regulation of an ingredient flow. All variables and

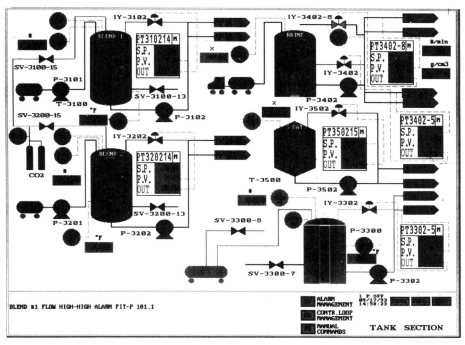

Fig. 17.4. Typical tank-farm screen layout for a saponification plant.

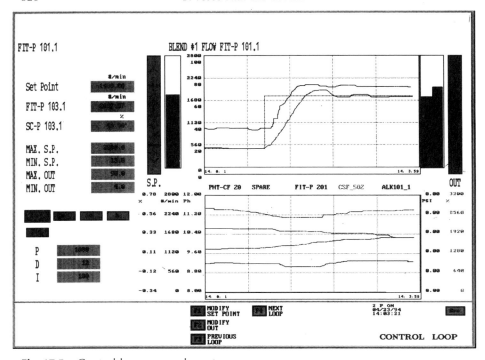

Fig. 17.5. Control loop screen layout.

parameters, as well as the operating modes (manual, automatic, cascaded, recipe), appear on the screen and can be modified at will. The screen also contains two diagram sets, one for the controlled variables and a second for up to five related parameters, used to evaluate in real time the effects of the implemented changes.

From a practical viewpoint, one can distinguish between slow and fast loops, PID algorithms, and threshold regulations. PID algorithms are accurate and react quickly to disturbances, whereas threshold regulations are less accurate but simpler and therefore less expensive.

In any control loop there are four variables:

- The *process variable,* defined as the instantaneous value of the physical parameter under control as measured by a sensor.
- The *set point,* defined as the value that the process variable should eventually reach and maintain.
- The *error variable,* defined as the difference of the previous two.
- The *control variable,* defined as the instantaneous value of the physical parameter used to control the loop (typically the power supplied to the actuator).

In a PID algorithm the control variable is a continuous signal computed as the sum of three contributions: the first proportional to the error variable (P), the second proportional to the integral of the error variable (I) and the third proportional to the derivative of the error variable (D). Tuning the loop consists of choosing (usually by means of a computer screen like the one shown in Fig. 17.5) the three coefficients P, I, and D so as to achieve an accurate, fast, and smooth response. In general, these three requirements tend to be mutually exclusive, so a compromise has to be sought. Without delving into this subject in detail, one can in practice keep in mind the following approximate rules: The P parameter provides a quick control, the I parameter is used to smooth the transition and enhance the steady-state accuracy, and the D parameter is seldom used, at least in typical soap and detergent plants.

In a threshold regulation, the control variable can only assume the statuses ON and OFF, according to a logic suitable to maintain the process variable within preassigned limits around the set point (for instance, if the temperature of a fluid becomes too high, a valve is automatically opened allowing cold water to cool it down).

Sequencing

Sequencing is the main function provided by automation. In fact, the essence of automatic control consists in providing the necessary commands to the machines and other parts of the plant at the exact time and in the exact order according to a preassigned process logic. In general, three main blocks of logic are implemented in the control system: (1) coordination and synchronization logic of the continuous normal production activity; (2) organization of the sometimes complex procedures for plant startup, shutdown, and changeover; and (3) immediate and proper response to emergency conditions. With the use of a friendly PC interface these procedures can become very informative and useful for operator training, leaving the human supervisor free to make high-level decisions concerning how to run production.

When the process is running steadily in automatic mode, the control system keeps the process variables within their limits by modulating valves and motors. This function is of paramount importance to ensure the constant quality of the process product. A well-designed system would, in this phase, monitor how accurately the process formula is being followed. Although special criteria can be implemented for each process, the rule normally applied is the following: If any formula variable de-rates from its assigned limits even for a short time, or the average of its square de-rates from much narrower limits for a sufficiently long time, then an alarm condition is set and an operator's decision is required.

Considering that typical values of these parameters could be no more than 3% de-rating for 5 s and no more than 0.5% average de-rating for 5 min, one can easily conclude that the accuracy in process tracking provided by modern control systems is more than adequate for practically all oil and detergents processes.

In case any warning or alarm condition is detected during normal production, a well designed system should be able to pinpoint the malfunction so to direct the maintenance people to take immediate and appropriate action.

The startup and shutdown procedures are usually organized in sequential steps. Consent to perform the next step of the sequence can either be given by the operator, when he or she determines that it is appropriate (semi-auto mode), or by the control system itself, when it detects that all step criteria have been entirely fulfilled (full automatic mode). In any case the process can be brought up and down very safely and in consistent way, preventing the waste of time required to correct abnormal conditions that too often occur when purely manual startup and shutdown procedures are followed.

The control system is usually designed so that safety rules prevail over any others. Although the most important commands are interlocked (that is, no operation can be done either manually or automatically unless the conditions are present that guarantee equipment and process safety), alarm situations can still occur. Such alarms are analyzed and assigned to one of the following three categories:

- *Process warnings.* The process has reached a critical condition and needs particular attention. No special action is usually taken by the control system, which will nevertheless continue its normal process steering in the attempt to recover.
- *Process failure.* An equipment failure has been detected that would jeopardize the process output quality or, in the long run, cause further and more severe equipment failure. The control system performs an orderly retreat to holding conditions to await manual repair.
- *Safety failure.* An equipment failure has been detected that would jeopardize the safety of the plant itself. The control system shuts down every item in the quickest possible way, disregarding effects on the product's quality and the conditions of the process.

Typical Computer Control Applications

Following are a few examples of automation applied to a few specific soap and detergent processes and a discussion as to what extent the concepts described above are applied in each case. It is not the purpose of this chapter to rehearse the processes themselves; this is done in depth elsewhere, and it is assumed that the reader is well apprised with the basic technologies. For easy understanding, each process has been divided in sections, and the most significant automation requirements are highlighted for each. Here are some common features that every computer control system must provide in any process:

- Remote command of valves, pumps, and motors
- Command logic of valves, pumps, motors, and their interlocks
- Continuous display on the screen of process variables and alarms
- On-line and historical trends
- On-line tuning of PID loops
- Data archiving
- Automatic startup and shutdown procedures and support to manual operations

Each process presents specific needs. The process control and computerization topics relevant to each are listed separately for each process section.

Fatty Acid Neutralization (Fig. 17.6)

Section A—Raw Material Storage
The raw materials for the main reaction, fatty acids of various blends, caustic soda, and salt are usually received from outside suppliers or from other plants in truckloads and are stored in tanks. The required physical storage conditions must be maintained and their inventory precisely reported. The salt must be dissolved in water to obtain a saturated solution. So the control system provides the following:

- Liquid raw material balance, tracking, and report
- Level control (high and low alarms)
- Temperature control (threshold regulations)
- Feeding pressure control

Section B—Dosing
The main purpose of the dosing section is to ensure that all ingredients required by the formula are fed at the exact rate and temperature. Temperatures are controlled by PID technique, adjusting the flow of hot or cold water, as required, in heat exchangers to maintain the feed temperatures within 1° or 2°C.

Dosing of the liquid components can be performed by one of the two methods:

1. By a volumetric pump adjusting the pump speed or its displacement so to maintain the volume rate. The accuracy achievable depends mainly on the pump quality and the mass rate is not controlled.
2. By a gear pump with mass flow meter, adjusting the pump speed with the PID technique so to maintain the mass rate. The overall accuracy depends mainly on the flow meter quality and can routinely reach 1%.

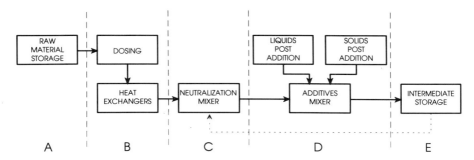

Fig. 17.6. Continuous fatty acids neutralization.

The system can be mechanically designed so that, if a problem occurs, feeding pumps are not stopped but valves are switched to interrupt the reactor feeding and recycle the materials still maintaining the rates required by the formula. A quick and accurate restart can be achieved in this way.

The computer accumulates the materials flows and provides the total materials consumption on a preassigned time base (for instance, daily consumption).

In summary, the following controls are in this section:

* Temperature control (PID)
* Dosing by volumetric pump with speed and/or displacement control or gear pump with flow meter and PID speed control
* Pump recycle logic
* Flow totalizers

Section C—Neutralization

In this section the neutralization reaction occurs; consequently, the main automation purpose is to monitor its main parameters: temperature, viscosity, and pH.

Temperature poses no problems. Viscosity can be directly measured by means of probes (which are rather expensive) or, more simply, by reading the electric power absorbed by the mixing process. The pH measure provides not only the most important reaction parameter but also a way to achieve constant neutralization degree. For this purpose, the reactor pH is constantly compared with a target set point. The error is processed by a carefully tuned PID controller that, in turn, adjusts the set points of the raw material pumps (this technique is called cascaded PID control). Consequently the reaction pH is brought and maintained to the desired value.

Both viscosity and pH are suitable for product quality on-line control, such as statistical process control (SPC).

In summary, this section includes the following:

* Alarms (neat soap viscosity and reactor temperature)
* pH control (PID control cascading on set points of raw material flows)

Figure 17.7 shows an example of the implementation of these concepts.

Section D—Additives Mixing

Most of the problems encountered here are similar to those already discussed for the dosing section. Usually solid components (powders) are added and mixed here in addition to the liquid ones. Solids powders are dosed with weigh belts with PID speed control to maintain the total rates to the desired targets.

If more salt is needed, it can be added here in the form of diluted brine. The brine concentration can be controlled by computer by adding water to saturated brine and checking, under PID control, the solution's specific gravity.

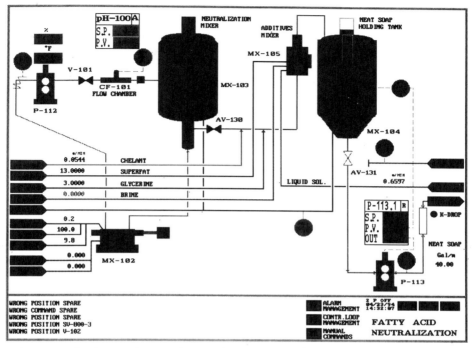

Fig. 17.7. Continuous fatty acids neutralization screen layout.

The level in the mixer is also PID controlled by adjusting the speed of the unloading pump so that, if the raw materials injection rate is changed or even stopped, the unloading rate will follow in an exact manner.

In summary, the following controls are in this section:

- Liquid and solid material balance, tracking, and report
- Level control (high and low alarms)
- PID dilution control for brine solution
- Liquid dosing by volumetric pump with speed control and/or displacement control or gear pump with mass flow meter and PID speed control
- Solid dosing by weigh belt (PID speed control)
- Pumps and motors recycle logic
- Flow totalizers
- Post addition mixer PID level or pressure control

Section E—Intermediate Storage
In this section the soap is brought either to intermediate storage tanks or, if the quality is not adequate enough, to an off-quality tank for reprocessing. The

switching between the two destinations is usually performed manually by the operator. When there are two or more tanks for soap intermediate storage, the computer selects (or helps the operator select) the appropriate tank, depending on the type of formula being produced and storage space availability. The rate at which soap is unloaded from the intermediate tanks for further processing (soap drying) is fed back to the control computer, which will accordingly set the rate at which soap is produced. Consequently, the computer recalculates the set point flow of each feeding pump in order to maintain the correct formula ratios.

In summary, this section includes the following:

- High-level control—feed back to production rate
- Storage and distribution logic
- Off-quality management

As already mentioned, the computer also takes care of the plant as a whole, notably plant safety, and provides the general facilities listed above as common features. In addition, this process requires the following computer services:

- Recipe management (editing, archiving, and downloading)
- Production rate on line control
- Changeover procedures
- Plant recycle logic

Sulfonation (Fig. 17.8)

Having previously discussed in some detail the continuous fatty acid neutralization process, we will assume that the reader is by now somewhat familiar with the applied concepts, so we will proceed with shorter descriptions for similar automation problems.

Section A—Raw Material Storage and Sulfur Melting

The only new automation subject here regards sulfur melting, the important parameter of which being its temperature. This is continuously measured, compared with limits to generate various levels of alarms, and brought to the operator's attention on the screen. No regulation is performed in general, because temperature melting is not critical.

Briefly, this is the list of the computer-provided services in this section:

- Liquid and solid raw material balance, tracking, and report
- Level control (high and low alarms)
- Sulfur melting (temperature alarm)
- Feeding pressure control

Section B—SO_2 and SO_3 Production

Air drying requires moisture and silica gel bed control. Usually there are two of

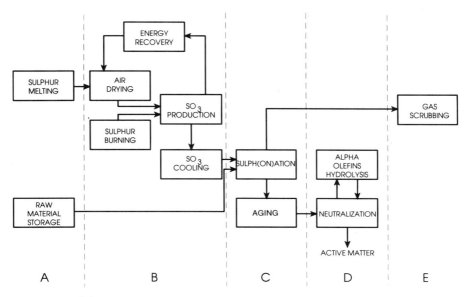

Fig. 17.8. Sulph(on)ation.

them, operating in alternate phases. While one is drying air, the other is being regenerated. The switching conditions are checked by computer.

The SO_2 and SO_3 production, although requiring highly specialized equipment, does not create any new problems for the computer control system, which provides the following functions:

- Moisture control on air drying
- Silica gel beds regeneration logic
- Threshold regulation of air pressure
- Sulfur dosing by volumetric pump with speed control and/or displacement control
- SO_2 temperature control (PID)
- Conversion tower temperatures control (PID)
- Flow totalizers

Section C—Sulfonation
In spite of the wide difference between the two chemical processes, the automation problems in this section are similar to the ones encountered in the neutralization of fatty acids. The reaction between two components must be controlled; thus they must be dosed according to the formula, and the rate of one of the reactants must be finely corrected in order to achieve the desired sulfonation degree. In this case, however, one component is gaseous (SO_3). The SO_3 flow can be calculated in the

computer as the square root of the differential pressure across a disk flow meter and used to adjust, under PID control, the feeding pressure until the calculated flow reaches the set point. The liquid dosing is achieved in one of the ways already discussed, and the degree of sulfonation is maintained by a PID cascade like the one for the pH control of the soap neutralization reaction.

In summary, the following are required:

• Liquid and gaseous dosing (PID)
• Degree of sulfonation control cascading on set point of reactant

Section D—Neutralization

Again, the neutralization reaction is controlled by the computer system as already discussed, providing the following functions:

• Dosing
• Reactor temperature control
• pH control cascading on set point of reactants

Section E—Gas Scrubbing

There is no need in this section for sophisticated control. Only the pH of the scrubbing solution needs to be controlled, so we list only this function:

• pH control of the scrubbing solution (PID)

As in the previous example, the sulfonation process can be assisted by computer driven startup, shutdown, and changeover procedures, although in some small installations they are performed manually by the operator at the computer screen for the sake of simplicity. A more significant automatic procedure is required in sulfonation plants in case of emergency shutdown. As was said, the SO_3 and the liquid material to be sulfonated must be fed into the reactor at the correct rate to ensure the right reaction degree, and we have shown how the computer takes care of that. If, however, the liquid flow de-rates from the set point for a preassigned time or, to a greater extent, if the flow stops completely, the SO_3 excess will cause severe damage to the reactor's surface, so an alarm must be generated and an emergency shutdown initiated. First, the SO_3 inlet valve must be closed; then the liquid emergency vessel valve should be opened to allow excess liquid to react with the remaining SO_3 and wash out any leftover trace. Finally, the orderly shut-down the whole plant can be completed.

This provides an example, of rather general significance, of how the detection of a local malfunction can sometimes generate a quite extensive action by the supervising computer, so we list this automatic procedure as the most significant in the plant:

• Emergency reactor washing logic

Soap Drying (Fig. 17.9)

Section A—Drying

The automation problem here is to achieve the desired moisture level in the dried soap. In most of the installed systems the soap is first pumped into a heat exchanger, where some moisture is preevaporated, the amount being a function of the steam pressure and back pressure. Subsequently, the soap is sprayed into a vacuum chamber, where the moisture removal takes place. The control system measures the moisture level at the output of the vacuum spray dryer, compares it with the set point, and modulates the steam input valve of the heat exchanger (by the PID technique) to achieve and maintain the target drying effect.

The flow of soap in the dryer is also controlled by a PID in order to synchronize its rate to the levels of the input and output storage tanks.

The summarized list of the automation services is as follows:

- PID moisture control acting on steam pressure of first-stage heat exchanger
- PID production rate control

Section B—Soap Fines Recovery

In the soap fines recovery cyclones only standard monitoring and alarming functions are required. The same applies to the vacuum systems for vapors condensing, where in some cases a steam jet booster is required (when the cooling water temperature is too high). The computer control provides the PID adjustment of the steam pressure, which we list as the only significant automation function in this part of the system:

- PID steam pressure control on booster

Section C—Dry Soap Transport and Storage

The dry soap, in the form of pellets, is taken by a pneumatic transport system to a number of storage tanks. Although the transport system itself is quite simple and does not need automation support, the levels in the tanks need to be controlled. The switching between the tanks is also done automatically because it must be quick and synchronized to avoid pressure waves in the system. The switching command can be automatically generated by computer according to the desired distribution logic or, in most cases, provided by the operator at the computer console. If the computer detects the reaching of critical available space in the storage tanks, an alarm is generated, and this information is fed back to the drying section for reducing the soap flow.

In summary, the computer control provides the following automation functions:

- High-level control—feed back to production rate
- Storage and distribution logic

Figure 17.10 shows the main control screen for a typical drying process.

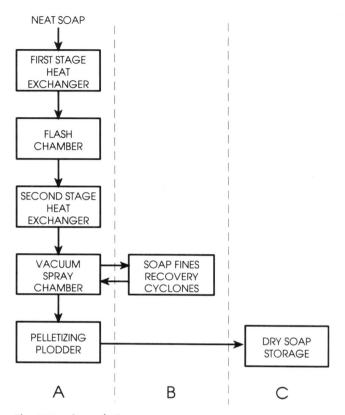

Fig. 17.9. Soap drying.

Bar Soap Finishing (Fig. 17.11)

Section A—Mixing

The dry soap base is mixed in this section with liquid and solid additives providing most of the attributes and qualities under which the final product will be marketed. Clearly, this generates the need for high operational flexibility, which, not surprisingly, can be achieved and enhanced by computer control, and indeed many automation functions are applicable to this operation. However, we will limit ourselves to list them in concise form, because the mixing problems have already been discussed in the previous examples.

- Liquid and solid raw material balance, tracking, and report
- Level control (high and low alarms)
- Recipes management (editing, archiving, and downloading)

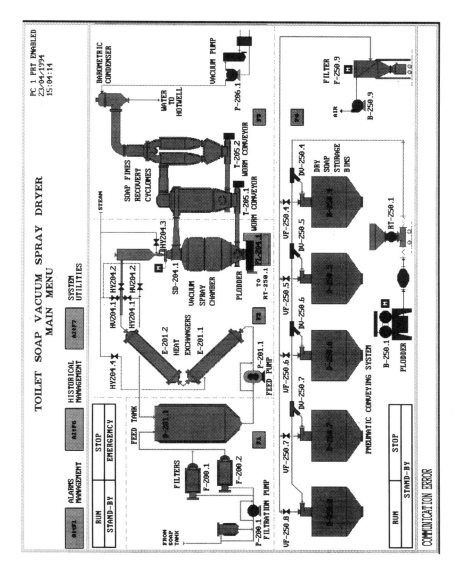

Fig. 17.10. Soap drying screen layout.

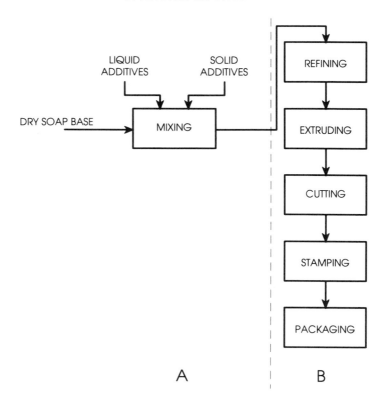

Fig. 17.11. Bar soap finishing.

- Liquid dosing by volumetric pump with speed control and/or displacement control
- Solid dosing by dosing belt (PID control on speed)
- Flow totalizers

Section B—Refining, Extruding, Cutting, Stamping, and Packaging
The operations grouped in this section are performed by a sequence of machines connected in cascade, so the main automation problem, at system level, is posed by their proper synchronization. This is provided by a computer logic that automatically stops and restarts (or, when allowed by the implemented technology, adjusts the rate of) each individual unit to allow a continuous and smooth average flow. This is made easier, in practice, by the constitution of the machines, which are normally designed and built to be highly automatic. Often each of them contains a dedicated microcomputer controlling the moving parts through logic equations implemented by relay networks and mechanical expedients. This way the machine

setup operations are substantially simplified, in some cases been reduced to the introduction of numeric parameters at the machine console. An example of a command screen for such automatic equipment is given in Fig. 17.12, in this case pertaining to a stamping machine.

In our condensed list we highlight the following:

- Flow synchronization logic between machines
- Temperature alarms
- Electronically controlled machines

Powdered Detergent Production (Fig. 17.13)

The automation problems encountered in this plant consist mainly of dosing and mixing liquid and solid components of various kinds and keeping the machines in working condition within appropriate limits. All of this has been considered at length in the previous examples, so we limit ourselves to listing them in a concise form.

Section A—Liquid and Solid Raw Material Storage

- Liquid and solid raw material balance, tracking, and report
- Level control (high and low alarms)

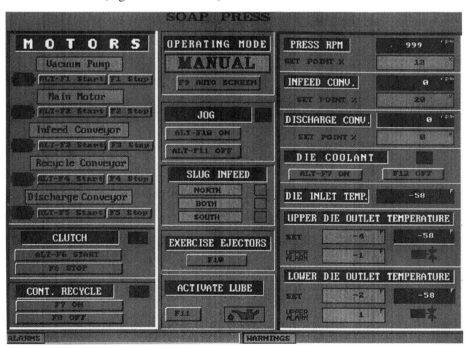

Fig. 17.12. Soap press screen layout.

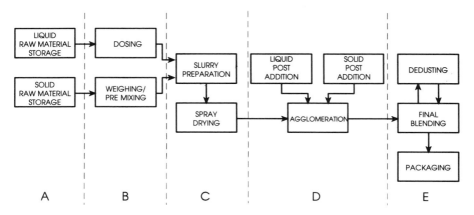

Fig. 17.13. Powdered detergent production.

- Temperature control (threshold regulations)
- Feeding pressure control

Section B—Liquid Dosing and Solid Weighing

- Liquid dosing by volumetric pump with speed control and/or displacement control
- Solid dosing by weigh belt (PID control on speed)
- Flow totalizers

Section C—Slurry Preparation and Spray Drying

- Alarms (powder density and moisture)

Section D—Post Addition and Agglomeration

- Liquid and solid material balance, tracking, and report
- Level control (high and low alarms)
- Liquid dosing by volumetric pump with speed control and/or displacement control
- Solid dosing by weigh belt (PID control on speed)
- Flow totalizers

Section E—Dedusting, Blending, and Packaging

- Dosing
- Storage and distribution logic

In addition, the control system provides the following plantwide services and automatic procedures:

- Recipe management (editing, archiving, and downloading)
- Production rate on-line control
- Changeover between both compatible and incompatible formulas

Control System Design Specifications

In Fig. 17.14 is a complete checklist of the functions that a modern computer control system makes available for implementation in the various soap and detergent projects, some in addition to those already cited in the previous chapters. These functions are to be specified in detail when designing a new system, because they will determine the cost of the automation. The given definitions briefly outline the computerization concepts and are intended to provide guidelines to the design of a new system. They also may provide a key to evaluate and rate the control packages available on the market. To this purpose we list the general characteristics to which any control system should comply, regardless the particular process to which it is applied.

Digital and Analog Signals and Control Devices

- At power-up all outputs and controllers should be set to a designated safe state.
- Switching from direct output setting to automatic loop control of the PIDs should be bumpless.
- Process variable, set point, control variable, and loop status (automatic/manual) should be displayed for each loop.
- Each loop should have a tag number and a description of its function.
- Each loop should have tuning facilities by means of the operator's console and trend diagrams to show the loop behavior.
- Analog inputs and outputs should be converted and scaled to engineering units.
- Analog inputs should be checked for validation limits.
- Access to a number of functions should be restricted via password.
- Safety-critical commands should be hardwire interlocked.
- All relevant motors and valves should be commanded by the control system.
- All motors and valves should transmit run/stop and position status to the control system.
- Modes of operation should include direct manual control and automatic mode. In direct manual control the operator should be enabled to control all individual devices and loop set points. In automatic mode all of them should be under the computer command.

Operator Interface

- Color animation should show the plant in P&ID form and be consistent throughout the different screens; symbols should be consistent with international standards.

PROCESS CONTROL
◊ Interlock Logic
◊ Plant Safety Procedures
◊ Start-up and Shut-down Sequences
◊ PID Control of Analog Loops
◊ Logic Control of Digital Signals
◊ Operation Mode Control (Manual/Automatic)

PROCESS SUPERVISION
◊ Acquisition of Instrument Readings
◊ Acquisition of Equipment Status
◊ Linearization, Scaling and Calculation of Process Variables
◊ Threshold Limit Checking and Logging for Warnings and Safety Alarms
◊ Numeric and Graphic Display of Variables and Alarms
◊ Long Term Archiving of Process Variables and Events
◊ Trends
◊ Remote Command of Motors, Valves and PID Regulators
◊ Materials and Thermal Flow Balance

PRODUCTION SCHEDULING
◊ Definition and Maintenance of Production Recipes
◊ Calculation of Required Mass Flows
◊ Scheduling of Production Formula or Rate Changes

MAINTENANCE
◊ Scheduling of Preventive Maintenance
◊ Logging and Tracing of Maintenance and Repair Interventions

PRODUCTION REPORTING
◊ Reports of Accumulated Production related to Formulas, Shifts, Calendar
◊ Raw Material, In Process and Finished Product Tracking
◊ Actual versus Scheduled Production and Reasons for Derating
◊ Equipment Down Time Reporting
◊ Quality Analysis and SPC/SQC Control

Fig. 17.14 Control system design checklist.

- Recipe handling should be friendly and allow manual entry by means of the system keyboard or downloading from host computers.
- Alarms should appear consistently in different screens and should provide date, time, tag ID, and description.

- Action of the keyboard keys and screen buttons should be consistent throughout different screen layouts.
- Current alarms should be displayed on every page.
- Alarms should be acknowledged individually or collectively at the operator's choice. Unless otherwise required by the process, the equipment should not restart automatically when alarms are accepted; a deliberate restart command should be given through the user interface.
- Whenever possible, messages pertaining to alarms that are direct and unavoidable consequences of a previous alarm situation should be suppressed.
- In case of alarm, related input tags and troubleshooting information should be displayed on the screen.

Hardware and Software Design

- Hardware should be modular and easily expandable.
- All inputs and outputs should be insulated and easily disconnectable from the control system by means of circuit breakers in the terminal board.
- Critical control circuits should have duplicated power supplies.
- Application software should be designed in modular blocks.
- Application software should be easily expandable both in number of devices controlled and in functionality.
- Numeric constants reflecting the process dimensions (such as number of controlled devices and number of inputs/outputs) should be in the form of parameters defined only once throughout the whole application.

Advanced Topics and Future Trends

The computerization of soap and detergent processes and machinery will continue to benefit from the improvements in the relevant technologies and will include application areas already available but not yet fully exploited. The developments listed below are expected to be at hand in the near future and provide a general idea of the direction in which computerization of processes will evolve.

- Development and availability of ever more powerful PLCs, specifically conceived to be used as nodes of integrated networks.
- Process supervision software packages that are even more graphics oriented than today, able to integrate video recorders and players, live TV cameras, sound, and computer-generated images (multimedia systems).
- Integration of the above packages with standard data processing systems and data structures, so that the output from the process control computers will be directly transferred and further processed by any of today's common office

automation packages. The process control computer will therefore be integrated with the factory management procedures on higher-level computers.

- More extensive development of mathematical models of processes, with the purpose of increasing the operation efficiency, allowing the computer to adjust some or all of the set points whenever it detects de-rating of some significant product quality parameter from its optimum value.

Additional Suggested Readings

1. Astrom, K.J., *Automatic Tuning of PID Controllers,* Instrument Society of America, 1994.
2. Batten, G.L., Jr., *Programmable Controllers,* McGraw-Hill, New York, 1994.
3. DiStefano, J.J., III, A.R. Stubberud, and I.J. Williams, Schaum's Outline Series: *Feedback and Control Systems,* McGraw-Hill, New York, 1990.
4. Gottelieb, I.M., *Power Supplies, Switching Regulators, Inverters and Converters,* McGraw-Hill, New York, 1994.
5. Grant, E.L., and R.S. Leavenworth, *Statistical Quality Control,* McGraw-Hill, New York, 1988.
6. Kompass, E.J., *Man-Machine Interfaces for Industrial Control,* Cahiers des Plaines, 1983.
7. Lewis, J.W., *Modeling Engineering Systems,* HighText Publications, 1994.
8. Ott, E.R., and E.D. Schilling, *Process Quality Control,* McGraw-Hill, New York, 1990.

Chapter 18

Market Trends and Fragrance Technology

Thomas McGee

Givaudan-Roure Corporation, Teaneck, New Jersey, U.S.A.

Introduction

The 1980s saw changes in socioeconomic values. These changes have been influential in causing soap and detergent manufacturers to become much more conscious that innovation is required to participate successfully in the market. Before the mid-1980s innovation in soaps and detergents involved mainly marketing changes. New technologies were introduced only after very careful laboratory and consumer testing. The slow technological evolution of these markets is exemplified by the length of time it took to introduce the now commonly used protease enzymes into main-brand detergents. Similarly, although synthetic detergents (syndets) were first used in personal wash (PW) bars in the 1950s, it was only after the mid-1980s that they started to be widely used.

One of the major socioeconomic factors that has influenced the soap and detergent market's dynamics is the increased consumer awareness of the environment. Environmental concerns became a major preoccupation in the 1980s. Biodegradability, landfill, energy conservation, renewable sources, and the like entered the consumer's vocabulary, resulting in changes in use habits, formulations, packaging, and the "green" positioning of some brands.

Lifestyles also changed. People became more leisure oriented and colored leisure garments became more common in the wash load. Women also assumed the dual role of worker and homemaker, which left them with less time. Hence products were sought that were multifunctional and easy to use. Convenience became a desirable brand attribute.

The Northern Hemisphere's populations also aged. Products that softened the effects of aging were sought. This growing awareness of skin care started to influence the PW category. Technology that offered skin benefits was introduced.

Consumers became more value conscious as the rise in real incomes decelerated in the developed Northern Hemisphere markets. For example, in the United States the average annual growth in personal incomes in 1984 was 4.4%; today the growth rate has declined to 1.5%. The growth of megasupermarket chains in the Northern Hemisphere, and their buying power, lowered the price that the soap and detergent manufacturers could obtain for their brands. These factors increased the pressure to be cost-effective. Additional cost pressure came from the widespread introduction of the supermarket chain's own products, which competed with branded products at a significantly lower cost. Supermarket brands have been the fastest-growing segment and are approaching 20% of the soap and detergent market in Europe and 15% in the United States.

Another force driving innovation in the soap and detergent markets is the need for the manufacturers to differentiate their brands. This is necessary to maintain their market share in the highly competitive, mature markets of Europe and North America and increasingly to maintain or increase their share in the rapidly developing markets of South America and Asia Pacific. Today, for success, a brand must have a benefit that clearly justifies the price, and this requires continuous technological innovation.

Below is a brief review of how the changes in socioeconomic factors have affected the soap and detergent market trends around the world. The innovative approaches that Givaudan-Roure, as a fragrance supplier, has had to adopt to keep pace with the new market dynamics are also outlined.

Personal Wash Market Trends

The major change in the personal wash market has been the replacement of soap by syndet products, both in liquid and bar forms. The wider range of syndet-active systems that can be used greatly enhances the performance and sensory benefits that can be delivered. Fragrance innovation in the PW market has also been possible due to the introduction of PW liquid products, which allow a much wider range of fragrance materials to be used.

PW Liquids

The PW habit in Europe has changed dramatically. Showers have replaced baths as the most frequent bathing practice. The greater involvement of people in sports activities, time pressure on women, and the economic and environmental factors of the relatively lower cost and energy consumption of showers has driven this habit change. The frequency of showering in the major European countries, on average, is now 70% of the bathing occasions.

The change to showering has caused a dramatic increase in shower gels, to the detriment of soap bars. Figure 19.1 shows the growth of their value share of the European PW market. In 1992 this ranged from 85% in Spain to around 64% in the UK, France, and Italy. The average growth in 1992 was 12%.

The change from traditional bathing to showering highlighted the poor lathering of soap in the hard water that is prevalent in Europe. The less calcium-sensitive syndet-active systems have a perceivable rate of lathering performance benefit over soap. They also do not form unsightly calcium soap scum. These benefits met consumer needs and initially compensated for their higher cost per wash of shower gels.

Shower gel growth was sustained by additional benefits such as improved skin feel and high-quality fragrance. Innovative fragrances, based on fine fragrance

Value Growth

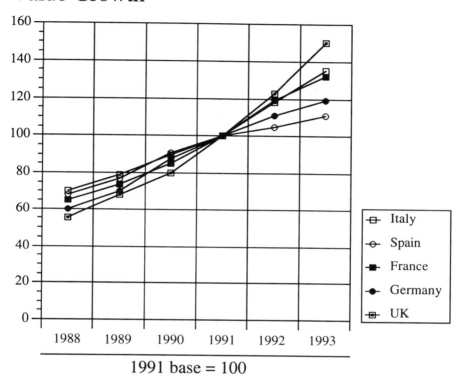

Fig. 18.1. The value growth of shower gels in Europe.

types, can be incorporated more easily into gels than bars. The processing requirements of soap greatly restrict the type of fragrance materials that can be used.

The gel market is much more fragmented than the bar market and has many
more brands and variants. A look at the leading brands across Europe shows that
there is little commonality among the countries. A Northern Europe–Southern
Europe split is evident. In the Southern countries skin care and family use categories
are important. In Northern Europe brand variants tend to have hedonistic value.
Currently there is not a common brand among the five market leaders in the major
European markets. However, there is an emerging trend for pan-European brands.

The seduction/beauty brands are evolving to reflect the growing concern with
aging and skin care. Major introductions have been Dove and Oil of Olay shower
gels. Technology support for these brands resides primarily on surfactant systems
designed to be milder for dry and delicate skin. Moisturizing, 2-in-1 shower and
lotion, light and soap-free variants have all been introduced on the market to provide

both cleaning and care for the skin. Marketing innovations are also addressing aging skin by claiming ingredients used in skin creams for shower gels, for example, alpha hydroxy acids, vitamins and deep moisturizers. The caring concept is also reflected in the trend of the natural/ecology brands to move to natural care and purity positionings. The "neutral" concept, which was specific to the Italian and Spanish markets, is also part of this care trend and is being adopted by other countries. Fragrance-driven freshness variants are moving toward unisex and natural freshness arenas.

The North American PW market is essentially a bar market. Liquid products in 1993 accounted for only approximately 17% of the value share of the PW market (Fig. 18.2). The majority of the liquid products are specifically for hand washing. In the liquid handwash market the main innovation has been antibacterial products that now account for 46% of the liquids market. Figure 18.3 shows the growth of antibacterial PW liquids; Dial was the first introduction in 1988, followed by other major brands. The sales value of antibacterial liquids grew in 1993 to $140 million.

The shower gel market accounts for only 3% value share of the U.S. PW market. Although this market is small it is seen by many companies as a growth area; however, several barriers to this growth exist:

1. U.S. water is relatively soft, and the rate of lather generation by soap bars is high.
2. Syndet bars have had major presence for many years on the U.S. market.
3. Lathering with showers gels under high-pressure showers frequently found in North America is difficult without using prelathered wash cloths or sponges, which are commonly used in the shower in Europe.

The most recent shower gel launches are trying to overcome the lathering problem by giving either a sponge or a puff to encourage prelathering. Early indications

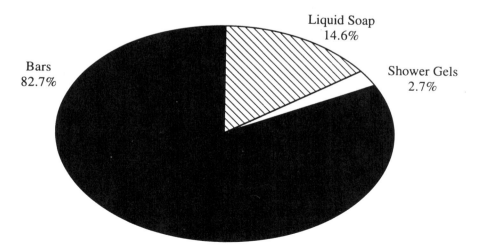

Fig. 18.2. Value shares of the 1993 U.S. personal wash market.

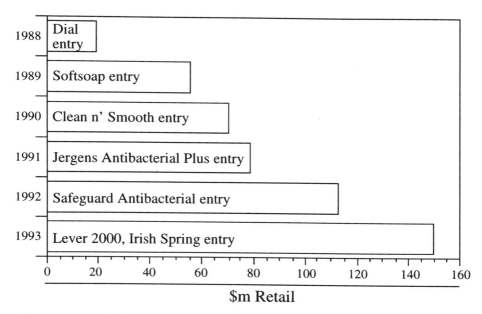

Fig. 18.3. Growth of antibacterial handwash liquids.

are that the increased market activity in this segment is creating growth in the shower gel market. Its full potential in the United States, however, remains to be seen.

In Asia Pacific the share of liquids of the PW market for the major countries is shown in Table 18.1. The acceptance of liquid products is country dependent. India and the Philippines, for example, are 100% bar markets. However, in countries such as Korea, Japan, and Taiwan shower gels are growing fast, and reached approximately a 35% market share in 1993.

In Southeast Asia facial wash liquids are an important PW category. Their use in the major Asia Pacific countries is shown in Table 18.2. Facial care is considered

TABLE 18.1
Personal Wash Products Used in Asia Pacific Countries, Bar Soaps vs. Shower Gels[a]

Country	Soap (%)	Shower Gels (%)
Australia	98	2
India	100	Negligible
Indonesia	95	5
Japan	70	30
Korea	65	35
Malaysia	75	25
Philippines	100	Negligible
Taiwan	60	40
Thailand	90	10

[a]1993 market value at RSP.

TABLE 18.2
Product Usage in Asia Pacific to Wash the Face

Country	Soap (%)	Facial Cleansers
Australia	30	70
India	100	0
Indonesia[a]	15	80
Japan	5	95
Korea	70	30
Malaysia	70	30
Philippines	95	5
Taiwan	20	80
Thailand	70	30

[a]5% of Indonesian women do not use any facial product for fear of irritation to skin.

to be very special in these countries, so much so that in Indonesia 5% of women use only water to wash their faces. Not surprisingly, mild active technology, such as Kao's monoalkyl phosphate, drives this category.

The South American liquids market is very small. Some introductions such as Lux shower gel have been made but as yet there is no significant movement in this segment.

PW Bar Markets

The bar market is fighting back. In Europe after years of continuous decline the bar market is growing in value terms. The reason for this is the introduction of syndet bars. We speculate that the consumer's experience of shower gels has set different standards of performance and price, making their acceptance in Europe much easier. They are also being positioned to reflect the skin care concerns outlined above.

Unilever's Dove was the pioneer on the European market. It was launched in Italy in 1989 and by 1993 had achieved 12% of the PW bar market. It was rolled out in Europe in 1992 and by 1993 it had gained leadership of the European PW bar market, with a value share of around 9%. Dove focuses on the consumer need of skin care.

Dove's success has stimulated several other manufacturers to launch syndet bars. It is too early to say how this will ultimately affect the shower gel market. Time pressure reduces the occasions when it is possible to indulge in pampering. The hedonic pleasure that the higher-quality fragrances possible in shower gels provide consumers may reduce the inroads that syndet bars will make into the shower gel market.

In South America the bar market is still a very traditional market. The major innovation on these markets has been the introduction of long-lasting soaps from the Chilean concept, executed by Le Sanci (Unilever) and Charmis (Colgate). Imported Dove was introduced in the major South American markets in 1992 and in value terms has made a good start. As in Europe and the United States, there are indications that mildness and care for aging skin will be the trends.

The Asia Pacific bar markets are still based on soap technology. Syndet bar acceptance is being tested in Asia Pacific. However, some doubt exists as to whether they will be successful. The current syndet technology leaves the skin feeling "moisturized and slippy," and Asian consumers apparently prefer the "squeaky-clean" feel of soap. A major segment in several Asian markets is the family health/antibacterial category. Brands such as Lifebuoy (Lever) , Safeguard (P&G), and Protex (Colgate) lead this category.

PW Emerging Technology

The patents that the major manufacturers, P&G, Unilever, and Colgate, have filed in the last year in the PW area were reviewed. This review shows what new and improved technologies the manufacturers are focusing on. Most of the patents are on new actives (1) or additives (2) that deliver skin care benefits. This reflects the market trend of products that will ameliorate problems of dryness and skin suppleness as they address perceived consumer needs.

New technology might be also stimulated if the proposed U.S. FDA tentative monograph on antiseptic drug products (3) is adopted. It would severely challenge antimicrobial PW products. This could require the development of ultra deodorant fragrances, which would allow strong freshness-all-day claims.

Fabric Wash Market Trends

The socioeconomic changes that occurred in the in the 1980s has had a greater impact on the detergent market than the PW markets. The growing awareness of environmental care resulted in a change of the fabric wash product active. The commonly used nonbiodegradable, branched chain dodecyl benzene sulphonate surfactant was replaced by the biodegradable, linear dodecyl benzene sulphonate. In the Northern Hemisphere, the builder changed from sodium tripolyphosphate to builders, such as zeolites, carbonates, and citrates, that do not cause eutrophication problems. Changes in surfactants in the 1990s, associated with care for the environment, are actives from renewable sources, such as coconut alcohol sulphate, coconut polyglucoside, coconut fatty acid ester sulphonate, and coconut methyl glucamide. Actives and builders are the workhorses of fabric wash product formulations, and any new product innovations will have to satisfy all the requirements of ecology, cost, and performance.

An outcome of the environmentalists' concern about conserving energy has been the move in Europe toward lower fabric-washing temperatures. The average wash temperature in Europe has dropped from 75°C in the 1970s to 50°C in the 1990s. This change in wash habit has reduced the wash performance of heavy-duty products. At the lower temperatures now prevalent, sodium perborate does not function efficiently as a bleach, and the detergent's ability to remove oily soils is greatly reduced.

The need to improve perborate bleaching at the lower wash temperatures caused the rediscovery of tetraacetyl ethylenediamine (TAED). It was invented over 30 years ago. It acts in combination with perborate to liberate peracetic acid, which generates active oxygen at lower temperatures. TAED was introduced in the late 1970s, and as the wash temperatures fell in the 1980s its use was greatly expanded. It is now the most commonly used bleach precursor in Europe.

In Europe, whiteness/stain removal remains the most powerful advertising claim, and intensive research has been devoted to the development of even better bleaching systems than TAED. A major innovation was the introduction in Europe by Unilever of the next-generation, low-temperature bleach. The bleach system is based on manganese activation of perborate. This gives good bleaching efficiency at temperatures as low as 20°C (4). This technology improves the spot bleaching and "pinholing" associated with metal-catalyzed perborate bleaching (5, 6). Its new formulation also includes a new type of zeolite and the more biodegradable primary alcohol sulphate surfactant. This technology allowed Unilever to make performance superiority claims. It was marketed under the names Persil Power, OMO Power, and the like.

However, P&G and some consumer product testing groups have shown that under some wash conditions this manganese-catalyzed bleaching system can damage the fibers of the clothes. Unilever initially reformulated its products, reducing the level of manganese activator to alleviate consumer concern, and the company has recently withdrawn it in Europe. The manganese activator technology is still being used in some ambient-wash-temperature countries and as a niche product in Europe.

P&G has also responded technologically to the performance challenge by its launch of Ariel Futur. This product is thought to have an innovative formulation mix that uses several safer routes to improved fabric wash performance at lower temperatures. The formula contains their sugar-based active methyl glucosamide, which in combination with conventional nonionic and anionic surfactants yields good oily soil removal (7). Its improved bleaching system is based on high levels of TAED and percarbonate. This in combination with a cocktail of five enzymes provides improved stain removal at low temperatures.

Unilever made a second attempt, launching a new formula under names such as New Generation Persil and OMO Total. The formula uses the base technology of the Power formula without the accelerator. Unilever also launched Persil Finesse, which is targeted at fine fabrics and wool.

In the United States the short wash time, cooler wash temperatures, and lower product concentrations, compared with Europe, have not favored the use of in-powder bleaches. To remove stains from white articles, a large amount of liquid hypochlorite bleach is used. However, the socioeconomic pressure for multifunctional convenience products prompted the introduction of in-powder bleach in 1985. The bleach was sodium perborate monohydrate, which dissolves faster than the tetrahydrate and is more suitable to the shorter wash times of the United States. In 1989 a bleach precursor, nonanoyloxybenzene sulfonates (NOBS), was used by P&G in combination with perborate monohydrate to improve bleaching performance. This is favored over TAED as it performs better at the lower concentrations, shorter wash times, and lower temperatures used in North America.

Improvements have been made to overcome the reduced cleaning efficiency of oily soils at the lower temperatures. Innovation in biotechnology has allowed the introduction of improved protease and lipase enzymes to help catalyze the breakdown and removal of oily soils at lower temperatures. Nonionic actives are now being used at higher levels in the active system, because of their greater oily soil solubilization properties.

South America and Asia Pacific, who have ambient-temperature wash conditions, have also benefited from the technological innovations of the Northern Hemisphere in response to the environmentalist pressure. Multienzyme, low-temperature bleaches and nonionic actives have all been introduced in specific markets in these regions. The photobleach aluminum phthalocyanine sulphonate has also been used in these markets.

The changes in 1980s lifestyles have caused a marked increase in the amount of easy-care and colored articles in the wash load. This has helped the introduction of heavy-duty fabric washing liquids, which are perceived as gentler than powders and more suitable for easy-care and colored fabrics. Liquids are also easier for spot treatment of specific oily soils, such as collar and cuff grime, that are not as easy to remove at lower wash temperatures.

The liquid formulations are either isotropic with high active levels or are structured to suspend solids such as builders. Structuring is achieved either by the formulation of the active system and electrolytes to form a suspending lamellar phase or by polymers. Normally the liquids contain high levels of enzymes as the stain-removal system. They were first introduced in the United States, and liquid products now account for around 40% of the heavy-duty fabric wash market.

The market share of liquids in the major Asia Pacific markets is shown in Table 18.3. They are relatively successful in Australia but not in the rest of Asia Pacific;

TABLE 18.3
Heavy-Duty Fabric Wash Liquid's Market in Asia Pacific

Country	Percentage of Fabric Wash Market
Australia	28
Indonesia	—
Japan	5
Korea	2
Malaysia	7
Philippines	0
Taiwan	4
Thailand[a]	15

[a]Mainly light-duty liquid, Givaudan-Roure 1992 share.

note that the apparently high level of usage of liquids in Thailand is due to fine-wash light-duty liquid products.

The patent literature (8) shows that the poorer stain removal performance of liquids may be resolved by using nonaqueous liquids in which bleach and activators can be suspended. These may be the next-generation products that will develop this category further.

The higher percentage of colored clothes in the wash load has also produced brand variants specifically targeted for washing colored articles. The first wash products for coloreds relied on being bleach free to reduce dye fading and fluorescer free to reduce color hue modification. Second-generation wash products for coloreds incorporated polymers, such as polyvinylpyrrolidone, which prevented dye in the wash solution from redepositing on other colored articles and modifying their color. A recent innovation has been the use of a cellulase enzyme to remove the cellulose fibrils and leave the colored articles more attractive. Originally offered in the 1970s as an in-wash softening ingredient, it has recently been found that colored articles washed in products containing cellulase look brighter. The mechanism may be that as luster is dependent on the ratio of specular to diffuse reflection, the removal of the microfibrils will decrease the diffuse reflectance and, therefore, enhance the brightness of the color. Interestingly, mindful of the prevailing value consciousness of consumers, P&G has formulated its color care product based on the cellulase Carezyme™ at no on-cost.

Environmentalist pressure to reduce unnecessary parts of the product mix, combined with the increasing cost pressures, has stimulated the introduction of concentrated products. The first successful introduction of a concentrated fabric wash powder was Kao's Attack in Japan in 1987. Following this success concentrated powders were introduced in Europe and the United States, respectively, in 1989. Concentrated powders have the following advantages that are powerful driving factors:

- Energy-efficient, low manufacturing costs
- Lower formulation and packaging costs
- Lower distribution costs
- Less shelf space
- Less disposable waste

The success of concentrated powders was then translated into concentrated heavy-duty fabric washing liquids in the early 1990s.

Concentrated powders have been extended around the Asia Pacific region (Table 18.4). In the Philippines, which has a small powders market, concentrates have renewed some interest in powders; the market increased from 10,000 tons to 30,000 tons in the 1990s. Elsewhere, in Asia Pacific they have not been as successful as in Japan; however, they have strengthened the main powder brands. For example, Unilever's recent launch in Indonesia of Rinso Ultra with bleach,

TABLE 18.4
Concentrated Heavy-Duty Fabric Wash Market in Asia Pacific

Country	Percentage of Powder Market
Australia	14
Indonesia	6
Japan	98
Korea	10
Malaysia	19
Philippines	100
Taiwan	50
Thailand[a]	23

[a]Givaudan-Roure 1992.

packaged in a plastic container, has attracted much attention and enhanced the Rinso franchise.

In Europe powder concentrates reached almost 40% of the market by 1992, and concentrated liquids attained almost 5% of the total market in one year. Unilever's new powder, containing the manganese-catalyzed bleaching system, uses a nontower route manufacturing process, which allows the company to produce a superconcentrated powder with large energy savings. In the nontower route perborate monohydrate has replaced the tetrahydrate, as its density is more compatible with the concentrated powder.

In the United States concentrated powders were launched in early 1992, followed by concentrated heavy-duty liquids. By 1993 they have reached 94% and 47% of the fabric wash powder and liquid markets, respectively. In a market in which concentrated products are already successful, further concentration has to be undertaken with care. The launch of Lever's superconcentrated liquids in the United States has not been successful even with heavy promotional expenditure. The established concentrate brands appear to have set the standard, and the consumers are skeptical that a slightly more concentrated form is value for money. Recently, Lever has reintroduced normal liquid concentrates.

Emerging FW Technology

A review of the patents filed on FW technology by Unilever, P&G, and Colgate indicates that the technological innovations most sought are even more effective low-temperature bleach system (9), enzyme systems (10), and color care (11). Other product forms, such as rapidly dissolving tablets (12), are also being researched.

Innovation has become a key to market success. The winners will be those who can exploit every factor to gain a performance benefit. Perhaps we will see product formulators extending performance by lateral thinking. For example, the recent use of genetic engineering to modify the properties of cellulose might be used to enhance the performance benefit of an FW formulation. Figure 18.4 illustrates some benefits that it might be possible to achieve from bioengineered cellulose in combination with a specifically formulated product.

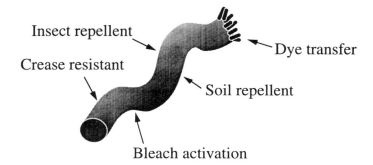

Fig. 18.4.   Genetic engineering of cellulose.

Fragrance Technology

The speed of innovation has required the development of fragrance technology, which can help in creating fragrances rapidly. They have to have excellent hedonics and perform well in innovative bases. The problems encountered most frequently with new bases are as follows:

1. The effect on fragrance performance of new surfactant systems
2. Malodor of new bases, wash solutions, and washed fabrics
3. Fragrance stability

Fragrance Performance in Surfactant Systems

Surfactant systems can significantly affect the perceived quality and impact of a fragrance. If new surfactants are used in a base, the perfumer must know how the fragrance materials interact with the new surfactant system to achieve the performance desired. We have developed a technology that we term Odor Value Technology; it quantifies changes in both the inherent quality and strength of a fragrance in a new base. This allows us to provide guidance to our perfumers. Odor Value Technology consists of two parts, dynamic headspace measurement and odor values.

The first part of this technology, the dynamic headspace technique, defines the complete spectrum of fragrance ingredients in the vapor phase and quantifies their concentration. The method developed by our research scientists (Fig. 18.5) involves placing the sample in an enclosed vessel. Volatiles in the headspace above the sample are slowly extracted through a special filter by a sucking device. This is designed to withdraw the headspace sample at a very constant rate, avoiding any temperature or pressure stress. The exact volume of vapor drawn through the filter is known. The fragrance trapped by the filter is thermally desorbed directly onto the gas chromatographic capillary column for analysis. This allows the accurate determination of the concentration of each fragrance component partitioned into the headspace.

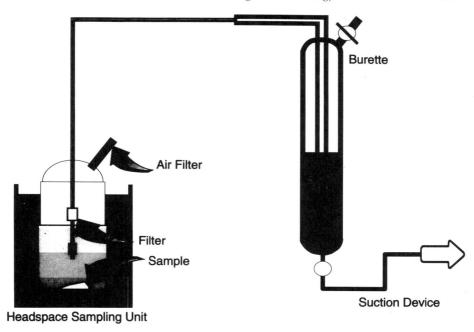

Headspace Sampling Unit

Fig. 18.5. Schematic of headspace sampling device.

If we were dealing only with an aqueous solution of fragrance, then the partitioning of the fragrance into the air phase could be calculated from Henry's law. However, in a surfactant solution the partitioning is more complex (13). Consider the partitioning of three fragrance molecules from an aqueous solution of sodium lauryl sulphate. We see (Fig. 18.6) that above the concentration at which the surfactant starts to form micelles, that is, above a critical micellar concentration, each of the three fragrance molecules partitions into the air phase differently. The surfactant micelle is acting as a third phase, which competes with the aqueous phase for the fragrance molecules. The stronger the partitioning into the micelle, the less fragrance is available to partition into the headspace.

Fragrance molecules can partition from the aqueous phase into the core of the micelle formed by the intertwined hydrocarbon chains. They can also partition into the palisade of the micelle. This is located just inside the hydrophilic layer, formed by the head group of the surfactant, and has a polarity intermediate between the hydrocarbon core and the aqueous solution. The extent to which a fragrance is solubilized by a specific micellar system, therefore, depends on the polarity of a perfume molecule. Nonpolar fragrance molecules will tend to partition strongly into the core of the micelle, whereas polar fragrance molecules will partition less strongly into the palisade layer.

For a given hydrocarbon chain the head group of the surfactant will affect how much fragrance will partition into the micellar phase. Figure 18.7 shows the

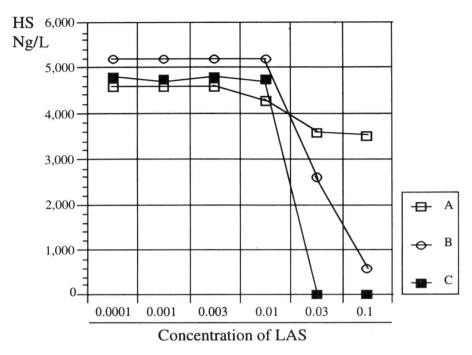

Fig. 18.6. Effect of surfactant on concentration of fragrance molecules in headspace.

quantity of dihydromyrcenol and phenyl ethyl alcohol that partitions into the air from 10% solutions of sodium lauryl sulphate and sodium lauryl ether sulphate, respectively. We see that the semipolar material dihydromyrcenol partitions more into the sodium lauryl sulphate micelle than into the sodium lauryl ether sulphate micelle. Consequently, less is available to partition into the air phase. The highly polar phenyl ethyl alcohol is affected to a much lesser degree by the head group of the surfactant. It strongly partitions into the aqueous phase regardless of the surfactant system and does not partition well into the air phase.

Micellar shape can also affect the partitioning of a fragrance molecule. The sodium lauryl sulphate micelle is spherical. Adding coco amine oxide forms mixed micelles that are rod shaped. The effect of this is to increase the lipophilic volume of the micelle. The partitioning of dihydromyrcenol, styrallyl acetate, benzyl acetate, and phenyl ethyl alcohol from a 6% sodium lauryl sulphate/4% amine oxide solution compared with that from a 10% sodium lauryl sulphate solution is shown in Fig. 18.8. The greater lipophilic volume of the rod-shaped micelles favors the partitioning of the semipolar materials. Dihydromyrcenol, benzyl acetate, and styrallyl acetate partition into the micellar phase more than into the simple spherical micelle. Thus the concentration of these materials in the air phase is reduced.

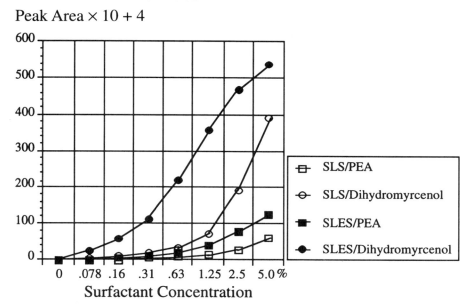

10% Surfactant Solution

Peak Area × 10 + 4

Legend:
- SLS/PEA
- SLS/Dihydromyrcenol
- SLES/PEA
- SLES/Dihydromyrcenol

Surfactant Concentration

Fig. 18.7. The effect of surfactant headgroup on the concentration of fragrance in the headspace.

If 5% sodium chloride is added to the above sodium lauryl sulphate/coco amine oxide surfactant system, it will form a lamellar micellar phase. This type of phase has a greater lipophilic volume than rod-shaped micelles. The effect of a lamellar phase on the partitioning of the four fragrance materials, used in the previous experiment, is shown in Fig. 18.9. The partitioning of the semipolar fragrance molecules into the lamellar phase is greater than that into the rod-shaped micelles. This effectively reduces the concentration of these molecules in the headspace.

As can be seen from the above examples, the effect of surfactants on the concentration of perfume that will be available in the headspace to be smelled is complex. The concentration of fragrance in the headspace will depend on how strongly it partitions into the micellar phase. This is dependent on several interacting factors:

- Size of the surfactant head group
- Nature of the head group
- Shape of the micelle
- Phase structure of the micelle
- Polarity of the fragrance molecule

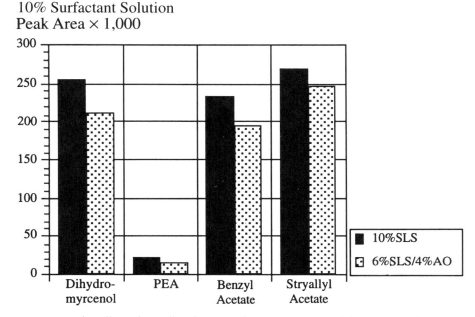

10% Surfactant Solution
Peak Area × 1,000

Fig. 18.8. The effect of micellar shape on the concentration of fragrance in the headspace.

The complexity of the effect of surfactant systems on fragrance partitioning can be seen from work we have done on PW bars. The phase structure of surfactants in highly concentrated solution, such as PW bars, is highly complex (14). In concentrated surfactant systems the surfactant can be present in many phases; the two most commonly formed are the liquid crystal hexagonal and lamellar phases, respectively. The type and quantity of the surfactant phases present, as well as being determined by formulation variables, can also be affected by process variables.

The partitioning of the four fragrance molecules into the air phase from several PW bars obtained from the Bradford Soap Company is shown in Fig. 18.10. It can be seen that phenyl ethyl alcohol and benzyl alcohol partitions well into the air phase from the soap bar but poorly from the translucent and syndet bars, respectively. The highest concentration of dihydromyrcenol in the headspace is obtained when it is incorporated into the syndet bar.

On dilution of these PW bars to a 1% (w/w) solution, the surfactant micelles will change in size, shape, and phase structure as the system moves from highly concentrated to dilute. The effect of this will be to change both the amount and the relative concentration of each of the fragrance materials in the headspace. The new composition of the fragrance in the headspace above the respective solutions is shown in Fig. 18.11. In the 1% solution the polar phenyl ethyl alcohol prefers to

10% Surfactant Solution
Peak Area × 1,000

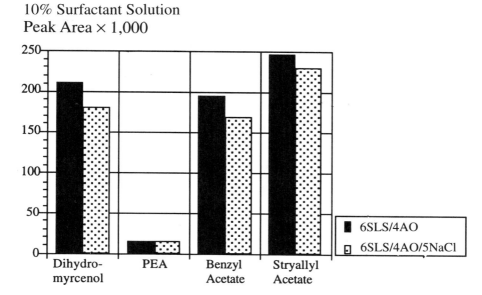

Fig. 18.9. The effect of micellar structure on the concentration of fragrance in the headspace.

reside in the aqueous phase irrespective of the PW bar formulation. The concentration of the semipolar fragrance materials in the headspace is greatest from Bradpride. The syndet formulation has the lowest amount of fragrance material partitioned into the air phase.

The above results show that the interaction of fragrances from surfactant systems is highly complex. We do not have a simplistic physicochemical model; however, we can provide guidance to the perfumer on how fragrance ingredients will partition from concentrated and dilute surfactant systems into the headspace.

The second part of our odor value technology is the conversion of the amount of fragrance that partitions into the headspace into odor character and impact. The fragrance available for smelling is determined quantitatively by headspace measurements. However, even though the quantity of each fragrance component can be reproducibly measured, it does not tell us about the sensation it produces, the probability of smelling the fragrance molecule, or its strength. In order to interpret the spectrum of volatile chemicals partitioned into the headspace for our perfumers, we convert the respective quantity of each fragrance material, in the headspace, into its odor values (OV) (15):

$$OV = \frac{\text{Concentration of odorant in headspace}}{\text{Average threshold concentration of odorant}}$$

Bradford Soap Bars
Peak Area × 1,000

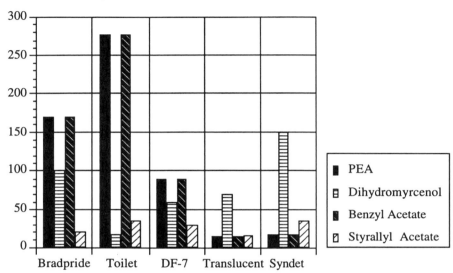

Fig. 18.10. The effect of PW bar formulation on the concentration of fragrance in the headspace above the bar.

The threshold is defined as the concentration of odorant at which a panelist can perceive it reliably. We use a panel of a minimum 25 people and determine their individual thresholds for the odorant. The geometric mean of the set of individual thresholds is calculated and used in the OV calculation.

The threshold is detected on our proprietary dynamic air dilution olfactometers that gives a very precise measure of odor intensity (Fig. 18.12). The saturated vapor of an odorant is accurately diluted with odorless air and randomly delivered to one of the three smelling ports. The same volume of odorless air is delivered to two other smelling ports. The panelist is asked to select the port that contains the odorant. If the port is identified correctly three times, the odorant is diluted by a factor of two. The binary dilution steps are repeated until the panelist can no longer correctly identify the port with the odorant three times. The concentration above this is the panelist's threshold for that odorant.

The OV converts the fragrance concentration into odor perception. The importance of OV is shown Fig. 18.13. The headspace concentration profile of the fragrance is shown in Fig. 18.13(a), in which component 5 appears to be an insignificant contributor to the smell. Whereas in the OV profile, Fig. 18.13(b), component 5 is a major olfactive component of the perceived odor because of its extremely low threshold.

Bradford Product 1% Solution
Peak Area × 1,000

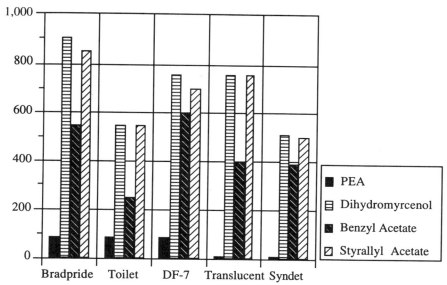

Fig. 18.11. The effect of PW bar formulation on the concentration of fragrance above a 1% solution.

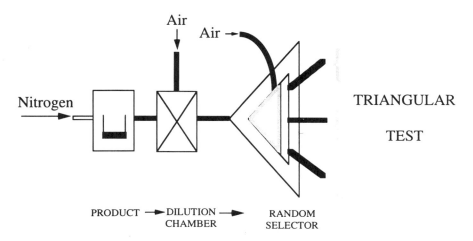

Fig. 18.12. Schematic of dynamic air dilution olfactometer.

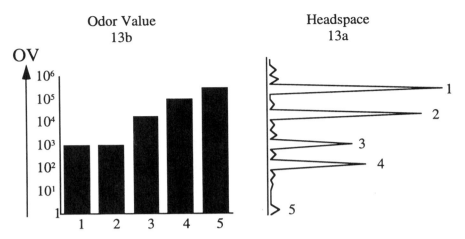

Fig. 18.13. (a) GC headspace contribution; (b) odor value profile.

In order to convert accurately the perception profile into odor strength it is necessary to know the dose response curve of each fragrance material studied. This is not available for all fragrance materials, and an approximation has to be used. Although the dose response curve is not linear for fragrance (16), it is assumed that it is approximately linear over the concentration ranges of interest. The impact is then calculated from Stevens' Power Law (17):

$$\text{Sensation} = k \, (\text{Stimulus})^n$$

where k and n are constants and are the intercept and the slope of the dose response curve plotted on log-log coordinates. For odor impact we can approximate (18) to

$$\text{Impact} = OV^n$$

where n is averaged from several dose response curves.

The combination of dynamic headspace and odor values has allowed us to study how formulation variables will affect the performance of fragrance molecules. This has proved extremely valuable and has allowed our perfumers to create the desired hedonics in new and often difficult bases. They can creatively design the hedonic and strength performance required at each stage of the use cycle. It also ensures that today's challenge of cost optimized performance is met.

Malodor Associated with Innovative Bases

The introduction of new chemicals into formulations can bring with it malodor problems either from new materials or from trace contaminants, such as solvents. The quantity of volatile components in the headspace of two unfragranced PW bases provided by the Bradford Soap Company is shown in Fig. 18.14. The syndet base has many more volatiles than the Bradpride soap base. Some of these, confirmed by GC smelling, are malodorous and show why it is more difficult to fragrance the syndet base than the Bradpride base.

Similarly, some new FW products have poor base odor. This malodor can be apparent in the wash/soak solution and can even be transferred onto the washed fabric. Some new ingredients also can themselves cause a malodor, such as isononanoyloxybenzene sulphonate (ISONOBS), which develops a malodor as the pH drops (19). Lipolase enzyme can also react with the residual oily soils on the fabric and leave the washed fabric with a malodor. Bleaches can also break down triglycerides and leave odorous fragments on the fabric. Unless the fragrance is engineered to mask these odors, the washed articles will have a malodor.

When we have sufficient time to study the new base, we can characterize the malodor. First, the malodorous peaks are identified by sniffing the peaks as they

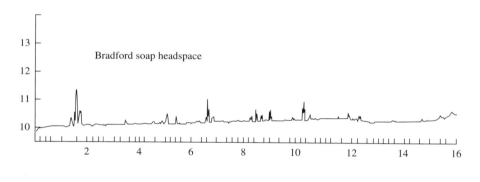

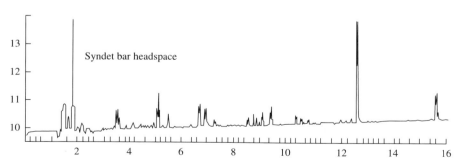

Fig. 18.14. The volatile components in the headspace of Bradpride soap and a syndet base.

elute on a GC column. These are then analyzed by multidimensional GC, mass spectrometry, and nuclear magnetic resonance, when sufficient material is available. The malodor peaks can be synthesized and the malodor reconstituted. Fragrance materials can then be screened to mask or blend in with the malodor.

We have speeded up the process of base odor coverage by developing a multichannel olfactive blender (20). This proprietary instrument, shown in Fig. 18.15, allows the perfumer to draw the volatile components of the base or the reconstituted malodor of the base into a vapor phase mixing chamber. The perfumer can then add accurately controlled amounts of fragrance vapors and combine them until the malodor is covered and a hedonically pleasing smell is perceived. This mixture is then converted by a computer program into a liquid formula, which the perfumer can use as a base to create the final fragrance.

Wash solutions and fabrics can be similarly examined. For malodor coverage on fabric we must ensure that the fragrance components used in the multichannel olfactive blender to mask the malodor are those able to be deposited from the wash solution onto the fabric. One key property that predicts deposition efficiency is the water solubility of a fragrance molecule (21). Our scientists have developed a dialysis technique that allows the determination of the solubility of odorant

Fig. 18.15.   Multichannel olfactive blender.

molecules with a wide range of water solubilities (22). This in combination with the malodor coverage properties of the odorant allows the perfumer to reverse-engineer the fragrance to provide a pleasant smell at all stages of the use cycle.

Fragrance Stability

Perfumes are a blend of chemicals, some of which can react with the base under certain conditions. Any potential reaction can be accelerated by exposure to heat, humidity, oxygen, and light, and by the presence of trace ingredients that can catalyze reactions. The new bleach systems can attack many of the fragrance components, especially unsaturated molecules. Enzymes, for example, can cause esters to break down into their constituent acids and alcohols.

To provide the perfumer with a list of fragrance materials that are stable, comprehensive stability testing of individual fragrance materials under standardized conditions is performed. This requires time. The growing urgency of the market often does not give sufficient time to develop a database on the stability of fragrance molecules in the new base. To help the perfumer, the chemical structure of the fragrance molecules in similar bases are being correlated by the perfume technologist and computer structure activity models are being developed to predict stability.

The computer-generated list of potentially stable molecules is used as the starting point of fragrance creation for new bases. This type of technology helps the perfumer meet the increasingly short deadlines to provide a fragrance that is compatible and stable with the new base.

Conclusion

The trend of the soap and detergent market is toward more rapid innovation. A recent article (23) pointed out that the lifetime of the old handwash, high suds powder was 60 or 70 years; the low suds powders of the 1960s lasted 20 years; detergent liquids 7 years; and the first phase of concentrates 5 years. The implication is that the lifetime of the second phase will be shorter.

This change in innovation dynamics is, by necessity, bringing new formulations into the market place with short lead times. In order to meet the challenges of the rapid product innovation, our fragrance research program has had to keep pace. Leading-edge fragrance technology has been developed that allows us to meet the demands of the new and exciting market dynamics. To be successful, the perfumer must design fragrances to be both hedonically preferred and technically designed to meet today's performance criteria.

Acknowledgment

Thanks is due to all my marketing colleagues in the regions for providing me with the marketing data and for my technical colleagues in the United States for carrying out the technical work.

References

1. U.S. Patent 5,236,710; European Patent 437,347; U.S. Patent 526,414; U.S. Patent 5,234,619.
2. International Patent WO 24,010; European Patent 589,407; U.S. Patent 5,290,471.
3. FDA Tentative Final Monograph, *Federal Register 59:* 116 (1994).
4. European Patent 509,787.
5. Comyns, A.E., Detergents Bleaching Catalyst Revealed, *Nature 369:* 609 (1994).
6. Hage, R., J.E. Iburg, J. Kerschner, J. Koek, E. Lempers, R. Martens, U. Racherla, S. Russell, T. Swarhoff, R. van Vliet, J. Warnaar, L. van der Wolf, and B. Krijnen, Manganese Catalyst with Low Temperature Bleaching Action, *Nature 369* (1994).
7. International Patent WO 9,222,629.
8. German Patent DE 3,829,087; European Patent 490,436.
9. European Patent 564,251; European Patent 572,724; International Patent WO 9,403,395; U.S. Patent 5,281,361; U.S. Patent 5,264,142.
10. International Patent 9,322,415; European Patent 544,359; International Patent WO 9,323,516; U.S. Patent 5,281,356/57.
11. European Patent 576,778; European Patent 579,295; European Patent 581,751/752/753.
12. U.S. Patent 5,225,100.
13. Labows, J.N., Surfactant Solubilization Behavior via Headspace Analysis, *J. Am. Oil Chem. Soc. 69:* 34 (1992).
14. Rosevear, F.B., The Microscopy of the Liquid Crystalline Neat and Middle Phases of Soaps and Synthetic Detergents, *J. Am. Oil Chem. Soc. 31:* 628 (1954).
15. Müller, P.M., and F. Etzweiler, Application of Odor Values in Perfuming Cosmetic Products, 15th IFSCC Congress, London, 359 (1988).
16. Müller, P.M., The Physiology of Perception of Flavors and Fragrances: Exploitable Findings?, Am. Chem. Soc. Fall National Meeting (1994) (unpublished).
17. Stevens, S.S., On the Psychological Law, *Psychol. Rev. 64:* 153 (1957).
18. Callan, B.T., Malodor Measurement and Control, *Chem. Ind. 21:* 845 (1993).
19. U.S. Patent 4,412,934.
20. Swiss Patent 2,517.
21. Müller, P.M., N. Neuner-Jehle, and F. Etzweiler, What Makes a Fragrance Substantive?, *Perfumer and Flavorist 18:* 45 (1993).
22. Etzweiler, F., E. Senn, and H.W.H. Schmidt, Method for Measuring Aqueous Solutions of Organic Components, *Analytical Chemistry 67:* 655–658 (1995).
23. Jackson, T., *Financial Times,* April 22, 1994.

Index